INTERNATIONAL SERIES OF MONOGRAPHS IN
NATURAL PHILOSOPHY
GENERAL EDITOR: D. TER HAAR

VOLUME 78

INTRODUCTION TO ELEMENTARY PARTICLE THEORY

INTRODUCTION
TO
ELEMENTARY PARTICLE THEORY

by

YU. V. NOVOZHILOV
Leningrad State University

Translated by

JONATHAN L. ROSNER
University of Minnesota

PERGAMON PRESS
OXFORD • NEW YORK
TORONTO • SYDNEY • BRAUNSCHWEIG

Pergamon Press Ltd., Headington Hill Hall, Oxford
Pergamon Press Inc., Maxwell House, Fairview Park, Elmsford
New York 10523
Pergamon of Canada Ltd., 207 Queen's Quay West, Toronto 1
Pergamon Press (Aust.) Pty. Ltd., 19a Boundary Street,
Rushcutters Bay, N.S.W. 2011, Australia
Pergamon Press GmbH, Burgplatz 1, Braunschweig 3300, West Germany

First edition 1975

Library of Congress Cataloging in Publication Data

Novozhilov, I͡Uriĭ Viktorovich.

Introduction to elementary particle theory.

(International series of monographs in natural philosophy, 78)

Translation of Vvedenie v teorii͡u elementarnykh chastit͡s.

1. Particles (Nuclear physics) I. Title.

QC793.2.N6813 1975 539.7'21 74-17036

ISBN 0-08-017954-1

Printed in Hungary

CONTENTS

PART IV

ELEMENT OF DYNAMICAL THEORY

[illegible] MENT OF DYNAMICAL THEORY

PREFACE

THE present stage of development in the physics of elementary particles began at the end of the 1950s. With the 1960s our view of the elementary (or fundamental) particles and their interactions broadened considerably. Instead of a score of stable or nearly stable particles and several resonances in pion–nucleon scattering, we now know of more than a couple of hundred particles and resonances. Our picture of space–time symmetry underwent still another change as a result of the discovery of the nonconservation of the combined parity CP (or noninvariance with respect to time reversal T), and, consequently, it is possible that the three basic types of interaction have been joined by a fourth—the superweak interaction. During the same time significant information on high-energy scattering and properties of resonances has accumulated.

In elementary particle theory the 1960s were an era of symmetry, Regge poles, current algebra, and duality. During this time fruitful schemes of internal symmetry were constructed, described by the groups SU_3 and SU_6, as well as by current algebra and chiral symmetry. At the same time, the basic features of the asymptotic behavior of scattering amplitudes were clarified, and the fundamental nature of the Regge trajectory was established. The notion of duality introduced in recent years may become a basic concept of future theory.

The vigorous development of elementary particle physics has made it evident that quantum field theory cannot assume the role of a theory of elementary particles. Nonetheless, the local quantum theory of fields remains the basis from which general principles are adopted and with whose help several models are constructed.

The contemporary theory of elementary particles may be called a constructive theory. Starting from well-founded principles of quantum field theory, from experiment, and from guesses, such a theory seeks to select and work out the ideas essential for the description of the elementary particles and resonances. Regge trajectories, unitary and chiral symmetries, current algebra, and duality are successful examples of the constructive approach; they depend on the local theory but cannot be unambiguously obtained from it or proven. The present book is meant as an introduction to such a constructive theory of elementary particles. The author hopes that such a book will be useful as a complement to other texts on elementary particle theory.[1–4]

The book consists of four parts. The introductory Part I acquaints the reader with the basic description of elementary particles. In Part II questions of relativistic quantum mechanics and kinematics are set forth; Part III is devoted to the problem of internal symmetry, and Part IV to those new dynamical approaches which are likely to have the greatest influence on the development of theory in the future. Quantum electrodynamics and renormalization are excluded from the present book, as these questions are contained in the standard quantum theory of fields.[5–7] The author does not give a systematic review of experimental data, but cites only the information essential to illustrate the pattern of phenomena and to connect theory with experiment. The Appendix contains tables of particles, but the reader's main reference on particle properties should be special annual reviews.[8]

The list of references contains only those works which, in the author's opinion, are basic. The reader may acquaint himself with a more complete list in books[9–20] and in reviews referred to here.

The plan of the book essentially follows the program of courses on elementary particle theory given in the Physics Faculty of Leningrad University.

The reader must be familiar with nonrelativistic quantum mechanics and classical relativity theory. It would also be very useful to have a preliminary acquaintance with the fundamentals of the Lagrangian formulation of quantum field theory and with Feynman diagrams.[5–7] A course in elementary particle field theory usually is preceded by a short course on group theory. We thus assume that the basic facts of group theory are known to the reader.[18, 19]

The author is grateful to A. A. Ansel'm, M. A. Braun, B. V. Medvedev, I. A. Terent'ev, and especially to L. V. Prokhorov, V. A. Frank, and Yu. P. Shcherbin for valuable advice.

The author thanks colleagues and students in the nuclear theory and elementary particle theory groups at Leningrad State University for help and discussions.

Several chapters of a first version of this book were written during the author's stay at the Faculty of Physics and Astronomy of Delhi University. The author wishes to use this occasion to express his gratitude to Professors D. S. Kothari and R. S. Majumdar for their hospitality and for pleasant working conditions.

AUTHOR'S PREFACE TO THE ENGLISH EDITION

In the time since this book was finished the center of gravity of interest in elementary particle physics has shifted to the weak and electromagnetic interactions. Gauge theories and parton models have become the subject of nearly universal attention.

The author, however, has decided to overcome the temptation to write a supplement on gauge fields and expand the chapter on weak interactions, since the book is primarily devoted to the phenomenological foundations of relativistic theory and to the strong interactions from the *S*-matrix standpoint. An exposition of gauge theories would have required an introduction to the Lagrangian formalism (which we do not consider separately) and the use of functional integration (which we do not even mention).

The author appreciates the careful and exacting work performed by Prof. Rosner in the process of translation. The changes he has made, of which I approve fully, can only serve to improve the treatment.

Yu. Novozhilov
Paris, July 1974

TRANSLATOR'S PREFACE

THE serious student of elementary particle theory must cope both with rather formal course work and with a highly specialized literature. The relation between the formalism and its application to practical problems often is not transparent at first glance. The present book, by presenting a unified treatment of both, helps in large measure to bridge this gap. The main emphasis of the book is on the strong interactions, where the need for such a "bridge" is greatest. The first two parts present a theoretical framework which is then used extensively in dealing with concrete problems in the last two parts. It was partly my appreciation for the particular fortunate combination of "theory" and "practice" in this book— which corresponded very closely to the way I happened to learn the subject—that led me to undertake the translation.

I would like to express deep gratitude to Professor Novozhilov for his constant interest and encouragement, for his patient help in clarifying points in the text, and for making available a list of misprints in the Russian edition. I would also like to ask his indulgence regarding the minor modifications (that a practising theorist cannot help making) which I felt would bring portions of the material more up to date. In general, however, the translation attempted to follow the text as closely as possible. This was not difficult owing to the exceptional clarity of the original prose.

Professor Morton Hamermesh provided much valuable technical advice regarding translation procedures. I am thankful to Mrs. Valerie Nowak, Mrs. Veronica Goidadin, Mlles M. Milligan and S. Williams, Mrs. Janace Ator, and Mrs. Sandra Smith for expert typing. Finally, my special appreciation is reserved for my wife, Joy, whose day-to-day moral support and whose assistance in the final correction of the typescript made a pleasant but protracted task seem shorter and sweeter.

NOMENCLATURE

This book uses units in which $\hbar = c = 1$.

Indices and abbreviations

Indices denoted by the Greek letters μ, ν, ϱ, ..., take on the values 0, 1, 2, 3.

Indices denoted by the Latin letters i, j, k, ..., take on the values 1, 2, 3.

Spinor and isospinor indices are denoted by $\alpha, \beta, \gamma, \ldots = 1, 2$, but for Dirac bispinors $\alpha, \beta, \gamma, \ldots = 1, 2, 3, 4$.

Indices denoted by the Latin letters a, b, c, ..., take on the values 1, 2, 3, ..., 8.

Contravariant components of a vector have the same sign as in the nonrelativistic theory: $a^\mu = (a^0, \boldsymbol{a})$.

The metric tensor $g^{\mu\nu} = g_{\mu\nu}$ has the signature $(+ - - -)$.

Scalar product: $a_\mu b^\mu = ab = a_0 b_0 - \boldsymbol{a} \cdot \boldsymbol{b}$.

Abbreviations: $\partial_\mu = \partial/\partial x^\mu$, $\Box = \partial_0^2 - \boldsymbol{\partial}^2$.

The antisymmetric pseudo-tensor $\varepsilon_{\mu\nu\lambda\sigma}$ is normalized according to $\varepsilon_{0123} = -1$ or $\varepsilon^{0ijk} = \varepsilon_{ijk}$.

Volume elements: $d^4a = da^0\, d^3a$, $d^3a = da^1\, da^2\, da^3$.

Commutators and anticommutators: $AB - BA = [A, B]_- = [A, B]$, $AB + BA = [A, B]_+ = \{A, B\}$.

Hermitian conjugation, complex conjugation, and transposition are denoted respectively by the superscripts +, *, and T.

State vectors and Lorentz group

Spin: J; spin-parity: J^{P}; spin projection on the z-axis: σ; helicity: λ.

Normalization of states:

$$\langle \boldsymbol{p}, \sigma \mid \boldsymbol{p}', \sigma' \rangle = 2p_0 \delta_{\sigma\sigma'} \delta(\boldsymbol{p} - \boldsymbol{p}').$$

Particle density corresponding to this normalization:

$$2p_0/(2\pi)^3.$$

Invariant phase space:

$$dR_n(p) = \delta\left(p - \sum_{r=}^{n} p_r\right) \delta(p_1^2 - m_1^2) \ldots \delta(p_n^2 - m_n^2)\, d^4p_1 \ldots d^4p_n.$$

The representation of the group SU_2 is $\mathcal{D}^J$.
Finite-dimensional representation of the Lorentz group: $\mathcal{D}^{(J_1, j_2)}$; notation: $\mathcal{D}^{(J, 0)} = D^J$.
2 by 2 matrices associated with the vector p^μ:

$$\mathrm{p} = \sigma_0 p^0 + \boldsymbol{\sigma} \cdot \boldsymbol{p}, \quad \tilde{\mathrm{p}} = \sigma_0 p^0 - \boldsymbol{\sigma} \cdot \boldsymbol{p}.$$

Dirac matrices:

$$\{\gamma_\mu, \gamma_\nu\} = 2g_{\mu\nu}, \quad \gamma_5 = \gamma_5^+ = -i\gamma^1\gamma^2\gamma^3\gamma^4,$$
$$\gamma_4 \equiv \gamma_0, \quad \sigma_{\mu\nu} = \tfrac{1}{2}[\gamma_\mu, \gamma_\nu], \quad \not{a} = a^\mu \gamma_\mu.$$

The matrix $\mathcal{C}$: $\gamma_\mu^{\mathrm{T}} = \mathcal{C}\gamma_\mu\mathcal{C}^{-1}$, $\mathcal{C}^{\mathrm{T}} = -\mathcal{C}$, $\mathcal{C}\mathcal{C}^+ = 1$.
Normalization of Dirac functions: $u^*(\boldsymbol{p})u(\boldsymbol{p}) = 2p_0$.
Geometrical (ordinary) spatial reflection: P.
Geometrical (Wigner) time reversal: T.
Charge conjugation: C.
Total reflection: $\theta =$ CPT.

Internal symmetry

Group: $\mathfrak{G}$; transformation: G; representation: $d(G)$; isospin: I_k, with $I_3 \equiv t$; generators of the group SU_3: F_a.
Hypercharge: Y; strangeness: $S = B+Y$.
Baryon number: B.
Lepton numbers: L_e, L_μ.

Scattering amplitude

The T-matrix is defined by

$$\langle a\,|S-1|\,b\rangle = i(2\pi)^4\,\delta^4(p_a - p_b)\,\langle a\,|T|\,b\rangle.$$

Partial wave: a_J or a_l; for equal masses below inelastic threshold these satisfy the two-particle unitarity conditon

$$\mathrm{Im}\; a_J = \sqrt{\frac{s-4m^2}{s}}\,|a_J|^2.$$

References to formulae from other chapters contain the chapter number first: (9.16) is equation (16) from Chapter 9.

Part I

INTRODUCTION: STATES OF ELEMENTARY PARTICLES

The description of states of elementary particles relies primarily on quantum mechanical regularities and concepts of symmetry.

Symmetry principles are more general than laws of motion; thus finding them is the first step in the formulation of laws. This is especially important in the theory of elementary particles where the equations of motion are unknown. Knowledge of the symmetry group allows one to find immediately a natural set of basis quantities (the generators of the group and their invariants) with which to describe particle states.

In elementary particle physics, where not all basic concepts are fully established, still another side of the symmetry method is important. The symmetry method is a general approach to the physical description of any phenomena. It allows one to introduce a logically closed physical description in which every stage of approximation to reality is endowed with precise symmetry properties. Thus at every stage of the approximation there exists an exact physical interpretation of the theory in terms of an idealized world. An elementary particle is thus defined as an object whose state forms the basis of an irreducible representation of a symmetry group.

Symmetries are divided into space–time (or geometrical) and internal (or dynamical), reflecting interaction properties.

CHAPTER 1

ELEMENTS OF RELATIVISTIC QUANTUM THEORY

THE theory of elementary particles is a relativistic theory. Quantum mechanics and the Einstein relativity principle are the basis of its apparatus.

The Einstein relativity principle determines the group of space–time symmetry in classical physics as the inhomogeneous Lorentz group, or the Poincaré group $\mathcal{P}^{\uparrow}_{+}$. In quantum mechanics the transformations of this group connect "equivalent" positions of classical observers performing a quantum experiment and defining quantum states. It is assumed that all transformations are in fact feasible, including those which correspond to passing to states of physical systems with infinite energy.

As is well known, a quantum state corresponds to not one but a number of state vectors ("unit ray") in Hilbert space. Because of this fact, and because the parameter space of the group $\mathcal{P}^{\uparrow}_{+}$ is doubly connected, the space–time symmetry group in quantum mechanics is not the Poincaré group $\mathcal{P}^{\uparrow}_{+}$ but its universal covering group $\overline{\mathcal{P}}^{\uparrow}_{+}$ or the quantum mechanical Poincaré group. The irreducible unitary representations of $\overline{\mathcal{P}}^{\uparrow}_{+}$ also describe elementary particle states. The invariants of the Poincaré group—mass and spin—characterize the invariant properties of particles, and the generators of the Poincaré group—the possible observables, i.e. the variables whose independent measurements fix the state of a relativistic particle.

The representations of the quantum mechanical Poincaré group $\overline{\mathcal{P}}^{\uparrow}_{+}$ contain both single- and double-valued representations of the classical group $\mathcal{P}^{\uparrow}_{+}$. This means that both integral and half-integral values of spin correspond to single-valued representations of the quantum mechanical group. The existence of physical objects with half-integral spin is thus a direct consequence of relativistic quantum theory.

After rules are established for the description of single-particle states, the problem of describing the scattering of particles arises.

In Chapter 1 we follow (without proofs) a chain of discussion leading to the quantum mechanical Poincaré group as a symmetry group, and also the simplest consequences relating to the nature of particle states. We recall also those original postulates which lie at the basis of the quantum theory of scattering.

§ 1.1. Homogeneity of space–time and the Poincaré group

In processes with elementary particles the gravitational field does not play an essential role because of the smallness of the gravitational constant; in any case, at present there are no experimental facts or weighty theoretical considerations indicating the need to account for the gravitational field of particles. Thus in a theory of elementary particles one can consider space–time homogeneous and isotropic and geometry pseudo-euclidean. Accordingly, the principle of relativity must hold; i.e. in frames of reference differing from one another with respect to orientation and position in space, any choice of initial time and relative speed (rectilinear and uniform) must be equivalent.

All such (inertial) systems are equivalent in that in any two systems respective physical phenomena proceed in the same way. Associating the coordinate x^μ ($\mu = 0, 1, 2, 3$) with a frame of reference, we thus postulate the invariance of physical laws relative to linear transformations from one reference frame to the other:

$$x^\mu \to x'^\mu = \Lambda^\mu{}_\nu x^\nu + a^\mu, \tag{1}$$

with the condition that the square length of the interval

$$(x-y)^2 = (x_\mu - y_\mu)(x^\mu - y^\mu) \tag{2}$$

is conserved. In defining (1) it is assumed that the translation is performed after the homogeneous transformation.

The linear transformations conserving (2) form the inhomogeneous Lorentz group or the Poincaré group. They contain not only translations and rotations (ordinary and hyperbolic) in four-dimensional pseudo-euclidean space, but also the space–time reflections P, T, and PT $= I$:

$$\left.\begin{array}{ll} \mathrm{P}x^k = -x^k, & Px^0 = x^0; \\ Tx^k = x^k, & \mathrm{T}x^0 = -x^0; \\ \mathrm{PT}x^\mu = -x^\mu & \\ (k = 1, 2, 3; & \mu = 0, 1, 2, 3). \end{array}\right\} \tag{3}$$

The interval (2) is invariant if the coefficients $\Lambda^\mu{}_\nu$ satisfy the condition

$$\Lambda^\nu{}_\mu \Lambda^\mu{}_\sigma = \delta^\sigma{}_\nu, \quad \Lambda^\nu{}_\mu = g_{\mu\varrho} \Lambda^\varrho{}_\beta g^{\beta\nu}, \tag{4}$$

where the nonzero $g_{\mu\nu}$ are equal to $g_{00} = -g_{11} = -g_{22} = -g_{33} = 1$. For the transformation matrix Λ with matrix elements $\Lambda^\mu{}_\sigma$ the condition (4) takes the form $\Lambda^{\mathrm{T}} g \Lambda = g$. From this relation it follows that the determinant of the matrix Λ in the general case is equal to

$$\det \Lambda = \pm 1. \tag{5}$$

From (4) it also follows that the coefficients $\Lambda^{\mu\nu}$ have the following property:

$$(\Lambda^{00})^2 - \Sigma(\Lambda^{0k})^2 = 1, \tag{6}$$

or $(\Lambda^{00})^2 \geqslant 1$. This means that there are two possibilities:

$$\Lambda^{00} \geqslant 1 \quad \text{or} \quad \Lambda^{00} \leqslant -1. \tag{7}$$

Thus the general transformations (1) may be divided into four classes differing with respect to the sign of det Λ and Λ^{00}. These classes of transformations correspond to the following connected components of the general Poincaré group:

1. $\mathcal{P}^{\uparrow}_{+}$: det $\Lambda = 1$, $\Lambda^{00} \geqslant 1$. The direction of time remains unchanged, and spatial reflection does not occur. The transformations (1) describe rotations and translations in pseudo-euclidean space, forming the proper or orthochronous Poincaré group $\mathcal{P}^{\uparrow}_{+}$.

2. $\mathcal{P}^{\downarrow}_{+}$: det $\Lambda = 1$, $\Lambda^{00} \leqslant -1$. The transformations (1) include time reflection. Since, however, the transformation is unimodular, it must also contain reflection of the spatial coordinates. Put differently, each transformation of this class may be represented in the form of the product of two operations: a transformation belonging to the class $\mathcal{P}^{\uparrow}_{+}$ and a reflection of all coordinates PT. In particular the operation PT itself is contained in $\mathcal{P}^{\downarrow}_{+}$. The operations P and T separately are not contained in $\mathcal{P}^{\downarrow}_{+}$ because of the condition det $\Lambda = 1$. Taken together, the transformations of the classes $\mathcal{P}^{\uparrow}_{+}$ and $\mathcal{P}^{\downarrow}_{+}$ form the proper Poincaré group $\mathcal{P}_{+}$.

3. $\mathcal{P}^{\uparrow}_{-}$: det $\Lambda = -1$, $\Lambda^{00} \geqslant 1$. Transformations of this class may be written as a product of a transformation from the class $\mathcal{P}^{\uparrow}_{+}$ and a reflection of spatial coordinates. Together with transformations of the class $\mathcal{P}^{\uparrow}_{+}$ they form the orthochronous Poincaré group.

4. $\mathcal{P}^{\downarrow}_{-}$: det $\Lambda = -1$, $\Lambda^{00} \leqslant -1$. The direction of time is changed. Each transformation of this class may be represented as the product of a transformation of the class $\mathcal{P}^{\uparrow}_{+}$ and a reflection of time.

The general Poincaré group may be symbolically depicted in the form of a sum

$$\mathcal{P} = \mathcal{P}^{\uparrow}_{+} + \mathrm{PT}\mathcal{P}^{\uparrow}_{+} + \mathrm{P}\mathcal{P}^{\uparrow}_{+} + \mathrm{T}\mathcal{P}^{\uparrow}_{+}, \tag{8}$$

where the respective terms correspond to the components 1–4.

Of all the components of the Poincaré group only the first $\mathcal{P}^{\uparrow}_{+}$ contains the unit transformation. For this reason transformations belonging to different classes cannot be connected with one another using any sort of continuous transformation belonging to $\mathcal{P}^{\uparrow}_{+}$. Transformations of the same class may be obtained from one another using transformations from $\mathcal{P}^{\uparrow}_{+}$.

The classification of particle states starts with the study of the proper orthochronous group $\mathcal{P}^{\uparrow}_{+}$. An element (a, Λ) of the group $\mathcal{P}^{\uparrow}_{+}$ is determined by a translation four-vector a^{μ} and an orthochronous unimodular transformation Λ. The multiplication law for two elements (a_1, Λ_1) and (a_2, Λ_2) is

$$(a_1, \Lambda_1)(a_2, \Lambda_2) = (a_1 + \Lambda_1 a_2, \Lambda_1 \Lambda_2). \tag{9}$$

In particular, $(a, \Lambda) = (a, 1)(0, \Lambda)$ in accord with the definition (1) of an inhomogeneous transformation as the product of a homogeneous transformation and a subsequent translation. The unit element of the group $\mathcal{P}^{\uparrow}_{+}$ is equal to

$$E = (0, 1), \tag{10}$$

and the inverse element is written as

$$(a, \Lambda)^{-1} = (-\Lambda^{-1}a, \Lambda^{-1}). \tag{11}$$

The multiplication law (9) for group elements may be represented by introducing the 5 by 5 matrices

$$(a, \Lambda) \rightarrow \begin{pmatrix} \Lambda & a \\ 0 & 1 \end{pmatrix}, \tag{12}$$

belonging to a nonunitary representation of the Poincaré group.

The transformation (a, Λ) is characterized by 10 parameters: four translations a^{μ}, three parameters of three-dimensional rotations, and three parameters connected with passage to another inertial system ("hyperbolic rotations"). It is convenient to introduce the six parameters of Lorentz transformations Λ using the fact that any Λ may be represented as a product of a pure Lorentz transformation H (without rotation of axes) and a three-dimensional rotation R:

$$\Lambda = RH. \tag{13}$$

Any pure Lorentz transformation $x' = Hx$ is fully characterized by a direction of motion $\boldsymbol{\nu}(\boldsymbol{\nu}^2 = 1)$ and a speed $v = \tanh \beta \; (0 < \beta < \infty)$:

$$\begin{aligned} x'^0 &= x^0 \cosh \beta + (\boldsymbol{\nu} \cdot \boldsymbol{x}) \sinh \beta, \\ \boldsymbol{x}' &= \boldsymbol{x} + x^0 \boldsymbol{\nu} \sinh \beta + \boldsymbol{\nu}(\boldsymbol{x} \cdot \boldsymbol{\nu}) (\cosh \beta - 1). \end{aligned} \tag{14}$$

If the direction $\boldsymbol{\nu}$ is kept fixed, then

$$H(\boldsymbol{\nu}, \beta_1)\, H(\boldsymbol{\nu}, \beta_2) = H(\boldsymbol{\nu}, \beta_1 + \beta_2).$$

Any rotation $y = Rx'$ may always be represented as a rotation around some axis $\boldsymbol{n}(\boldsymbol{n}^2 = 1)$:

$$\begin{aligned} y^0 &= x'^0, \\ \boldsymbol{y} &= R(\boldsymbol{n}, \alpha)\boldsymbol{x}' = \boldsymbol{x}' \cos \alpha + \boldsymbol{n}(\boldsymbol{n} \cdot \boldsymbol{x}') (1 - \cos \alpha) + \boldsymbol{x}' \times \boldsymbol{n} \sin \alpha \qquad (0 \leqslant \alpha \leqslant \pi). \end{aligned} \tag{15}$$

The normals $\boldsymbol{n}$ and $\boldsymbol{\nu}$ may be expressed in terms of angles in polar coordinates:

$$\begin{aligned} \boldsymbol{n} &= (\sin \vartheta \cos \varphi,\ \sin \vartheta \sin \varphi,\ \cos \vartheta) \qquad (0 \leqslant \varphi \leqslant 2\pi,\ 0 \leqslant \vartheta \leqslant \pi), \\ \boldsymbol{\nu} &= (\sin \gamma \cos \delta,\ \sin \gamma \sin \delta,\ \cos \gamma) \qquad (0 \leqslant \delta \leqslant 2\pi,\ 0 \leqslant \gamma \leqslant \pi). \end{aligned}$$

Then as the six parameters of the transformation Λ it is convenient to choose ξ_k and η_k $(k = 1, 2, 3)$:

$$\boldsymbol{\xi} = \alpha \boldsymbol{n}, \quad \boldsymbol{\eta} = \beta \boldsymbol{\nu}. \tag{16}$$

The unit transformation corresponds to $\boldsymbol{\xi} = \boldsymbol{\eta} = 0$.

The space of parameters $\boldsymbol{\xi}$, $\boldsymbol{\eta}$ is doubly connected. In fact, since by (14) there exists a one-to-one correspondence between a Lorentz transformation and the parameters η_1, η_2, and η_3, the sets of parameters $\boldsymbol{\xi} = \pi \boldsymbol{n}$ and $\boldsymbol{\xi} = -\pi \boldsymbol{n}$, by (15), describe the same rotation. Consequently, these points of $\boldsymbol{\xi}$-space must be identified with one another, making $\boldsymbol{\xi}$-space doubly connected. The space of $\boldsymbol{\xi}$ is a sphere of radius π with center at $\boldsymbol{\xi} = 0$, inside of which every point $\boldsymbol{\xi} = \alpha \boldsymbol{n}$ is associated with a rotation $R(\boldsymbol{n}, \alpha)$ while points on opposite ends of diameters are identical (Fig. 1), so that the diameter should be considered a closed path. Each point $(\boldsymbol{n}, \alpha)$ may thus be reached from the center by two inequivalent paths 1 and 2, which cannot be made to coincide by continuous deformations: for example, the path l_1 proceeds directly from the center along a line $\xi_1(t) = n\alpha t$ $(0 < t < 1)$, while the path l_2 involves a discontinuity, e.g. along the line $\xi_2(t) = -n\pi t$ $(0 < t < 1)$ and then along a line $\xi_2'(t') = n(\alpha-\pi)t' + n\pi$ $(0 < t' < 1)$.

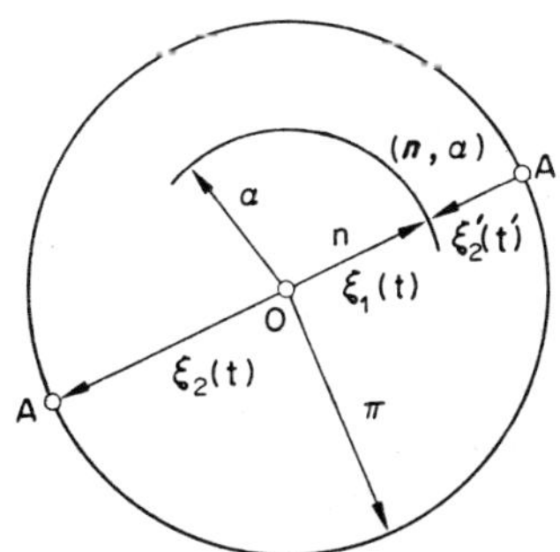

FIG. 1. Space of points $\xi = \alpha n$ associated with rotation by an angle α about an axis n. Points on opposite sides of a diameter are identified with one another.

In quantum mechanics the single-connectedness of parameter space becomes important. There is a well-known way of passing from a multiply-connected topological group $\mathfrak{G}$ to a singly connected covering group $\overline{\mathfrak{G}}$, whose element $\bar{g} = \{g, l(t)\}$ is characterized by an element of the initial group g and a path $l(t)$ from the unit element to g. In the case of rotations R via paths 1 and 2, the covering group $\bar{R}$ has elements

$$\left.\begin{aligned} \bar{R}_1(n, a) &= \{R(n, \alpha),\ l_1 = R(n, t\alpha)\}, \\ \bar{R}_2(n, \alpha) &= \{R(n, \alpha),\ l_2 = R(-n, (2\pi - t\alpha))\}, \end{aligned}\right\} \tag{17}$$

since formally $R(n, \alpha) = R(-n, 2\pi-\alpha)$ for $\pi < \alpha < 2\pi$.

Combining (14) and (17), one can easily write an element of the covering Lorentz group.

§ 1.2. Quantum mechanics and relativity

Let us recall some characteristic features of quantum mechanical description. In quantum mechanics physical quantities are represented by self-adjoint operators α which act on state vectors $|\gamma\rangle$ forming a Hilbert space $\mathcal{H}$. The state vector $|\gamma\rangle$ describes a physical system in a probabilistic way: if the system is in the state γ, then the quantity $W_{\varrho\gamma} = |\langle\varrho|\gamma\rangle|^2$ is equal to the probability of observing the system in the state ϱ, where $\sum_\varrho W_{\varrho\gamma} = 1$. The vectors $|\gamma\rangle$ and $u|\gamma\rangle$ differing by the phase factor $u(|u| = 1)$ thus describe the same state, so that the correspondence between the physical state γ and the vector $|\gamma\rangle$ is not one-to-one.

The set of Hilbert space vectors $|\gamma\rangle$ differing from one another only by a relative phase is called a unit ray γ; all vectors of a unit ray may be obtained by multiplying one of them by a phase. Thus each state is associated in a one to one way with a unit ray.

When α has a definite value α' in some state $|\alpha'\rangle$, then

$$\alpha|\alpha'\rangle = \alpha'|\alpha'\rangle. \tag{18}$$

If the operators α_1 and α_2 commute ($[\alpha_1, \alpha_2] = 0$), then there exists a state $|\alpha_1', \alpha_2'\rangle$, arising as a consequence of the simultaneous measurement of these quantities. In quantum mechanics it is assumed that for every physical system there exists a full set of mutually commuting operators $\alpha_1 \ldots \alpha_n$, whose eigenvalues $\alpha_1' \ldots \alpha_n'$ may be used to fully characterize the state $|\alpha_1' \ldots \alpha_n'\rangle$. States belonging to different eigenvalues are orthogonal:

$$\langle \alpha_1'' \ldots \alpha_n'' | \alpha_1' \ldots \alpha_n' \rangle = \delta_{\alpha_1''\alpha_1'} \ldots \delta_{\alpha_n''\alpha_n'}. \tag{19}$$

One may obviously choose different sets of mutually commuting operators. Any physical quantity must be contained in at least one of the complete sets since there always exists a state in which it has a definite value (the postulate of the existence of "pure states"). According to this postulate, the state of the system may be prepared with arbitrary accuracy using a single measurement of each of the quantities in a complete set.

Moreover, in quantum mechanics it is postulated that the set of vectors $|\alpha_1' \ldots \alpha_n'\rangle$ with various values $\alpha_1' \ldots \alpha_n'$ exhausts all possible states of the system (the postulate of completeness).

Any state of a system may be represented in the form of a superposition of basis states (the principle of superposition):

$$|\gamma\rangle = \sum_{\alpha_1' \ldots \alpha_n'} \psi_\gamma(\alpha_1' \ldots \alpha_n') |\alpha_1' \ldots \alpha_n'\rangle. \tag{20}$$

The expansion coefficients $\psi_\gamma(\alpha_1' \ldots \alpha_n') = \langle \alpha_1' \ldots \alpha_n' | \gamma \rangle$ are sometimes called wave functions in the $(\alpha_1 \ldots \alpha_n)$ representation.

The states we investigate are those of the Heisenberg type. Such a state $|\alpha_1' \ldots \alpha_n'\rangle$ may be obtained as the result of a series of consecutive measurements $\alpha_1 \ldots \alpha_n$, which, in general, may be performed at any time in the study of the system. It describes properties of the system, including, for example, interactions among its separate parts. The operators α in this case may depend on time. This means that observation of the quantities α at the time x^0 may demand a different experiment from that used for observing α at the moment $x^{0\prime}$.

The states $|\alpha_1' \ldots \alpha_n'\rangle$ differ from states of the Schroedinger type, which are based on multiple measurements of all quantities $\alpha_1 \ldots \alpha_n$ at the same time x^0. The set of Schroedinger states at different times contains the same information as the state $|\alpha_1' \ldots \alpha_n'\rangle$.

Superselection rules

Not all self-adjoint operators may be observable, just as not all vectors $|\varrho\rangle$ correspond to physical states. This situation arises, for example, when the observables include quantities Q_i contained in all complete sets. The operators Q_i then commute with all remaining operators corresponding to physical quantities. In this case the right-hand side of the expansion (20) of the physical state $|\gamma\rangle$ must contain only vectors referring to the same set of eigenvalues Q_i'.

In fact, if $|\gamma\rangle$ is a physical state, then the projection operator on this state $P(\gamma) = |\gamma\rangle\langle\gamma|$ must be observable as well. Since by definition Q_i commutes with all observables, $[Q_i, P(\gamma)] = 0$. The equality $Q_i|\gamma\rangle\langle\gamma| - |\gamma\rangle\langle\gamma|Q_i = 0$ is satisfied only if $|\gamma\rangle$ is an eigenvector of Q_i. The Hilbert space of physical states in this case may be expanded as a direct sum of coherent subspaces each of whose states correspond to the same set of values Q_i'. The expansion of the physical state $|\gamma\rangle$ in terms of basis states is then written in the form

$$|\gamma\rangle = \sum_{\alpha_1' \dots \alpha_n'} \psi_\gamma(\alpha_1' \dots \alpha_n') \, |\alpha_1' \dots \alpha_n', (Q_i')\rangle, \tag{21}$$

where the values $(Q_i') = (Q_1' \dots Q_n')$ are the same for all basis vectors in the given coherent subspace. This restriction on the superposition principle is called a superselection rule.[21, 22] Quantities of the type Q_i have strictly conserved additive quantum numbers: electric, baryonic and leptonic charges.

According to a superselection rule, operators corresponding to observable quantities cannot have matrix elements between the states belonging to different coherent subspaces. If a self-adjoint operator B had matrix elements $\langle Q'' | B | Q' \rangle \neq 0$, connecting different coherent spaces, then it could not commute with Q. But the operator B would then not be contained in any of the complete sets of observables (which always include Q).

Symmetry and invariance

To clarify these concepts, let us turn to the role of the measuring apparatus† in quantum mechanical description. The measuring apparatus participates in quantum mechanical description in two ways. Firstly, a state of a quantum mechanical object is prepared using one apparatus; then a measurement on this object is carried out using another. These devices are called preparing and measuring devices respectively. As a result of the preparation of the state $|\gamma\rangle$, there arises, as we have mentioned earlier, a unit ray γ in Hilbert space. The prepared quantum mechanical object then interacts with the measuring device. If the process of measurement determines whether the object is in the state ϱ, then the measuring device corresponds in Hilbert space to the unit ray ϱ.

The probability of finding the object in state ϱ, if it was prepared in state γ, is equal to $W_{\varrho\gamma} = |\langle \varrho | \gamma \rangle|^2$, where $|\varrho\rangle$ and $|\gamma\rangle$ are any vectors belonging respectively to the rays ϱ and γ. The probability of transitions $W_{\varrho\gamma}$ together with the expectation values of physical quantities $\bar{\alpha} = \langle \gamma | \alpha | \gamma \rangle$ are direct results of experimental study of the physical system. It is by the values of $W_{\varrho\gamma}$ and $\bar{\alpha}$ that we judge the accuracy of model representations of the system.

Two systems a and b observed using the same devices (laboratories) X will be identical if the set of their physical states is the same: $\{\gamma(a)\} = \{\gamma(b)\}$. The states $\gamma(a)$ and $\gamma(b)$ of these systems will, in general, be different.

Since the sets $\{\gamma(a)\}$ and $\{\gamma(b)\}$ are the same, there exists a one-to-one correspondence $\gamma(a) \leftrightarrow \gamma(b)$ between their members. Changing the order of observations of a, we can set up such a correspondence in another way as well: $\gamma(a) \leftrightarrow \gamma'(b)$, etc. All such correspondences form a group of transformations $\gamma \leftrightarrow \gamma'$ of the states of the set $\{\gamma\}$ into themselves. The product of its elements is the sequence of transformations $\gamma \to \gamma' \to \gamma'' = \gamma \to \gamma''$, which corresponds to a unique transformation of the initial observation into the final observation, and the unit element is $\gamma \to \gamma$. The inverse element to $\gamma \to \gamma'$ is obviously $\gamma' \to \gamma$.

† As usual, we mean by a measuring apparatus an instrument whose change in state upon interaction with the physical systems being studied can be detected. A measuring apparatus must thus have at the end a macroscopic indicator fixing the result of the observation in terms of classical physics.

The general group of transformations of physical states $\gamma \to \gamma'$ contains a subgroup of transformations leaving invariant the probability of the transition $1 \to 2$:

$$|\langle\gamma_2|\gamma_1\rangle|^2 = |\langle\gamma_2'|\gamma_1'\rangle|^2. \tag{22}$$

The properties of such transformations are characterized by the following general theorem of Wigner,[18, 23] which we present without proof.

Any one-to-one transformation $\gamma \to \gamma'$ of a set of unit rays γ in a Hilbert space $\mathcal{H}$ into a collection of unit rays γ' in a Hilbert space $\mathcal{H}' = \mathcal{H}$, which satisfies the condition

$$|\langle\gamma_2|\gamma_1\rangle|^2 = |\langle\gamma_2'|\gamma_1'\rangle|^2,$$

generates either a unitary or an antiunitary operator, determined up to a phase factor.

A unitary transformation U has the properties:

$$\mathrm{U}(c_1|\alpha_1'\rangle + c_2|\alpha_2'\rangle) = c_1\mathrm{U}|\alpha_1'\rangle + c_2\mathrm{U}|\alpha_2'\rangle,$$
$$\mathrm{UU}^+ = \mathrm{U}^+\mathrm{U} = 1,$$

where c_1 and c_2 are any complex numbers. Under the transformation

$$|\gamma'\rangle = \mathrm{U}|\gamma\rangle, \quad \boldsymbol{\alpha}' = \mathrm{U}\alpha\mathrm{U}^{-1} \tag{23}$$

the probability amplitude and the matrix elements of observables do not change:

$$\langle\gamma_2'|\gamma_1'\rangle = \langle\gamma_2|\gamma_1\rangle, \quad \langle\gamma_2'|\alpha'|\gamma_1'\rangle = \langle\gamma_2|\alpha|\gamma_1\rangle.$$

An antiunitary transformation A contains the operation of complex conjugation:

$$\mathrm{A}(c_1|\alpha_1\rangle + c_2|\alpha_2\rangle) = c_1^*\mathrm{A}|\alpha_1\rangle + c_2^*\mathrm{A}|\alpha_2\rangle;$$

it changes a probability amplitude into its complex conjugate:

$$\langle\mathrm{A}\gamma_1|\mathrm{A}\gamma_2\rangle = \langle\gamma_1|\gamma_2\rangle^* = \langle\gamma_2|\gamma_1\rangle. \tag{24}$$

Thus symmetry operations are described by unitary or antiunitary operators. However, not every transformation leaving the transition probability invariant forms a symmetry operation. For example, some transformations may describe features of the interaction in a given particular problem. A unitary transformation U may also describe a "renumbering" of basis vectors of the state $|\alpha'\rangle$.

Symmetry principles are usually related to rather universal properties of physical systems. Space–time symmetry expresses the relativistic invariance of laws of motion. Internal symmetry characterizes the properties of interactions of a given class. Processes with elementary particles may differ also with respect to operations of discrete symmetry (space–time reflections).

Among symmetries of various sorts, space–time symmetry occupies a special place: only in this case can one give an exact recipe for carrying out any symmetry transformation. It is impossible to give an actual prescription for carrying out a measurement on an apparatus with the direction of time reversed. Internal symmetry transformations, as a rule, carry a state vector out of a coherent space, rotating this state into an unphysical one (see § 9.1).

Relativistic invariance

In quantum mechanics an "inertial reference frame X" is associated with a uniformly moving set X of macroscopic devices (laboratories) essential for the full description of the physical system. Using these devices one can prepare and measure any states of a given physical system.

A transformation of the Poincaré group characterized by an element $g = (a, \Lambda)$ may be interpreted in two ways. On the one hand it may describe the difference between the position of two identical physical systems with respect to the given frame of reference ("active point of view"). On the other hand, this transformation may characterize the difference between two frames of reference in which the same physical system is being studied ("passive point of view").

In the "active" interpretation of a relativistic transformation with element $g = (a, \Lambda)$, each physical state γ is associated with a transformed state γ_g. The state γ_g differs from the state γ by the location of the preparing devices. In preparing the state γ_g the apparatus with the preparing devices is translated, rotated, or moved uniformly relative to the previous location (according to the geometrical sense of g). Consequently, the transformation g generates a one-to-one transformation of the set of all those unit rays in Hilbert space which correspond to physical states. From this it follows that every set of state vectors forming a Hilbert space is invariant with respect to transformations of the Poincaré group.

Analogously, with every state ϱ created by the measuring devices one can associate a transformed state ϱ_g. To find the probability of the transition $\gamma_g \to \varrho_g$, we have to repeat the experiment $\gamma \to \varrho$ (on the same or on an identical physical system), but with all devices (both preparing and measuring) translated or moving relative to the initial position in the sense of $g = (a, \Lambda)$. The existence of a space–time symmetry is manifested in the equality of the transition probabilities for $\gamma \to \varrho$ and $\gamma_g \to \varrho_g$, i.e. in the invariance of the theory relative to transformations of the Poincaré group:

$$|\langle \varrho | \gamma \rangle|^2 = |\langle \varrho_g | \gamma_g \rangle|^2. \tag{25}$$

Here $|\gamma\rangle$ and $|\varrho\rangle$ are any vectors in the rays γ and ϱ.

In the "passive" interpretation, the transformations $g = (a, \Lambda)$ compare descriptions of the same physical system in two reference frames X and Y, i.e. in two laboratories X and $Y = gX$. These laboratories are considered equivalent in the sense that the set of possible observations in X and Y are the same. Elements of the Poincaré group describe the respective location of equivalent laboratories, and all such equivalent laboratories may be enumerated relative to some fixed laboratory using the properties of the group. Equivalent laboratories need not be thought of as copies of one laboratory, since the concept of the group of motion in space–time has already been used.

The equivalence of laboratories X and Y means that there is a one-to-one correspondence $\alpha_i(X) \leftrightarrow \alpha_i(Y)$ between all operators $\alpha_i(X)$ and $\alpha_i(Y)$ comprising a complete set of observables in X and Y. The operators $\alpha(X)$ and $\alpha(Y)$ describe the same observables from the point of view of the systems X and Y; $\alpha(Y)$ is that operator for some time x^0 in the system X which from the point of view of system Y has the same form as the operator $\alpha(X)$ in the system X

at the same time x^0. The transformation $\alpha(X) \to \alpha(Y)$ is a "translation" of the language of the reference frame X into the language of the equivalent system Y.

Passive "translation" transformations may, in general, not even coincide with symmetry transformations; the "translation" group guarantees only the possibility of recalculating the observables of one reference frame in another, i.e. the covariance of the description. If, for example, the effects of gravity are taken into account, then the "translation" group will exist, while the invariance relative to transformation of the Poincaré group will no longer be present.

"Translation" transformations are symmetry transformations when the principle of relativity, according to which physical phenomena proceed in the same manner in all inertial reference frames, is fulfilled. In this case the transition probability between the states $\gamma(a, X)$ and $\varrho(a, X)$ of the system a in the reference frame X will be equal to the transition probability between the states $\gamma(a, Y)$ and $\varrho(a, Y)$ of the same physical system a described from the point of view of the reference frame Y.

Unitarity and continuity of representations of the group $\mathcal{P}^{\uparrow}_{+}$

We shall adhere to the active interpretation of transformations of the Poincaré group. Let us consider the transformation of state vectors by an element of the proper orthochronous group $g = (a, \Lambda)$:

$$|\gamma\rangle \to |\gamma_g\rangle = \mathrm{U}(g)|\gamma\rangle, \tag{26}$$

which conserves the probability (25), and let us find the properties of the operator $\mathrm{U}(g)$. The definition of an operator performing the transformation in (26) assumes that the operator $\mathrm{U}(g)$ depends only on the properties of the transformation g, and not on the properties of the state $|\gamma\rangle$. If the opposite were true, the influence of properties of the specific states would be impossible to separate from physical laws. We see that for the transformations $g = (a, \Lambda)$ the operator $\mathrm{U}(g)$ is unitary and continuous in g. Moreover, we see that the indeterminancy in the choice of the phase factor for the state vector together with the double-connectedness of the parameter space of the Lorentz group leads to the fact that the state vectors transform according to unitary representations of the universal covering group $\overline{\mathcal{P}}^{\uparrow}_{+}$ of the Poincaré group $\mathcal{P}^{\uparrow}_{+}$.

Superselection rules demand that the unit rays γ and ϱ in (25) lie in the same coherent subspace. The respective rays γ_g and ϱ_g must also lie in the same coherent subspace. Otherwise the vector $(|\gamma_g\rangle + |\varrho_g\rangle)/\sqrt{2}$ will not describe a physical state, while the vector $(|\gamma\rangle + |\varrho\rangle)/\sqrt{2}$ describes such a state. Thus under the transformation $\gamma \to \gamma_g$ induced by transformations of the Poincaré group (1), separate coherent subspaces are taken either into themselves or into each other.

The act of measurement in quantum mechanics is carried out using classical devices. The transformation of reference frames connected with measuring devices is macroscopic. We thus postulate that the transition probability

$$W_{\varrho\gamma_g} = |\langle\varrho|\gamma_g\rangle|^2 = |\langle\varrho|\mathrm{U}(g)|\gamma\rangle|^2 \tag{27}$$

is a continuous function of parameters on which the transformation $g = (a, \Lambda)$ depends. Here γ is a unit ray describing the physical space in the reference frame X, while γ_g is the "corresponding" ray referring to the reference frame Y. The reference frame Y is obtained from X by the transformation $g = (a, \Lambda)$. Finally, ϱ denotes any fixed ray.

From the continuity of (27) it follows that the matrix elements of the operators $U(g)$ are continuous. Moreover, the group $\mathcal{P}_+^\uparrow$ contains the identity transformation, and the different coherent subspaces are mutually orthogonal. This means that the postulated continuity may hold only when each coherent subspace goes into itself.

The successive application of two transformations (26) gives

$$|\gamma_{g_1 g_2}\rangle = U(g_1)\, U(g_2)\, |\gamma\rangle. \tag{28}$$

The result must be equivalent to the direct transformation $|\gamma\rangle \to |\gamma_{g_3}\rangle$ with $g_3 = g_1 g_2$. However, the state vector $|\gamma\rangle$ is determined only up to a phase factor ω, so that $|\gamma_g\rangle$ and $\omega|\gamma_g\rangle$ describe the same state. Thus, in the general case, one must write

$$|\gamma_{g_1 g_2}\rangle = \omega(g_1, g_2)\, U(g_1)\, U(g_2)\, |\gamma\rangle$$

and the multiplication rule for transformations is

$$U(g_1 g_2) = \omega(g_1, g_2)\, U(g_1)\, U(g_2). \tag{29}$$

The unitarity of the operator $U(g)$ for transformations of the group $\mathcal{P}_+^\uparrow$ follows from the fact that $\mathcal{P}_+^\uparrow$ contains the identity transformation $E = (0, 1)$. Every small (but finite) transformation g close to E may be represented as a square $g = g'^2$ of some other transformation g'. This means, by (29), that $U(g) = \omega(g', g')\, U^2(g')$. The square of such a unitary U, as well as of an antiunitary operator A, is a unitary operator:

$$A^2(c_1|\alpha_1\rangle + c_2|\alpha_2\rangle) = c_1 A^2|\alpha_1\rangle + c_2 A^2|\alpha_2\rangle,$$
$$\langle A^2\varrho | A^2\gamma\rangle = \langle \varrho | \gamma\rangle.$$

Thus $U(g)$ is always unitary in a finite neighborhood of E. Any transformation may be obtained by an application of a finite number of operators $U(g)$ and is thus always unitary.

The complex factor $\omega(g_1, g_2)$ in (29) depends on the choice of the form of the representation $U(g)$. A unitary operator $U(g)$ may be multiplied by a phase factor $U'(g) = \varphi(g)\, U(g) \times (|\varphi(g)| = 1)$ without changing the unitary nature of $U(g)$. The factor $\omega(g_1, g_2)$ is then changed to another:

$$\omega'(g_1, g_2) = \varphi(g_1)\, \varphi(g_2)\, \omega(g_1, g_2)\, \varphi^{-1}(g_1 g_2). \tag{30}$$

Both factors ω and ω' are equally valid from the point of view of unitary representations $U(g)$. This circumstance may be used to get rid of the phase arbitrariness in the multiplication rule (29) for the case of transformations g_2 close to the unit E, when $g_1 = E$ in (30). However, this result may be carried over to finite transformations only in the case that the region of variation of the parameters is simply connected. In the case of Lorentz transformations the space of parameters ξ_k [formula (17)] is doubly connected. Thus if one makes the space ξ simply connected and passes from $\mathcal{P}_+^\uparrow$ to the covering group $\bar{\mathcal{P}}_+^\uparrow$, then the factor ω in (29) vanishes, but one must take U to mean a unitary representation of the group $\bar{\mathcal{P}}_+^\uparrow$.[23] Denoting an element of the group $\bar{\mathcal{P}}_+^\uparrow$ by $\bar{g}$,

$$U(\bar{g}_1)\, U(\bar{g}_2) = U(\bar{g}_1, \bar{g}_2), \tag{31}$$

i.e., the usual multiplication rule for group representations. The proof of formula (31) was given by Bargmann.[(24)]

Thus the condition of relativistic invariance in quantum mechanics is equivalent to the demand that the state vectors transform according to continuous unitary representations of the universal covering group $\bar{\mathcal{P}}^{\uparrow}_{+}$ of the Poincaré group without reflections. We will call the group $\bar{\mathcal{P}}^{\uparrow}_{+}$ the quantum mechanical Poincaré group.†

In passing to another reference frame, the observables α transform according to the rule (23)

$$\alpha \rightarrow \alpha_{g} = \mathrm{U}(\bar{g})\,\alpha\mathrm{U}^{-1}(\bar{g}),$$

so that their matrix elements remain unchanged:

$$\langle \varrho_{\bar{g}} \,|\alpha_{\bar{g}}|\, \gamma_{\bar{g}}\rangle = \langle \varrho\,|\alpha|\,\gamma\rangle.$$

The physical quantities Q_i corresponding to superselection rules are invariant with respect to transformations of the group $\bar{\mathcal{P}}^{\uparrow}_{+}$.

§ 1.3. Basis quantities

In the previous paragraph it was assumed implicitly that we already knew the quantities serving as dynamical variables. In fact, the choice of these quantities is determined by symmetry considerations, i.e. by the choice of invariance group. For the purposes of the previous section it was sufficient that these dynamical variables existed. In this section we shall determine the observables associated with the Poincaré group $\bar{\mathcal{P}}^{\uparrow}_{+}$, i.e. with the properties of homogeneity and isotropy of space–time. These quantities (momentum, orbital angular momentum, and spin) are known from experiment and nonrelativistic theory.

Let us turn to infinitesimal transformations. Formula (1) may be rewritten in the form

$$x'^{\mu} = x^{\mu} + \varepsilon^{\mu} + \omega^{\mu}{}_{\nu} x^{\nu}, \tag{32}$$

where the infinitesimal parameters ε^{μ} and $\omega_{\mu\nu}$ are real [$\omega_{\mu\nu} = -\omega_{\nu\mu}$ is antisymmetric by virtue of the condition (4)]. The infinitesimal coordinate transformation (32) entails a change of state vectors $|\gamma\rangle \rightarrow (1+\delta\mathrm{U})|\gamma\rangle$ which, as a consequence of the continuity of the transformation, will also be infinitesimal:

$$\delta\mathrm{U} = iP_{\mu}\varepsilon^{\mu} - i\tfrac{1}{2}M_{\mu\nu}\omega^{\mu\nu}. \tag{33}$$

In formula (33) we introduce the operators P_{μ} and $M_{\mu\nu} = -M_{\nu\mu}$, which are Hermitian since $1+\delta\mathrm{U}$ is unitary. By virtue of (31) these operators commute with all charges Q_i associated with superselection rules. The quantities P_{μ} and $M_{\mu\nu}$, consequently, can be observables. Let us discuss their meaning.

According to (32) and (33) the operator

$$1 + iP_{\mu}\varepsilon^{\mu} \tag{34}$$

† A review of geometrical invariance principles is given in ref. 25.

describes an infinitesimal translation generated by motion of the reference frame:

$$x_\mu \to x'^\mu = x^\mu + \varepsilon^\mu .$$

Infinitesimal three-dimensional rotations and Lorentz transformations are described by the operator

$$1 - i\tfrac{1}{2}M_{\mu\nu}\omega^{\mu\nu} . \tag{35}$$

For example, the transformation

$$1 - iM_{12}\omega^{12} \tag{36}$$

describes the change of a state vector caused by the rotation

$$x'^0 = x^0, \qquad x'^1 = x^1 - \omega^{12}x^2,$$
$$x'^2 = x^2 + \omega^{12}x^1, \quad x'^3 = x^3 .$$

The Lorentz transformation to a moving inertial system

$$x'^0 = x^0 - x^1\omega^{01}, \quad x'^1 = x^1 - x^0\omega^{01},$$
$$x'^2 = x^2, \qquad x'^3 = x^3$$

causes a transformation of the state vector by the operator

$$1 - iM_{01}\omega^{01} . \tag{37}$$

Thus the quantity iP_μ is a generator of translations along the axis x^μ, $-iM_{jk}$ is a generator of rotations in the (jk) plane, and the quantity $-iM_{0j}$ is a generator of a pure Lorentz transformation. Ten generators have been enumerated, corresponding to the 10 single-parameter transformations of the Poincaré group, and these are the basis quantities in relativistic quantum mechanics. The quantity P_μ is called the energy-momentum or four-momentum vector; the three-vector

$$\boldsymbol{M} = (M_{23}, M_{31}, M_{12})$$

is the orbital angular momentum. Instead of the component M_{0j} one frequently introduces the three-vector

$$\boldsymbol{N} = (M_{01}, M_{02}, M_{03}).$$

If $\Phi(x)$ is an operator of some physical quantity, depending on the coordinate x, then from (23) and (34) it follows that

$$i[P_\mu, \Phi(x)] = \frac{\partial \Phi(x)}{\partial x^\mu} .$$

The development in time of the dynamical variables of the system is determined by the operator P_0: the Hamiltonian of the system. Specification of the Hamiltonian (i.e. P_0) as a function of the dynamical variables allows the complete characterization of its development.

Let us find the commutation relations between the operators P_μ and $M_{\lambda\nu}$. For this we shall use the group multiplication law (9) and the fact that the commutation relations are the same for the generators of the group as for the generators of the covering group. Thus in the multiplication law for unitary representations we shall set $\omega(g_1, g_2) = 1$. Then, according to (9) and (29),

$$\left.\begin{aligned} U(a_1, \Lambda_1)\, U(a_2, \Lambda_2) &= U(a_1+\Lambda_1 a_2, \Lambda_1\Lambda_2), \\ U(a, \Lambda)^{-1} &= U(-\Lambda^{-1}a, \Lambda^{-1}). \end{aligned}\right\} \tag{38}$$

Here $g_1 = (a_1, \Lambda_1)$ and $g_2 = (a_2, \Lambda_2)$.

We shall first introduce the transformation rule for generators under a Lorentz transformation Λ. In the case of momentum, we shall start from the relation

$$U(0, \Lambda^{-1})\, U(a, 1)\, U(0, \Lambda) = U(\Lambda^{-1}a, 1),$$

which, after inserting the infinitesimal form (34),

$$U(\varepsilon, 1) \approx 1+iP_\mu \varepsilon^\mu,$$

may be rewritten in the form

$$U^{-1}(0, \Lambda)\, P_\mu \varepsilon^\mu U(0, \Lambda) = P_\mu(\Lambda^{-1}\varepsilon)^\mu.$$

From this, by virtue of (4), we find the transformation of the momentum operator P_μ:

$$U^{-1}(0, \Lambda)\, P^\mu U(0, \Lambda) = \Lambda^\mu{}_\nu P^\nu. \tag{39}$$

The transformation of the generator $M_{\mu\nu}$ may be introduced starting from the relation

$$U(0, \Lambda^{-1})\, U(0, \Lambda')\, U(0, \Lambda) = U(0, \Lambda^{-1}\Lambda'\Lambda),$$

in which $U(0, \Lambda')$ is expressed in terms of $M_{\mu\nu}$ by (35). The right-hand side can be written in the form

$$U(0, \Lambda^{-1}\Lambda'\Lambda) \approx 1-i\tfrac{1}{2}(\Lambda^{-1})^\mu{}_\varrho \Lambda^\nu{}_\lambda \omega^{\varrho\lambda} M_{\mu\nu}.$$

Comparing coefficients of $\omega^{\mu\nu}$ and using (4), we obtain a formula for the transformation of $M_{\mu\nu}$:

$$U(0, \Lambda^{-1})\, M^{\mu\nu} U(0, \Lambda) = \Lambda^\mu{}_\varrho \Lambda^\mu{}_\lambda M^{\varrho\lambda}. \tag{40}$$

In order to obtain the commutation relations between generators of the Poincaré group, it is now sufficient to insert (35) for infinitesimal transformations $U(0, \Lambda)$ associated with the coordinate transformation $x^\mu \to (\Lambda x)^\mu = x^\mu + \omega^\mu{}_\nu x^\nu$ into formulae (39) and (40).

To summarize, the commutation relations have the form

$$\left.\begin{aligned} [M_{\mu\nu}, M_{\lambda\sigma}] &= i[g_{\mu\sigma}M_{\nu\lambda}+g_{\nu\lambda}M_{\mu\sigma}-g_{\mu\lambda}M_{\nu\sigma}-g_{\nu\sigma}M_{\mu\lambda}], \\ [M_{\mu\nu}, P_\lambda] &= i[g_{\nu\lambda}P_\mu - g_{\mu\lambda}P_\nu], \quad [P_\mu, P_\nu] = 0. \end{aligned}\right\} \tag{41}$$

The relation between components of the momentum is obtained directly from the group multiplication law (38) as a consequence of the commutativity of translations in different directions.

The commutation relations for $M_{\mu\nu}$ are equivalent to the following relations for the orbital angular momentum $\boldsymbol{M}$ and the Lorentz generator $\boldsymbol{N}$:

$$\begin{aligned} [M_i, M_j] &= i\varepsilon_{ijl}M_l, \quad [N_i, N_j] = -i\varepsilon_{ijl}M_l, \\ [M_i, N_j] &= i\varepsilon_{ijl}N_l \qquad (i, j, l = 1, 2, 3), \end{aligned}$$

where ε_{ijl} is totally antisymmetric ($\varepsilon_{123} = 1$). The combinations of these operators

$$\mathcal{L}_j^{\pm} = \tfrac{1}{2}(M_j \pm iN_j)$$

satisfy the commutation relations for angular momentum:

$$[\mathcal{L}_k, \mathcal{L}_j] = i\varepsilon_{kjl}\mathcal{L}_l, \quad [\mathcal{L}_i^+, \mathcal{L}_j^-] = 0,$$

where $\mathcal{L}_i^+$ and $\mathcal{L}_j^-$ are independent.

Using the infinitesimal transformation (33), one can construct a finite transformation. Thus the unitary operators U(a, 1) and U(0, Λ) have the following general form:

$$\mathrm{U}(a, 1) = e^{iP_\mu a^\mu}, \quad \mathrm{U}(0, \Lambda) = e^{-i(1/2)M_{\mu\nu}\omega^{\mu\nu}}. \tag{42}$$

Let us now construct quantities commuting with all generators using $M_{\mu\nu}$ and P_μ. Such quantities do not change under transformations of the group and will be proportional to the unit matrix in each irreducible representation. One may classify the irreducible representations by the values of these quantities.

One can form only one invariant quantity

$$m^2 = P_\mu P^\mu. \tag{43}$$

from the components of momentum. Being a scalar, m^2 commutes with $M_{\lambda\sigma}$; as a function of momentum, m^2 commutes with P_λ.

For $m^2 \geqslant 0$ m^2 has the meaning of the square of the rest mass. In this case there exists a second invariant quantity constructed from the momenta: the operator of the sign of the energy

$$\varepsilon = \frac{P_0}{|P_0|} \tag{44}$$

with eigenvalues $\varepsilon' = \pm 1$. Since proper orthochronous transformations $\mathcal{P}_+^\uparrow$ do not change the sign of the components of a time-like vector, ε is invariant.

Using only one of the quantities $M_{\lambda\nu}$ it is impossible to form an invariant operator. There is only one combination of generators $M_{\lambda\nu}$ and P_μ which commutes with the momentum P_λ, namely the pseudo-vector

$$w_\mu = \tfrac{1}{2}\varepsilon_{\mu\nu\lambda\sigma}M^{\nu\lambda}P^\sigma \tag{45}$$

with the property

$$w_\mu P^\mu = 0, \quad [w_\nu, P_\lambda] = 0. \tag{46}$$

The square of the pseudo-vector w_μ is a scalar, and, consequently, commutes with $M_{\mu\nu}$. Thus $w^2 = w_\mu w^\mu$ commutes with all 10 generators. An explicit expression for w^2 is

$$w^2 = \tfrac{1}{2}M_{\mu\nu}M^{\mu\nu}P_\lambda P^\lambda - M_{\mu\sigma}M^{\nu\sigma}P^\mu P_\nu. \tag{47}$$

The physical meaning of the invariant w^2 may be easily seen when $m^2 > 0$. Let us take the four-momenta P_μ diagonal. Then in the rest system, i.e. when acting on the state vector

$|\boldsymbol{p} = 0, m\rangle$ the invariant w^2 is equal to†

$$w^2|\boldsymbol{p} = 0, m\rangle = -m^2(M_{12}^2+M_{23}^2+M_{31}^2)|\boldsymbol{p} = 0, m\rangle = -m^2\boldsymbol{M}^2|\boldsymbol{p} = 0, m\rangle. \tag{48}$$

In other words, the quantity $-w^2/m^2$ is equal to the square of the angular momentum $\boldsymbol{M}^2$, i.e. the square of the spin, in the rest frame. The eigenvalues of $\boldsymbol{M}^2$, as is well known in quantum mechanics, have the form $J(J+1)$, where $J = 0, \frac{1}{2}, 1, \frac{3}{2}, \ldots$ Since w^2 is an invariant operator in any reference frame, applying it to a state $|\boldsymbol{p}, m\rangle$ and transforming according to an irreducible representation of $\mathcal{P}_+^\uparrow$, gives

$$w^2|\boldsymbol{p}; m, J\rangle = -m^2J(J+1)|\boldsymbol{p}; m, J\rangle. \tag{49}$$

To establish a covariant relation between w_μ and the spin operator J_k, we introduce the space-like normals $n_\mu^{(r)}$ $(r = 1, 2, 3)$: $n_\mu^{(r)}n^{(r')\mu} = -\delta_{rr'}$, $n_\mu^{(r)}p^\mu = 0$. Together with the velocity $v_\mu = p_\mu/m \equiv n_\mu^{(0)}$, they form a complete system of four-normals: $n_\mu^{(\alpha)}g^{\mu\nu}n_\nu^{(\beta)} = g^{\alpha\beta}$.

Since $w^\lambda n_\lambda^{(0)} = 0$, there are only three independent space-like components of w_μ; in covariant form:

$$w_\lambda = m\sum_{k=1}^{3} J^k n_\lambda^{(k)}, \qquad J^k = -\frac{1}{m}\, w^\lambda n_\lambda^{(k)}. \tag{50}$$

Introducing the vector $\boldsymbol{J} = (J^1, J^2, J^3)$, we find by direct calculation:

$$w^2 = -m\boldsymbol{J}^2, \qquad [J^i, J^j] = i\varepsilon^{ijk}J^k, \tag{51}$$

i.e. the vector $\boldsymbol{J}$ indeed has the properties of a spin operator. The values of J_3 take on the $2J+1$ values $J_3' \equiv \sigma = -J, -J+1 \ldots J-1, J$.

Along with the basic invariants w^2, m^2, and ε (for $m^2 \geqslant 0$), there exists still one operator in the Poincaré group:

$$z = e^{i2\pi M_3} = (-1)^{2M_3} = (-1)^{2J}, \tag{52}$$

which commutes with all generators of the group $\mathcal{P}_+^\uparrow$. The quantity z takes the value $z = 1$ in states with integral angular momentum or spin and $z = -1$ in states with half-integral angular momentum or spin. A rotation by 360° is physically indistinguishable from the identity transformation, and thus the physical properties of the system cannot change under such a rotation. Consequently the value of z must be associated with a superselection rule (see § 1.2), so that a superposition of states with half-integral and integral angular momenta is impossible, and the operators of all observable quantities must commute with z.[21, 22]

Thus an irreducible unitary representation of the Poincaré group is distinguished by the values of spin J, mass m, and the sign of the energy (for $m^2 > 0$). The energy of physical states is always positive, i.e. $\varepsilon > 0$. Particles with integral spin $J = 0, 1, 2, 3, \ldots$, are called bosons; particles with half-integral spin $J = \frac{1}{2}, \frac{3}{2}, \ldots$, are fermions. State vectors forming a basis of an irreducible representation may be distinguished by the eigenvalues of the operators in one of the complete sets. These operators are functions of the generators P_λ and $M_{\mu\nu}$. As an example, we introduce one of the bases: the canonical basis, for $m^2 > 0$.

† We shall designate the eigenvalues of the operator P_μ in one-particle states by p_μ.

In the canonical basis the components of the momentum p_j and the spin projection j_3, on a given fixed axis orthogonal to the four-momentum p_μ are diagonal. In other words, the complete set of mutually commuting quantities contains the operators

$$\begin{gathered} m^2,\ J^2,\ p_j,\ J_3 = \frac{1}{m}\omega\, n_\lambda^{(3)} = \sigma, \\ (n^{(3)})^2 = -1, \quad p^\lambda n_\lambda^{(3)} = 0, \end{gathered} \tag{53}$$

while the state vector is written in the form $|\boldsymbol{p}, \sigma; m, J\rangle$.

Representations of the Poincaré group will be examined in detail in Chapter 4.

§ 1.4. Description of scattering. The S-matrix

The previous section dealt with the quantities describing an isolated system in relativistic quantum mechanics. However, the concept of an isolated system plays only an auxiliary role since particles do not exist without interactions and may be observed only as a result of these interactions. The theory of elementary particles is primarily a theory of interactions. Let us turn to the description of processes with elementary particles.

Information on elementary particles is provided by experiments on their collisions with one another. In all known particle-scattering experiments one observes either the initial particles (elastic scattering) or other particles. In decay experiments, the observed decay products are also only particles. No experiments observe quantum objects other than particles (elementary or composite). We thus assume that particles are the only manifestation of matter in the quantum régime. Consequently the possible states ("complete set") which we shall examine are all possible particle states.

Before and after interaction, particles are detected at large distances from one another (in comparison with the interaction region), and may be considered essentially free. In fact, in the absence of particles with zero mass, the force between particles falls exponentially with increasing distance between them, so that for infinite distances they become free. Asymptotic states of this type are also taken as initial and final states in the description of collisions; they are designated respectively by α_{in} and β_{out} and determine unit rays in the Hilbert spaces $\mathcal{H}_{\text{in}}$ and $\mathcal{H}_{\text{out}}$ in a one-to-one fashion. The vector states in these rays are related by a phase to $|\alpha, \text{in}\rangle$ and $|\beta, \text{out}\rangle$.

The set of all vectors $|\alpha, \text{in}\rangle$, referring to the initial configuration of particles, is complete as a consequence of the assumption of the corpuscular nature of matter (the axiom of asymptotic completeness). Analogously, the set of all possible vector states $|\beta, \text{out}\rangle$ referring to the final configuration of particles also will be complete. Obviously, the Hilbert spaces $\mathcal{H}_{\text{in}}$ and $\mathcal{H}_{\text{out}}$ formed by the vector states $|\alpha, \text{in}\rangle$ and $|\beta, \text{out}\rangle$ will coincide by virtue of the axiom of asymptotic completeness with the full space of states $\mathcal{H}$: $\mathcal{H}_{\text{in}} = \mathcal{H}_{\text{out}} = \mathcal{H}$.

At high energies the colliding particles can produce new particles. In every experiment we observe a definite number of particles. A state with n particles may be characterized by a wave function $\varphi_n(k_1, \ldots k_n)$, depending on the variables $k_1 \ldots k_n$ describing these particles. Since the number of particles is arbitrary (but finite), the state vector arising in collisions may be represented in the form of a superposition of states with a definite number of particles, i.e. in the form of a Fock column vector:[26]

$$|\Psi\rangle = \begin{bmatrix} \varphi_0 \\ \varphi_1(k_1) \\ \cdots\cdots\cdots\cdots \\ \varphi_n(k_1 \ldots k_n) \\ \cdots\cdots\cdots\cdots \end{bmatrix}. \tag{54}$$

If one assumes for simplicity that the variables $k_1 \ldots k_n$ take on discrete values, then the column vector (54) with 1 in the nth place (n particles with variables $k_1 \ldots k_n$) and the remaining $\varphi_i = 0$ is a state vector $|k_1 \ldots k_n\rangle$, so that

$$|\Psi\rangle = \sum_{n=0}^{\infty} \sum_{k_1 \ldots k_n} \varphi_n(k_1 \ldots k_n) \,|\, k_1 \ldots k_n\rangle. \tag{55}$$

The space of states $\mathcal{H}$ is thus the direct sum of spaces $\mathcal{H}_n$ formed by the state vectors with a definite number of particles n.

The scalar product of the vectors $|\Psi\rangle$ and $|\Psi'\rangle$ has the form

$$\langle\Psi' | \Psi\rangle = \sum_{n=0}^{\infty} \sum_{k_1 \ldots k_n} \varphi_n'^{*}(k_1 \ldots k_n)\, \varphi_n(k_1 \ldots k_n)\mu_n, \tag{56}$$

where μ_n is a relativistically invariant measure.

As a consequence of its asymptotic character, the function $\varphi_n\,(k_1 \ldots k_n)$ has the following properties.†

1. In the case of identical particles, $\varphi_n\,(k_1, \ldots k_n)$ is a symmetrized or antisymmetrized product of single-particle wave functions:

$$\varphi_n(k_1 \ldots k_n) = \{\psi_1(k_1) \ldots \psi_n(k_n\}_{\pm},$$

or

$$|\, k_1 \ldots k_n\rangle = \{|\, k_1\rangle \ldots |\, k_n\rangle\}_{\pm}, \tag{57}$$

where the symbol $\{\}_{\pm}$ denotes symmetrization or antisymmetrization together with normalization factors. This symmetry property of the wave function for identical particles is well known in quantum mechanics.

The symmetry character is always the same for particles of a given type; if a wave function for two protons is antisymmetric, then also the wave function of the system of three protons will be antisymmetric, etc. If a wave function of a system of identical particles is symmetric in the in-state, then this symmetry property is preserved in the out-state. The symmetry character of a wave function of a system of identical particles is related to spin: particles with half-integral spin (fermions) are described by antisymmetric functions, and particles with integral spin (bosons) by symmetric functions. This statement ("spin-statistics-theorem") has been proven using various approaches.[28–33] An illustration of this theorem will be presented in § 4.4.

† The foundations of the theory of particle scattering are investigated in ref. 27.

2. Under the symmetry transformation $|\alpha\rangle \to |\alpha_g\rangle$ (where g may refer to transformations in addition to those of the Poincaré group) every single-particle state transforms separately:

$$|(k_1 \ldots k_n)_g\rangle = \mathrm{U}(g)\,|k_1 \ldots k_n\rangle = \{\mathrm{U}(g)\,|k_1\rangle \ldots \mathrm{U}(g)\,k_n\rangle\}_{\pm}.$$

This property of asymptotic states depends on the possibility of representing each state in the form of a sum of products of single-particle states.

From the property of factorizability of transformations it follows that the single-particle states $|k\rangle$ transform according to unitary representations of the quantum mechanical Poincaré group $\overline{\mathcal{P}}^{\uparrow}_{+}$. A mathematical expression of the simplicity of a representation is its irreducibility. For this reason, elementary particles are described by the irreducible representations of $\overline{\mathcal{P}}^{\uparrow}_{+}$, which contain the smallest set of states invariant with respect to transformations of the Poincaré group.

Consequently the classification of elementary particles arising from the theory of relativistic invariance consists of the classification of unitary representations of the quantum mechanical Poincaré group. The quantities characterizing irreducible representations (mass and spin) describe invariant properties of particles. The quantum numbers characterizing the basis of an irreducible representation are variables describing particle states (e.g. momentum and spin projection in the canonical basis).

Under transformations of the Poincaré group the state vector with n particles thus transforms as a direct product of unitary irreducible representations of this group corresponding to the spins $J_1 \ldots J_n$ and masses $m_1 \ldots m_n$ of those particles.

Let us now turn to the definition of the S-matrix. We shall investigate massive particles, since for massless particles the introduction of asymptotic states and of the S-matrix involves difficulties (the infrared divergence). Let us expand the state vector $|\alpha, \text{in}\rangle$, corresponding to the initial configuration of particles, in the system $|\beta, \text{out}\rangle$:

$$|\alpha, \text{in}\rangle = \sum_{\beta} S_{\beta\alpha}\,|\beta, \text{out}\rangle, \tag{58}$$

assuming that both systems are orthonormal:

$$\langle\alpha, \text{in (out)}\,|\,\beta, \text{in (out)}\rangle = \delta_{\alpha\beta}. \tag{59}$$

$S_{\beta\alpha}$ is equal to the probability amplitude for the transition $\alpha_{\text{in}} \to \beta_{\text{out}}$. The matrix with elements $S_{\beta\alpha}$ is called the S-matrix:

$$S_{\beta\alpha} = \langle\beta, \text{out}\,|\alpha, \text{in}\rangle = \langle\beta, \text{in}\,|\,S\,|\,\alpha, \text{in}\rangle = \langle\beta, \text{out}\,|S|\,\alpha, \text{out}\rangle; \tag{60}$$

it depends on the parameters of the initial and final state. The S-matrix is unitary since it connects two orthonormal bases in the same space $\mathcal{H}$:

$$SS^+ = S^+S = 1. \tag{61}$$

In the presence of superselection rules the S-matrix subdivides into blocks corresponding to coherent subspaces, since by virtue of the superselection rules transitions between different coherent subspaces are forbidden. If, for example, we write the S-matrix in a representation with baryon number B diagonal, then the S-matrix will have block structure:

$$S = \begin{bmatrix} \cdots & & & \\ & S(-1) & & \\ & & S(0) & \\ & & & S(1) \\ & & & & \cdots \end{bmatrix}.$$

Here $S(B)$ denotes a block of the S-matrix referring to baryon number B. This form of the S-matrix expresses the conservation of the charges Q_i associated with the superselection rule

$$[S, Q_i] = 0 \quad \text{or} \quad (Q_i(\alpha)-Q_i(\beta))\langle\beta\,|S|\,\alpha\rangle = 0. \tag{62}$$

The Q_i, hence, are those charges which are conserved exactly for all transformations. According to experimental data, the absolutely conserved quantum numbers include the electric charge Q, the baryon number B, and the leptonic numbers L_e and L_μ (see Chapter 2).

The probability amplitude $S_{\beta\alpha}$ is relativistically invariant:

$$S_{\beta\alpha} = \langle\beta, \text{out}\,|\,\alpha, \text{in}\rangle = \langle\beta_g, \text{out}\,|\,\alpha_g, \text{in}\rangle. \tag{63}$$

It does not change in passing to a transformed state in a new reference frame, translated, rotated, or uniformly moving with respect to the old system. Since, according to (26), $|\beta_g\rangle = \mathrm{U}(g)|\beta\rangle$, the invariance of the S-matrix is equivalent to the condition

$$\mathrm{U}^{-1}(g)\,S\mathrm{U}(g) = S. \tag{64}$$

Passing to the infinitesimal transformation $\mathrm{U}(g)$ and inserting (33) in (64), we obtain the connection, familiar from classical mechanics, between invariance and conservation laws. The invariance of the S-matrix with respect to translations in space and time is expressed by the law of conservation of four-momentum p_μ:

$$[P_\mu, S] = 0 \quad \text{or} \quad (p_\mu(\alpha)-p_\mu(\beta))S_{\alpha\beta} = 0. \tag{65}$$

The invariancc of the S-matrix with respect to rotations and Lorentz transformations entails the law of conservation of angular momentum or

$$[M_{\mu\nu}, S] = 0. \tag{66}$$

As a particular case, let us investigate the invariance of the S-matrix (63) with respect to rotations around the third axis by an angle 2π; in this case $\mathrm{U} = e^{2\pi i M_3} = z = \mathrm{U}^{-1}$ [see (52)] and

$$\langle\beta\,|S|\,\alpha\rangle = \langle\beta\,|\,e^{i2\pi M_3}\,Se^{i2\pi M_3}\,|\alpha\rangle = (-1)^{2\Sigma J_i}\langle\beta\,|S|\,\alpha\rangle.$$

Here ΣJ_i is the sum of spins of all particles in the states α and β. Particles with integral spin do not contribute to the definition of the sign, so that $2\Sigma J_i$ is equal ("modulo 2") to the sum of the number of fermions ($J=\frac{1}{2}, \frac{3}{2}, \ldots$) in states α and β. Consequently, in any process the total number of fermions in the initial and final states must be even.

If the four-momentum p_μ is diagonal, then according to (65) $S_{\beta\alpha}$ is proportional to a δ-function of $p_\mu(\alpha)-p_\mu(\beta)$. Let us separate out from the S-matrix a $\delta_{\alpha\beta}$ part corresponding to the absence of scattering; then one may write, introducing the scattering matrix T:

$$\langle\beta|S-1|\alpha\rangle = i(2\pi)^4\,\delta^4(p(\alpha)-p(\beta))\,\langle\beta|T|\alpha\rangle. \tag{67}$$

The unitarity condition (61) turns into the relation

$$\frac{1}{i}\langle\beta|(T-T^+)|\alpha\rangle = (2\pi)^4\sum_n \delta^4(p(\alpha)-p(\beta))\,\langle\beta|T^+|n\rangle\langle n|T|\alpha\rangle, \tag{68}$$

where the sum is over a complete set of states n, subject to the conservation laws, where $p_\mu(\alpha) = p_\mu(\beta)$. The transition probability for $\alpha \to \beta$ is equal to

$$\begin{aligned}|\langle\beta|S-1|\alpha\rangle|^2 &= (2\pi)^8\,\delta^4(0)\,\delta^4(p(\alpha)-p(\beta))\,|\langle\beta|T|\alpha\rangle|^2\\ &= \lim_{V,\,t\to\infty}(2\pi)^4\,Vt\delta^4(p(\alpha)-p(\beta))\,|\langle\beta|T|\alpha\rangle|^2.\end{aligned} \tag{69}$$

where V is the volume and t is a time interval.

Thus the transition probability for $\alpha \to \beta$ per unit time and unit volume is

$$w(\alpha\to\beta) = \lim_{V,\,t\to\infty}\frac{1}{V}\frac{d}{dt}|\langle\beta|S-1|\alpha\rangle|^2 = (2\pi)^4\,\delta^4(p(\alpha)-p(\beta))\,|\langle\beta|T|\alpha\rangle|^2. \tag{70}$$

The total transition probability W from the state α to any other state is obtained by summing (and integrating) the function $w(\alpha\to\beta)$ over all final states β:

$$W(\alpha) = \sum_\beta w(\alpha\to\beta) = (2\pi)^4\sum_\beta \delta^4(p(\alpha)-p(\beta))\,|\langle\beta|T|\alpha\rangle|^2. \tag{71}$$

Detailed properties of the S-matrix are examined in Chapter 7.

CHAPTER 2

FOUNDATIONS OF PHENOMENOLOGICAL DESCRIPTION

RELATIVISTIC quantum theory determines fully only the kinematics of a process, i.e. that part of the scattering matrix which depends on the properties of free particles. The interaction must be introduced in addition: for the same kinematics various dynamical models are possible.

In classical physics there exist only two types of force between particles—electromagnetic and gravitational. The capacity of particles for interaction is characterized by charges—electric and mass. In the physics of elementary particles there exist several types of interactions besides electromagnetism whose explicit nature (in Lagrangian or other form) is still unknown. The general properties of these interactions are studied primarily via the group theoretic approach and by introducing internal symmetries of various types. This means of approximating phenomena (i.e. the construction of approximations endowed with definite symmetry properties) has the virtue that it is always logically self-consistent.

However, specific internal symmetry groups do not have the universality that are inherent in space–time symmetry. With the exception of the simplest symmetries connected with superselection rules, internal symmetries are approximate in nature. Several groups have been found describing (in various approximations) the symmetry of interactions: the groups SU_2, SU_3, SU_6, $SU_2\times SU_2$, . . . ; but up to now the principle has still not been discovered which would indicate the necessity for approximate internal symmetry groups. For this reason, model representations, guesses, and hypotheses are an important part of the contemporary group-theoretic approach to the physics of elementary particles.

An overwhelming number of strongly interacting particles are unstable and are observed only as resonances in reactions with other particles. Meanwhile the concept of the quantum mechanical Poincaré group, of which we spoke in Chapter 1, refers only to absolutely stable particles. In a consistent relativistic quantum theory, unstable particles would appear when studying dynamics and approximations. In the physics of elementary particles, the concept of an unstable particle must be considered just as fundamental as the concept of a stable particle. In the (approximate) application of internal symmetries, a distinction is not made between stable and unstable particles. This difference is also not very important in the basic problem of the physics of elementary particles—the explanation of the mass spectrum. Thus, in contrast to Chapter 1, we shall investigate only approximate concepts in this chapter. This blend of exact and approximate concepts is specific to the present-day theory of elementary particles.

§ 2.1. Interactions and internal symmetry

The experimental values of transition probabilities show that the interactions of elementary particles may be subdivided into the following three basic classes: strong, electromagnetic, and weak. Dimensionless constants, appearing linearly in the *S*-matrix element at low energies, characterize the interactions numerically. The comparison of these constants gives an idea of the relative size of interactions of various classes:

Strong interactions, $g_s^2 \approx 1\text{–}10$.
Electromagnetic interactions, $\alpha \equiv e^2 = 1/137$.
Weak interactions, $Gm_p^2 \equiv g_w^2 = 10^{-5}$, where m_p is the proton mass.

In the case of the gravitational interaction, the dimensionless constant is equal to $\gamma m_p^2 \approx 10^{-20}$, which justifies the neglect of gravitational phenomena in elementary particle physics. Among strong interactions one may single out the medium strong interaction with constant $g^2/10$; moreover, it is possible that there exists a superweak interaction with constant $G_{sw} \sim 10^{-9}G$.

With respect to interactions, the various classes of particles fall into three groups.

(1) Hadrons: particles which can participate in all interactions. Hadrons include the majority of particles.
(2) Leptons: particles which participate in electromagnetic and weak interactions (the charged leptons e and μ) or only in weak (the neutral leptons ν_e and ν_μ).
(3) Photon: a single particle participating only in electromagnetic interactions.

All the bosons known at present with mass $m \neq 0$ are hadrons. Fermions may be either hadrons or leptons. Fermionic hadrons are called baryons.

Interactions are characterized not only by their "strength", determined by their intrinsic dimensionless constants of the type g^2 or α, but also by symmetry properties. The fact that interactions may have symmetry properties is well known from nuclear physics: nuclear forces do not depend on the electric charge of the proton and neutron. The development of the idea of charge independence of nuclear forces led to the introduction of a new approximately conserved quantum number—isotopic spin, or isospin, and to the classification of particles with respect to isospin.

In elementary particle physics, only the theory of interactions of leptons with the electromagnetic field—quantum electrodynamics (which is not considered in the present book)—may be considered as well established. The study of symmetries of the strong and weak interactions is one of the few effective means of revealing properties of these interactions, and thus is of prime importance.

In nonrelativistic quantum mechanics and nuclear physics, a symmetry of the interaction potential may entail invariance of the *S*-matrix with respect to transformations of the given symmetry, as well as the appearance of new conserved quantum numbers. One may denote the states of the system by these quantum numbers.

Analogously, the establishment of an interaction symmetry in elementary particle physics reveals new conservation laws and quantum numbers for classifying particles; moreover, an interaction symmetry entails relations between transition amplitudes for various processes.

Description of internal symmetry

Let us assume that the interaction between particles possesses some symmetry and the full description of each particle now includes the new quantum numbers ζ_i', so that the state vector may be written in the form

$$|\ldots; m, J; \zeta_i'\rangle. \tag{1}$$

Here the dots denote the variables of the Poincaré group (e.g. the momentum and spin projection). In contrast to space–time (or geometrical) symmetry, the additional symmetry associated with the nature of the interaction is called an internal or dynamical symmetry.

The starting point of the description of internal symmetry is the concept of complete independence of both types of symmetries—space–time and internal (exact internal symmetry). The notation (1) already assumes that the quantities ζ_i commute with the mass, spin, and variables of the Poincaré group. This is equivalent to the fact that the operators ζ_i commute with all generators P_μ and $M_{\mu\nu}$;

$$[\zeta_i, P_\lambda] = 0, \quad [\zeta_i, M_{\mu\nu}] = 0, \tag{2}$$

and, consequently, the coordinate transformations $x \to x' = \Lambda x + a$ do not change the ζ_i: $\mathrm{U}(a, \Lambda)\,\zeta_i \mathrm{U}^{-1}(a, \Lambda) = \zeta_i$. It is obvious that the ζ_i do not depend on coordinates. In particular, the ζ_i do not depend on time.

The transformations $\tilde{\zeta} = G\zeta$ of the variables ζ_i, which do not change probability amplitudes, are internal symmetry transformations. The corresponding transformation $u(G)$ of the state vector must be unitary:

$$|\ldots; mJ; \tilde{\zeta}'\rangle' = u(G)|\ldots; mJ; \tilde{\zeta}'\rangle = \sum_{\zeta'} |\ldots; mJ; \zeta'\rangle\langle\ldots; mJ; \zeta'\,|u(G)|\ldots; mJ; \tilde{\zeta}'\rangle. \tag{3}$$

The unitary matrices $u_{\zeta'\zeta''} = \langle\ldots; mJ; \zeta'\,|u(G)|\,\ldots; mJ; \zeta''\rangle$ form a group, since u_1u_2 is also unitary and $u = 1$ for $G = E$.

The invariance of the S-matrix for internal symmetry transformations entails

$$u(G)S = Su(G). \tag{4}$$

Internal symmetry groups may be discrete or continuous. The most interesting are the continuous groups, since they lead to additive conservation laws. Continuous groups may be compact or noncompact. The unitary representations of compact groups are finite-dimensional and correspond to multiplets with a finite number of particles. The unitary representations of noncompact groups are infinite-dimensional, i.e. a multiplet in this case includes an infinite number of particles. In what follows we shall study only compact groups.

Let $\mathfrak{G}$ be a compact internal symmetry group. Let us write an infinitesimal unitary transformation in the form

$$u = 1 + iF_r\varepsilon_r \qquad (r = 1, 2, \ldots, N), \tag{5}$$

where ε_r is an infinitesimal real parameter, while the generators F_r are Hermitian as a result of the unitarity of u. A finite transformation is then

$$u = e^{iF_r\alpha_r}, \tag{6}$$

where α_r is real and finite.

The group $\mathfrak{G}$ is characterized by a multiplication law of the transformations (6), or by the commutation relations between the generators

$$[F_r, F_t] = ic_{rts}F_s \qquad (r, t, s = 1, 2, \ldots, N). \tag{7}$$

The structure constants c_{rts} are real and may be chosen antisymmetric in all indices. They satisfy the Jacobi identity

$$c_{rtp}c_{psu} + c_{tsp}c_{pru} + c_{srp}c_{ptu} = 0. \tag{8}$$

From this it follows, in particular, that the N matrices c_r with matrix elements $(c_r)_{tp} = -ic_{rtp}$ satisfy the commutation relations (7) and, consequently, form a unitary N-dimensional representation of the generators (the adjoint or regular representation).

The generators F_r are physical quantities introduced by the internal symmetry group. The quantities ζ_i, by which states are denoted, must be constructed from F_r. The dimension of irreducible representations of the group (or the number of rows in the matrices of the generators F_r) is the number of particles in multiplets described by the group $\mathfrak{G}$. In particular, using the adjoint representation, one describes a multiplet with N particles, i.e. with a number of particles equal to the number of parameters of the group.

A multiplet or irreducible representation is characterized by the eigenvalues of the invariant quantities (the Casimir operators) z_l which commute with all generators F_r. Individual particles in a multiplet may be denoted by the eigenvalues of all generators F_r which can be diagonalized simultaneously, so that $\{\zeta_i\} = \{h_\alpha, z_l\}$. The eigenvalues h_α are additive quantum numbers, i.e. the eigenvalue $h(1+2)$ in the two-particle state $1+2$ is equal to the sum of the eigenvalues $h(1)+h(2)$ in the single-particle states 1 and 2. The additivity of the action of the generators on asymptotic multi-particle states follows from property 2 of asymptotic states (see § 1.4). In fact, using (5)

$$u(G)\,|\zeta_1, \zeta_2\rangle \sim \{1 + i(F_r(1) + F_r(2))\varepsilon_r\}\,|\zeta_1\rangle\,|\zeta_2\rangle, \tag{9}$$

where the tilde ($\sim$) denotes: "transforms as ...", while $F_r(i)$ is a generator applied to the state of the ith particle. The number of diagonal generators is equal to the rank of the group.(19)

The Casimir operators z_l, being at least bilinear quantities, are not additive quantum numbers, with the exception of the simplest case $N = 1$.

The internal symmetry transformation (6) commutes with transformations of the inhomogeneous Lorentz group. Thus all generators F_r of internal symmetries commute with generators of the Poincaré group:

$$[F_r, P_\lambda] = 0, \quad [F_r, M_{\nu\mu}] = 0. \tag{10}$$

so that all the F_r are relativistically invariant. As a consequence of (4) they commute with the S-matrix:

$$[F_r, S] = 0, \tag{11}$$

and, consequently, in processes involving interactions one may observe conservation laws for the diagonal operators h_α and for the Casimir operators z_l. For example, in the case of the reaction $1+2 \to 3+4$ one will have

$$\begin{aligned} h_\alpha(1)+h_\alpha(2) &= h_\alpha(3)+h_\alpha(4), \\ z_l(1+2) &= z_l(3+4), \end{aligned} \tag{12}$$

where $z_l(1+2)$ is the eigenvalue of z_l in the two-particle state $1+2$.

Additive quantities may be diagonalized simultaneously both in single- and multi-particle states, and thus, if single-particle states are denoted using h_α, then a superselection rule may exist corresponding to h_α. In the case of the nonadditive quantities z_l the diagonality of z_l in the single-particle states does not lead, in general, to the diagonality of z_l in multi-particle states, and, consequently, a superselection rule with respect to z_l does not exist.

Additive quantum numbers, or charges h_α, allow one to distinguish a particle from an antiparticle. An antiparticle (with respect to the particle a) is a particle $\bar{a}$ which differs from a only by the sign of its charge: $h_\alpha(\bar{a}) = -h_\alpha(a)$. The mass, spin, and nonadditive quantum numbers of a particle and its antiparticle are the same. In all cases subject to experimental test, each particle may be associated with its antiparticle. For this reason the existence of antiparticles is a strict regularity and the table of particles does not contain antiparticles separately (with the exception of special cases, when this is called for by additional considerations).

Using the concept of charges, one can introduce the idea of a neutral particle. Generalizing the definition of an electrically neutral particle, we shall call a neutral particle (i.e. genuinely neutral) one whose charges are all equal to zero: $h_\alpha = 0$. It is obvious that a neutral particle coincides with its antiparticle. In particular, the photon, the pion π^0, and the mesons η, ϱ^0, ω, and φ are all neutral particles.

Let us now turn to the properties of multiplets determined by the Casimir operators z_l. Each particle in a multiplet is characterized by a set of quantum numbers h_α. The internal symmetry transformations [according to (3)] connect states with different values of h_α in a given multiplet, i.e. with different particles of the multiplet. As a result of (10) the generators F_i of these transformations will commute with the operators for the mass $m^2 = P_\mu P^\mu$ and the invariant spin w^2, where $w_\mu = \frac{1}{2}\varepsilon_{\mu\nu\lambda\sigma}M^{\nu\lambda}P^\sigma$. This means that in the case of exact symmetry all particles of a multiplet will have the same mass and spin.

Let us now consider the transition amplitude for the process $1+2 \to 3+4$, in which all participating particles belong to the same multiplet. We perform an internal symmetry transformation and use the invariance (4) of the scattering matrix. Then, using (3), we arrive at the relation

$$\begin{aligned} \langle h(3), h(4)|S|h(1), h(2)\rangle &= \langle h(3), h(4)|u^{-1}Su|h(1), h(2)\rangle \\ &= \sum_{h'(i)} \langle h'(3), h'(4)|S|h'(1), h'(2)\rangle \\ &\quad \times d_{h'(1)\,h(1)} d_{h'(2)\,h(2)} d^*_{h(3)\,h'(3)} d^*_{h(4)\,h'(4)}, \end{aligned} \tag{13}$$

where $d_{h'h} \equiv d_{h'h}(G)$. This relation connects the amplitude for the initial process with the amplitude for the process $1'+2' \to 3'+4'$, in which, generally speaking, other particles of the given multiplet [with quantum numbers $h'(i)$] participate.

Thus an internal symmetry may manifest itself by a multiplet structure of the mass spectrum, by transformation laws for additive and nonadditive quantum numbers, and also by the existence of relations between cross-sections for different processes.

§ 2.2. Symmetry and particle classification

One must distinguish between exact and approximate internal symmetries. Exact symmetries are satisfied for all interactions. Approximate symmetries describe the properties of the stronger interactions.

Exact symmetries

The difference between leptons and baryons is expressed using the concept of baryonic B and leptonic L_e, L_μ charges. The quantities B, L_e, L_μ may take on the values $0, \pm 1, \pm 2, \ldots$, for physical systems; the values of these quantities for elementary particles are limited to the numbers 0, 1. Fermions may participate in strong interactions when their baryonic charge differs from 0, so that for elementary baryons $B = \pm 1$, while for leptons $B = 0$. Analogously, the leptonic charges of the leptons are equal to $L_e = \pm 1$, $L_\mu = \pm 1$, while for the baryons $L_e = L_\mu = 0$. All mesons and the photon have $B = L_e = L_\mu = 0$.

Experiment shows that the baryonic B and the leptonic L_e, L_μ charges are additive quantities, conserved in all interactions, as in the case of the electric charge Q.

The conservation of baryonic charge expresses the fact that transformations of baryons into leptons, photons, and mesons are not observed. The absence of transformations of leptons into mesons and photons of the type $e^\pm + \mu^\pm \rightarrow \pi^\pm + \pi^\pm$ is explained by the conservation of leptonic charge $L = L_e + L_\mu$. To explain the absence of the reactions $\mu^\pm \rightarrow 2e^\pm + e^\mp$ or $\mu^\pm \rightarrow e^\pm + \gamma$, it is necessary to have two different leptonic charges L_e and L_μ with separate conservation laws. In turn, the difference between the charges L_e and L_μ leads to two sorts of neutrinos: ν_e and ν_μ. The existence of two neutrinos follows from the experiments

$$\nu_\mu + n \rightarrow p + \mu^-, \quad \nu_\mu + n \nrightarrow p + e^-$$

in which only muons, but not electrons, were detected. Usually leptons are assigned the following charges:

$$\begin{aligned}
&\text{neutrino } \nu_e: && Q = 0, && L_e = 1, && L_\mu = 0;\\
&\text{neutrino } \nu_\mu: && Q = 0, && L_e = 0, && L_\mu = 1;\\
&\text{electron } e^-: && Q = -1, && L_e = 1, && L_\mu = 0;\\
&\text{muon } \mu^-: && Q = -1, && L_e = 0, && L_\mu = 1.
\end{aligned}$$

Mathematically the conservation of the charges B, L_e, L_μ, and Q is equivalent to the invariance of the probability amplitude (for the scattering matrix) with respect to phase transformations u_B, u_{L_e}, u_L, u_Q, where, for example,

$$u_B = e^{iB\alpha} \tag{14}$$

(α is a real parameter). The transformations (14) form a one-dimensional unitary group $U_B(1)$. The groups U_{L_e}, U_{L_μ}, and U_Q are determined analogously. The charges B, L_e, L_μ, and Q are conserved quantities. These are just the quantum numbers associated with superselection rules (see § 1.2). Since the charges B, L_e, L_μ are assumed independent, only the operator z may be a function of other quantities, i.e. of charges. A study of the particle table shows that the following relation holds:

$$z = (-1)^{2J} = (-1)^{B+L_e+L_\mu}.$$

The representations of one-parameter groups of the type $U_B(1)$ are one-dimensional, i.e. exact symmetries do not imply the existence of multiplets of particles (and consequently, of relations between the probabilities for various processes). Thus the exact symmetries known at present are manifested only in conservation laws for the charges B, Q, L_e, and L_μ.

Approximate symmetries

Every class of interactions is characterized by its own particular symmetry properties; thus the stronger interactions display a greater degree of symmetry. This (experimental) relation between the strength of the interaction and its symmetry allows one to introduce the important concept of approximate (or broken) symmetry, as the symmetry of some group of the stronger interactions. An exact symmetry, as a symmetry of all interactions, is lower in comparison with an approximate one.

Let us see which of the results in § 2.1 changes if the symmetry described by the group $\mathfrak{G}$ [formulae (4)–(7)] is in fact approximate.

Let us suppose that the compact group $\mathfrak{G}$ is a symmetry group of the stronger part of the interaction. Then formulae (10) and (11), which express the independence of internal and space–time symmetries, hold only approximately in the absence of all weaker interactions. When these interactions are taken into account, the generators F_r may depend on time:

$$[F_r, P_0] \neq 0, \quad [F_r, M_{0k}] \neq 0, \tag{15}$$

while the full S-matrix becomes noninvariant with respect to transformations of the group $\mathfrak{G}$, i.e.

$$[F_r, S] \neq 0 \tag{16}$$

for some r.

The approximate nature of the symmetry does not change the group $\mathfrak{G}$, which, as before, is characterized by the same commutation relations (7) for the generators F_r; the irreducible representations of the group do not depend on whether the symmetry is exact. Consequently, if the relation

$$\begin{matrix}\text{basis of an irreducible representation} \\ \text{of the internal symmetry group } \mathfrak{G}\end{matrix} \leftrightarrow \text{particle multiplet} \tag{17}$$

holds when the symmetry is approximate, the classification of particles with respect to multiplets of the group $\mathfrak{G}$ will be the same as in the case of the exact symmetry. Here the value of the spins of all particles in the multiplet is assumed to be the same.

Equation (17) is the physical interpretation of an approximate symmetry. Together with characteristics borrowed from the Poincaré group and from exact symmetry, it defines the concept of a particle in the case of an approximate symmetry.

In comparison with the purely kinematic definition in § 1.4, which related elementarity to the irreducibility of a representation of the Poincaré group, this definition of an elementary particle has additional features. First of all, a dynamical element is introduced by which particles are characterized both by charges Q_i (including the electric charge) and the quantum numbers z_l and h_α denoting particles in a multiplet of the symmetry group of the interaction. Secondly, the concept of a particle has now become approximate. A particle which quantum numbers J, Q_i', z_l, h_α is defined only up to weaker interactions; at the same time it is assumed that such a particle owes its existence only to that interaction whose symmetry group is $\mathfrak{G}$. It is produced in processes caused by this interaction.

According to (1) and (17) the state vector of a particle may be written approximately in the form

$$|\ldots; m, J; Q_i', z_l, h_\alpha\rangle, \tag{18}$$

where, as in (1), the variables of the Poincaré group are omitted. The approximate nature of the state vector (18) is clear, for example, from the fact that as a consequence of (15) the mass will not in general commute with the operators of the quantities z_l and h_α, and in such a case cannot have a definite value in the state (18) in which z_l and h_α are fixed. But then the particle must be described by a mass distribution, which entails its instability (see § 2.3). The quantity m in (18) here must be interpreted as an average mass.

Moreover, in the presence of weaker interactions, which break the symmetry and cause transitions between states with different z_l and h_α, it may be impossible to distinguish one of the states of the type (18) from some superposition of it with other states.

The definition (17) in the case of an approximate symmetry does not depend on whether the symmetry is strongly or weakly broken. It acts as long as one can follow any "genetic" connection of particles observed in experiments with an assumed symmetry group. For a strongly broken symmetry it is more difficult to establish quantitatively the consequences of breaking (mass splittings, etc.) and to justify the proposed symmetry group.

Thus approximate symmetries involve more information on particles and interactions than exact symmetries. The introduction of such symmetries is the first step in the theoretical treatment of data on elementary particles. It is precisely using approximate symmetries that one can understand the multiplet structure of the mass spectrum. Approximate symmetries are a convenient (and sometimes natural) language for describing the strongly interacting particles. Thus the manifestation of successive degrees of symmetry, corresponding to interactions of ever-increasing strength, is one of the main problems of classification of particles. The approximate symmetry groups SU_2, SU_3, and SU_6 will be studied in Part III, while the groups $SU_2 \times SU_2$ and $SU_3 \times SU_3$ will be treated in Chapter 15.

§ 2.3. Unstable particles

Of the many unstable particles there are only five absolutely stable ones—both neutrinos, the photon, the electron, and the proton, as well as their antiparticles. The remaining particles are unstable. Unstable particles may be subdivided into two classes—metastable particles and resonances.

Metastable particles decay as a result of weak or electromagnetic interactions. As a result of the weakness of these interactions, the lifetime of metastable particles is large so that the inverse lifetime Γ is small in comparison with the particle mass m. For metastable particles, the ratio Γ/m does not exceed 5×10^{-6} (the η-meson). Metastable particles are usually included in the group of stable particles. The table of stable particles (Table A.1, p. 359) contains both absolutely stable and metastable particles. All these particles are stable with respect to decays caused by the strong interactions.

Resonant particles are produced and decay via strong interactions. These particles are not observed in a "free" state; they appear in scattering in the form of a quasi-stationary state of two or more strongly interacting particles. Resonances have the properties of particles, specifically: (a) the spin of a resonance has a definite value; and (b) resonances may be characterized by internal quantum numbers conserved by strong interactions (e.g. isospin and hypercharge). However, in contrast to stable particles, resonances do not have a definite or nearly definite value of mass. Resonances are described by a mass spectrum of a dispersion type, and the maximum of this spectrum is usually called the mass of the resonance. The width Γ of the mass spectrum of the resonance may be comparable to the mass of the particle. In the case of the ϱ-meson, the ratio Γ/m attains the value 0.2. For the S-wave $K\pi$ resonance observed in several experiments,[34, 35] this ratio may be as large as 0.3.

The nature of an unstable particle is evidently described most exactly by introducing the idea of a quasi-stationary state, which is well known from nonrelativistic quantum mechanics. When considering an unstable particle a, we may write the total operator for the energy H in the form of a sum of two parts: $H = H_a+H'$. When H' is neglected, particle a becomes stable, while the state of the (stable) particle $|a\rangle$ is an eigenstate of the operator H_a. The "inclusion" of H' causes the decay of the particle a. In the case of a metastable particle, H' contains weak and electromagnetic interactions, while for resonances H' contains a part of the strong interactions as well. When H' is taken into account, the state for the particle a becomes quasi-stationary.

This picture, however, may be checked only in its most general features. The operator H' is known only for the electromagnetic interaction. For weak interactions the only expression for H' which is established at present is that which may be used to first order in perturbation theory. In the more interesting case of strong interactions, there are neither reliable expressions for H' nor methods for solving the problem outside the framework of perturbation theory. Thus the definition of a resonance as a physical object must rely on only those properties of stationary states which do not depend on the concrete form of interactions. One such property is the form of the mass spectrum of an unstable particle.

Relation between the mass spectrum and the decay law

The relation between the decay law of a quasi-stationary state and the energy distribution function is based on the Krylov–Fock theorem.[36]

Let $|\psi(\alpha, x^0 = 0)\rangle$ be the initial state of the system. Here α denotes the set of variables labeling the states of the system. Represent $|\psi(\alpha, 0)\rangle$ in the form of an integral over the eigenvalues of the energy operator E:

$$|\psi(\alpha, 0)\rangle = \int c(E)|\alpha, E\rangle\, dE. \tag{19}$$

Then the state of the system at the time x^0 will be

$$|\psi(\alpha, x^0)\rangle = \int e^{-iEx^0}c(E)|\alpha, E\rangle\, dE. \tag{20}$$

The probability that after a time x^0 the system will still be in the initial state is equal to

$$L(x^0) = |p(x^0)|^2,$$

where

$$p(x^0) = \int \langle\psi(\alpha, 0)\,|\,\psi(\alpha, x^0)\rangle\, d\alpha = \int e^{-iEx^0} |c(E)|^2\, dE.$$

But

$$|c(E)|^2\, dE = w(E)\, dE = dW(E) \tag{21}$$

is the energy distribution function of the initial state (and, consequently, also of the state $|\psi(\alpha, x^0)\rangle$). Then

$$L(x^0) = \left|\int e^{-iEx^0}\, dW\,(E)\right|^2, \tag{22}$$

i.e. the decay law for the state $|\psi(\alpha, 0)\rangle$ depends only on the energy distribution function in this state (the Krylov–Fock theorem).

For an unstable particle at rest the energy distribution $dW(E)$ is just the particle's mass spectrum. Here the state $|\alpha, E\rangle$ includes the states of the decay products. A necessary and sufficient condition for the decay to occur is the continuity of the integral distribution function $W(E)$, which excludes a discrete mass spectrum. In fact, only when $W(E)$ is continuous does the probability $L(x^0)$ vanish for $x^0 \to \infty$.

The Krylov–Fock theorem (22) is very general; it is applicable to the decay of any non-stationary state whether an unstable particle, or, for example, a wave packet for a stable particle. The usual (exponential) form of the decay law for a stationary state

$$L(x^0) \approx e^{-\Gamma x^0} \tag{23}$$

is obtained by assuming the meromorphic character of $w(E)$ as a function of the complex variable E. Since for real values of E the function $w(E)$ will be real, its poles will be distributed symmetrically with respect to the real axis, so that the residues at the poles will be complex conjugates of one another. If the function $w(E)$ has only one pair of poles, $E = E_0 \pm i(\Gamma/2)\,(\Gamma > 0)$, then one may set

$$w(E) = \frac{1}{\pi}\,\frac{(\Gamma/2)^2}{(E-E_0)^2+\Gamma^2/4}\,. \tag{24}$$

When one inserts this expression in (22) one obtains the exponential decay law (23) for large times and $\Gamma \ll E_0$.

If the function $w(E)$ has several pairs of poles, then the main contribution to the asymptotic behavior of $L(x^0)$ for large times x^0 is governed by the pair of poles closest to the real axis. Let the next pair of poles have an imaginary part $\Gamma' > \Gamma$. Then the exponential law (23) with width Γ describes the decay for times x^0 satisfying

$$(\Gamma' - \Gamma)x^0 \gg 1. \tag{25}$$

Thus in the simplest case, when the decay of the particle is determined by one exponent (23), the mass distribution of an unstable particle at rest must be of the form (24) and will be determined by two parameters—the mass m (the position of the maximum) and the width Γ of the peak. A stable particle is obtained by passing to the limit $\Gamma \to 0$.

Unstable particles in relativistic quantum theory

The description of unstable particles in relativistic quantum theory relies primarily on the Poincaré group and the scattering matrix.

Since unstable particles are characterized by a mass spectrum, they should be described by reducible representations of the Poincaré group (since irreducible representations correspond to definite values of mass). But the group-theoretic definition of elementarity assumes the irreducibility of a representation; the space of states of elementary particles must be irreducible. This means that unstable particles cannot be considered elementary in the group-theoretic sense.

However, such an approach to the role of unstable particles is hard to justify physically. Unstable particles would have to be treated as less fundamental than stable particles if there were some indication of a special role for stable particles (e.g. if all particles were constructed only of stable particles). But in all phenomena known to us, stable and unstable particles participate nearly identically in interactions. Unstable particles are more like the excited states of a system whose ground states are stable particles. For this reason, with respect to interactions, unstable particles are just as fundamental as stable particles, while the group-theoretic definition of their elementarity is inconsistent.

The existence of a difference between the physical and group-theoretic ideas of elementarity means that within the context of the Poincaré group one has not yet managed to construct an effective description of unstable particles which would broaden the mathematical concept of elementarity.

The conclusions of relativistic quantum theory may be used approximately for unstable particles as well (when the interactions causing the decay are neglected). For example, in the case of muons, we must neglect the weak interaction; in the case of neutral pions, both electromagnetic and weak interactions; and in the case of resonances, also a part of the strong interaction. Such an approximation, of course, is also necessary when introducing the scattering matrix for unstable particles. The asymptotic in- and out-states cannot contain unstable particles (since unstable particles would already have decayed by the time asymptotic separations had been reached). When factoring many-particle asymptotic states into single-particle ones, we are not allowed to include the states of unstable particles among the single-particle states. For this reason, the scattering matrix may be considered, strictly speaking, as connecting only stable particles with stable particles. If, however when one defines the asymptotic states one neglects the decay interaction, all results of S-matrix theory continue to hold.

From the standpoint of the S-matrix (from stable to stable particles), unstable particles correspond to poles in the partial wave-scattering amplitude reflecting the spin and other quantum numbers of the unstable particle (see Chapter 12).

Observation of resonances

The experimental detection of a resonant particle is based on its having a mass spectrum of the type (24).

Consider the scattering of stable particles a and b, in which the resonant particle Y is formed together with other stable particles $c_1 \ldots$:

$$a+b \to c_1+ \ \ldots \ +\text{Y} \to c_1+ \ \ldots \ +d+f.$$
$$\hookrightarrow d+f$$

Let the resonance Y decay into the stable particles d and f so that the final products of the reaction are $c_1 \ldots d, f$. In this scattering process the energy and momentum of particles a and b are divided among the groups of particles $c_1+ \ \ldots$ and $d+f$ in the final state. If the resonance Y had been a stable particle, its rest-system energy E (i.e. its mass) would have had a definite value. But the resonance Y is unstable and is characterized by the distribution $w(E) = |c(E)|^2$. The function $w(E)$ is directly connected with the mass spectrum of the products of the decay $\text{Y} \to d+f$ when

$$E = E_d+E_f, \quad \mathbf{p}_d+\mathbf{p}_f = 0.$$

Let us assume for simplicity that the particles d and f are the only possible products of the decay of resonance Y. Then in the centre of mass system a complete set of states in (19) is the set of those two-particle states $|d, f; E\rangle$ which has the same spin and internal quantum numbers as the particle Y. Here the coefficient $c(E)$ in (19) will be equal to $\langle d, f; E|\text{Y}\rangle$, while the mass spectrum of the decay products d and f coincides with the mass spectrum of Y.

Thus by studying the mass distribution of complexes of particles in the final states of reactions, one can obtain directly the mass distribution of unstable particles which are "resonant states" of such complexes, i.e. having the same internal quantum numbers, but higher masses and spins. This does not mean, of course, that every peak in mass distributions of reaction products may be associated with an unstable resonant particle; kinematic peaks existing for a given reaction are also possible. The concept of a resonant particle is universal: it must appear in all strong interaction reactions for which states with the quantum numbers of the resonance are accessible.

Peaks corresponding to resonant particles are also detected both in the elastic-scattering amplitude $d+f \to \text{Y} \to d+f$ and in the total cross-section for reactions initiated by the particles d and f. This case differs from the one considered earlier in that the energy interval ΔE defining the state $d+f$ is significantly smaller than the width of the resonance Γ_Y: $\Delta E \ll \Gamma_\text{Y}$. For this reason the system $d+f$ will be localized in a volume with linear dimensions of order $\Delta x \approx 1/\Delta p \approx v/\Delta E$, which significantly exceeds the distance v/Γ_Y over which the resonant particle Y may pass in its own mean lifetime. Consequently, the resonance Y in this case is a virtual intermediate state in the scattering process.

If the scattering $d+f$ is dominated by formation of the virtual resonance Y, then the total cross-section for df scattering will depend on the energy E close to the resonance via the same law (24) as the mass distribution $w(E)$ (up to a kinematic factor). The total cross-section will have a maximum in the region of energy E of the colliding particles (in the c.m.s.) equal to the resonance mass, and the particles in the final state will appear with a delay in comparison with the scattering without formation of the resonance.

For example, the total cross-section for scattering of pions π^+ on protons has a prominent peak in the energy region $E=p^0(\pi^+)+p^0(\text{p}) \approx 1150–1300$ MeV with width $\Gamma \approx 120$ MeV,

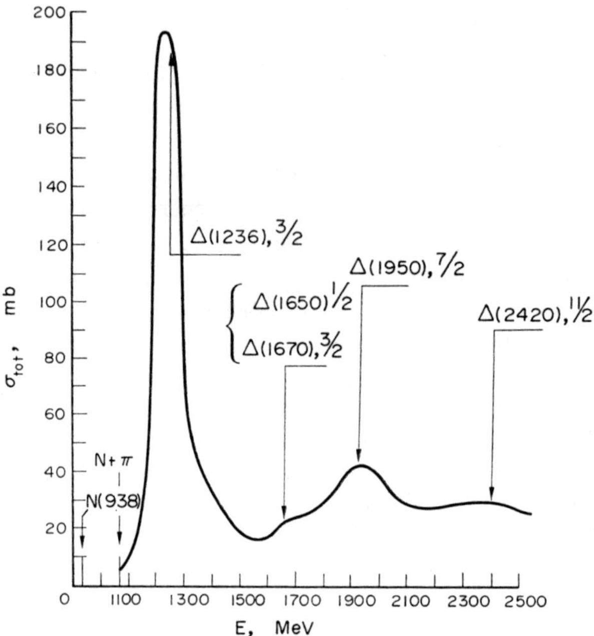

FIG. 2. Total π^+p cross-section σ_{tot} as a function of energy E in the center of mass. (Adapted from ref. 37.)

corresponding to a state with total spin $J = \frac{3}{2}$ (Fig. 2).[37] This peak is interpreted as evidence for a resonant particle Δ with mean mass $m_\Delta = 1236$ MeV, spin $J = \frac{3}{2}$ and the same remaining quantum numbers as in the π^+p system. In this energy region π^+p scattering consists of the successive processes of formation of the particle Δ and its decay:

$$\pi^+ + \text{p} \to \Delta \to \pi^+ + \text{p}.$$

The study of the total cross-section for π^+p scattering at higher energy allows one to establish the existence of several more resonances with the internal quantum numbers of the π^+p system, but with different spins and masses: Δ (1670) with spins $\frac{1}{2}$ and $\frac{3}{2}$ [these may be distinguished from one another using phase shift analyses], Δ (1950) with spin $\frac{7}{2}$, and Δ (2420) with spin $\frac{11}{2}$. The number in parentheses after the particle symbol denotes the mass of the resonance.

Decay probability

Decays may be subdivided into rapid, electromagnetic, and slow. Rapid decays are caused by strong interactions and slow decays byweak interactions. Let the Hamiltonian be $H = H_a + H'$, where H' is the decay interaction. We shall write a formula for the decay probability to first order in H'.

Consider the decay of the particle Y into the multi-particle state $|a, b, \ldots; \text{out}\rangle$: $\text{Y} \to a + b + \ldots$. Since the decay is considered to first order, the state $|\text{Y}\rangle$ is an eigenstate of the operator H_a. The multi-particle states are also defined in the absence of the interaction $H'(x^0)$. Since energy and momentum are conserved, the decay amplitude may be written in the form

$$S(\mathrm{Y} \rightarrow a+b+\ \ldots) = i(2\pi)^4\,\delta^4(p_{\mathrm{Y}}-p_a-p_b-\ \ldots)\,\langle \mathrm{Y}\,|T|\,a, b, \ldots;\ \mathrm{out}\rangle \tag{26}$$

and to first order

$$\langle \mathrm{Y}\,|T|\,a, b, \ldots\rangle = -i\langle \mathrm{Y}\,|H'(0)|\,a, b, \ldots\rangle. \tag{27}$$

The decay probability $w(\mathrm{Y} \rightarrow a+b+\ \ldots)$ may be computed in the standard fashion described in § 1.4:

$$\begin{aligned} w(\mathrm{Y} \rightarrow a+b+\ \ldots) &= \frac{1}{Vt}\,|S(\mathrm{Y} \rightarrow a+b+\ \ldots)|^2 \\ &= (2\pi)^4\,\delta^4(p_{\mathrm{Y}}-p_a-p_b-\ \ldots)\,|\langle \mathrm{Y}\,|T|\,a, b, \ldots\rangle|^2. \end{aligned} \tag{28}$$

Formula (28) gives the decay probability if the decaying particle and the decay product are in definite states. In an experiment one usually observes the decay of a particle in an unspecified spin state. Expression (28) must thus be averaged over the spin states of the initial particle.

The total probability for the decay with given products $a, b, \ldots$, is obtained by summing (28) over all possible final states of the particles $a, b \ldots$.

PART II

RELATIVISTIC KINEMATICS AND REFLECTIONS

RELATIVISTIC kinematics is a direct consequence of the invariance of a theory with respect to transformations of the Poincaré group. Relativistic kinematics deals with transformation properties of the in- and out-states and of the scattering matrix. Its purpose is to give a simple and convenient expansion of the scattering matrix in terms of irreducible amplitudes in order to serve as a starting point for dynamical theories.

This separation of the kinematic aspects from the dynamical ones may be made, in general, in various ways. We shall examine a parametrization of the amplitude using spinor states or $\mathscr{M}$-functions, as well as an expansion in terms of helicity amplitudes. However, wave functions for particles with higher spins are given in various forms, which allows the reader to try other representations for the amplitude in which it is easy to take account of the invariance of the theory with respect to reflections.

The relativistic invariance of a theory does not by itself imply its invariance with respect to separate reflections of space and time. But since the theory depends on local quantum field theory it is automatically invariant with respect to a combined reflection in space and time. Moreover, strong and electromagnetic interactions are invariant with respect to reflections, and consequently, in these cases, reflections are part of the kinematics.

CHAPTER 3

THE LORENTZ GROUP AND THE GROUP $SL(2, c)$

As was explained in § 1.2, relativistic invariance in quantum mechanics means that physical states must transform under Lorentz transformations according to unitary representations not of the Poincaré group $\mathcal{P}^{\uparrow}_{+}$, but of its universal covering group $\overline{\mathcal{P}}^{\uparrow}_{+}$ or the quantum mechanical Poincaré group. The universal covering group of the homogeneous Lorentz group is the group of complex unimodular 2 by 2 matrices of second order $SL(2, c)$.

In classical physics, covariant equations are formulated in the language of tensor quantities transforming according to irreducible (nonunitary) representations of the Lorentz group. In relativistic quantum theory, an analogous role belongs to quantities which transform according to irreducible representations of the group $SL(2, c)$. These quantities are called spinors (of various rank).[(38–41)] Only spinors of even rank may be expressed in terms of tensors. Spinors of odd rank are quantities which cannot be obtained in the framework of single-valued representations of the Lorentz group; they transform according to double-valued representations of this group.

Spinor analysis is presented in the present chapter. Moreover, in §§ 3.4 and 3.5, Clebsch–Gordan coefficients and representations (single-valued and double-valued) of the representation group will be discussed briefly.

§ 3.1. Second-order unimodular matrices and the Lorentz transformation

Let us consider the complex second-order matrix

$$A = \begin{pmatrix} \alpha & \beta \\ \gamma & \delta \end{pmatrix}, \tag{1}$$

where the complex parameters α, β, γ, and δ are restricted by the unimodularity condition

$$\det A = \alpha\delta - \gamma\beta = 1. \tag{2}$$

The matrix A thus depends on six real parameters.

The product A_1A_2 of two unimodular matrices A_1 and A_2 will also be a unimodular matrix: $\det(A_1A_2) = 1$. The unit matrix E is

$$E = \begin{pmatrix} 1 & 0 \\ 0 & 1 \end{pmatrix}. \tag{3}$$

The inverse matrix always exists, since $\det A$ is not equal to 0:

$$A^{-1} = \begin{pmatrix} \delta & -\beta \\ -\gamma & \alpha \end{pmatrix}. \tag{4}$$

We see that matrices of type A form a binary group. In standard notation this group is $SL(2, c)$.

Among the matrices A, a special role is played by the matrix C:

$$C = \begin{pmatrix} 0 & -1 \\ 1 & 0 \end{pmatrix} = -C^{+} = -C^{-1}. \tag{5}$$

Using this matrix one can establish a connection between the transpose matrix A^{T} and the inverse A^{-1}:

$$A^{-1} = C^{-1}A^{\mathrm{T}}C. \tag{6}$$

In other words, the matrices $A\,C^{-1}\,A^{\mathrm{T}}\,C$ and $C^{-1}\,A^{\mathrm{T}}\,C\,A$ are equivalent to the unit matrix.

The basis matrices for the matrices of second order are the matrices σ_μ, including the Pauli matrices σ_k and the unit matrix

$$\sigma_0 = E, \quad \sigma_1 = \begin{pmatrix} 0 & 1 \\ 1 & 0 \end{pmatrix}, \quad \sigma_2 = \begin{pmatrix} 0 & -i \\ i & 0 \end{pmatrix}, \quad \sigma_3 = \begin{pmatrix} 1 & 0 \\ 0 & -1 \end{pmatrix}. \tag{7}$$

The Pauli matrices have the following commutation properties:

$$[\sigma_k, \sigma_q] = i2\varepsilon_{kqp}\sigma_p, \quad \{\sigma_k, \sigma_q\} = 2\delta_{kq}, \tag{8}$$

where ε_{kqp} is the totally antisymmetric tensor with $\varepsilon_{123} = 1$. The Pauli matrices are orthogonal in the sense that

$$\mathrm{Tr}\,(\sigma_\mu\sigma_\nu) = 2\delta_{\mu\nu}. \tag{9}$$

The relation

$$\sum_{k=1}^{3} (\sigma_k)_{ij}\,(\sigma_k)_{ln} = 2\delta_{in}\delta_{jl} - \delta_{ij}\delta_{ln}, \tag{10}$$

which may be easily checked by a direct calculation, holds between the matrix elements of σ_k.

As a result of the commutation relations (8) the product of any number of matrices σ_μ may be expressed as a linear combination of these matrices. Thus any matrix of second-order O may be represented in the form of a linear combination of the basis matrices:

$$\mathrm{O} = \sum_\mu O(\mu)\sigma_\mu, \tag{11}$$

with complex coefficients $O(\mu)$ which by the orthogonality (9) of the matrices σ_ν may be determined unambiguously in terms of the matrix O:

$$O(\mu) = \tfrac{1}{2}\,\mathrm{Tr}\,(\sigma_\mu \mathrm{O}) \tag{12}$$

If the matrix O is unimodular, then condition (2) holds, or

$$O^2(0) - \sum_{k=1}^{3} O^2(k) = 1. \tag{13}$$

Let us now establish a relation between matrices and Lorentz four vectors. For each real four-vector x^μ, we form the Hermitian matrix X, according to the rule

$$\mathrm{X} = x^\mu \sigma_\mu = \begin{pmatrix} x^0+x^3 & x^1-ix^2 \\ x^1+ix^2 & x^0-x^3 \end{pmatrix}. \tag{14}$$

The relation between x^μ and X is one-to-one; the inverse formula is

$$x^\mu = \tfrac{1}{2}\,\mathrm{Tr}\,(\sigma_\mu\,\mathrm{X}). \tag{15}$$

Thus every point in space x^μ leads to its corresponding Hermitian matrix X. The determinant of the matrix X

$$\det \mathrm{X} = (x^0)^2 - x^k x^k = x^2 \tag{16}$$

is just the square of the four-vector x^μ.

Let us consider the transformation $\mathrm{X} \to \mathrm{X}'$ induced by the unimodular matrix A:

$$\mathrm{X}' = A\mathrm{X}A^+. \tag{17}$$

It is a linear transformation of the four-vector x^μ for which the Hermitian nature of the matrix X is conserved, i.e. real x^μ transform into real x'^μ. Since $\det A = \det A^+ = 1$, the determinant of the matrix X does not change under the transformation (17): $\det \mathrm{X} = \det \mathrm{X}'$, which by (16) implies the invariance of x^2. But a linear transformation of the four-vector $x^\mu \to x'^\mu$, for which its length does not change ($x'^2 = x^2$), is a transformation of the Lorentz group. Thus (17) describes a transformation of the Lorentz group on a four-vector x^μ.

Let us now obtain the relation between a transformation of the matrix X by (17) and a transformation of a four-vector $x'^\mu = \Lambda^\mu{}_\nu x^\nu$. In detailed notation, formula (17) reads

$$\mathrm{X}' = \sigma_\mu x'^\mu = \sigma_\mu \Lambda^\mu{}_\nu x^\nu = A\mathrm{X}A^+ = A\sigma_\nu A^+ x^\nu, \tag{18}$$

from which, by (15), we obtain $x'_\mu = \frac{1}{2}\,\mathrm{Tr}\,\sigma_\mu(\mathrm{X}')$, or

$$\Lambda^\mu{}_\nu(A) = \tfrac{1}{2}\,\mathrm{Tr}\,(\sigma_\mu A \sigma_\nu A^+). \tag{19}$$

From (19) we conclude that every transformation $\Lambda^\mu{}_\nu$ gives rise to two matrices $\pm A$, differing by a sign: $\Lambda^\mu{}_\nu(A) = \Lambda^\mu{}_\nu(-A)$.

Lowering the upper index in (19), we obtain the coefficients of a Lorentz transformation with lower indices:

$$\Lambda_{\mu\nu}(A) = \tfrac{1}{2}\,\mathrm{Tr}\,(\tilde{\sigma}_\mu A \sigma_\nu A^+), \tag{20}$$

which may be expressed in terms of the matrix

$$\tilde{\sigma}_\mu = (\sigma_0, -\boldsymbol{\sigma}) = C^{-1}\sigma_\mu^{\mathrm{T}} C. \tag{21}$$

It is sometimes convenient to use the matrices σ_μ and $\tilde{\sigma}_\nu$ simultaneously for transformations, since then the relations may be written in a relativistically covariant form. Instead of equations (8) and (9) one may use the formulae

$$\sigma_\mu \tilde{\sigma}_\nu = g_{\mu\nu} + i\tfrac{1}{2}\varepsilon_{\mu\nu\lambda\tau}\sigma^\lambda \tilde{\sigma}^\tau, \tag{22}$$

$$\tfrac{1}{2}\,\mathrm{Tr}\,(\tilde{\sigma}_\mu\,\sigma_\nu) = g_{\mu\nu}. \tag{23}$$

Formula (10) may be transcribed in the form

$$(\sigma_\nu)_{ij}(\tilde{\sigma}_\mu)_{ln}\, g_{\mu\nu} = 2\delta_{in}\delta_{jl}. \tag{24}$$

We shall show that the quantities $\Lambda_{\mu\nu}(A)$ really have all properties of coefficients of a Lorentz transformation. We obtain

$$\Lambda^\mu{}_\nu(A)\,\Lambda_\varrho{}^\nu(A) = \tfrac{1}{4}\,\mathrm{Tr}\,(\sigma_\mu A\sigma_\nu A^+)\,\mathrm{Tr}\,(\tilde{\sigma}_\varrho A\tilde{\sigma}_\nu A^+) = \delta^\mu_\varrho,$$

or the orthogonality condition for the Lorentz transformation (19). Analogously it is easy to verify that $\Lambda(A_1)\,\Lambda(A_2) = \Lambda(A_1A_2)$, and in particular $\Lambda(A^{-1}) = \Lambda^{-1}(A)$. All coefficients $\Lambda^\mu{}_\nu$ are real:

$$\Lambda^{*\mu}{}_\nu(A) = \tfrac{1}{2}\,\mathrm{Tr}\,(\sigma_\mu A\sigma_\nu A^+)^+ = \tfrac{1}{2}\,\mathrm{Tr}\,(\sigma_\mu A\sigma_\nu A^+) = \Lambda^\mu{}_\nu(A).$$

The condition that the transformation be orthochronous, $\Lambda^{00} \geqslant 1$, imposes the condition

$$\Lambda^{00} = \tfrac{1}{2}\,\mathrm{Tr}\,(AA^+) \geqslant 1. \tag{25}$$

on the matrix A.

Besides the matrix X with the transformation law (17) one may also consider the matrix

$$\tilde{\mathrm{X}} = x^\mu\tilde{\sigma}_\mu = C^{-1}\mathrm{X}^*C.$$

From this and from (17), we find the transformation properties of $\tilde{\mathrm{X}}$:

$$\tilde{\mathrm{X}}' = C^{-1}A^*\mathrm{X}^*A^{+*}C = A^{-1+}\tilde{\mathrm{X}}A^{-1}.$$

In other words, under $\mathrm{X} \to \tilde{\mathrm{X}}$ one must make the substitution $A \to A^{-1+}$. The product of matrices X and $\tilde{\mathrm{Y}}$ transforms according to the rule $(\mathrm{X}\tilde{\mathrm{Y}})' = A\mathrm{X}\tilde{\mathrm{Y}}A^{-1}$, so that if in some reference frame this product is proportional to the unit matrix, then this property holds in all systems. It may be easily seen that for any X and Y the following relation is true:

$$\tfrac{1}{2}\,\mathrm{Tr}\,(\mathrm{X}\tilde{\mathrm{Y}}) = x_\mu y^\mu, \tag{26}$$

giving the scalar product of two four-vectors.

Let us return to the transformations A. The matrix A may be represented in the general case as a product of a Hermitian matrix H and a unitary matrix R:

$$A = RH \tag{27}$$

As a result of the unitarity of R the matrix H is defined in terms of A: $A^+A = HR^{-1}RH = H^2$ Thus $R = AH^{-1} = A(A^+A)^{1/2}$. As we see, the Hermitian matrices $A^+ = A$ describe pure Lorentz transformations, while the unitary matrices $A^+ = A^{-1}$ describe spatial rotations.

Let A_H be a Hermitian matrix. The general expression for it in terms of the matrices σ_μ is

$$A_H = \sum_\mu A(\mu)\sigma_\mu,$$

where the coefficients $A(\mu)$ are now real. To satisfy the unimodularity condition, we set

$$A(0) = \cosh\frac{\theta}{2}, \quad A(k) = n_k \sinh\frac{\theta}{2} \quad (n_1^2+n_2^2+n_3^2 = 1).$$

Then A_H takes the form

$$A_H = \sigma_0 \cosh\frac{\theta}{2}+(\boldsymbol{\sigma}\cdot\boldsymbol{n})\sinh\frac{\theta}{2} = \exp\left[(\boldsymbol{\sigma}\cdot\boldsymbol{n})\frac{\theta}{2}\right]. \tag{28}$$

Transformation of the matrix $\mathrm{X} \to \mathrm{X}' = A_H \mathrm{X} A_H^+$ corresponds to the transformation of the four-vector $x^\mu \to x'^\mu$:

$$x_0' = x_0 \cosh\theta+(\boldsymbol{n}\cdot\boldsymbol{x})\sinh\theta, \quad [\boldsymbol{n}\times\boldsymbol{x}'] = [\boldsymbol{n}\times\boldsymbol{x}], \tag{29}$$
$$(\boldsymbol{n}\cdot\boldsymbol{x}') = (\boldsymbol{n}\cdot\boldsymbol{x})\cosh\theta+x_0\sinh\theta,$$

where $\boldsymbol{x}$ is a space-like vector with components x^1, x^2, x^3. This is a Lorentz transformation of the four-vector x^μ from the system O to the system O′ moving with respect to O in the direction $\boldsymbol{n}$ with velocity $v = \tanh\theta$.

Let A_R now be unitary. In this case, instead of (28), one may write the following general expression for A_R:

$$A_R = \sigma_0 \cos\frac{\omega}{2} + i(\boldsymbol{\sigma}\cdot\boldsymbol{n})\sin\frac{\omega}{2} = \exp\left[i(\boldsymbol{\sigma}\cdot\boldsymbol{n})\frac{\omega}{2}\right]. \tag{30}$$

Formula (19) allows one to write the transformation of x^μ at once:

$$\boldsymbol{x}' = \boldsymbol{x}\cos\omega+(\boldsymbol{x}\cdot\boldsymbol{n})\,\boldsymbol{n}(1-\cos\omega)+[\boldsymbol{n}\times\boldsymbol{x}]\sin\omega,$$
$$x'^0 = x^0.$$

This is a rotation around the axis $\boldsymbol{n}$ by an angle ω.

The unitary transformations (30) form the unitary subgroup $SU(2)$ of the group $SL(2, c)$, holomorphic to the group R_3 of spatial rotations. In this case,

$$x_k' = R_{kl}x_l, \quad x_0' = x_0$$

and the matrix R satisfies the orthogonality condition

$$R^{\mathrm{T}}R = 1.$$

Instead of the full matrix X it is sufficient to limit oneself to transformations of its spatial part alone:

$$\mathrm{X}_R = x^k\sigma_k,$$

which transforms by the rule

$$\mathrm{X}_R' = A_R\mathrm{X}_R A_R^{-1}, \qquad A_R^+ = A_R^{-1},$$

where $\det A_R = 1$ in the absence of reflections. The relation between the coefficients R_{kl} of a transformation of the rotation group R_3 and a transformation matrix A_R of the group $SU(2)$ is given by [see (19)]

$$R_{kl}(A) = -\tfrac{1}{2}\,\mathrm{Tr}\,(\sigma_k A_R \sigma_l A_R^{-1}). \tag{31}$$

From (31) it is clear that this relation is double-valued: $R(A) = R(-A)$. In particular, the unit rotation, i.e. the rotation by an angle $\omega = 2n\pi$ ($n = 0, 1, 2, \ldots$), corresponds to the matrices $\pm E$.

If we are interested in a general Lorentz transformation, including both a pure Lorentz transformation and a rotation, the matrix A may be taken as

$$A = \exp\left(\tfrac{1}{2}\boldsymbol{\sigma}\cdot\boldsymbol{\alpha}\right), \tag{32}$$

where $\boldsymbol{\alpha}$ is a complex three-vector.

Another parametrization of a unimodular matrix A, which will be used frequently in applications, has the form

$$A = RBN, \tag{33}$$

where R describes any rotation (in Euler angles):

$$R = e^{-i\frac{1}{2}\sigma_3\varphi}e^{-i\frac{1}{2}\sigma_2\vartheta}e^{-i\frac{1}{2}\sigma_3\psi}$$
$$(0 \leqslant \varphi+\psi \leqslant 4\pi, \quad -2\pi \leqslant \varphi-\psi \leqslant 2\pi). \tag{34}$$

The matrix B refers to the Abelian subgroup:

$$B = e^{-\frac{1}{2}\sigma_3\beta} \qquad (-\infty < \beta < \infty); \tag{35}$$

the matrix N is triangular:

$$N = 1+\tfrac{1}{2}(\sigma_1 - i\sigma_2)z \tag{36}$$

(here z is any complex number). The matrix elements of A are:

$$\begin{aligned}
A_{11} &= e^{-\beta/2}\cos(\vartheta/2)e^{-i[(\varphi+\psi)/2]} - e^{\beta/2}z\sin(\vartheta/2)e^{-i[(\varphi-\psi)/2]},\\
A_{12} &= -e^{\beta/2}\sin(\vartheta/2)e^{-i[(\varphi-\psi)/2]},\\
A_{21} &= e^{-\beta/2}\sin(\vartheta/2)e^{i[(\varphi-\psi)/2]} + e^{\beta/2}z\cos(\vartheta/2)e^{i[(\varphi+\psi)/2]},\\
A_{22} &= e^{\beta/2}\cos(\vartheta/2)e^{i[(\varphi+\psi)/2]}.
\end{aligned}$$

To conclude this section, we verify that the group $SL(2, c)$ of unimodular 2 by 2 matrices really is a covering group. For this it is necessary to establish an isomorphism between the elements defined in § 1.1 of the covering group $\bar{L}$ and $SL(2, c)$. By (1.13) and (1.17) the relation between $\bar{L}$ and 2 by 2 matrices A is

$$\begin{aligned}
\bar{L}_0 &= [H\bar{R}]_0 = HR(\boldsymbol{n}, \theta) \leftrightarrow A(\boldsymbol{n}, \theta),\\
\bar{L}_1 &= [H\bar{R}]_1 = HR_1(\boldsymbol{n}, \theta) \leftrightarrow -A(\boldsymbol{n}, \theta),
\end{aligned}$$

so that by (30)

$$R(\boldsymbol{n}, \theta) = -R(-\boldsymbol{n}, 2\pi-\theta).$$

The matrices $A(\boldsymbol{n}, \theta)$ have the same multiplication law as HR, and, consequently, the matrices A form the covering group $\bar{L}$.

§ 3.2. Spinors

As we showed in the previous section, the group $SL(2, c)$ associates every four-vector x^{μ} with a Hermitian matrix X and in the space of these matrices the Lorentz transformation $x' = \Lambda x$ is given as a linear transformation $\mathrm{X}' = A\mathrm{X}A^{+}$ induced by the unimodular matrix A.

Let us now consider other quantities whose transformation law is determined by the matrix A.

The two-component spinor (or a spinor of the first rank)

$$\xi = \begin{pmatrix} \xi_1 \\ \xi_2 \end{pmatrix} \tag{37}$$

is a pair of complex numbers ξ_1, ξ_2, which transform by the rule

$$\xi'_\alpha = A_\alpha{}^\beta \xi_\beta \quad \text{or} \quad \xi' = A\xi. \tag{38}$$

The spinor ξ may be treated as a vector in the linear complex two-dimensional space associated with some reference frame in the space of Lorentz four-vectors. When passing to another reference frame, ξ goes into ξ' by (38). Since the relation $\Lambda \leftrightarrow \pm A$ is two-valued, a spinor may be given only up to a sign in any system.

We shall also denote the components of the spinor ξ by $\xi_1 = \xi_{1/2}$ and $\xi_2 = \xi_{-1/2}$, keeping in mind the effect of the matrix $(1/2)\sigma_3$ on ξ.

We define the spinor with an upper index η^α by the condition of invariance of the product $\eta^\alpha \xi_\alpha$ with respect to group transformations. From (38) it follows that η^α must transform by the matrices $A^{-1\mathrm{T}}$.

$$\eta'^\alpha = A^{-1\mathrm{T}\alpha}{}_\beta \eta^\beta. \tag{39}$$

Setting $A^{-1\mathrm{T}\alpha}{}_\beta = (C^{-1}AC)^\alpha{}_\beta$ by (6), we verify that in order that the contraction rule in upper and lower indices be satisfied, the matrix elements of C and C^{-1} must be written as $C_{\alpha\beta}$ and $C^{-1\alpha\beta}$. The spinor $\xi^\alpha = C^{-1\alpha\beta}\xi_\beta$ transforms by the rule (39), while the spinor $\eta_\alpha = C_{\alpha\beta}\eta^\beta$ transforms by formula (38). In other words, spinors with upper indices ξ^α and spinors with lower indices ξ_α transform via unitarily equivalent representations.

The bilinear form

$$\eta^\alpha \xi_\alpha = \eta^\alpha C_{\alpha\beta}\xi^\beta = -\eta_\beta C^{-1\beta\alpha}\xi_\alpha = -\eta_\beta \xi^\beta \tag{40}$$

is invariant with respect to Lorentz transformations. C and C^{-1} play the role of metric matrices. Here the rule for raising and lowering indices refers only to contraction with the right-hand index in C and C^{-1} since $C^\mathrm{T} = C^{-1}$, i.e. the properties of C and C^T are different in this respect.

With each matrix A one may associate a complex conjugate matrix A^* which is also unimodular. Under a Lorentz transformation, the complex conjugate pair of numbers $\xi^*_{1/2}$ and $\xi^*_{-1/2}$, forming the complex conjugate spinor ξ^*_α, will transform via the matrix A^*:

$$\xi^*_\alpha = A^{*\beta}_\alpha \xi^*_\beta.$$

Instead of ξ^*_α one usually writes $\xi_{\dot\alpha}$, while the matrix elements of A^* are written as $A^{*\dot\beta}_{\dot\alpha}$. Thus the complex conjugate spinor $\xi_{\dot\alpha}$ transforms via

$$\xi_{\dot\alpha} = A^{*\dot\beta}_{\dot\alpha} \xi_{\dot\beta}. \tag{41}$$

The matrix A^* cannot be obtained from A using any linear transformation since such a transformation would involve changing all parameters α, β, γ and δ in (1) into their complex conjugates, which is not a linear operation. Consequently, a representation by A^*

is not equivalent to a representation by the matrices A. The spinors ξ_α and $\xi_{\dot\alpha}$ transform in an inequivalent way. If, however, A is unitary, then $A^* = A^{-1\text{T}}$. This means, by (39) and (41), that under spatial rotations the spinors ξ^α and $\xi_{\dot\alpha}$ transform in the same way, i.e. in the unitary group SU_2 there is no difference between a spinor with an upper index and a complex conjugate spinor.

In analogy with (39) one may introduce the contravariant complex conjugate spinors $\xi^{\dot\alpha}$ having an upper dotted index:

$$\xi'^{\dot\alpha} = A^{-1+\dot\alpha}{}_{\dot\beta}\xi^{\dot\beta}. \tag{42}$$

The metric matrices here have the same form

$$\xi_{\dot\alpha} = C_{\dot\alpha\dot\beta}\xi^{\dot\beta}, \quad \xi^{\dot\alpha} = C^{-1\dot\alpha\dot\beta}\xi_{\dot\beta}, \tag{43}$$

$$C_{\dot\alpha\dot\beta} = -C^{-1\dot\alpha\dot\beta} = -C^{-1\alpha\beta} = C_{\alpha\beta} = \begin{pmatrix} 0 & -1 \\ 1 & 0 \end{pmatrix} = -i\sigma_2. \tag{44}$$

It is easy to see that the quantity

$$\eta^{\dot\alpha}\xi_{\dot\alpha} = -\eta_{\dot\alpha}\xi^{\dot\alpha} = \eta^{\dot\alpha}C_{\dot\alpha\dot\beta}\xi^{\dot\beta}$$

is invariant. The matrix elements of the unit matrix E will be written as $E_{\dot\alpha}^{\dot\delta} = \delta_{\dot\alpha}^{\dot\delta}$ and $E_\alpha^\beta = \delta_\alpha^\beta$.

As is obvious from the way that they were introduced, the dotted and undotted indices are independent, i.e. it is impossible, for example, to contract $\eta_{\dot\alpha}\xi^\alpha$ with respect to indices of different type. From this it follows that using spinors of only one type it is impossible to form an invariant bilinear Hermitian form.

To see how one must write the matrix elements of X, we turn to the transformation formula $X' = AXA^+$. The matrices A and A^+ have matrix elements $A_\alpha{}^\beta$ and $A^{+\dot\lambda}{}_{\dot\tau}$, so that the matrix elements of X must be denoted by $X_{\beta\dot\lambda}$, i.e. $X_{\beta\dot\lambda}$ is a mixed spinor of second rank:

$$X = \begin{pmatrix} X_{1/2\,\dot{1/2}} & X_{1/2-\dot{1/2}} \\ X_{-1/2\,\dot{1/2}} & X_{-1/2-\dot{1/2}} \end{pmatrix}. \tag{45}$$

The transformation properties of $X_{\beta\dot\lambda}$ coincide with the properties of $\xi_\beta\xi_{\dot\lambda}$:

$$X_{\beta\dot\lambda} \sim \xi_\beta\xi_{\dot\lambda},$$

where the tilde means "transforms as ...". But the matrix X is constructed from the Pauli matrices σ_μ. Thus the Pauli matrices which appear in X (or, of course, in any other matrix Y with the same transformation law) have the matrix elements $(\sigma_\mu)_{\alpha\dot\beta}$.

The form of the matrix elements of $\tilde\sigma_\mu$ may be obtained from the definition $\tilde\sigma_\mu = C^{-1}\sigma_\mu^{\text{T}}C$. Using C^{-1} to raise the dotted index and writing the matrix $C_{\alpha\beta}$ in the form $C^{-1\beta\alpha}$ (one may contract only the right index of C and C^{-1}!), we finally obtain

$$\tilde\sigma_\mu \equiv (\sigma_\mu)^{\dot\alpha\beta}. \tag{46}$$

Consequently, the contravariant coordinate matrix $\tilde{X}$ is $\tilde{X} = X^{\dot\alpha\beta}$ and

$$X_{\dot\alpha\beta} \sim \xi^{\dot\alpha}\xi^\beta. \tag{47}$$

The orthogonality condition (24) may now be written in the form

$$(\sigma_\mu)_{\alpha\dot\beta}\,(\tilde\sigma^\mu)^{\dot\beta'\alpha'} = 2\delta_\alpha^{\alpha'}\delta_{\dot\beta}^{\dot\beta'}. \tag{48}$$

Using (45) and the definition (14) of the matrix X, it is easy to obtain the properties of the components of the four-vector x^μ with respect to transformations of the binary group:

$$\left.\begin{aligned} x^0 &\sim \tfrac{1}{2}(\xi_{1/2}\xi_{\dot 1/2}+\xi_{-1/2-\dot 1/2}), & x^1 &\sim \tfrac{1}{2}(\xi_{1/2}\xi_{-\dot 1/2}+\xi_{-1/2}\xi_{-\dot 1/2}), \\ x^2 &\sim \tfrac{1}{2}i(\xi_{1/2}\xi_{-\dot 1/2}-\xi_{-1/2}\xi_{-\dot 1/2}), & x^3 &\sim \tfrac{1}{2}(\xi_{1/2}\xi_{\dot 1/2}-\xi_{-1/2}\xi_{-\dot 1/2}). \end{aligned}\right\} \tag{49}$$

Spinors of higher rank (spin-tensors) with lower indices are quantities transforming as the product of (different) spinors:

$$\Psi_{\alpha_1 \ldots \alpha_n \dot\beta_1 \ldots \dot\beta_m} \sim \xi_{\alpha_1} \ldots \eta_{\alpha_n}\zeta_{\dot\beta_1} \ldots \chi_{\dot\beta_m}. \tag{50}$$

Indices may be raised using the matrices $C^{-1\alpha\beta}$ and $C^{-1\dot\alpha\dot\beta}$. Thus the definition (50) also allows one to introduce the concept of spinors of higher rank with upper indices or the concept of mixed spin-tensors. Thus under Lorentz transformations the spinor $\psi_{\alpha\ldots}{}^{\beta\ldots}{}_{\dot\alpha\ldots}{}^{\dot\beta\ldots}$ transforms according to the unimodular matrix in every index, specifically: the lower undotted indices with the matrices $A_\alpha{}^{\alpha'}$, the lower dotted ones using $A_{\dot\alpha}^{*\dot\alpha'}$, the upper undotted ones using $A^{-1\mathrm{T}\beta'}{}_\beta$, and the upper dotted ones by $A^{-1+\dot\beta'}{}_{\dot\beta}$:

$$\Psi'_{\alpha'\ldots}{}^{\beta'\ldots}{}_{\dot\alpha'\ldots}{}^{\dot\beta'\ldots} = A_{\dot\alpha}{}^{\alpha} \ldots A^{-1\mathrm{T}\beta'}{}_\beta \ldots A^{*\dot\alpha}_{\dot\alpha'} \ldots A^{-1+\dot\beta'}{}_{\dot\beta}\Psi_{\alpha\ldots}{}^{\beta\ldots}{}_{\dot\alpha\ldots}{}^{\dot\beta\ldots}. \tag{51}$$

A spinor ψ of higher rank may be formed of spinors of lower rank φ and χ by multiplication: $\psi = \varphi\chi$. Here the spinor ψ has all indices of the spinors φ and χ. The product of a spinor with (n, m) indices (n undotted and m dotted) by a spinor with (n', m') indices gives a spinor with $(n+n', m+m')$ indices, or a spinor of $(n+n'+m+m')$th rank. For example, multiplying two spinors of second rank by one another

$$\varphi_{\alpha\dot\varrho}\chi_{\beta\dot\sigma} = \psi_{\alpha\beta\dot\varrho\dot\sigma}, \tag{52}$$

we obtain a spinor of fourth rank.

One may obtain spinors of lower rank from spinors of higher rank by contraction (summation) in lower and upper indices of the same type. For example, raising the index β in (52) and summing, we form a dotted spinor of second-rank

$$\eta_{\dot\varrho\dot\sigma} = \varphi_{\alpha\dot\varrho}\chi^{\alpha}{}_{\dot\sigma}.$$

Raising one of the dotted indices and summing, we obtain an invariant

$$\eta = \eta_{\dot\varrho}^{\dot\varrho} = \varphi_{\alpha\dot\varrho}\chi^{\alpha\dot\varrho}.$$

Spinors of higher rank which cannot be simplified (i.e. which give zero upon contraction) transform according to irreducible representations of the binary group. Since the metric matrices C and C^{-1} are antisymmetric ($C = -C^{\mathrm{T}}$), it is impossible to simplify any spinor of higher rank which is separately symmetric both in dotted and in undotted indices of the same character (whether all indices are upper or lower). In fact, $C^{-1\alpha\beta}\psi_{\alpha\beta\ldots} = 0$ and

$C_{\dot{\alpha}\dot{\beta}}\psi^{\dot{\alpha}\dot{\beta}\cdots} = 0$, if $\psi_{\alpha\beta\ldots}$ and $\psi^{\dot{\alpha}\dot{\beta}\cdots}$ are symmetric in $\alpha\beta$ and in $\dot{\alpha}\dot{\beta}$. An example of a spinor symmetric in indices of both types is the product of single spinors

$$\xi_{\alpha_1}\xi_{\alpha_2}\ldots\xi_{\alpha_n}\xi_{\dot{\beta}_1}\xi_{\dot{\beta}_2}\ldots\xi_{\dot{\beta}_m}. \tag{53}$$

Any spinor (50) has, in general, 2^{m+n} components. The symmetric spinor (53) has $(2m+1)(2n+1)$ components, since each of them takes only two values: $\pm\frac{1}{2}$ for $\alpha, \beta, \ldots,$ and $\pm 1\dot{/}2$ for $\dot{\alpha}, \dot{\beta}, \ldots$

§ 3.3. Irreducible representations and generalized spinor analysis

Symmetric spinors describe the higher representations of the group $SL(2, c)$. Since the order of the indices has no meaning for symmetric spinors, a spinor of the type (53) is defined only by the number of indices having the values $\pm\frac{1}{2}$ or $\pm 1\dot{/}2$. Thus we write (53) in the form

$$\xi_{1/2}^{\;n-s}\xi_{-1/2}^{\;s}\eta_{1\dot{/}2}^{\;m-r}\eta_{-1\dot{/}2}^{\;r},$$

where n and m are the number of undotted and dotted spinors ξ and η, while s and r are the number of components of these spinors with projection $\frac{1}{2}\sigma_3' = -\frac{1}{2}$. It is obvious that $0 \leqslant r \leqslant m$, $0 \leqslant s \leqslant n$.

To pass to the usual notation, we set $n = 2j_1$, $m = 2j_2$, $n-s = j_1+\sigma$, $s = j_1-\sigma$, $m-r = j_2+\dot{\varrho}$, $r = j_2-\dot{\varrho}$. Then j_1 and j_2 take on integer and half-integer nonnegative values, while σ and $\dot{\varrho}$ run respectively through the $2j_1+1$ and $2j_2+1$ values $\sigma = j_1, j_1-1, \ldots, -j_1+1, -j_1$ and $\dot{\varrho} = j_2, j_2-1, \ldots, -j_2+1, -j_2$. In other words, the quantum numbers σ and $\dot{\varrho}$ have the properties of spin projections on the third axis, while j_1 and j_2 are the eigenvalues of the spin.

Consequently, we characterize a spinor of rank $(2j_1, 2j_2)$ by the numbers j_1 and j_2 having the properties of eigenvalues of angular momentum. The components of this spinor are denoted by the projections σ, $\dot{\varrho}$ on the third axis. Passing to the new notation, we write a symmetric spinor $F(j_1, j_2)$ with lower indices in the form

$$F_{\sigma\dot{\varrho}}(j_1, j_2) = a(j_1, j_2; \sigma, \dot{\varrho})\xi_{1/2}^{\;j_1+\sigma}\xi_{-1/2}^{\;j_1+\sigma}\eta_{1\dot{/}2}^{\;j_2-\dot{\varrho}}\eta_{-1\dot{/}2}^{\;j_2-\dot{\varrho}}, \tag{54}$$

where $a(j_1, j_2; \sigma, \dot{\varrho})$ is a number.

We define analogously the spinor $\Phi(j_1, j_2)$ with upper indices:

$$\Phi^{\sigma\dot{\varrho}}(j_1, j_2) = a(j_1, j_2; \sigma, \dot{\varrho})\xi'^{\frac{1}{2}\,j_1+\sigma}\xi'^{-\frac{1}{2}\,j_1-\sigma}\eta'^{\dot{\frac{1}{2}}\,j_2+\dot{\varrho}}\eta'^{-\dot{\frac{1}{2}}\,j_2-\dot{\varrho}}, \tag{55}$$

where ξ' and η' are spinors of first rank different from ξ and η.

Expressions (54) and (55) agree with each other only for a definite choice of their common factor $a(j_1, j_2; \sigma, \dot{\varrho})$. In fact, the notation (54) and (55) with indices in different positions assumes that the quantity

$$F_{\sigma\dot{\varrho}}(j_1, j_2)\,\Phi^{\sigma\dot{\varrho}}(j_1, j_2) = \text{invt} \tag{56}$$

is invariant with respect to transformations.

If one inserts the expressions (54) and (55) into (56), one obtains a polynomial of degree $(4j_1, 4j_2)$ in the spinors ξ, ξ', η, η'. This polynomial can be invariant only when ξ and ξ' occur in the invariant combination $\xi_\alpha \xi'^\alpha$, and η and η' via the invariant $\eta_{\dot\alpha}\eta'^{\dot\alpha}$. Consequently, (56) must have the form

$$(\xi_\alpha \xi'^\alpha)^{2j_1} (\eta_{\dot\beta}\eta'^{\dot\beta})^{2j_2} c(j_1, j_2), \tag{57}$$

where $c(j_1, j_2)$ is an invariant factor, depending only on j_1 and j_2, i.e. on the number of spinors of various types. Comparing (57) with the results of putting (54) and (55) into (56),

$$a^2(j_1, j_2; \sigma, \dot\varrho) = \frac{(2j_1)!\,(2j_2)!\,c(j_1, j_2)}{(j_1+\sigma)!\,(j_1-\sigma)!\,(j_2+\dot\varrho)!\,(j_2-\dot\varrho)!}. \tag{58}$$

The factors depending only on j_1 and j_2 are identical for all components of the symmetric spinor of higher rank, and therefore may be discarded.

Thus the basis symmetric spinors of higher rank may be written in the form

$$F_{\sigma\dot\varrho}(j_1, j_2) = \frac{\xi_{1/2}^{\;j_1+\sigma}\xi_{-1/2}^{\;j_1-\sigma}\eta_{\dot{1/2}}^{\;j_2+\dot\varrho}\eta_{-\dot{1/2}}^{\;j_2-\dot\varrho}}{[(j_1+\sigma)!\,(j_1-\sigma)!\,(j_2+\dot\varrho)\,(j_2-\dot\varrho)!]^{1/2}}, \tag{59}$$

where $-j_1 \leqslant \sigma \leqslant j_1$, $-j_2 \leqslant \dot\varrho \leqslant j_2$; the quantum numbers j_1, j_2 may be nonnegative integers and half-integers, while neighboring values of σ and $\dot\varrho$ differ by unity. The spinors ξ and η are written as $F_\sigma(\frac{1}{2}, 0)$ and $F_{\dot\varrho}(0, \frac{1}{2})$.

Let us turn to the transformation properties of the spinors (59). Under Lorentz transformations, the $(2j_1+1)(2j_2+1)$ components of the spinor (59) transform among one another; thus there must exist a representation of the group $SL(2, c)$ by the square matrix $\mathcal{D}^{(j_1, j_2)}(A)$ with $(2j_1+1)(2j_2+1)$ rows, by which the spinor (59) transforms:

$$F'_{\sigma\dot\varrho}(j_1, j_2) = \{\mathcal{D}^{(j_1, j_2)}(A)\}_{\sigma\dot\varrho}^{\;\;\lambda\dot\tau} F_{\lambda\dot\tau}(j_1, j_2). \tag{60}$$

The matrices $\mathcal{D}(A)$ must be defined uniquely in terms of the matrices A.

The construction of the matrices $\mathcal{D}(A)$ proceeds from the definition of the basis spinor (59). The transformed spinor $F'(j_1, j_2)$ is given by the same formula (59) but with the transformed spinors $\xi' = A\xi$ and $\eta' = A^*\eta$ in place of the spinors ξ and η in the initial reference frame:

$$F'_{\sigma\dot\varrho}(j_1, j_2) = \frac{(A\xi)_{1/2}^{\;j_1+\sigma}(A\xi)_{-1/2}^{\;j_1-\sigma}(A^*\eta)_{\dot{1/2}}^{\;j_2+\dot\varrho}(A^*\eta)_{-\dot{1/2}}^{\;j_2-\dot\delta}}{[(j_1+\sigma)!\,(j_1-\sigma)!\,(j_2+\dot\varrho)!\,(j_2-\dot\varrho)!]^{1/2}}. \tag{61}$$

Comparison of (60) and (61) allows one to find the matrices $\mathcal{D}(A)$. To do this it is necessary to expand each factor in (61) of the type

$$(A\xi)_\alpha^{\;j+\sigma} = (A_\alpha^{\;1}\xi_{1/2} + A_\alpha^{\;2}\xi_{-1/2})^{j+\sigma}$$

via the binomial formula, then to construct a polynomial in $\xi_{1/2}, \eta_{\dot{1/2}}, \xi_{-1/2}$, and $\eta_{-\dot{1/2}}$. Collecting terms in $F_{\lambda\dot\tau}(j_1, j_2)$ and then comparing with (61), one may write the matrix elements of $\mathcal{D}^{(j_1, j_2)}(A)$. Since the polynomials in ξ and η are independent, we may find separately the matrices with dotted and undotted indices.

The spinors (59) and (61) may be represented in the form of a product of the spinors $f_\varrho(j_1)$ and $\tilde{f}_{\dot\sigma}(j_2)$, each of which is associated only with dotted or undotted indices:

$$F_{\varrho\dot\sigma}(j_1, j_2) = f_\varrho(j_1)\tilde{f}_{\dot\sigma}(j_2).$$

It is obvious that

Setting $j_2 = 0$, we obtain the matrix $\mathcal{D}^{(j_1,0)}(A)$, transforming only spinor components with undotted lower indices:

$$f_\varrho(j_1) = [(j_1+\varrho)!(j_1-\varrho)!]^{-1/2}\,\xi_{1/2}{}^{j_1+\varrho}\xi_{-1/2}{}^{j_1-\varrho}. \tag{62}$$

Setting $j_1 = 0$, we find the matrix $\mathcal{D}^{(0,j_2)}(A^*)$, transforming only components with dotted indices:

$$\tilde{f}_{\dot\sigma}(j_2)[(j_2+\dot\sigma)!\,(j_2-\dot\sigma)!]^{-1/2}\,\eta_{\dot{1/2}}{}^{j_2+\dot\sigma}\eta_{-\dot{1/2}}{}^{j_2-\dot\sigma}.$$

In the general case the matrix $\mathcal{D}^{(j_1,j_2)}(A)$ is a direct product of the matrices $\mathcal{D}^{(j_1,0)}(A)$ and $\mathcal{D}^{(0,j_2)}(A)$:

$$\mathcal{D}^{(j_1,j_2)}(A) = \mathcal{D}^{(j_1,0)}(A)\times\mathcal{D}^{(0,j_2)}(A). \tag{63}$$

Let us now compare the matrices $\mathcal{D}^{(0,j)}(A)$ and $\mathcal{D}^{(j,0)}(A)$. The spinor components $f_\varrho(j)$ and $\tilde{f}_{\dot\varrho}(j)$ are formed by the same rule respectively from the spinors ξ_α and $\eta_{\dot\alpha}$. The difference in the matrices $\mathcal{D}^{(j,0)}(A)$ and $\mathcal{D}^{(0,j)}(A)$ thus arises from the fact that ξ transforms by the matrix A and η by the complex conjugate matrix A^*. In other words, if we perform the substitution $A \to A^*$, then from $\mathcal{D}^{(j,0)}(A)$ we obtain $\mathcal{D}^{(0,j)}(A)$:

$$\mathcal{D}^{(0,j)}(A) = \mathcal{D}^{(j,0)}(A^*).$$

But $f_\varrho(j)$ contains the real coefficients $[(j+\varrho)!(j-\varrho)!]^{-1/2}$ times powers of the complex numbers $\xi_{1/2}$ and $\xi_{-1/2}$. Thus [as one may easily verify directly from (61)], one will have

$$\mathcal{D}^{(j,0)}(A^*) = [\mathcal{D}^{(j,0)}(A)]^* = \mathcal{D}^{(0,j)}(A). \tag{64}$$

From this it follows that the matrices $\mathcal{D}^{(j_1,j_2)}(A)$ and $\mathcal{D}^{(j_2,j_1)}(A)$ are complex conjugates of one another:

$$\mathcal{D}^{(j_1,j_2)}(A) = [\mathcal{D}^{(j_2,j_1)}(A)]^*. \tag{65}$$

This means that the transformation is real if $j_1 = j_2$.

On the basis of relations (62)–(64) we may ascertain the properties of $\mathcal{D}^{(j_1,j_2)}(A)$ by studying only

$$\mathcal{D}^{(j,0)}(A) \equiv D^j(A) \tag{66}$$

or the matrices $\mathcal{D}^{(0,j)}(A)$.

Let us examine the matrices $D^j(A)$. They form a representation of the binary group. The identity transformation for $A = E$ is

$$D^{j\beta}_{\alpha}(E) = \delta^{\beta}_{\alpha}. \tag{67}$$

Performing two successive transformations A_1 and A_2, we obtain from (60) and

$$(A_2A_1\xi)_{\frac{1}{2}}^{j+\alpha}(A_2A_1\xi)_{-\frac{1}{2}}^{j-\alpha} = [(j+\alpha)!\,(j-\alpha)!]^{\frac{1}{2}}\, D^j(A_2A_1)_\alpha^{\ \beta} f_\beta(j)$$
$$= [(j+\alpha)!\,(j-\alpha)!]^{\frac{1}{2}}\, D^j(A_2)_\alpha^{\ \varrho}(A_1\xi)_{\frac{1}{2}}^{j+\varrho}(A_1\xi)_{-\frac{1}{2}}^{j-\varrho}[(j+\varrho)!\,(j-\varrho)!]^{-\frac{1}{2}},$$

or

$$D^j(A_2)\,D^j(A_1) = D^j(A_2A_1). \tag{68}$$

As a consequence of (68),

$$D^j(A^{-1}) = [D^j(A)]^{-1} \equiv D^{j-1}(A). \tag{69}$$

In an analogous way, one may show that

$$D^j(A^{\mathrm{T}}) = [D^j(A)]^{\mathrm{T}} \equiv D^{j\mathrm{T}}(A). \tag{70}$$

The relation between the inverse A^{-1} and the transposed A^{T} 2 by 2 matrices was written earlier in the form (6), or $A^{-1} = C^{-1}A^TC$. The corresponding relation between the $(2j+1)\times(2j+1)$-matrices $D^{j-1}(A)$ and $D^{j\mathrm{T}}(A)$ may be obtained if one uses the multiplication rule (68) and formulae (69) and (70):

$$D^{j-1}(A) = D^j(C^{-1})\,D^{j\mathrm{T}}(A)\,D^j(C). \tag{71}$$

Thus this relation is completely analogous to the relation between the 2 by 2 matrices.

If we limit ourselves to unitary matrices $A^+ = A^{-1}$, the representation $D^j(A)$ coincides with the representations $\mathcal{D}^j(A)$ of the group of spatial rotations R_3 or the unitary group SU_2, which, as is well known, are irreducible. The group SU_2 is a subgroup of the binary group, and, consequently, the representations $D^j(A)$ are also irreducible. From this it follows as well that the representations $\mathcal{D}^{(j_1,j_2)}(A)$ are irreducible.

The matrices $D^j(A)$, in contrast to $\mathcal{D}^j(A)$, also are determined for nonunitary transformations A. The matrices D^j and $\mathcal{D}^j$ have in common the rule for construction of a representation in the space of basis spinors $f_\sigma(j)$. Thus in the case of the matrices D_j one can use those results obtained for $\mathcal{D}^j$ which are based only on the form of $f_\sigma(j)$ and the transformed spinor $f_{\sigma'}(j)$.

Let us now multiply each spinor in formula (62) for $f_\sigma(j)$ by a constant (complex) number a, thus obtaining

$$f'_\sigma(j;\,\xi) \equiv f_\sigma(j;\,a\xi) = a^{2j}f_\sigma(j,\,\xi).$$

The transformation $\xi \to a\xi$ may be treated as one induced by the nonunimodular 2 by 2 matrices $a\cdot E(\det(aE) = a^2)$. If one maintains the previous definition of $D^j(A)$ [formulae (61) and (62)], given there for unimodular matrices A, one may then write

$$D^j(aE) = a^{2j}D^j(E) = a^{2j}E^{(j)},$$

where $E^{(j)}$ denotes the unit matrix in the space of the $2j+1$ functions $f_\sigma(j)$.

Analogously, one may see that

$$D^j(aB) = a^{2j}D^j(B), \tag{72}$$

where B is any 2 by 2 matrix not limited by the unimodularity condition. From (72) one may easily conclude as well that D_j form a representation of the group of 2 by 2 matrices (all complex 2 by 2 matrices B). For this we note that if $\det B = b^2 \neq 0$, then one may set $B = bA$, where $\det A = 1$. Then

$$D^j(B_1)\, D^j(B_2) = D^j(b_1A_1)\, D^j(b_2A_2) = (b_1b_2)^{2j}\, D^j(A_1A_2) = D^j(b_1b_2A_1A_2) = D^j(B_1B_2), \quad (73)$$

so $D^j(B)$ is a representation of the group of matrices B with nonzero determinant (nonsingular matrices). But the representations $D^j(B)$ are continuous functions of the matrix elements B [see, for example, (61)]. Thus (73) remains valid for singular matrices B, when $\det B = 0$.

Up to now we have considered spinors with lower indices. In analogy with (55) and (52) we may define a spinor with an upper index

$$f^\sigma(j) = [(j+\sigma)!\,(j-\sigma)!]^{-1/2}\, \xi^{\frac{1}{2}\,j+\sigma} \xi^{-\frac{1}{2}\,j-\sigma}. \quad (74)$$

The passage from spinors f_σ to spinors f^σ (raising of indices) is carried out, according to the definition (74), using the operation in which the index of every spinor ξ_α in (62) is raised by the metric matrix C^{-1}. In other words, raising indices of the spinor f_σ is performed using the matrix $D^j(C^{-1})$:

$$f^\varrho(j) = D^{j\varrho\sigma}(C^{-1})\, f_\sigma(j). \quad (75)$$

Since $(C^{-1}\xi)^{-1/2} = -\xi_{1/2}$, $(C^{-1}\xi)^{1/2} = \xi_{-1/2}$, via the definition of $D^j(A)$, we may find the matrix elements

$$D^{j\varrho\sigma}(C^{-1}) = (-1)^{j-\varrho}\, \delta_{\varrho}^{-\sigma}. \quad (76)$$

An index is lowered using the matrix $D^j(C)$. From (76) we find

$$D^j_{\varrho\sigma}(C) = (-1)^{j+\varrho}\, \delta_{\varrho}^{-\sigma}. \quad (77)$$

The matrices $D^j(C)$ and $D^j(C^{-1})$ are symmetric for even values of j and antisymmetric for odd j:

$$D^j_{\varrho\sigma}(C) = (-1)^{2j}\, D^{j\varrho\sigma}(C^{-1}), \quad D^j_{\varrho\sigma}(C) = (-1)^{2j}\, D^j_{\sigma\varrho}(C). \quad (78)$$

From this, in particular, one obtains a generalization of (40) for the contraction of two spinors:

$$f^\beta \varphi_\beta = (-1)^{2j} f_\beta \varphi^\beta. \quad (79)$$

For spinors with dotted indices $\psi^{\dot\varrho}(j)$ and $\chi_{\dot\sigma}(j)$ and representations $D^j(A^*) = \mathcal{D}^{(0,j)}(A)$ we could repeat all operations beginning with (64), finally obtaining all the above formulae but with dotted indices, including

$$D^j_{\dot\varrho\dot\sigma}(C) = (-1)^{2j}\, D^{j\dot\varrho\dot\sigma}(C^{-1}) = (-1)^{j+\dot\varrho}\, \delta_{\varrho}^{-\dot\sigma}. \quad (80)$$

Thus spinor algebra in the case of spinors of the type $f_\sigma(j)$ is a direct generalization of the algebra for the simplest spinors ξ, η. Instead of two values of $\varrho, \dot\sigma = \pm\frac{1}{2}$, the indices now take on $2j+1$ values $\varrho, \dot\sigma = j, j-1, \ldots, -j+1, -j$. These spinors transform by the matrices $D^j(A)$ with $(2j+1)^2$ matrix elements; here the type of indices in D^j exactly corresponds to the type of indices in its argument A. We write $D^{j\beta}_{\alpha}(A)$, $D^{j\beta}{}_{\alpha}(A^{-1\mathrm{T}})$, $D^{j\dot\beta}_{\dot\alpha}(A^*)$, $D^{j\dot\beta}{}_{\dot\alpha}(A^{-1+})$, and instead of the metric tensors $C_{\alpha\beta}$, $C_{\dot\beta\dot\alpha}$ and $C^{-1\alpha\beta}$, $C^{-1\dot\alpha\dot\beta}$ use the matrices $D^j_{\alpha\beta}(C)$, $D^j_{\dot\alpha\dot\beta}(C)$, $D^{j\alpha\beta}(C^{-1})$, and $D^{j\dot\alpha\dot\beta}(C^{-1})$. Of course, numerically all four matrices $D^{j\beta}_{\alpha}(B)$, $D^{j\alpha}{}_{\beta}(B)$

$D^{j\dot\beta}{}_{\dot\alpha}(B)$, and $D^{j\dot\alpha}{}_{\dot\beta}(B)$ will be the same for the same numerical matrix B. According to (64), (69), and (70) the matrices $D^j(A)$ are unitary if the matrix A is unitary. The matrices $D^j(A)$ are Hermitian if the matrix A is Hermitian.

The one-to-one connection between the properties of the matrices $D^j(B)$ and B allows one to simplify the writing of the indices; we can, for example, write only the lower indices, easily reestablishing their real meaning via properties of B. For example, $D^j_{\sigma\sigma'}(A) \equiv D^{j\sigma'}_{\sigma}(A)$, $D^j_{\sigma\sigma'}(A^*) \equiv D^{j\dot\sigma'}_{\dot\sigma}(A^*)$, $D^j_{\sigma\sigma'}(C^{-1}) \equiv D^{j\sigma\sigma'}(C^{-1})$, $D^j_{\sigma\sigma'}(\mathrm{X}) \equiv D^j_{\sigma\dot\sigma'}(\mathrm{X})$, $D^j_{\sigma\sigma'}(\tilde{\mathrm{X}}) = D^j_{\dot\sigma\sigma'}(\tilde{\mathrm{X}})$. This form of writing indices will be used later on in Chapter 4.

Let us sum up. An irreducible finite-dimensional representation $\mathcal{D}^{(j_1, j_2)}(A)$ of the unimodular group $SL(2, c)$ is characterized by two numbers j_1, j_2, each of which may take on integer or half-integer nonnegative values. The dimension of a representation (j_1, j_2) is equal to $(2j_1+1)(2j_2+1)$. The matrices $\mathcal{D}^{(j_1, j_2)}(A)$ may be written as a direct product of matrices $D^{j_1}(A)$ and $D^{j_2*}(A)$: $\mathcal{D}^{(j_1, j_2)} = D^{j_1} \times D^{j_2*}$. The representation $\mathcal{D}^{(j_1, j_2)}$ is nonunitary.

The representation (j_1, j_2) of the group $SL(2, c)$ for integral j_1+j_2 is independent of the substitution $A \to -A$ [see (62) and (72)], and corresponds to (single-valued) representations of the Lorentz group. If j_1+j_2 is a half-integer, the representation (j_1, j_2) is a double-valued representation of the Lorentz group.

§ 3.4. Direct products of representations and covariant Clebsch–Gordan coefficients

The product of two spinors $f_\sigma(j_1)$ and $f_\varrho(j_2)$ transforms according to a reducible representation $D^{j_1}D^{j_2}$. Using (62) it is easily established that $D^{j_1}D^{j_2}$ may be expanded in a sum of representations D^j with

$$j = |j_1-j_2|, \quad |j_1-j_2|+1, \ \ldots, \quad |j_1+j_2| \tag{81}$$

or

$$f_{\sigma_1}(j_1)\, f_{\sigma_2}(j_2) = \sum_j \langle j_1 j_2 \sigma_1 \sigma_2 \,|\, j_1 j_2 j \sigma\rangle f_\sigma(j), \qquad \sigma = \sigma_1+\sigma_2. \tag{82}$$

The expansion coefficients $\langle j_1 j_2 \sigma_1 \sigma_2 | j_1 j_2 j \sigma\rangle$ are defined for $\sigma = \sigma_1+\sigma_2$ and for the values of j enumerated in (81). For fixed j_1, j_2, j these coefficients are rectangular matrices, where σ_1 varies from $-j_1$ to $+j_1$ while σ_2 takes on the values $-j_2$ to $+j_2$. The coefficients do not depend on the form of the matrix A and thus may be found for unitary A when the representations D^j coincide with the representations $\mathcal{D}^j$ of the rotation group SU_2. Consequently, the coefficients in (82) coincide with the usual Clebsch–Gordan coefficients:

$$\langle j_1 j_2 \sigma_1 \sigma_2 \,|\, j_1 j_2 j \sigma\rangle = C(j_1 j_2 j;\, \sigma_1 \sigma_2 \sigma), \tag{83}$$

and are the matrix elements of the unitary transformation from the basis $f(j_1)f(j_2)$ to the basis $f(j)$ with values j in (81).

From the unitarity of the transformation, the orthogonality of the coefficients (83) follows:

$$\left.\begin{aligned} &\sum_{\sigma, j} \langle j_1 j_2 \sigma_1 \sigma_2 \,|\, j_1 j_2 j \sigma\rangle \langle j_1 j_2 \sigma_1' \sigma_2' \,|\, j_1 j_2 j \sigma\rangle = \delta_{\sigma_1}^{\sigma_1'} \delta_{\sigma_2}^{\sigma_2'}, \\ &\sum_{\sigma_1, \sigma_2} \langle j_1 j_2 \sigma_1 \sigma_2 \,|\, j_1 j_2 j' \sigma'\rangle \langle j_1 j_2 \sigma_1 \sigma_2 \,|\, j_1 j_2 j \sigma\rangle = \delta_{jj'} \delta_\sigma^{\sigma'}. \end{aligned}\right\} \tag{84}$$

Let us perform the Lorentz transformation A on both sides of formula (82). The independence of the Clebsch–Gordan coefficients of the form of the matrices A means that

$$\langle j_1 j_2 \sigma_1 \sigma_2 \,|\, j_1 j_2 j \sigma\rangle = \sum_{\sigma_1,\,\sigma_2,\,\sigma} D^{j_1\sigma_1}_{\ \ \sigma_1'}(A^{-1})\, D^{j_2\sigma_2}_{\ \ \sigma_2'}(A^{-1})\, D^{j\sigma'}_{\ \ \sigma}(A)\, \langle j_1 j_2 \sigma_1' \sigma_2' \,|\, j_1 j_2 j \sigma'\rangle. \tag{85}$$

must hold. Formula (85) determines the transformation properties of the coefficients $\langle j_1 j_2 \sigma_1 \sigma_2 | j_1 j_2 j \sigma\rangle$ for Lorentz transformations: it transforms as $f^{\sigma_1}(j_1)\, f^{\sigma_2}(j_2)\, f_\sigma(j)$. In short, the Clebsch–Gordan coefficients have the same numerical value for all reference frames.

Passing from the matrices A to the matrices A^*, we obtain from the representation $\mathcal{D}^{(j,\,0)}$ the representation $\mathcal{D}^{(0,\,j)} = D^{j*}$ with dotted indices. Because of their independence of the matrices A, the Clebsch–Gordan coefficients are the same both for D^j and for D^{j*}. From this it follows also that they are real.

To emphasize the spinor properties of the coefficients (82) one sometimes introduces the notation

$$\langle j_1 j_2 \sigma_1 \sigma_2 \,|\, j_1 j_2 j \sigma\rangle = \begin{bmatrix} j j_1 j_2 \end{bmatrix}_{\sigma}^{\ \sigma_1\sigma_2} = \begin{bmatrix} j j_1 j_2 \end{bmatrix}_{\dot\sigma}^{\ \dot\sigma_1\dot\sigma_2}. \tag{86}$$

The covariant Clebsch–Gordan coefficient has only upper spinor indices:(18, 42)

$$\begin{pmatrix} j_1 & j_2 & \big| & j \\ \sigma_1 & \sigma_2 & \big| & \sigma \end{pmatrix} \equiv [j j_1 j_2]^{\sigma\sigma_1\sigma_2} = D^{j\sigma\sigma'}(C^{-1}) \begin{bmatrix} j j_1 j_2 \end{bmatrix}_{\sigma'}^{\ \sigma_1\sigma_2} = (-1)^{j-\sigma} \langle j_1 j_2 \sigma_1 \sigma_2 \,|\, j_1 j_2 j {-\sigma}\rangle$$

$$= (-1)^{j-j_2-j_1} \sqrt{2j+1} \begin{pmatrix} j_1 & j_2 & j \\ \sigma_1 & \sigma_2 & \sigma \end{pmatrix}, \qquad (\sigma = \sigma_1 + \sigma_2), \tag{87}$$

where $\begin{pmatrix} j_1 & j_2 & j \\ \sigma_1 & \sigma_2 & \sigma \end{pmatrix}$ is the usual Wigner $3j$ symbol.

The Clebsch–Gordan coefficients have the following symmetry properties:

$$\left.\begin{aligned} \langle j_1 j_2 \sigma_1 \sigma_2 \,|\, j_1 j_2 j \sigma\rangle &= (-1)^{j_1+j_2-j} \langle j_1 j_2 {-\sigma_1} {-\sigma_2} \,|\, j_1 j_2 j {-\sigma}\rangle, \\ \langle j_1 j_2 \sigma_1 \sigma_2 \,|\, j_1 j_2 j \sigma\rangle &= (-1)^{j_1+j_2-j} \langle j_2 j_1 \sigma_2 \sigma_1 \,|\, j_2 j_1 j \sigma\rangle. \end{aligned}\right\} \tag{88}$$

We note also that:

$$\left.\begin{aligned} \begin{bmatrix} j j_1 j_2 \end{bmatrix}^{\sigma}_{\ \sigma_1\sigma_2} &= (-1)^{2j} \begin{bmatrix} j j_1 j_2 \end{bmatrix}_{\sigma}^{\ \sigma_1\sigma_2} = (-1)^{2j} \begin{bmatrix} j_1 j_2 j \end{bmatrix}_{\sigma_1\sigma_2}^{\ \ \sigma}, \\ \begin{bmatrix} 0 j j \end{bmatrix}_0^{\ \sigma_1\sigma_2} &= \frac{1}{\sqrt{2j+1}} D^{j\sigma_1\sigma_2}(C^{-1}) = \frac{1}{\sqrt{2j+1}} (-1)^{j-\sigma_1} \delta_{\sigma_1}^{-\sigma_2}, \\ \begin{bmatrix} j j 0 \end{bmatrix}_{\sigma_1\sigma_2}^{\ \ 0} &= \frac{1}{\sqrt{2j+1}} D^{j}_{\sigma_1\sigma_2}(C) = \frac{1}{\sqrt{2j+1}} (-1)^{j+\sigma_1} \delta_{\sigma_1}^{-\sigma_2}. \end{aligned}\right\} \tag{89}$$

§ 3.5. Representations of the unitary group SU_2

The representations of the unitary group SU_2 may be obtained from the representations $(J, 0)$ of the group $SL(2, c)$ if we restrict ourselves to unitary 2 by 2 matrices R. Denoting a matrix of the representation of SU_2 by $\mathcal{D}^j$, we have

$$D^J(R) = \mathcal{D}^J(R).$$

The matrices $\mathcal{D}^J$ are usually studied in a course on nonrelativistic quantum mechanics. In relativistic quantum theory they are also important, since, as we shall see below in Chapter 4, it is precisely the matrices $\mathcal{D}^J$ that are involved in the transformation law of state vectors of massive particles. For this reason, we shall dwell on the properties of these matrices.

The infinitesimal form of a unitary 2 by 2 matrix R describing a rotation may be written as

$$R = 1 - i\tfrac{1}{2}(\boldsymbol{\sigma}\cdot\boldsymbol{\omega}),$$

where the rotation parameters ω_k are real. In the Jth representation of the group SU_2, this form corresponds to the infinitesimal form of the matrix $\mathcal{D}^J(R)$:

$$\mathcal{D}^J(R) = 1 - i(\boldsymbol{J}\cdot\boldsymbol{\omega}). \tag{90}$$

The generators J_k introduced in formula (90) are self-adjoint operators and satisfy the same commutation relations

$$[J_i, J_j] = i\varepsilon_{ijk}J_k,$$

as in the particular case of the representation $J = \frac{1}{2}$, when $J_k = \frac{1}{2}\sigma_k$. For finite transformations $R = \exp\left[-i\frac{1}{2}\boldsymbol{\sigma}\cdot\boldsymbol{\omega}\right]$ we must write in place of (90)

$$\mathcal{D}^J(R) = \exp\{-i(\boldsymbol{J}\cdot\boldsymbol{\omega})\}.$$

It is obvious that if R is parametrized using Euler angles [expression (34)], then

$$\mathcal{D}^J(R) \equiv \mathcal{D}^J(\varphi, \vartheta, \psi) = e^{-iJ_3\varphi}e^{-iJ_2\vartheta}e^{-iJ_3\psi}. \tag{91}$$

The basis of a representation $\mathcal{D}^J$ is a set of $2J+1$ functions Φ_σ^J distinguished by the rank $2n = 2J$ of an irreducible multi-spinor or by the eigenvalue of the invariant $\boldsymbol{J}^2$:

$$(J_1^2 + J_2^2 + J_3^2)\Phi_\sigma^J = J(J+1)\Phi_\sigma^J.$$

The basis functions Φ_σ^J differ from one another with respect to the eigenvalue of the operator J_3:

$$J_3\Phi_\sigma^J = \sigma\Phi_\sigma^J;$$

the quantity σ may take the following values: $-J, -J+1, \ldots, J-1, J$. For a transformation given by a 2 by 2 matrix R, the basis functions Φ_σ^J transform by the rule

$$\Phi_\sigma^J \xrightarrow{R} \Phi_\sigma'^J = \sum_{\sigma'} \mathcal{D}_{\sigma'\sigma}^J(R)\Phi_{\sigma'}^J. \tag{92}$$

Since the transformations of the group SU_2 are unitary, we may introduce a quantum mechanical notation for the basis functions:

$$\Phi_\sigma^J \equiv |\sigma, J\rangle,$$

and write (92) in the form

$$|\sigma, J\rangle' = \mathcal{D}^J(R)\,|\sigma, J\rangle = \sum_{\sigma'} |\sigma', J\rangle\, \mathcal{D}_{\sigma'\sigma}^J(R).$$

The matrix $\mathcal{D}_{\sigma'\sigma}^J$ is thus formed from the matrix elements of the unitary operator $\mathcal{D}^J$:

$$\mathcal{D}_{\sigma'\sigma}^J(R) = \langle\sigma', J\,|\,\mathrm{U}(R)\,|\,\sigma, J\rangle = \langle\sigma', J\,|\,e^{-iJ_3\varphi}e^{-iJ_2\vartheta}e^{-iJ_3\psi}\,|\,\sigma, J\rangle \equiv \mathcal{D}_{\sigma'\sigma}^J(\varphi, \vartheta, \psi). \tag{93}$$

The second relation (93) is satisfied if the parametrization (34) is chosen for R, or the expression (91) is chosen for the operator $U(R)$. This parametrization is convenient since the operator J_3 is usually assumed diagonal.

The functions $\mathcal{D}^J_{\sigma'\sigma}(\varphi, \vartheta, \psi)$ introduced in (93) are the matrix elements of any rotation $U(R)$ in an irreducible representation J. They are those fundamental functions which are connected with the group SU_2 and completely characterize it. The functions $\mathcal{D}^J$ satisfy the equation

$$\left\{\frac{1}{\sin\vartheta}\frac{\partial}{\partial\vartheta}\left(\sin\vartheta\frac{\partial}{\partial\vartheta}\right)+\frac{1}{\sin^2\vartheta}\left(\frac{\partial^2}{\partial\psi^2}+\frac{\partial^2}{\partial\varphi^2}-2\cos\vartheta\frac{\partial^2}{\partial\varphi\,\partial\psi}\right)\right\}\mathcal{D}^J_{\sigma'\sigma}(\varphi,\vartheta,\psi)$$
$$= -J(J+1)\,\mathcal{D}^J_{\sigma'\sigma}(\varphi,\vartheta,\psi) \quad (94)$$

where

$$\sigma'\mathcal{D}^J_{\sigma'\sigma}(\varphi,\vartheta,\psi) = -i\frac{\partial}{\partial\varphi}\mathcal{D}^J_{\sigma'\sigma}(\varphi,\vartheta,\psi),$$

$$\sigma\mathcal{D}^J_{\sigma'\sigma}(\varphi,\vartheta,\psi) = -i\frac{\partial}{\partial\psi}\mathcal{D}^J_{\sigma'\sigma}(\varphi,\vartheta,\psi).$$

The orthogonality condition for $\mathcal{D}^J_{\sigma'\sigma}(\varphi, \vartheta, \psi)$ has the form

$$\frac{1}{8\pi^2}\int_0^{2\pi}\int_0^{\pi}\int_0^{2\pi}\mathcal{D}^{J*}_{\sigma\gamma}(\varphi,\vartheta,\psi)\,\mathcal{D}^{J'}_{\sigma'\gamma'}(\varphi,\vartheta,\psi)\sin\vartheta\,d\vartheta\,d\varphi\,d\psi = \frac{1}{2J+1}\,\delta_{JJ'}\delta_{\sigma\sigma'}\delta_{\gamma\gamma'}. \quad (95)$$

If the angles φ and ψ are connected in such a way that $\varphi+\psi = 0$, then

$$\int\mathcal{D}^{J'*}_{\sigma'\gamma}(\varphi,\vartheta,-\varphi)\,\mathcal{D}^J_{\sigma\gamma}(\varphi,\vartheta,-\varphi)\sin\vartheta\,d\vartheta\,d\varphi = \frac{4\pi}{2J+1}\,\delta_{JJ'}\delta_{\sigma\sigma'}. \quad (96)$$

holds.

The functions $\mathcal{D}^J_{\sigma\sigma'}(\varphi, \vartheta, \psi)$, as functions of the Euler angles for various values of J, σ, σ', are a complete set, or

$$\frac{1}{8\pi^2}\sum_{J\sigma\sigma'}(2J+1)\,\mathcal{D}^{J*}_{\sigma'\sigma}(\varphi,\vartheta,\psi)\,\mathcal{D}^J_{\sigma'\sigma}(\varphi',\vartheta',\psi') = \delta(\varphi-\varphi')\,\delta(\cos\vartheta-\cos\vartheta')\,\delta(\psi-\psi'). \quad (97)$$

The relation (97) is satisfied separately for integral and for half-integral values of J.

Since σ is an eigenvalue of J_3, by (93) one may isolate the dependence of the function $\mathcal{D}^J_{\sigma'\sigma}$ on the angles φ and ψ:

$$\mathcal{D}^J_{\sigma'\sigma}(\varphi,\vartheta,\psi) = e^{-i\sigma'\varphi-i\sigma\psi}d^J_{\sigma'\sigma}(\vartheta), \quad (98)$$

where

$$d^J_{\sigma'\sigma}(\vartheta) = \langle\sigma', J|e^{-iJ_2\vartheta}|\sigma, J\rangle.$$

The functions $d^J(\vartheta)$ have been calculated by Wigner.[18] They have the following symmetry properties:

$$\left.\begin{aligned} d^J_{\sigma'\sigma}(\vartheta) &= (-1)^{\sigma'-\sigma}\,d^J_{\sigma\sigma'}(\vartheta) = (-1)^{\sigma'-\sigma}\,d^J_{-\sigma'-\sigma}(\vartheta),\\ d^J_{\sigma-\sigma}(\vartheta) &= (-1)^{J+\sigma'}\,d^J_{\sigma'\sigma}(\pi-\vartheta). \end{aligned}\right\} \quad (99)$$

Formulae (95) and (97) are equivalent to the following relations for $d^J(\vartheta)$:

$$\left.\begin{aligned}\int_0^\pi d^{J'}_{\sigma'\sigma}(\vartheta)\, d^J_{\sigma'\sigma}(\vartheta)\sin\vartheta\, d\vartheta &= \frac{2}{2J+1}\,\delta_{JJ'},\\ \sum_J (J+\tfrac{1}{2})\, d^J_{\sigma\sigma'}(\vartheta')\, d^J_{\sigma\sigma'}(\vartheta) &= \delta(\cos\vartheta'-\cos\vartheta).\end{aligned}\right\} \tag{100}$$

The explicit expression for $d^J(\vartheta)$ is rather cumbersome:

$$\begin{aligned} d^J_{\sigma'\sigma}(\vartheta) = \sum_l (-1)^{l+\sigma'-\sigma} &\frac{\sqrt{(J+\sigma)!\,(J-\sigma)!\,(J+\sigma')!\,(J-\sigma')!}}{(J-\sigma'-l)!\,(J+\sigma-l)!\,l!\,(l+\sigma'-\sigma)!}\\ &\times\left(\cos\frac{\vartheta}{2}\right)^{2J+\sigma-\sigma'-2l}\left(\sin\frac{\vartheta}{2}\right)^{2l+\sigma'-\sigma}. \end{aligned} \tag{101}$$

CHAPTER 4

THE QUANTUM MECHANICAL POINCARÉ GROUP

THE relativistic dynamics of a free particle is fully determined by the behavior of its state vector under transformations $\bar{g}$ of the quantum mechanical Poincaré group $\bar{\mathcal{P}}^{\uparrow}_{+}$. Knowledge of the explicit form of the operator $U(\bar{g})$ acting on the particle state vector is equivalent to the solution of the equation of motion for a free particle. In particular, after constructing $U(\bar{g})$ the transition to local fields becomes obvious.

In studying a scattering process, effects connected with the motion of free particles make up the kinematics of the process. Thus for interacting particles knowledge of $U(\bar{g})$ allows one to distinguish kinematics from dynamics.

In the case of elementary particles, the operators $U(\bar{g})$ form an irreducible unitary representation of the quantum mechanical Poincaré group (see § 1.2). It is thus necessary for us to construct an explicit expression for $U(\bar{g})$ for masses $m > 0$ and $m = 0$.

§ 4.1. Introductory remarks

As was shown in §§ 1.3 and 1.4, elementary particles are classified in the Poincaré group by two invariants—the mass m and the spin J($m^2 = P_\mu P^\mu$ and $J(J+1) = -w^2/m^2$). We consider only the case $m^2 \geqslant 0$. The polarization vector w_μ (or the Pauli–Lyubanski vector) is defined here by the formula

$$w_\mu = \tfrac{1}{2}\varepsilon_{\mu\nu\lambda\sigma}M^{\nu\lambda}P^\sigma .$$

The third invariant—the sign of the eigenvalue of the operator P^0—is always positive for physical states.

In the canonical basis introduced in § 1.3, as quantities forming a complete set of mutually commuting operators, one chooses (along with mass and spin) the momentum $\boldsymbol{p}$ and one of the components of the spin vector

$$J^k = -\frac{1}{m}\, w^\lambda n_\lambda^{(k)},$$

where $n_\lambda^{(k)}$ is a unit four-vector orthogonal to the four-momentum p_μ. Thus the full notation for a particle state vector is

$$|\,\boldsymbol{p}, \sigma;\, m, J;\, \zeta\rangle, \tag{1}$$

where σ is the eigenvalue of the operator J^3, while ζ refers to all remaining variables. We shall sometimes omit for brevity the letters m, J, and ζ from (1) (if this does not cause misunderstandings) and write instead of (1) simply

$$|\boldsymbol{p}, \sigma\rangle.$$

Our problem consists in finding an explicit form of the unitary transformation of the state vector (1)

$$|\boldsymbol{p}, \sigma\rangle \to \mathrm{U}(\bar{g})\,|\boldsymbol{p}, \sigma\rangle, \tag{2}$$

induced by an element of the quantum mechanical Poincaré group $\bar{g}$.

A transformation of the classical Poincaré group $g = (a, \Lambda)$ is given by a translation four-vector a^μ and a real 4 by 4 Lorentz transformation matrix Λ. In the quantum mechanical group, a homogeneous transformation may be written in the form of a unimodular complex 2 by 2 matrix A, while the four-vector a^μ may be associated with the Hermitian matrix $\mathrm{a} = \sigma_\mu a^\mu$, so that the element $\bar{g}$ of this group is $\bar{g}(\mathrm{a}, A)$. The Lorentz matrix Λ is defined in terms of the matrix A; specifically, the transformation of a matrix $\mathrm{b} = \sigma_\mu b^\mu$

$$\mathrm{b}' = A\mathrm{b}A^+$$

corresponds to the Lorentz transformation

$$b'^\mu = \Lambda^\mu{}_\nu b^\nu, \tag{3}$$

where $\Lambda(\pm A) = \Lambda(A)$ (see Chapter 3). Instead of the multiplication law of the classical Poincaré group

$$(a_1, \Lambda_1)(a_2, \Lambda_2) = (a_1 + \Lambda_1 a_2, \Lambda_1 \Lambda_2)$$

the multiplication law

$$(\mathrm{a}_1, A_1)(\mathrm{a}_2, A_2) = (\mathrm{a}_1 + A_1 \mathrm{a}_2 A_1^+, A_1 A_2). \tag{4}$$

holds in the quantum mechanical Poincaré group. The unit element is (0, 1); the inverse element is equal to

$$(\mathrm{a}, A)^{-1} = (-A^{-1}\mathrm{a}A^{-1+}, A^{-1}). \tag{5}$$

An inhomogeneous transformation (a, A) was defined as a homogeneous transformation (0, A) followed by a translation (a, 1); in fact, in this case (4) then reads

$$(\mathrm{a}, A) = (\mathrm{a}, 1)(0, A) = (0, A)(A^{-1}\mathrm{a}A^{-1+}, 1). \tag{6}$$

Thus, to find the unitary representations U(a, A) it is sufficient to find separately the unitary representations U(a, 1) of the translation and U(0, A) of the homogeneous subgroups.

The unitary representations U(a, A) of the quantum mechanical Poincaré group were found by Wigner.(23) The method of induced representations used to construct U(a, A) rests on the concept of the little, or stationary, group. We shall not touch on the mathematical questions connected with this method and its application to the Poincaré group, referring the reader to the specialized literature.(43)

§ 4.2. Transformations and momenta. The little group and the Wigner operator

Let us consider the subgroup of translations. Its representations satisfy the commutative multiplication law

$$\mathrm{U}(\mathrm{a}_1, 1)\,\mathrm{U}(\mathrm{a}_2, 1) = \mathrm{U}(\mathrm{a}_1+\mathrm{a}_2, 1), \quad (\mathrm{U}(0, 1) = 1), \tag{7}$$

reflecting the fact that translations in different directions do not influence one another. Since the relation between the matrix a and the four-vector a^μ is one-to-one, the representations (7) coincide with the representations of the translation subgroup of the classical Poincaré group. Translations form an algebra of the subgroup with four continuous parameters a^μ. The operator of an infinitesimal transformation $(\delta \mathrm{a}, 1)$, by (1.34), is equal to $\mathrm{U}(\delta\mathrm{a}, 1) = 1+iP_\mu\,\delta a^\mu$, so that a finite transformation U(a, 1) is

$$\mathrm{U}(\mathrm{a}, 1) = e^{iP_\mu a^\mu}. \tag{8}$$

Applying (8) to the states (1), when the momentum is diagonal,

$$\mathrm{U}(\mathrm{a}, 1)\,|\,\boldsymbol{p}, \sigma\rangle = e^{ipa}\,|\,\boldsymbol{p}, \sigma\rangle.$$

Apart from variables not connected with space–time symmetry, the states (1) exhaust the possible states of a free elementary particle with mass m and spin J. The completeness condition for the system of states (1) for given m and J may be written in the form

$$\sum_\sigma \int d\mu(\boldsymbol{p})\,|\,\boldsymbol{p}, \sigma\rangle\langle \boldsymbol{p}, \sigma\,| = 1, \tag{9}$$

where $d\mu(\boldsymbol{p})$ is the invariant volume element in momentum space. Any state in (9) may be obtained from some chosen state using a Lorentz transformation, since by definition elementary particle states transform via irreducible representations. Bearing in mind that the value of the mass $m^2 = p^2$ is given, we find

$$d\mu(\boldsymbol{p}) = \mu_0\delta(p^2-m^2)\,d^4p = \mu_0\frac{d^3p}{2p_0} \qquad (p_0 > 0), \tag{10}$$

where μ_0 is an invariant factor which we shall set equal to 1. For this choice of $d\mu(\boldsymbol{p})$ the orthogonality condition for the states (1) must be written in the form

$$\langle \boldsymbol{p}', \sigma'\,|\,\boldsymbol{p}, \sigma\rangle = 2p_0\delta(\boldsymbol{p}-\boldsymbol{p}')\delta_{\sigma\sigma'} \qquad (p_0 > 0). \tag{11}$$

Any elementary particle state $|\,\Psi\rangle$ may be represented in the form of an expansion in basis states $|\,\boldsymbol{p}, \sigma\rangle$:

$$|\,\Psi\rangle = \sum_\sigma \int \frac{d^3p}{2p_0}\,|\,\boldsymbol{p}, \sigma\rangle\,\psi(\boldsymbol{p}, \sigma), \tag{12}$$

where the function $\psi(\boldsymbol{p}, \sigma)$ may be called the Wigner wave function of the particle. The scalar product of two states $|\,\Phi\rangle$ and $|\,\Psi\rangle$ may be expressed in terms of their Wigner wave functions:

$$\langle \Phi\,|\,\Psi\rangle = \sum_\sigma \int \frac{d^3p}{2p_0}\,\varphi^*(\boldsymbol{p}, \sigma)\,\psi(\boldsymbol{p}, \sigma). \tag{13}$$

As is usual in quantum mechanics, for the existence of the scalar product (13) it is essential that the wave functions φ and ψ be quadratically integrable. The four-momentum matrix $\mathrm{p} = \sigma_\mu p^\mu$ transforms via the formula (see (3.17))

$$\mathrm{p}' = A\mathrm{p}A^+, \quad p'^\mu = \Lambda_\nu{}^\mu(A)p^\nu. \tag{14}$$

Let us distinguish among the transformations A those $\tilde{A}(p)$ which leave the momentum matrix invariant:

$$\tilde{A}(p)\,\mathrm{p}\tilde{A}^+(p) = \mathrm{p}. \tag{15}$$

The form of the matrix $\tilde{A}(p)$ will, of course, depend on the choice of p. The product $\tilde{A}_1(p)\times\tilde{A}_2(p)$ of two matrices $\tilde{A}_1(p)$ and $\tilde{A}_2(p)$ with the property (15) also does not change p, i.e. the set of matrices $\tilde{A}(p)$ forms a subgroup of the unimodular group. This subgroup is called the stationary subgroup $L(p)$ of the momentum p or the little group belonging to the momentum p.

Thus for every value of the momentum p one may define the little group $L(p)$ and, consequently, considering different momenta, we must introduce the set of little groups $L(p_1)$, $L(p_2)$. ...

It is natural to expect that the little groups $L(p)$ and $L(\mathring{p})$ belonging to the momenta p and $\mathring{p}$ have the same form (i.e. are isomorphic), if the momenta p and $\mathring{p}$ may be connected by a Lorentz transformation. In fact, let us set

$$\mathrm{p} = \alpha(p, \mathring{p})\,\mathring{\mathrm{p}}\alpha^+(p, \mathring{p}). \tag{16}$$

It is easy to check that if $\tilde{A}(p)$ refers to the little group $L(p)$, i.e. $\tilde{A}(p)\,\mathrm{p}\tilde{A}^+(p) = \mathrm{p}$, the matrix

$$\tilde{A}(\mathring{p}) = \alpha^{-1}(p, \mathring{p})\,\tilde{A}(p)\alpha(p, \mathring{p})$$

belongs to the little group $L(\mathring{p})$ of the momentum $\mathring{p}$, i.e. the relation $\tilde{A}(\mathring{p})\,\mathring{\mathrm{p}}\tilde{A}^+(\mathring{p}) = \mathring{\mathrm{p}}$ holds. Fixing $\alpha(p, \mathring{p})$ we may form from every matrix $\tilde{A}(p)$ the matrix $\tilde{A}(\mathring{p})$, and vice versa. In other words, there exists a one-to-one relation between the sets of matrices $\tilde{A}(p)$ and $\tilde{A}(\mathring{p})$, while the groups $L(p)$ and $L(\mathring{p})$ have identical structure. As a consequence of this, it is sufficient to consider the little group belonging to a particular chosen momentum $\mathring{p}$ for momenta p with the same value of the invariant p^2. The momentum $\mathring{p}$ is called the standard momentum.

The transformation (16) taking the standard momentum $\mathring{p}$ into the momentum p is not unique. It is obvious that two transformations $\alpha(p, \mathring{p})$ and $\alpha(p, \mathring{p})\,\tilde{A}(\mathring{p})$ differing by a matrix $\tilde{A}(\mathring{p})$ in the little group $L(\mathring{p})$ of the standard momentum will lead to the same result:

$$\alpha(p, \mathring{p})\,\tilde{A}(\mathring{p})\,\mathring{\mathrm{p}}\tilde{A}^+(\mathring{p})\alpha^+(p, \mathring{p}) = \alpha(p, \mathring{p})\,\mathring{\mathrm{p}}\alpha^+(p, \mathring{p}).$$

We may, however, fix $\alpha(p, \mathring{p})$ by demanding that it always be chosen in a definite way. The operator α is called the Wigner operator. Setting the form of the Wigner operator, we may find the connection between any unimodular transformation A and the corresponding transformation of the little group $\tilde{A}(\mathring{p})$.

Let the transformation A take the momentum p into p':

$$A\mathrm{p}A^+ = \mathrm{p}'. \tag{17}$$

The momenta p and p' may be obtained from the standard momentum $\mathring{p}$ using $\alpha(p, \mathring{p}) \equiv \alpha(p)$:

$$\alpha(\boldsymbol{p})\,\mathring{\mathrm{p}}\alpha^{+}(\boldsymbol{p}) = \mathrm{p}, \quad \alpha(\boldsymbol{p}')\,\mathring{\mathrm{p}}\alpha^{+}(\boldsymbol{p}') = \mathrm{p}'. \tag{18}$$

Inserting expression (18) into (17) and multiplying from the left by $\alpha^{-1}(\boldsymbol{p}')$ and on the right by $\alpha^{-1+}(\boldsymbol{p}')$,

$$\alpha^{-1}(\boldsymbol{p}')\,A\alpha(\boldsymbol{p})\,\mathring{\mathrm{p}}\alpha^{+}(\boldsymbol{p})\,A^{+}\alpha^{-1+}(\boldsymbol{p}') = \mathring{\mathrm{p}}.$$

This means that the matrix

$$\alpha^{-1}(\boldsymbol{p}')\,A\alpha(\boldsymbol{p}) = \tilde{A}(\mathring{p}, A) \tag{19}$$

belongs to the little group of the standard momentum $\mathring{p}$. From this we conclude that any unimodular transformation A may be represented in the form of a product of three matrices:

$$A = \alpha(\boldsymbol{p}')\,\tilde{A}(\mathring{p}, A)\,\alpha^{-1}(\boldsymbol{p}). \tag{20}$$

Fixing the Wigner operator α, we have thereby established a one-to-one relation betwen transformations of the homogeneous group A and transformations of the little group $\tilde{A}(\mathring{p}, A)$. This fact is crucial in the construction of unitary representations of the Poincaré group.

The little group and the Wigner operator for $m^2 > 0$

In this case it is convenient to choose the momentum in the rest state: $\mathring{p}^{\mu} = (m, 0, 0, 0)$ as a standard momentum $\mathring{p}$. Then the matrix of the standard momentum is $\mathring{\mathrm{p}} = mE$, i.e. proportional to the unit matrix. Transformations of the little group $L(\mathring{p})$ belonging to this standard momentum must thus, by (15), satisfy the condition

$$\tilde{A}(\mathring{p})\,\tilde{A}^{+}(\mathring{p}) = E, \tag{21}$$

which is just the unitarity condition for the matrix $\tilde{A}(\mathring{p})$. Consequently, the little group in the case of timelike momenta is the subgroup of unitary transformations SU_2, which is the covering group of the rotation group.

The Wigner operator $\alpha(\boldsymbol{p})$ is defined for $m^2 > 0$ by (16) up to unitary transformations R in SU_2; specifically, the matrices α and αR lead to the same results. We shall use this freedom and choose as a "standard" operator α the operator $\alpha(\boldsymbol{p})$ which takes the four-momentum at rest $\mathring{p}^{\mu} = (m, 0, 0, 0)$ into a given four-momentum p^{μ} using a pure Lorentz transformation without rotations. As we saw in § 3.1, pure Lorentz transformations are described by Hermitian matrices A; for this reason our choice of the operator α is equivalent to the imposition of Hermiticity on α:

$$\alpha^{+}(\boldsymbol{p}) = \alpha(\boldsymbol{p}). \tag{22}$$

The transformation $\alpha(\mathrm{p})\,\mathrm{k}\alpha(\mathrm{p}) = \mathrm{K}$, where $k\mathring{p} = 0$, hence takes the three-dimensional plane orthogonal to the vector $\mathring{p}$ (and consisting of the vectors k) into the three-dimensional plane orthogonal to the vector p (and consisting of the vectors K). As one may show, the condition (22) means that those vectors k which are simultaneously orthogonal to the vectors $\mathring{p}$ and p (i.e. which belong to both three-dimensional planes), do not change under the transformation.

Equation (16) for α may be rewritten in the form

$$\alpha(\boldsymbol{p})\,\mathring{\mathrm{p}}\alpha^{+}(\boldsymbol{p}) = m\alpha^2(\boldsymbol{p}) = \mathrm{p}. \tag{23}$$

Calculation of the operator α thus amounts to the extraction of the square root of the momentum matrix p. To find α, we set, as in (3.28),

$$\alpha(\boldsymbol{p}) = \cosh\frac{\beta}{2} + (\boldsymbol{\sigma}\cdot\boldsymbol{n})\sinh\frac{\beta}{2} \qquad (\boldsymbol{n}^2 = 1).$$

From (23) we find $\cosh\beta = p_0/m$, $\boldsymbol{n} = \boldsymbol{p}/|\boldsymbol{p}|$, from which we obtain the desired expression for the Wigner operator:

$$\alpha(\boldsymbol{p}) = \frac{m+p_0+(\boldsymbol{\sigma}\cdot\boldsymbol{p})}{[2m(m+p_0)]^{1/2}}\,. \tag{24}$$

If, as a standard momentum, we choose a momentum k differing from the rest momentum $\mathring{p}^{\mu} = (m, 0, 0, 0)$, the most general expression for α in (16) is

$$\alpha(\boldsymbol{p}, \boldsymbol{k}) = \alpha(\boldsymbol{p})\,R\alpha^{-1}(\boldsymbol{k}), \tag{25}$$

where R is any unitary 2 by 2 matrix, while $\alpha(\boldsymbol{p})$ has the former meaning (24). The operator (25) in general is not Hermitian. The choice of the operator α in the form (24) [(i.e. with the limitation (22)] is convenient for writing state vectors in the canonical basis. In the case of the helicity and E_2 bases a different structure is convenient (see § 4.3).

The little group and the Wigner operator for $m^2 = 0$

The momentum of a zero-mass particle is light-like; $p^2 = 0$. Let us choose the standard momentum in the form $\mathring{p}^{\mu} = k(1, 0, 0, 1)$ so that the matrix of the standard momentum is $\mathring{p} = k(1+\sigma_3)$. The little group belonging to this momentum may be found from (15), i.e.

$$\tilde{A}(1+\sigma_3)\tilde{A}^{+} = 1+\sigma_3\,. \tag{26}$$

Let us obtain the unimodular transformation $\tilde{A}$ in the form

$$\tilde{A} = \sum_{\mu} c(\mu)\sigma_{\mu}, \quad \det\tilde{A} = 1.$$

We may find the coefficients $c(\mu)$ from (26). In short,

$$\tilde{A} = e^{-i\frac{1}{2}\sigma_3\theta} + \tfrac{1}{2}(\sigma_1+i\sigma_2)ze^{i\frac{1}{2}\theta}, \tag{27}$$

where z is any complex number while θ is a real parameter $(0 \leqslant \theta \leqslant 4\pi)$. The triangular matrix (27) is a product of a diagonal unitary matrix $u(\theta) = e^{-i\frac{1}{2}\sigma_3\theta}$ and a triangular unimodular matrix $t(z) = 1+\frac{1}{2}(\sigma_1+i\sigma_2)z$:

$$\tilde{A} = t(z)\,u(\theta). \tag{28}$$

The unitary transformations $u(\theta)$ form the Abelian subgroup U(1) isomorphic to the rotation group in two-dimensional Euclidean space (rotations around the third axis):

$$u(\theta_1)\,u(\theta_2) = u(\theta_1+\theta_2). \tag{29}$$

The transformations $t(z)$ also form an Abelian subgroup:

$$t(z_1)\,t(z_2) = t(z_1+z_2). \tag{30}$$

The transformations $t(z)$ and $u(\theta)$ do not commute with one another:

$$u(\theta)\,t(z) = t(ze^{-i\theta})\,u(\theta). \tag{31}$$

If one introduces the real variables $y_1, y_2, (z = y_1+iy_2)$, then $u(\theta)$ will describe the rotation in the coordinate plane (y_1, y_2) by an angle θ, while $t(z)$ will describe a translation in the (y_1, y_2) plane. Formula (31) is the multiplication rule (4) for the case of two-dimensional Euclidean space.

Thus the little group (27) of a light-like momentum is isomorphic to the inhomogeneous group of transformations in two-dimensional Euclidean space. This group is the group of "generalized" rotations of the three-dimensional plane tangent to the light cone, which has the degenerate metric

$$g_{ik} = \begin{pmatrix} 1 & & \\ & 1 & \\ & & 0 \end{pmatrix}.$$

Let us now find the Wigner operator α for a light-like momentum. For a standard momentum matrix $\mathring{\mathrm{p}} = k(1+\sigma_3)$ the definition (16) may be written in the form

$$\alpha(\boldsymbol{p})\,k(1+\sigma_3)\,\alpha^+(\boldsymbol{p}) = |\boldsymbol{p}|\,\sigma_0+(\boldsymbol{\sigma}\cdot\boldsymbol{p}).$$

We note that when acting on the standard momentum matrix $\mathring{\mathrm{p}} = k(1+\sigma_3)$ the matrices $\frac{1}{2}(\sigma_1+i\sigma_2)$ and $(1-\sigma_3)$ give 0. For this reason the Wigner operator may also be written as

$$\alpha(\boldsymbol{p}) = \sigma_0 b_0+\tfrac{1}{2}(1+\sigma_3)b_3+\tfrac{1}{2}(\sigma_1-i\sigma_2)b \tag{32}$$

or, because of the unimodularity of $\alpha(\boldsymbol{p})$, in the form

$$\alpha = (a_{11})^{-1/2}\begin{pmatrix} a_{11} & 0 \\ a_{21} & 1 \end{pmatrix} \tag{33}$$

with real a_{11} and complex a_{21}. The relation between the matrix elements of (33) and the components of the momentum may be found using $\alpha k(1+\sigma_3)\alpha^+ = \mathrm{p}$, so that

$$\mathrm{p} = k\begin{pmatrix} a_{11} & a_{21}^* \\ a_{21} & \dfrac{a_{21}a_{21}^*}{a_{11}} \end{pmatrix} = \begin{pmatrix} p^0+p^3 & p^1-ip^2 \\ p^1+ip^2 & p^0-p^3 \end{pmatrix}; \tag{34}$$

thus

$$a_{11} = \frac{p^0+p^3}{k}, \quad a_{21} = \frac{p^1+ip^2}{k}. \tag{35}$$

Formulae (33) and (35) give the most economical expression for the operator α. Another frequently used form for α may be obtained if one chooses α as the product of a pure Lorentz transformation $H(|\boldsymbol{p}|)$ along the three-axis and a three-dimensional rotation $V(\boldsymbol{p})$ from the three-axis to the direction of $\boldsymbol{p}$:

$$\alpha(\boldsymbol{p}) = V(\boldsymbol{p})\,H(|\boldsymbol{p}|). \tag{36}$$

The rotation matrix $V(\boldsymbol{p})$ has the form

$$\left.\begin{aligned} V(\boldsymbol{p}) &= \exp\{i\tfrac{1}{2}\beta(\boldsymbol{\sigma}\cdot\boldsymbol{n})\}; \\ \cos\beta &= \frac{|p^3|}{|\boldsymbol{p}|}, \quad \boldsymbol{n} = -\frac{[\boldsymbol{p}\times\boldsymbol{e}_3]}{|[\boldsymbol{p}\times\boldsymbol{e}_3]|}, \end{aligned}\right\} \tag{37}$$

while the Lorentz transformation $H(|\boldsymbol{p}|)$ is

$$H(|\boldsymbol{p}|) = H^+(|\boldsymbol{p}|) = \frac{1}{2}\sqrt{\frac{k}{|\boldsymbol{p}|}}\left[\frac{|\boldsymbol{p}|}{k}+1+\sigma_3\left(\frac{|\boldsymbol{p}|}{k}-1\right)\right]. \tag{38}$$

Formulae (36)–(38) may be checked by the reader as an exercise.

§ 4.3. Unitary representations. Case $m^2 > 0$

Let us turn to the construction of the special class of irreducible unitary representations $\mathrm{U}(0, A)$ of the homogeneous subgroup of the quantum mechanical Poincaré group for $m^2 > 0$. By (6) and (8) this allows one to obtain the unitary representations of the Poincaré group itself. Under a Lorentz transformation A a particle state vector in the canonical basis $|\boldsymbol{p}, \sigma\rangle$ undergoes the unitary transformation

$$|\boldsymbol{p}, \sigma\rangle \to \mathrm{U}(0, A)|\boldsymbol{p}, \sigma\rangle,$$

where the state is normalized according to (11). Since the momentum p is transformed into $p' = \Lambda(A)p$, the transformed state must belong to a new eigenvalue p'_μ of the momentum operator P_μ:

$$P_\mu \mathrm{U}(0, A)|p, \sigma\rangle = \mathrm{U}(0, A)\Lambda_\mu^{\ \nu} P_\nu|\boldsymbol{p}, \sigma\rangle = p'_\mu \mathrm{U}(0, A)|\boldsymbol{p}, \sigma\rangle. \tag{39}$$

We have used here the definition (1.39) of the transformed operator P_μ:

$$\mathrm{U}^{-1}(0, A)P_\mu \mathrm{U}(0, A) = \Lambda_\mu^{\ \nu}(A)P_\nu.$$

A Lorentz transformation A is accompanied not only by a change in momentum but also by a transformation in the space of spin variables σ. The transformed state vector $\mathrm{U}(0, A)\,|\boldsymbol{p}, \sigma\rangle$ thus may be written in the following form:

$$\mathrm{U}(0, A)|\boldsymbol{p}, \sigma\rangle = \sum_{\sigma'}|\boldsymbol{p}', \sigma'\rangle\, V_{\sigma'\sigma}(p, A), \quad p' = \Lambda(A)p, \tag{40}$$

where the matrix $V(p, A)$ acts on the spin variables σ. Let us find this matrix. First of all, we find the conditions imposed by the unitarity of the transformation $\mathrm{U}(0, A)$. Because of (9) and (40), we find

$$\begin{aligned} \mathrm{U}(0, A)\,\mathrm{U}^+(0, A) &= \int d\mu(\boldsymbol{p})\sum_{\sigma}\mathrm{U}(0, A)|\boldsymbol{p}, \sigma\rangle\langle\boldsymbol{p}, \sigma|\,\mathrm{U}^+(0, A) \\ &= \int d^4p\,\delta(p^2-m^2)\sum_{\sigma, \sigma', \sigma''} V_{\sigma'\sigma}(\Lambda^{-1}p, A)|\boldsymbol{p}, \sigma'\rangle\langle\boldsymbol{p}, \sigma''|\,V^*_{\sigma''\sigma}(\Lambda^{-1}p, A). \end{aligned}$$

Consequently, $\mathrm{U}(0, A)\,\mathrm{U}^+(0, A) = 1$, if the matrix V is unitary:

$$V(p, A)\,V^+(p, A) = 1. \tag{41}$$

The meaning of the matrix $V(p, A)$ may be seen if one uses as A one of the matrices $\tilde{A}$ of the little group of the momentum p, so that $\tilde{A}(p)\,\mathrm{p}\tilde{A}^{+}(p) = \mathrm{p}$. Then (40) becomes

$$\mathrm{U}(0, \tilde{A}(p))\,|\,\boldsymbol{p}, \sigma\rangle = \sum_{\sigma'} |\,\boldsymbol{p}, \sigma'\rangle\, V_{\sigma'\sigma}(p, \tilde{A}(p)). \tag{42}$$

The matrices $V(p, \tilde{A}(p)) = V(\tilde{A}(p))$ form a representation of the little group $L(p)$ belonging to the momentum p.

As we verified in the previous section, the little groups $L(p)$ belonging to various momenta are isomorphic to one another, and the relation between their elements may be established using the Wigner operator α:

$$\tilde{A}(p) = \alpha(\boldsymbol{p})\,\tilde{A}(\mathring{p})\,\alpha^{-1}(\boldsymbol{p}).$$

As a consequence, it is sufficient to consider the matrices $V(\tilde{A}(\mathring{p}))$ only for a standard momentum $\mathring{p}$, fixing the operator $\alpha(\boldsymbol{p})$ from the start. We thus define the basis vectors of the states for momentum p differing from the standard one in terms of the state vector for the standard momentum, i.e. we shall fix the transformation $\mathrm{U}(0, \alpha(\boldsymbol{p}))$ as one in which (by definition) the internal variables do not change. For a different choice of $\alpha(\boldsymbol{p})$ we would arrive, obviously, at a unitarily equivalent representation.

Thus we set

$$|\,\boldsymbol{p}, \sigma\rangle = \mathrm{U}(0, \alpha(\boldsymbol{p}))\,|\,\mathring{p}, \sigma\rangle, \tag{43}$$

choosing $\alpha(\boldsymbol{p})$ in a definite way.

Using the condition (43) we may express the matrix $V(p, A)$ for any transformation A in terms of the matrix $V(\tilde{A}(\mathring{p}))$ for the standard momentum $\mathring{p}$. Multiplying (40) by $\mathrm{U}^{-1}(0, \alpha(\boldsymbol{p}'))$ we find, in correspondence with (1.29), (4) and (43),

$$\begin{aligned}\mathrm{U}^{-1}(0, \alpha(\boldsymbol{p}'))\,\mathrm{U}(0, A)\,|\,\boldsymbol{p}, \sigma\rangle &= \mathrm{U}(0, \alpha^{-1}(\boldsymbol{p}')\,A\alpha(\boldsymbol{p}))\,|\,\mathring{p}, \sigma\rangle \\ &= U(0, \tilde{A}(\mathring{p}))\,|\,\mathring{p}, \sigma\rangle = \sum_{\sigma'} |\,\mathring{p}, \sigma'\rangle\, V_{\sigma'\sigma}(\tilde{A}(\mathring{p})).\end{aligned} \tag{44}$$

Comparison with (42) shows that the matrices $V(p, A)$ form a representation of the little group of the standard momentum $\mathring{p}$.

The irreducible representations of the quantum mechanical group will be obtained if we choose the matrices V corresponding to an irreducible representation of the little group. Here each transformation A for a fixed p corresponds to a transformation of the little group $\tilde{A}(\mathring{p})$ of the standard momentum $\mathring{p}$:

$$\left.\begin{aligned}V(p, A) &\equiv V(\tilde{A}(\mathring{p})),\\ \tilde{A}(\mathring{p}) &= \alpha^{-1}(\boldsymbol{p}')\,A\alpha(\boldsymbol{p}) \equiv \tilde{A}(\mathring{p}, A),\\ p' &= \Lambda\,(A)p.\end{aligned}\right\} \tag{45}$$

As was shown in § 4.2, the little group of the momentum $\mathring{p}^{\mu} = (m, 0, 0, 0)$ is the unitary group SU_2 considered in Chapter 3. The unitarity of the transformation $\tilde{A}(\mathring{p}, A)$ is easily checked:

$$\tilde{A}\tilde{A}^{+} = \alpha^{-1}(\boldsymbol{p}')\,A\alpha^{2}(p)\,A^{+}\alpha^{-1+}(\boldsymbol{p}') = \alpha^{-1}(\boldsymbol{p}')A\,\frac{\mathrm{p}}{m}\,A^{+}\alpha^{-1+}(\boldsymbol{p}') = \tilde{A}^{+}\tilde{A} = 1.$$

The matrix $V(\tilde{A})$ thus coincides in the case of spin J with the unitary $(2J+1)\times(2J+1)$ matrices $\mathcal{D}^J(\tilde{A})$ of the group SU_2, studied in Chapter 3:

$$V_{\sigma\sigma'}(\tilde{A}) = \mathcal{D}^J_{\sigma\sigma'}(\tilde{A}). \tag{46}$$

One may now obtain the unitary representations of the quantum mechanical Poincaré group. Performing the translation (8) on (44) and substituting (45) and (46), we obtain the final formula for a unitary irreducible representation U(a, A) of the quantum mechanical Poincaré group for spin J in the canonical basis:

$$\mathrm{U}(\mathrm{a}, A)\,|\,\boldsymbol{p}, \sigma\rangle = e^{ip'a}\sum_{\sigma'}|\,\boldsymbol{p}', \sigma'\rangle\,\mathcal{D}^J_{\sigma'\sigma}(\tilde{A}(\mathring{p})), \tag{47}$$

where $\mathcal{D}^J$ is an irreducible (unitary) representation of the group SU_2, $\tilde{A}$ is given by (45), and σ is equal to the spin projection on the third axis:

$$M_3\,|\,\mathring{p}, \sigma\rangle = \sigma\,|\,\mathring{p}, \sigma\rangle. \tag{48}$$

In the rest state the spin operator $\boldsymbol{J}$ is equal to the angular momentum $\boldsymbol{M}$, i.e. it may be found from the polarization three-vector $\boldsymbol{J} = -(1/m)\boldsymbol{w}$. Using (43) to pass to state vectors with momentum $\boldsymbol{p}$, we find a general expression for the spin vector in the canonical basis:[44]

$$J = -\frac{1}{m}\left[\boldsymbol{w}-\boldsymbol{p}\frac{w^0}{p^0+m}\right], \tag{49}$$

where

$$w_\mu = \tfrac{1}{2}\varepsilon_{\mu\nu\lambda\sigma}M^{\nu\lambda}p^\sigma.$$

To find the generators $M_{\mu\nu}$ in terms of known operators of a finite transformation is not especially difficult. The Hermitian generators $M_{\mu\nu}$ or M_k, N_k in a unitary representation were introduced by formula (1.33) so that in infinitesimal form the transformation U(0, A) was written as

$$\mathrm{U}(0, A) = 1-i\tfrac{1}{2}M_{\mu\nu}\omega^{\mu\nu} = 1+i\boldsymbol{M}\cdot\boldsymbol{\omega}+i\boldsymbol{N}\cdot\boldsymbol{\beta}. \tag{50}$$

The parameters $\boldsymbol{\omega}$ and $\boldsymbol{\beta}$ are the same in (50) as in the infinitesimal matrix A:

$$A = 1+i\tfrac{1}{2}\boldsymbol{\sigma}\cdot\boldsymbol{\omega}+\tfrac{1}{2}\boldsymbol{\sigma}\cdot\boldsymbol{\beta}. \tag{51}$$

To find an expression for $\boldsymbol{M}$ and $\boldsymbol{N}$ we must write the right-hand side of (47) for $a^\mu = 0$ and infinitesimal A, and the left-hand side must be rewritten using (50).

Let us first consider spinless particles ($J = 0$). In the absence of internal variables σ the right-hand side of (47) depends on A only via the momentum $p' = \Lambda(A)p$. We thus obtain the following expression for the generators of the spinless representation:

$$\left.\begin{aligned} \boldsymbol{M}^0\,|\,\boldsymbol{p}, J = 0\rangle &= -i\left[\boldsymbol{p}\times\frac{\partial}{\partial\boldsymbol{p}}\right]|\,\boldsymbol{p}, J = 0\rangle, \\ \boldsymbol{N}^0\,|\,\boldsymbol{p}, J = 0\rangle &= -ip^0\frac{\partial}{\partial\boldsymbol{p}}\,|\,\boldsymbol{p}, J = 0\rangle, \\ p^0 &= (p^2+m^2)^{1/2}. \end{aligned}\right\} \tag{52}$$

The quantities $\boldsymbol{M}^0$ and $\boldsymbol{N}^0$ are the relativistic orbital angular momenta.

If the spin of the particle is nonzero, the state vector also changes as a result of the rotation of the spin basis which is described in (47) by the matrix $\mathcal{D}^J(\tilde{A})$. Let us write the infinitesimal form of $\tilde{A}(\mathring{p}, A)$ as

$$\tilde{A}(\mathring{p}, A) = 1+i\tfrac{1}{2}\boldsymbol{\sigma}\cdot\boldsymbol{\theta}(\boldsymbol{\omega}, \boldsymbol{\beta}). \tag{53}$$

Then an infinitesimal transformation $\mathcal{D}^J$ is

$$\mathcal{D}^J(\tilde{A}) = 1+i\mathfrak{M}\cdot\boldsymbol{\theta}(\boldsymbol{\omega}, \boldsymbol{\beta}) = 1+i\mathfrak{M}\cdot\boldsymbol{\omega}+i\mathfrak{N}\cdot\boldsymbol{\beta}. \tag{54}$$

Inserting the matrices α and A from formulae (24) and (51) into $\tilde{A}(\mathring{p}, A) = \alpha^{-1}(\boldsymbol{p}')\,A\alpha(\boldsymbol{p})$, we find by a simple calculation,

$$\tilde{A}(\mathring{p}, A) = 1+i\frac{1}{2}\left\{\boldsymbol{\sigma}\cdot\boldsymbol{\omega}+\frac{(\boldsymbol{p}\times\boldsymbol{\sigma}\cdot\boldsymbol{\beta})}{m+p^0}\right\}. \tag{55}$$

Consequently, the angular momentum in the canonical basis for a unitary representation has the following form:[(44–46)]

$$\boldsymbol{M} = \boldsymbol{M}^0+\mathfrak{M} = \boldsymbol{M}^0+\boldsymbol{J}, \tag{56}$$

$$\boldsymbol{N} = \boldsymbol{N}^0+\mathfrak{N} = \boldsymbol{N}^0+\frac{\boldsymbol{p}\times\boldsymbol{J}}{p_0+m}. \tag{57}$$

One may verify by a direct check that the operators $\boldsymbol{M}$ and $\boldsymbol{N}$ can satisfy the commutation relations (1.41).

We present without proof a formula for a finite spin rotation $\tilde{A}(\mathring{p}, A)$ induced by a Lorentz transformation. If A describes a rotation ($A = A^{-1+} = R$), then [as follows from (56) and (57)] the rotation $\tilde{A}(\mathring{p}, A)$ coincides with the rotation R: $\tilde{A}(\mathring{p}, A) = R$. When A is a pure Lorentz transformation ($A = A^+ = e^{[\frac{1}{2}(\sigma\cdot n)\beta]}$), characterized by a relative four-velocity $u^\mu = (\cosh\beta, \boldsymbol{n}\sinh\beta)$, then the rotation of the spin basis is determined by the matrix

$$\tilde{A}(\mathring{p}, A) = \frac{(p^0+m)(u^0+1)-(\boldsymbol{u}\cdot\boldsymbol{p})+i(\boldsymbol{\sigma}\cdot[\boldsymbol{p}\times\boldsymbol{u}])}{[2(p^0+m)(u^0+1)((\boldsymbol{u}\cdot\boldsymbol{p})+m)]^{1/2}} \equiv \exp\left\{i\frac{1}{2}\Omega(p, A)\frac{(\boldsymbol{\sigma}\cdot[\boldsymbol{p}\times\boldsymbol{u}])}{|[\boldsymbol{p}\times\boldsymbol{u}]|}\right\}. \tag{58}$$

The matrix (58) describes a rotation around the axis $\boldsymbol{p}\times\boldsymbol{u}$ by an angle Ω:

$$\sin\Omega = \frac{2b}{b^2+|\boldsymbol{p}\times\boldsymbol{u}|^2}|\boldsymbol{p}\times\boldsymbol{u}|,$$

where

$$b = (u^0+1)(p^0+m)+\boldsymbol{u}\cdot\boldsymbol{p}.$$

As is clear from (58), the spin basis does not undergo a rotation if $\boldsymbol{p}$ and $\boldsymbol{u}$ are parallel. This result also follows directly from the definition (43) of the spin basis for the momentum $\boldsymbol{p}$, since in this case $A\alpha(\boldsymbol{p}) = \alpha(\boldsymbol{p}')$.

Helicity basis

In a number of applications it is convenient to use the helicity basis, which differs from the canonical one in the replacement of $J^3 = (1/m)w^\lambda n_\lambda^{(3)}$ by the helicity

$$\lambda = -\frac{w^0}{|\boldsymbol{p}|} = \frac{\boldsymbol{M}\cdot\boldsymbol{p}}{|\boldsymbol{p}|},$$

i.e. by the projection of the total angular momentum on the direction of motion. It is clear that the helicity is invariant with respect to three-dimensional rotations (and translations). The covariant expression for the helicity in terms of the pseudo-vector w^λ or the spin J has the form

$$\lambda = \frac{1}{m} w^\lambda \tilde{n}_\lambda^{(3)} = \boldsymbol{J} \cdot \tilde{\boldsymbol{n}}^{(3)}, \tag{59}$$

in which the unit vector $\tilde{n}_\lambda^{(3)}$ is orthogonal to the momentum, $\tilde{n}_\lambda^{(3)} p^\lambda = 0$, $(\boldsymbol{n}^{(3)})^2 = -1$, and belongs to a two-dimensional plane passing through the vectors p^μ and $\mathring{p}^\mu = (m, 0, 0, 0)$. It is obvious that the eigenvalues of λ are the same as those of the spin projection: $\lambda = -J, -J+1, \ldots, J$.

Let the momentum form an angle ϑ with the z-axis, and φ be the angle between the projection of p on the (x, y) plane and the x-axis. We shall choose α in the form

$$\alpha(\boldsymbol{p}) = R(\varphi, \vartheta, -\varphi) e^{\frac{1}{2}\sigma_3 \beta} \equiv h(\boldsymbol{p}) \equiv h(p, \vartheta, \varphi), \tag{60}$$

where

$$\sinh \beta = |\boldsymbol{p}|/m \qquad (0 \leqslant \varphi \leqslant \pi,\ 0 \leqslant \vartheta \leqslant \pi).$$

The operator (60) corresponds to that transformation of the standard momentum $\mathring{p}$ which first "boosts" the momentum $|\boldsymbol{p}| \equiv p$ along the positive z-axis, using a pure Lorentz transformation $e^{\frac{1}{2}\sigma_3 \beta}$, and then takes it by the rotation

$$R(\varphi, \vartheta, -\varphi) = e^{-i\frac{1}{2}\sigma_3 \varphi} e^{-i\frac{1}{2}\sigma_2 \vartheta} e^{i\frac{1}{2}\sigma_3 \varphi}$$

into the momentum p. Here the parametrization of the momentum has the form

$$p^\mu = m(\cosh \beta,\ \sinh \beta \sin \vartheta \cos \varphi,\ \sinh \beta \sin \vartheta \sin \varphi,\ \sinh \beta \cos \vartheta).$$

In the case of motion with momentum $|\boldsymbol{p}|$ along the negative z-axis, R must be chosen in the form $R(0, -\pi, 0)$.

Thus the helicity basis consists of the $2J+1$ states with $-J \leqslant \lambda \leqslant J$ defined by formula (43) with the operator $\alpha(\boldsymbol{p})$ in the form (60), or

$$|\boldsymbol{p}, \lambda\rangle \equiv |p, \vartheta, \varphi, \lambda\rangle = \mathrm{U}(h(p, \vartheta, \varphi))\, |\mathring{p}, \lambda\rangle, \tag{61}$$

where $\lambda = J^3 = \sigma$ in the rest state. The general formulae (43), (45), and **(47)** also apply in the helicity basis, but the operator $\alpha(\boldsymbol{p})$ is now chosen in a different way than in the canonical basis.

The rotations (45) belonging to the little group must be written in the helicity basis using the operator $h(\boldsymbol{p})$ [see (60)]:

$$\tilde{A}(\mathring{p}, A) = h^{-1}(\boldsymbol{p}')\, A h(\boldsymbol{p}), \tag{62}$$

where $p' = \Lambda(A)p$. If A describes a rotation, $A^+ = A^{-1}$, then $\tilde{A}(\mathring{p}, A)$ corresponds to a rotation of the spin basis around the direction of the momentum. Here the state $|p, \vartheta, \varphi; \lambda\rangle$ is multiplied by a phase factor (the helicity does not change).

If A is a pure Lorentz transformation along the negative z-axis, so that $\boldsymbol{p} = |\boldsymbol{p}|\, \boldsymbol{e}_z \to -\boldsymbol{p}$, then $V = e^{i\frac{1}{2}\sigma_2 \pi}$. Here the helicity basis (43) undergoes a rotation

$$\mathcal{D}^J_{\lambda'\lambda}(e^{i\frac{1}{2}\sigma_2 \pi}) = \mathcal{D}^J_{\lambda'\lambda}(i\sigma_2) = (-1)^{J+\lambda}\, \delta_{\lambda', -\lambda},$$

i.e. the helicity changes sign.

The choice of $\alpha(\boldsymbol{p})$ in the form (60) singles out the positive direction along the z-axis. Passing to the rest state by (61) the particle helicity λ always stays equal to the helicity projection $\boldsymbol{J}\cdot\boldsymbol{e}_z$ on the positive z-axis. Instead of (60) we could have defined a different operator $h^-(p, \vartheta, \varphi)$ for which the negative z-axis would have been singled out:

$$h^-(p, \vartheta, \varphi) = R(\varphi, \vartheta, -\varphi)e^{-\frac{1}{2}\sigma_3\beta}.$$

Acting with it on a state vector we would have obtained a particle with momentum $-\boldsymbol{p}$, while the helicity λ would have corresponded to the projection of $-J^3$ in the rest state:

$$|-\boldsymbol{p}, \lambda\rangle^- = U(h^-(p, \vartheta, \varphi))\,|\mathring{p}, \sigma\rangle\,|_{\sigma=-\lambda}. \tag{63}$$

We note that the operators h and h^- contain the same rotations $R(\varphi, \vartheta, -\varphi)$. For $\varphi = \vartheta = 0$ he relation between the states (61) and (63) has the form

$$|-\boldsymbol{p}, \lambda\rangle^- = (-1)^{J-\lambda}e^{-i\pi J_2}\,|\boldsymbol{p}, \lambda\rangle. \tag{64}$$

Consequently, the transformation from the state (61) to the state (63) includes not only a rotation by π around the y-axis, but also the introduction of an additional phase factor. States of the type (63) along with (61) are used in describing two-particle states (see § 4.6).

Other bases

In scattering experiments particles are characterized by their momenta. For this reason, the canonical and helicity bases, in which the momenta are diagonal, are most frequently used.

The expansion of a state in angular momenta corresponds to the angular momentum basis.(47) As variables describing the particle state one chooses the square of the angular momentum M^2, its projection M_3, the helicity λ, and the energy P_0.

In another angular momentum basis—the $\boldsymbol{L}\cdot\boldsymbol{J}$ basis(48)—the states are characterized by the eigenvalues of the operators $\boldsymbol{L}^2$, L_3, and J_3, where

$$\boldsymbol{L} \equiv \boldsymbol{M}^0 = -i\,\boldsymbol{P}\times\frac{\partial}{\partial \boldsymbol{P}}$$

is the orbital part of the generator $\boldsymbol{M}$ [see (56)].

In some relativistic applications, and also in the theory of infinite dimensional multiplets, the basis of the homogeneous Lorentz group is used.(49, 50) In this basis the operators $M_{\mu\nu}M^{\mu\nu}$ and $\varepsilon_{\mu\nu\lambda\sigma}M^{\mu\nu}M^{\lambda\sigma}$, (the invariants of the homogeneous group), and also M^2 and M_3, are diagonal.

In applications connected with the use of infinite-momentum rest frames, the $E(2)$ basis(51) is convenient, in which the operators

$$\left.\begin{aligned} E_1 &= M_{01}+M_{31} = N_1+M_2,\\ E_2 &= M_{02}+M_{32} = N_2-M_1. \end{aligned}\right\} \tag{65}$$

are diagonal. Together with $M_{12} = M_3$ the operators E_1 and E_2 form the system of generators of the Euclidean subgroup $E(2)$:

$$[E_1, E_2] = 0, \quad [M_3, E_1] = iE_2, \quad [M_3, E_2] = -iE_1. \tag{66}$$

Transformations of the subgroup $E(2)$ leave invariant the plane $x^0 - x^3 = 0$ tangent to the light cone.

Depending on the choice of the complete system of diagonal operators there are two different $E(2)$ bases (for given J and m):

$$\left.\begin{array}{lll} E_1, E_2, P_E = P^0 - P^3, & J_3^E = \dfrac{w^0 - w^3}{P^0 - P^3} & (E\text{-basis}), \\ E_1, E_2, \ F = -\frac{1}{4} M_{\mu\nu} M^{\mu\nu}, & G = \frac{1}{4} \varepsilon_{\mu\nu\lambda\sigma} M^{\mu\nu} M^{\lambda\sigma} & (\tilde{E}\text{-basis}). \end{array}\right\} \tag{67}$$

In the E-basis the operator J_3^E has the meaning of a spin projection. In fact, on the one hand, $[J_3^E, P_\mu] = 0$, while, on the other hand, in the rest system J_3^E coincides with the projection M_3 of the angular momentum $\boldsymbol{M}$. The eigenvalues of J_3^E are thus $-J, -J+1, \ldots, J-1, J$. An expression for the components J_1^E and J_2^E may be found using the commutation relations (1.41) and (1.51), specifically:

$$J_{1,2}^E = \frac{1}{m} (w_{1,2} - P_{1,2} J_3^E). \tag{68}$$

The operators J_k^E (k $= 1, 2, 3$) form an "angular momentum vector", i.e. satisfy the relations

$$[J_i^E, J_j^E] = i\varepsilon_{ijk} J_k^E.$$

In contrast to the spin operator J_k in the canonical basis, the operator J_k^E commutes with the generators E_1 and E_2. Since

$$w^2 = -m^2 (J^E)^2 = -m^2 J(J+1)$$

the eigenvalues of $(J^E)^2$ coincide with the eigenvalues of the spin, i.e. the vector J^E determines the spin in the E-basis.

§ 4.4. Spinor functions and quantum fields for $m^2 > 0$

The representation U(a, A) of the quantum mechanical Poincaré group (47) depends on momentum. The same Lorentz transformation A in states with different momenta and spin $J \neq 0$ may be associated with the unitary matrices $\tilde{A} = \alpha^{-1}(\boldsymbol{p}') A \alpha(\boldsymbol{p})$ describing rotations in spin space. If we had passed to coordinate space using a Fourier transformation we would have obtained a nonlocal transformation law. This dependence on momentum is, of course, purely kinematical in nature.

Let us pass to a new spinor basis of states whose transformation law does not depend on momentum. We consider first particles with spin $\frac{1}{2}$ and mass $m \neq 0$. The unitary matrix $\tilde{A}$ in this case is equal to $\tilde{A} = \alpha^{-1}(\boldsymbol{p}') A \alpha(\boldsymbol{p})$ $(\boldsymbol{p}' = \Lambda(A)p)$, i.e. the product of three unimodular matrices.

Let us introduce the spinor states[52, 53] †

$$|\boldsymbol{p}, \sigma) = \sum_{\sigma'} |\boldsymbol{p}, \sigma'\rangle \alpha_{\sigma'\dot{\sigma}}^{-1}(\boldsymbol{p}). \tag{69}$$

† Let us recall that the nature of the indices of the matrices B and $D^J(B)$ is defined by properties of the matrix B (see the end of § 3.3).

From the transformation formula (47) or

$$U(0, A)\,|\boldsymbol{p}\,,\sigma\rangle = \sum_{\sigma'} |\boldsymbol{p}', \sigma'\rangle\,(\alpha^{-1}(\boldsymbol{p}')\,A\alpha(p))_{\sigma'\sigma}$$

we conclude that the transformation of the states (69) does not depend on momentum

$$U(0, A)\,|\boldsymbol{p}, \sigma) = \sum_{\sigma'} |\boldsymbol{p}', \sigma')A_{\sigma'\sigma}, \tag{70}$$

and is given directly by the matrix A, as in the case of the spinor (fundamental) representation of the homogeneous group.

The normalization condition for the states (69) is

$$\left.\begin{aligned}(\boldsymbol{p}', \sigma'\,|\,\boldsymbol{p}, \sigma) &= \langle \boldsymbol{p}', \sigma'\,|\,\boldsymbol{p}, \varrho\rangle \alpha^{-1+}_{\sigma'\varrho'}\alpha^{-1}_{\varrho\sigma} \\ &= 2p_0\delta(\boldsymbol{p}-\boldsymbol{p}')\,(\alpha^{-2})_{\sigma'\sigma} = 2p_0\delta(\boldsymbol{p}-\boldsymbol{p}')\,\frac{(\sigma_0 p_0-(\boldsymbol{\sigma}\cdot\boldsymbol{p}))_{\sigma'\sigma}}{m}, \\ \alpha^{-2} = m\mathrm{p}^{-1} &= \frac{\tilde{\mathrm{p}}}{m} = (\alpha\alpha^+)^{-1},\end{aligned}\right\} \tag{71}$$

i.e. in passing to simple transformation properties one has lost orthogonality. The completeness condition is now

$$\sum_{\sigma,\,\sigma'} \int |\boldsymbol{p}, \sigma)\,\frac{\mathrm{p}_{\sigma\sigma'}}{m}\,(\boldsymbol{p}, \sigma'|\,\frac{d^3p}{2p_0} = 1. \tag{72}$$

In the case of particles with spin $J > \frac{1}{2}$ and mass $m > 0$ the unitary matrix $V(\tilde{A})$ is equal to the matrix $\mathcal{D}^J(\tilde{A})$ of the rotation group. The matrix $\mathcal{D}^J(\tilde{A})$ is defined only with respect to the unitary matrices $\tilde{A}$; thus in calculating $\mathcal{D}^J(\alpha^{-1}(\boldsymbol{p}')\,A\alpha(\boldsymbol{p}))$ with nonunitary α and A we cannot use the group multiplication law. However, as was shown in Chapter 3, the matrix $\mathcal{D}^J(\tilde{A})$ coincides with the matrix $\mathcal{D}^{(J,\,0)}(A) \equiv D^J(\tilde{A})$ for unitary $\tilde{A}$. The matrices $D^J(\tilde{A})$, being representations of the group $SL(2, c)$, are defined here also for nonunitary unimodular matrices A. Thus, using the group multiplication law, we may decompose $D^J(\tilde{A})$ into three factors:

$$\mathcal{D}^J(\tilde{A}) = D^J(\tilde{A}) = D^J(\alpha^{-1}(\boldsymbol{p}')\,A\alpha(\boldsymbol{p})) = D^J(\alpha^{-1}(\boldsymbol{p}'))\,D^J(A)\,D^J(\alpha(\boldsymbol{p})).$$

Acting further in analogy with the case of spin $J = \frac{1}{2}$, we may introduce the new basis states[52, 53].

$$|\boldsymbol{p}, \sigma; J) = \sum_{\sigma'} |\boldsymbol{p}, \sigma'; J\rangle\, D^J_{\sigma'\sigma}(\alpha^{-1}(\boldsymbol{p})). \tag{73}$$

The transformation law of these states does not depend on momentum; for $p \to p' = \Lambda(A)p$ instead of (70),

$$U(0, A)\,|\boldsymbol{p}, \sigma; J) = \sum_{\sigma'} |\boldsymbol{p}', \sigma'; J)\,D^J_{\sigma'\sigma}(A). \tag{74}$$

The normalization condition (71) is replaced in the general case by the condition

$$(\boldsymbol{p}, \sigma; J\,|\boldsymbol{p}', \sigma'; J) = 2p_0\delta(\boldsymbol{p}-\boldsymbol{p}')\,D^J_{\sigma\sigma'}\left(\frac{\tilde{\mathrm{p}}}{m}\right), \tag{75}$$

while the completeness condition for a given mass $m > 0$ and spin J is

$$\sum_{\sigma, \sigma} \int |\boldsymbol{p}, \sigma\rangle D^J_{\sigma\sigma'}\left(\frac{\mathrm{p}}{m}\right) \langle \boldsymbol{p}, \sigma' | \frac{d^3p}{2p_0} = 1. \tag{76}$$

The role of the momentum matrix p in the case of higher spins is played by the matrix $D^J(\mathrm{p})$, while

$$D^{1/2}(\mathrm{p}) = \mathrm{p}. \tag{77}$$

The generators of infinitesimal transformations M_k and N_k in the spinor basis may be calculated directly from (74):

$$M_k = M_k^0 + J_k, \quad N_k = N_k^0 - iJ_k, \tag{78}$$

where J_k is the $(2J+1)$-row spin matrix, $J^2 = J(J+1)$, and M_k^0 and N_k^0 describe the orbital angular momentum. If $A = e^{\frac{1}{2}\sigma\cdot(i\omega+\beta)}$ is a 2 by 2 Lorentz transformation matrix, the $(2J+1)\times(2J+1)$ matrix $D^J(A)$ is

$$D^J(A) = e^{\frac{1}{2}J\cdot(i\omega+\beta)}. \tag{79}$$

In the particular case when $A = \mathrm{p}/m = e^{\frac{1}{2}\sigma\cdot\theta}$, formula (79) gives

$$D^J\left(\frac{\mathrm{p}}{m}\right) = \exp\frac{\beta(\boldsymbol{J}\cdot\boldsymbol{p})}{|\boldsymbol{p}|}, \quad \theta^i = \beta\frac{p^i}{|\boldsymbol{p}|}, \tag{80}$$

where $\sinh\beta = |\boldsymbol{p}|/m$ and $\boldsymbol{J}$ is the spin operator.

Equations (73)–(76) describe properties of spinor states. The local nature of the transformations (70) and (74) of these states makes them very convenient as a starting point for the introduction of local quantum fields $\varphi(x)$ depending on the coordinates x as parameters.

Let us now introduce creation and annihilation operators. The single-particle state $|\boldsymbol{p}, \sigma; J, m\rangle$ may be thought of as the result of applying a creation operator $a^+(\boldsymbol{p}, \sigma; J, m)$ to the vacuum state $|0\rangle$:

$$|\boldsymbol{p}, \sigma; J, m\rangle = a^+(\boldsymbol{p}, \sigma; J, m)|0\rangle. \tag{81}$$

Since only particles with mass m and spin J will be considered below, one may omit m and J from the identification of the state.

The vacuum $|0\rangle$ is a homogeneous and isotropic state of lowest energy:

$$P^\mu|0\rangle = 0, \quad M_{\mu\nu}|0\rangle = 0,$$

and, consequently, is invariant with respect to relativistic transformations:

$$\mathrm{U}(\mathrm{a}, A)|0\rangle = |0\rangle. \tag{82}$$

Knowing the transformation properties of the states, it is not hard using (81) and (82) to find the way a^+ transforms. In order that (47) hold:

$$\mathrm{U}(\mathrm{a}, A)|\boldsymbol{p}, \sigma\rangle = \sum_{\sigma'} e^{ip'a}|\boldsymbol{p}', \sigma'\rangle \mathcal{D}^J_{\sigma'\sigma}(\alpha^{-1}(\boldsymbol{p}')A\alpha(\boldsymbol{p})), \quad p' = \Lambda(A)p,$$

a^+ must transform according to

$$\mathrm{U}(\mathrm{a}, A)a^+(\boldsymbol{p}, \sigma)\mathrm{U}^{-1}(\mathrm{a}, A) = \sum_{\sigma'} e^{ip'a}a^+(\boldsymbol{p}', \sigma')\mathcal{D}^J_{\sigma'\sigma}(\alpha^{-1}(\boldsymbol{p}')A\alpha(\boldsymbol{p})). \tag{83}$$

The annihilation operator a is the Hermitian conjugate of a^+ As a result of the normalization $\langle \boldsymbol{p}', \sigma | \boldsymbol{p}, \sigma \rangle = 2p_0 \delta_{\sigma\sigma'} \delta(\boldsymbol{p}-\boldsymbol{p}')$, the commutation relation between a and a^+ is

$$[a(\boldsymbol{p}, \sigma), a^+(\boldsymbol{p}', \sigma')]_{\mp} = 2p_0 \delta_{\sigma\sigma'} \delta(\boldsymbol{p}-\boldsymbol{p}'), \tag{84}$$

where the sign $\mp$ refers to commutation (anticommutation) in the case of Bose–Einstein (Fermi–Dirac) statistics. We shall assume that the statistics of the particles is known. In what follows we shall dwell on the connection between spin and statistics.

The Hermitian conjugate of formula (83) describes the transformation properties of the annihilation operator $\boldsymbol{a}$:

$$\mathrm{U}(\mathrm{a}, A)\, a(\boldsymbol{p}, \sigma)\, \mathrm{U}^{-1}(\mathrm{a}, A) = \sum_{\sigma'} e^{-ip'a} \mathcal{D}^J_{\sigma\sigma'}(\alpha^{-1}(\boldsymbol{p})\, A^{-1} \alpha(\boldsymbol{p}'))\, a(\boldsymbol{p}', \sigma'). \tag{85}$$

In (85) the fact that $\mathcal{D}^J$ is a unitary matrix has been used.

In what follows it will be convenient to have another way of writing the formula (83) for a^+. Since $A^{-1} = C^{-1} A^{\mathrm{T}} C$ (see Chapter 2), one may write instead of (83) the expression

$$\begin{aligned} &\sum_{\sigma'} \mathrm{U}(\mathrm{a}, A)\, D^J_{\sigma\sigma'}(C^{-1})\, a^+(\boldsymbol{p}, \sigma')\, \mathrm{U}^{-1}(\mathrm{a}, A) \\ &= \sum_{\sigma', \sigma''} \mathcal{D}^J_{\sigma\sigma'}(\alpha^{-1}(\boldsymbol{p})\, A^{-1} \alpha(\boldsymbol{p}'))\, D^J_{\sigma'\sigma''}(C^{-1})\, a^+(\boldsymbol{p},' \sigma''). \end{aligned} \tag{86}$$

This means that $D^J(C^{-1})\, a^+(\boldsymbol{p}) \equiv \sum_{\sigma'} D^J_{\sigma\sigma'}(C^{-1})\, a^+(\boldsymbol{p}, \sigma')$ transforms in the same way as the annihilation operator.

One may pass from the operators a and a^+ to the spinor operators $D^J(\alpha)a$ and $D^J(\alpha C^{-1})a^+$, which transform locally according to the $(J, 0)$ representation of the Lorentz group; for example:

$$\mathrm{U}(0, A)\, [D^J(\alpha(\boldsymbol{p}))\, a(\boldsymbol{p})]\, \mathrm{U}^{-1}(0, A) = D^J(A^{-1})\, [D^J(\alpha(\boldsymbol{p}'))\, a(\boldsymbol{p}')].$$

As a basis for the spinor operators $D^J(\alpha)a$ and $D^J(\alpha C^{-1})a^+$, one may construct the quantum fields $\varphi^{(+)}_a$ and $\varphi^{(-)}_a$:[54, 55]

$$\varphi^{(+)}_{a\sigma}(x) = \frac{1}{(2\pi)^{3/2}} \int \frac{d^3p}{2p^0} \sum_{\sigma'} D^J_{\sigma\sigma'}(\alpha)\, a(\boldsymbol{p}, \sigma') e^{-ipx}, \tag{87}$$

$$\varphi^{(-)}_{a\sigma}(x) = \frac{1}{(2\pi)^{3/2}} \int \frac{d^3p}{2p^0} \sum_{\sigma'} D^J_{\sigma\sigma'}(\alpha C^{-1})\, a^+(\boldsymbol{p}, \sigma') e^{ipx}, \tag{88}$$

where the form of the exponent is determined by formulae (83) and (85).

A unitary transformation $\mathrm{U}(\mathrm{a}, A)$, when acting on the fields $\varphi^{(+)}_a(x)$ and $\varphi^{(-)}_a(x)$, transforms them into the fields

$$\mathrm{U}(a, A)\, \varphi^{(\pm)}_{a\sigma}(x)\, \mathrm{U}^{-1}(\mathrm{a}, A) = \sum_{\sigma'} D^J_{\sigma\sigma'}(A^{-1})\, \varphi^{(\pm)}_{a\sigma'}(\Lambda x + a), \tag{89}$$

i.e. $\mathrm{U}(\mathrm{a}, A)$ produces translations and rotations in x-space corresponding to the element $g = (\mathrm{a}, A)$ of the Poincaré group $\mathcal{P}^{\uparrow}_{+}$.

Up to now, when considering the Poincaré group, we have not been interested in the difference between particles and antiparticles since their relativistic properties are the same.

However, when constructing a local field operator $\varphi(x)$ containing both creation and annihilation operators, it is natural to demand that the field $\varphi(x)$ have a simple phase transformation law $\varphi(x) \to e^{i\delta}\varphi(x)$. Under phase transformations the operators for creating particles a^+ and antiparticles b^+ transform in an opposite fashion: if $a^+ \to a^+e^{-i\delta}$, then $b^+ \to b^+e^{i\delta}$. We must, then, write the field φ in the form

$$\varphi(x) = \varphi_a^{(+)}(x) + \eta\varphi_b^{(-)}(x), \tag{90}$$

where $\varphi_b^{(-)}$ is given by the same formula (88) but with the operator $b^+(\boldsymbol{p}, \sigma)$ in place of $a^+(\boldsymbol{p}, \sigma)$. Here η is a phase factor.

Because of (89), the field $\varphi(x)$ transforms locally under Lorentz transformations. Let us now see if the field may be considered local in the sense that the commutator or anticommutator $[\varphi(x), \varphi^+(y)]_\mp$ vanishes for space-like intervals $(x-y)^2$, i.e. let us see whether the causality principle holds for $\varphi(x)$. Forming the anticommutator (commutator) and using (87), (88), and (90), we find

$$\begin{aligned}[\varphi_\sigma(x), \varphi_{\sigma'}^+(y)]_\pm &= \frac{1}{(2\pi)^3}\int \frac{d^3p}{2p_0} D^J_{\sigma\sigma'}\left(\frac{\mathrm{p}}{m}\right)\{e^{ip(x-y)} \pm \eta\eta^* e^{-ip(x-y)}\} \\ &= \frac{1}{(2\pi)^3} D^J_{\sigma\sigma'}\left(\frac{-i\partial}{m}\right)\int \frac{d^3p}{2p_0}\{e^{ip(x-y)} \pm (-1)^{2J}\eta\eta^* e^{-ip(x-y)}\},\end{aligned}$$

where we have used $m\alpha\alpha^+ = \mathrm{p}$ and $D^J(-1) = (-1)^{2J}$.

The right-hand side of $[\varphi(x), \varphi^+(y)]_\pm$ is equal to zero for $(x-y)^2 < 0$ if the bracket in the integrand contains the difference of exponents, i.e. if

$$\pm(-1)^{2J}\eta\eta^* = -1. \tag{91}$$

Equation (91) is possible only for

$$\eta\eta^* = 1, \quad \pm(-1)^{2J} = -1. \tag{92}$$

Here the upper sign refers to the anticommutator or Fermi–Dirac statistics, while the lower sign refers to the commutator or Bose–Einstein statistics.

Formula (92) expresses the well-known connection of spin with statistics: particles with half-integral spin $J = \frac{1}{2}, \frac{2}{3}, \ldots$, obey Fermi–Dirac statistics, while particles with integral spin, equal to 0, 1, ..., obey Bose–Einstein statistics. Thus the $(2J+1)$-component fields

$$\varphi(x) = \frac{1}{(2\pi)^{3/2}}\int \frac{d^3p}{2p_0}\{D^J(\alpha)\,a(\boldsymbol{p})e^{-ipx} + D^J(\alpha C^{-1})\,b^+(\boldsymbol{p})e^{ipx}\} \tag{93}$$

not only transform locally by virtue of (89),

$$\mathrm{U}(\mathrm{a}, A)\,\varphi(x)\,\mathrm{U}^{-1}(\mathrm{a}, A) = D^J(A^{-1})\,\varphi(\Lambda x + a), \tag{94}$$

but also satisfy the locality condition in the form

$$[\varphi(x), \varphi^+(y)]_\pm = 0, \quad (x-y)^2 < 0, \text{ for } \begin{Bmatrix}\text{fermions}\\ \text{bosons}\end{Bmatrix} \tag{95}$$

This condition holds if the normal connection between spin and statistics exists.

We note that the field $\varphi(x)$ may not even have an energy-momentum density $T_{\nu\mu}(x)$ or other local observables. In the case of higher spins $J \geqslant \frac{3}{2}$ the definition of such local quantities meets with difficulties.

To conclude this section, we discuss the method used above for describing a system with $m^2 \neq 0$ and spin J. In this method we chose the rest state $|0, \sigma_0; m, J\rangle$ as a standard state. States with arbitrary momentum $|\boldsymbol{p}, \sigma_0; m, J\rangle$ and the same spin projection σ_0 were obtained from the standard state by the Lorentz transformation $\mathrm{U}(0, \alpha(\boldsymbol{p}))$. This corresponds to the fact that the same state $|0, \sigma_0; J, m\rangle$ given in a stationary system of coordinates may be observed from the coordinate origin of various frames of reference moving with velocities $\boldsymbol{v} = -\boldsymbol{p}/p_0$, where $\boldsymbol{v}$ takes on all values $0 \leqslant |\boldsymbol{v}| < 1$ and all directions $\boldsymbol{v}/|\boldsymbol{v}|$. Rotating the frame of reference, we can obtain states $|\boldsymbol{p}, \sigma; m, J\rangle$ with various $\sigma(-J \leqslant \sigma \leqslant J)$.

Thus to obtain a complete system of states $|\boldsymbol{p}, \sigma; m, J\rangle$, it is sufficient to subject the standard state $|0, \sigma_0; J, m\rangle$ to all possible transformations of the homogeneous Lorentz group.

Passing from measuring devices placed at the coordinate origin of moving systems to devices in the same system at some point x, we perform the translation $\mathrm{U}(x, 1)$. For any creation operator $a^+(\boldsymbol{p}, \sigma)$ or annihilation operator $b(\boldsymbol{p}, \sigma)$, we may then define the translated operators $a^+(\boldsymbol{p}, \sigma)\, e^{ipx}$ and $b(\boldsymbol{p}, \sigma)\, e^{-ipx}$. The set of all translated creation and annihilation operators gives us a full description of a field. The invariant sum of all these translated operators $\mathrm{U}(x, 1)\, a(\boldsymbol{p}, \sigma)\, \mathrm{U}^{-1}(x, 1)$, taken with weight $D(\alpha(\boldsymbol{p}))$, is just the field operator $\varphi(x)$. Consequently, the existence of the field operator $\varphi(x)$ is closely connected with the Poincaré group.

§ 4.5. Unitary representations in the case $m = 0$. Equations of motion

The Wigner method for constructing unitary representations[(23)] set forth in § 4.3 [formulae (39)–(45)] holds as well for the case of particles of zero rest mass. The irreducible unitary representations for $p^2 = 0$ and $p_0 > 0$ are also defined by the irreducible representations of the little group of the standard momentum. In § 4.2 it was established that the little group of a light-like momentum $p^2 = 0$ is isomorphic to the group $E(2)$ of rotations and translations in the Euclidean plane [see formulae (28)–(31)]. If we find its irreducible unitary representations, we will at the same time obtain the behavior of a state vector of a massless particle under transformations of the Poincaré group.

A transformation of the group $E(2)$ is given by a unimodular 2 by 2 matrix K of the following form [see (28)]:

$$K = t(z)\, u(\theta) = \left(1 + \tfrac{1}{2}(\sigma_1 + i\sigma_2)z\right) e^{-i\frac{1}{2}\sigma_3\theta}, \tag{96}$$

where $z = y_1 + iy_2$, the parameters y_1, y_2, and θ are real, and the standard momentum is equal to $\mathring{p}^\mu = k(1, 0, 0, 1)$.

Let $Y^{\varrho}_{\xi\xi'}$ be the desired matrix of the irreducible unitary representation ϱ of the group $E(2)$ corresponding to the transformation (96) with parameters y_1, y_2, and θ. The quantities ξ are variables characterizing the components of the representation of the group $E(2)$. As a result of the noncompactness of this group, its representations may be infinite-dimensional and the variables ξ may take on a continuous set of values. The state vector of a particle with $m = 0$ may then be written in the form $|\boldsymbol{p}, \xi; \varrho\rangle$.

The transformation law of a state vector $|\boldsymbol{p}, \xi; \varrho\rangle$ induced by the transition to a new frame of reference $x \to x' = \Lambda(A)x+a$ is obtained directly from the general formula (40) by making the substitutions $\sigma \to \xi$ and

$$V_{\sigma'\sigma}(p, A) \equiv V_{\sigma'\sigma}(\tilde{A}(\mathring{p})) \to Y^{\varrho}_{\xi'\xi}(K), \tag{97}$$

specifically:

$$\mathrm{U}(0, A)\,|\boldsymbol{p}, \xi; \varrho\rangle = \sum_{\xi'} |\boldsymbol{p}', \xi'; \varrho\rangle\, Y^{\varrho}_{\xi'\xi}(K). \tag{98}$$

The form of the matrix K is given by (96), while the dependence of K on momentum and the matrix A may be found from (19) or (45):

$$K(p, A) = \tilde{A}(p, A)\Big|_{m=0} = \alpha^{-1}(\boldsymbol{p}')\,A\alpha(\boldsymbol{p}), \quad p' = \Lambda p.$$

Since the group $E(2)$ is similar to $\mathcal{P}^{\uparrow}_{+}$ in the sense that $E(2)$ is also an inhomogeneous group, one may use the apparatus developed in § 4.3 to study its matrices.

To make the analogy more distinct, we denote a translation in $E(2)$ by $t(z) \equiv [y, 1]$, and a rotation in the (y_1, y_2) plane by $u(\theta) \equiv [0, u]$. An element of the group $E(2)$ is $[y, u] = [y, 1][0, u]$, while the matrix Y may be written in the form

$$Y(y, u) \equiv \mathrm{U}^E[y, u] = \mathrm{U}^E[y, 1]\,\mathrm{U}^E[0, u]. \tag{99}$$

The operator $\mathrm{U}^E[y, u]$ acts only on the internal variables ξ of the state vector $|\boldsymbol{p}, \xi; \varrho\rangle$, which we shall write for brevity simply as $|\xi\rangle$, and is the analog of the operator $\mathrm{U}(a, A)$. The variables ξ may be formally subdivided into two parts: $\xi = (t, \gamma)$; the quantities t_1, t_2 are connected with the representation of the translation subgroup $\mathrm{U}^E[y, 1]$ in (99) and take the place of momenta, while the variables γ characterize the "internal" state of the vector $|t, \gamma\rangle$ and are connected with the little group of the standard "momentum" $\mathring{t}$.

In analogy with (47) we may write at once the general form of the unitary representation $\mathrm{U}^E[y, u]$ of the group $E(2)$:

$$\mathrm{U}^E[y, u]\,|t, \gamma\rangle = e^{iyt'} \sum_{\lambda'} |t', \gamma'\rangle\, V^E_{\gamma'\gamma}(t, u), \tag{100}$$

where the matrix $V^E_{\gamma'\gamma}$ acts only on the variables γ and forms a representation of the little group belonging to the standard "momentum" $\mathring{t}$; $t' = \Lambda(u)t$, $yt = y_1t_1+y_2t_2$.

Under rotations in the Euclidean plane, the quantity $t^2 = t_1^2+t_2^2$ remains invariant, and thus the representations (100) fall into two classes: $t^2 > 0$ and $t^2 = 0$ (or $t_1 = t_2 = 0$). If $t^2 > 0$, the "momenta" are different from 0 and take on a continuous set of values corresponding to the points of a circle of radius $|\boldsymbol{t}|$ in the (t_1, t_2) plane.

If $t^2 = 0$, then $\mathrm{U}^E[y, 1] = 1$ in (99), and the matrices $\mathrm{U}^E[0, u] = Y(0, u)$ form a representation of the rotation group in the plane, or, more precisely, of the group U(1). These representations, as in well known, are one-dimensional and are characterized by integers or half-integers λ, so that if $u(\theta) = e^{-i(1/2)\sigma_3\theta}$, then

$$Y^{\lambda}(0, u) = e^{-i\lambda\theta} \quad (0 \leqslant \theta \leqslant 4\pi;\ \lambda = 0, \pm\tfrac{1}{2}, \pm 1, \pm\tfrac{3}{2}, \ldots). \tag{101}$$

It remains to clarify the meaning of the invariant quantity t^2, and, consequently, the meaning of the quantities t_1 and t_2. Since the second invariant of the Poincaré group is w^2, one must first of all find the polarization vector $w_\mu = (\frac{1}{2})\,\varepsilon_{\mu\nu\lambda\sigma} M^{\nu\lambda} P^\sigma$ for the state $|\mathring{p}, t; \lambda\rangle$ with standard momentum $\mathring{p}^\mu = k(1, 0, 0, 1)$:

$$\left.\begin{aligned} w^\mu(\mathring{p}) &= k(-M_{12}, E_2, -E_1, -M_{12}),\\ E_1 &= M_{01}+M_{31}, \quad E_2 = M_{02}+M_{32}. \end{aligned}\right\} \tag{102}$$

The operators E_1, E_2, and M_{12} satisfy the commutation relations of the group $E(2)$:

$$[E_1, E_2] = 0, \quad [M_{12}, E_2] = -iE_1, \quad [M_{12}, E_1] = iE_2,$$

commute with the momentum p, and thus may be identified with the generators of the little group of a light-like momentum $\mathring{p}$. This means that t_1 and t_2 are the eigenvalues of the generators E_1 and E_2. The invariant spin w^2 in the state $|\mathring{p}, t; \lambda\rangle$ is equal to

$$w^2\,|\mathring{p}, t; \lambda\rangle = -t^2k^2\,|\mathring{p}, t; \lambda\rangle, \quad \mathring{p}^\mu = k(1, 0, 0, 1).$$

Consequently, t^2 is expressed in terms of the invariant spin operator w^2.

Thus in contrast to the case of particles with mass $m \neq 0$, for which only finite-dimensional irreducible representations $\mathcal{D}^J$ are possible, for mass $m = 0$ the representations of the little group contain both finite-dimensional (for $t^2 = 0$) and infinite-dimensional (for $t^2 > 0$) irreducible unitary representations. The infinite-dimensional representations should be interpreted as a manifestation of infinite spin. But particles with infinite spin are not observed in nature, and thus we may discard such representations. The remaining physical representations are one-dimensional: $Y^\theta_{\lambda'\lambda} = e^{-i\lambda\theta}\delta_{\lambda'\lambda}$, and may be characterized by the quantum number λ, which is, according to (102) the helicity: $\lambda = \boldsymbol{M}\cdot\boldsymbol{P}/|\boldsymbol{P}|$ ((102) is written in a system for which $\mathring{p}^\mu = k(1, 0, 0, 1)$ and $\lambda = M_{12}$).

Since the eigenvalues of the operators E_1 and E_2 are equal to zero for physical representations, $t_1 = t_2 = 0$, formula (102) implies $w_0 = -\lambda p_0$, $w_3 = -\lambda p_3$, i.e. in covariant form

$$w_\mu = -\lambda p_\mu.$$

A change in sign of λ in general means passing to another particle (which may not even exist). The absolute value of λ is called the spin of the massless particle.

The physical states with zero mass ($m = 0$) are thus classified according to their helicity λ. Each value of the helicity corresponds to one independent state (one polarization state). The helicity λ may take on the values $0, \pm\frac{1}{2}, \pm 1, \pm\frac{3}{2}, \ldots$

Under transformations of the Poincaré group, $x \to x' = \Lambda(A)\,x+a$ the state vector $|\boldsymbol{p}, \lambda\rangle$ of a zero-mass particle acquires a phase factor, and the momentum changes:

$$\mathrm{U}(a, \boldsymbol{A})\,|\boldsymbol{p}, \lambda\rangle = e^{ip'a}e^{-i\lambda\theta}\,|\boldsymbol{p}', \lambda\rangle. \tag{103}$$

Here the state with momentum p is defined in terms of the state with standard momentum $\mathring{p}$ by (43):

$$|\boldsymbol{p}, \lambda\rangle = \mathrm{U}(0, \alpha(\boldsymbol{p}))\,|\mathring{p}, \lambda\rangle,$$

without the additional phase θ. The angle $\theta = \theta(p, A)$ depends on the momentum p and the transformation A. This dependence may be established in accord with (27) from the relation

$$e^{-i\frac{1}{2}\theta}\tfrac{1}{2}(1+\sigma_3) = \tfrac{1}{2}(1+\sigma_3)\,\tilde{A}(p, A)\,\tfrac{1}{2}(1+\sigma_3)$$
$$= \tfrac{1}{2}(1+\sigma_3)\,\alpha^{-1}(\boldsymbol{p}')\,A\alpha(\boldsymbol{p})\,\tfrac{1}{2}(1+\sigma_3), \qquad (p' = \Lambda(A)p), \tag{104}$$

since $e^{-i\frac{1}{2}\theta}$ is equal to the matrix element $\tilde{A}_{11}$. When the Wigner operator $\alpha(\boldsymbol{p})$ is (32), a direct calculation of (104) gives

$$e^{i\frac{1}{2}\theta} = \frac{(Ap)_{11}}{|(Ap)_{11}|}, \tag{105}$$

where $B_{ij} = (Ap)_{ij}$ denotes the elements of the 2 by 2 matrix B.

In (104) $\frac{1}{2}(1+\sigma_3)$ plays the role of a projection operator on the representation with $w^2 = t^2 = 0$. Indeed, multiplication of the triangular matrix K by $\frac{1}{2}(1+\sigma_3)$ gives the matrix $u\frac{1}{2}(1+\sigma_3)$, and, consequently, "translations" in Euclidean space of the little group $t(z) = 1+\frac{1}{2}(\sigma_1+i\sigma_2)\,z$ do not change the state:

$$t(z)\,\tfrac{1}{2}(1+\sigma_3) = \tfrac{1}{2}(1+\sigma_3).$$

Let us compare the representation of the Poincaré group for a particle with $m \neq 0$ (in the helicity basis) and for a particle with zero mass. For $m \neq 0$ a particle with spin J has $2J+1$ independent states differing in helicity λ $(-J \leqslant \lambda \leqslant J)$; in the case $m = 0$ a particle with spin J has only one independent state with either left-hand helicity $(\lambda = -J)$, or with right-hand helicity $(\lambda = J)$. A wave function for a particle with spin J and $m \neq 0$ may be represented in the form of a $(2J+1)$-dimensional column vector formed from $\varphi^J_\lambda(p)$; for $m = 0$ there remains only one function $\Phi^J_J(\boldsymbol{p})$ or $\Phi^J_{-J}(\boldsymbol{p})$. Under Lorentz transformations the $2J+1$ components of φ^J_λ for $m \neq 0$ transform among one another, since the helicity of a particle with mass is not an invariant quantity. In the case $m = 0$, the helicity is a relativistically invariant operator, and the only form of the transformation of $\Phi^J_{\pm J}$ consists of multiplication by a phase factor.

We note that the one-dimensional helicity representation for $m = 0$ may be obtained via the limit $m \to 0$ from a representation with $m \neq 0$ and spin $J = |\lambda|$.

The photon, as is well known, has two independent polarization states and not one, as would have followed from the classification via the irreducible representations of the Poincaré group. This means that the photon is described by a reducible representation. In fact, the left and right circular polarizations of the photon correspond to the helicities $\lambda = \pm 1$. Under space reflection the left (right) polarization transforms into the right (left); the two polarization states of a photon will transform according to an irreducible representation of the Poincaré group with reflection (see Chapter 6).

Let us find the generators $\boldsymbol{M}$ and $\boldsymbol{N}$ in a unitary representation for $m = 0$. Under an infinitesimal transformation

$$A = 1+\delta A = 1+i\tfrac{1}{2}(\boldsymbol{\sigma}\cdot\boldsymbol{\omega})+\tfrac{1}{2}(\boldsymbol{\sigma}\cdot\boldsymbol{\beta})$$

the right-hand side of (103) varies as a result of both the variation of $\theta(p, A)$ and the variation of the momentum in $|\boldsymbol{p}, \lambda\rangle$. The spin parts $\mathfrak{M}$ and $\mathfrak{N}$ of the operators $\boldsymbol{M}$ and $\boldsymbol{N}$ are defined by the variation of $\theta(p, A)$:

$$\tfrac{1}{2}(1+\sigma_3)\,\delta(e^{-i\lambda\theta}) = 2\lambda(\delta\alpha^{-1}\alpha+\alpha^{-1}\,\delta A\alpha)\,\tfrac{1}{2}(1+\sigma_3)$$
$$\equiv i(\boldsymbol{\mathfrak{M}}\cdot\boldsymbol{\omega}+\boldsymbol{\mathfrak{N}}\cdot\boldsymbol{\beta})\,\tfrac{1}{2}(1+\sigma_3).$$

After inserting eqn. (32) for α we obtain[44]

$$\left.\begin{aligned}\mathfrak{M} &= \left(\frac{p^1}{p^0+p^3}, \frac{p^2}{p^0+p^3}, 1\right)\lambda,\\ \mathfrak{N} &= \left(-\frac{p^2}{p^0+p^3}, \frac{p^1}{p^0+p^3}, 0\right)\lambda.\end{aligned}\right\} \tag{106}$$

All the operators $\mathfrak{M}_k$ and $\mathfrak{N}_k$ commute with one another; the generators $\boldsymbol{M} = \boldsymbol{M}^0+\mathfrak{M}$ and $\boldsymbol{N} = \boldsymbol{N}^0+\mathfrak{N}$, which include the orbital parts $\boldsymbol{M}^0$ and $\boldsymbol{N}^0$ [see (52)] satisfy the required commutation relations (see § 1.3).

Spinor states and equations of motion

In the case of particles of zero mass, one may also construct spinor states whose transformation law does not depend on momenta.[56, 57] However, for $m = 0$ there is an additional condition distinguishing physical states, namely: only the one-dimensional representations of the little group $E(2)$ characterized by the invariant $w^2 = t^2 = 0$ correspond to physical states. When one passes to spinor states, unphysical $(2J+1)$-dimensional non-unitary representations are introduced. One must thus isolate the physical part of this set of spinor states. The resulting supplementary condition is equivalent to an equation of motion for the spinor function.

Let $\lambda = \frac{1}{2}$. To use the analogy with the case $m \neq 0$ we will consider the state vector $|\boldsymbol{p}, \frac{1}{2}\rangle$ as a component of the row-vector $\varphi^+_{\frac{1}{2}}(\boldsymbol{p})$ with $\varphi^+_{-\frac{1}{2}}(\boldsymbol{p}) = 0$:

$$\varphi^+_\sigma(\boldsymbol{p}) = (|\boldsymbol{p}, \tfrac{1}{2}\rangle, 0), \quad \sigma = \pm\tfrac{1}{2}. \tag{107}$$

Let us introduce an auxiliary row-vector Φ_σ whose first component coincides with $\varphi^+_{\frac{1}{2}}(\boldsymbol{p})$:

$$\varphi^+(\boldsymbol{p}) = \Phi(\boldsymbol{p})\,\tfrac{1}{2}(1+\sigma_3). \tag{108}$$

The spinor state $|\boldsymbol{p}, \sigma; \frac{1}{2})$ for particles with zero-mass may now be defined in analogy with (69):

$$|\boldsymbol{p}, \sigma; \tfrac{1}{2}) = \sum_{\sigma'} \Phi_{\sigma'}(\boldsymbol{p})\,\alpha^{-1}_{\sigma'\sigma}(\boldsymbol{p}). \tag{109}$$

Adhering to the analogy with the case $m \neq 0$, we postulate that the spinor states (109) transform by the rule (70) or

$$|\boldsymbol{p}', \sigma; \tfrac{1}{2})' \equiv \mathrm{U}(0, A)|\,\boldsymbol{p}, \sigma; \tfrac{1}{2}) = \sum_{\sigma'} |\boldsymbol{p}', \sigma'; \tfrac{1}{2})A_{\sigma'\sigma}, \tag{110}$$

These transformation properties of spinor states must be consistent with formula (103) for the physical state vector $|\boldsymbol{p}, \frac{1}{2}\rangle$, which may be rewritten in matrix form for the quantities $\varphi^+_\sigma(\boldsymbol{p})$ using (104) and (107):

$$\mathrm{U}(0, A)\,\varphi^+_\sigma(\boldsymbol{p}) = \sum_{\sigma'} \varphi^+_{\sigma'}(\boldsymbol{p}')\left\{\tfrac{1}{2}(1+\sigma_3)\,\alpha^{-1}(\boldsymbol{p}')\,A\alpha(\boldsymbol{p})\,\tfrac{1}{2}(1+\sigma_3)\right\}_{\sigma'\sigma}. \tag{111}$$

Formulae (110) and (111) may hold simultaneously only when the spinor state (109) satisfies the relation

$$\left.\begin{aligned} |\boldsymbol{p}, \sigma; \tfrac{1}{2}) &= \sum_{\sigma'} |\boldsymbol{p}, \sigma'; \tfrac{1}{2})\, \Delta_{\sigma'\sigma}(\boldsymbol{p}), \\ \Delta(\boldsymbol{p}) &= \alpha(\boldsymbol{p})\, \tfrac{1}{2}(1+\sigma_3)\, \alpha^{-1}(\boldsymbol{p}). \end{aligned}\right\} \tag{112}$$

The matrix $\Delta(\boldsymbol{p})$ is a covariant projection operator on the component $\sigma = \frac{1}{2}$. Equation (112) defines a light-like spinor.

The matrix $\Delta(p)$ is easily found if the whole calculation is performed using the Wigner operator $\alpha(\boldsymbol{p}) = VH$ [formula (36)]; then

$$\Delta(\boldsymbol{p}) = VH\frac{1}{2}(1+\sigma_3)\,(VH)^{-1} = V\frac{1}{2}(1+\sigma_3)V^{+} = \frac{1}{2}\left(1+\frac{\boldsymbol{\sigma}\cdot\boldsymbol{p}}{|\boldsymbol{p}|}\right). \tag{113}$$

Equation (112) is equivalent to the equation of motion for $|\boldsymbol{p}, \sigma; \frac{1}{2})$ when $\lambda = \frac{1}{2} > 0$. Multiplying (112) by $\sigma_0 p^0 - \boldsymbol{\sigma}\cdot\boldsymbol{p}$ and using (113),

$$\sum_{\sigma'} |\boldsymbol{p}, \sigma'; \tfrac{1}{2})\,(\sigma_0 p^0 - \boldsymbol{\sigma}\cdot\boldsymbol{p})_{\sigma'\varrho} = 0. \tag{114}$$

The generalization of this result to arbitrary helicity $\lambda = \pm J (J > 0)$ is simple if one notes that the matrix $D^J_{\sigma\sigma'}(\frac{1}{2}(1 \pm \sigma_3))$ has only matrix elements with $\sigma = \sigma' = \pm J$. (The matrices $D^J(B)$ depending on the matrix B with $\det B = 0$ were defined in Chapter 3.) We replace $|\boldsymbol{p}, \lambda\rangle$ by the $(2J+1)$-component row vector

$$\varphi^{+}_{\sigma}(\boldsymbol{p}, J) = (|\boldsymbol{p}, \lambda\rangle, 0, \ldots, 0), \qquad (\lambda = J > 0), \tag{115}$$

and then introduce the spinor state with $2J+1$ components

$$\varphi^{+}_{\sigma}(\boldsymbol{p}, J) = \sum_{\sigma'} |\boldsymbol{p}, \sigma'; J)\, D^J_{\sigma'\sigma}(\alpha \tfrac{1}{2}(1+\sigma_3)),$$

whose transformation law is postulated to be

$$\mathrm{U}(0, A)\,|\,\boldsymbol{p}, \sigma; J) = \sum_{\sigma'} |\,\boldsymbol{p}', \sigma'; J)\, D^J_{\sigma'\sigma}(A).$$

Instead of (111) the transformation properties of the physical states (115) according to (103) and (104) are described by

$$\mathrm{U}(0, A)\,\varphi^{+}_{\sigma}(\boldsymbol{p}, J) = \left\{\varphi^{+}(\boldsymbol{p}', J)\, D^J(\tfrac{1}{2}(1+\sigma_3)\, \tilde{A}\tfrac{1}{2}(1+\sigma_3))\right\}_{\sigma}.$$

The covariant operator for projection on the component $\sigma = J$ of a spinor state is now equal to $D^J(\Delta(\boldsymbol{p}))$, while the condition that the spinor state be light-like reads

$$|\boldsymbol{p}, \sigma; J) = \sum_{\sigma} |\boldsymbol{p}, \sigma'; J)\, D^J_{\sigma'\sigma}(\Delta(\boldsymbol{p})). \tag{116}$$

Equation (116) is equivalent to the equation of motion for a light-like spinor of helicity J:

$$\sum_{\sigma'} |p, \sigma; J)\,(\boldsymbol{J}\cdot\boldsymbol{p} - Jp^0)_{\sigma\sigma'} = 0. \tag{117}$$

Here J is the spin matrix in the D^J representation.

We now introduce the conjugate states. Omitting steps which repeat the derivation from (115) to (117) we give the result for helicity $\lambda = J$:

$$(\boldsymbol{p}, \sigma; J | \mathrm{U}^{-1}(0, A) = \sum_{\sigma'} D^J_{\sigma\sigma'}(A^{-1}) (\boldsymbol{p}', \sigma'; J |,$$

$$\varphi_\sigma(\boldsymbol{p}, J) = \sum_{\sigma'} D^J_{\sigma\sigma'}(\tfrac{1}{2}(1+\sigma_3)\alpha^{-1}) (\boldsymbol{p}, \sigma; J |,$$

$$\sum_{\sigma'} D^J_{\sigma\sigma'}(\Delta(\boldsymbol{p})) (\boldsymbol{p}, \sigma'; J | = (\boldsymbol{p}, \sigma; J |,$$

where $\varphi(\boldsymbol{p}, J)$ is the state conjugate to the state (115).

Introducing creation $a^+(\boldsymbol{p}, \lambda)$ and annihilation $a(\boldsymbol{p}, \lambda)$ operators via the same formulae (81) and (84) as for massive particles, one may obtain the rule for the transformation of these operators under the action of U(a, A) from (103) and the conjugate equation. The next step—the construction of quantum fields—also turns out to be trivial once one knows the procedure for constructing fields with $m \neq 0$ (see § 4.4) and light-like spinor states.

The neutrino equation

Experiments on β-decay of pions show that the neutrinos ν_e and ν_μ have negative, or left-handed, helicity ($\lambda(\nu) = -\frac{1}{2}$), while the antineutrinos $\bar{\nu}_e$ and $\bar{\nu}_\mu$ have positive, or right-handed, helicity ($\lambda(\bar{\nu}) = \frac{1}{2}$). The states (107) thus describe antineutrinos. A two-component neutrino field $\nu(x)$ may be written in the form [see (93)]

$$\nu_\sigma(x) = \frac{1}{(2\pi)^{3/2}} \int \frac{d^3p}{2p_0} \{(\alpha(\boldsymbol{p}))_{\sigma-J}\, a(\boldsymbol{p}) e^{-ipx} + (\alpha(\boldsymbol{p}))_{\sigma-J}\, b^+(\boldsymbol{p}) e^{ipx}, \quad \left(J = \frac{1}{2}\right), \tag{118}$$

where $a(\boldsymbol{p})$ is the annihilation operator for a neutrino ($\lambda = -\frac{1}{2}$), while $b^+(\boldsymbol{p})$ is the creation operator for an antineutrino ($\lambda = \frac{1}{2}$). Under transformations $x \to x' = \Lambda x + a$ the field $\nu_\sigma(x)$ undergoes a transformation

$$\mathrm{U}(a, A)\, \nu_\sigma(x)\, \mathrm{U}^{-1}(a, A) = \sum_{\sigma'} A^{-1}_{\sigma\sigma'} \nu_{\sigma'}(\Lambda x + a).$$

Equations of the type (114) for spinor states entail the Weyl equation for the field $\nu(x)$:

$$(\boldsymbol{\sigma} \cdot \nabla - \partial_0)\, \nu(x) = 0. \tag{119}$$

We note that the choice (36) of $\alpha(\boldsymbol{p})$ allows one to rewrite the "wave function" $(\alpha(p))_{\sigma-J}$ in the form

$$(\alpha(\boldsymbol{p}))_{\sigma-J} = (2|\boldsymbol{p}|)^J\, V(\boldsymbol{p}).$$

As in the case of fields with $m \neq 0$, one can check that for fields with $m = 0$ the general connection between spin and statistics must hold, i.e. particles with zero mass are bosons for $\lambda = 0, \pm 1, \ldots$, and fermions for $\lambda = \pm\frac{1}{2}, \pm\frac{3}{2}, \ldots$

§ 4.6. Multi-particle states

The asymptotic in- and out-states have properties of the states of free particles. Each particle r in the in- or out-complex is on the mass shell: $p_r^2 = m_r^2$, while the state vector of n particles $|\alpha_1 \ldots \alpha_n\rangle$ is the symmetrized or antisymmetrized (for identical particles) product of single-particle states (see § 1.4). Consequently the in- and out-states may be constructed using creation and annihilation operators (for free particles).

Let $a^+(\boldsymbol{p}\lambda, mJ)$ and $a(\boldsymbol{p}\lambda, mJ)$ be the creation and annihilation operators for particles with spin J and mass m, having momentum $\boldsymbol{p}$ and helicity λ. These operators satisfy the following commutation relations:

In the case of bosons:

$$[a(\boldsymbol{p}\lambda, mJ), a^+(\boldsymbol{p}'\lambda', m'J')]_- = \delta_{mm'}\delta_{JJ'}2p_0\delta_{\lambda\lambda'}\delta(\boldsymbol{p}-\boldsymbol{p}'), \tag{120}$$

In the case of fermions:

$$[a(\boldsymbol{p}\lambda, mJ), a^+(\boldsymbol{p}'\lambda', m'J')]_+ = \delta_{mm'}\delta_{JJ'}2p_0\delta_{\lambda\lambda'}\delta(\boldsymbol{p}-\boldsymbol{p}'). \tag{121}$$

These relations assume that fermion operators for different mass and spin anticommute. (This result has been proven in axiomatic quantum field theory.[(32)] Then the product of n creation operators applied to the vacuum $|0\rangle$ gives a state with n particles:

$$|\boldsymbol{p}_1\lambda_1, m_1J_1; \ldots; \boldsymbol{p}_n\lambda_n, m_nJ_n\rangle = Aa^+(\boldsymbol{p}_1\lambda_1, m_1J_1) \ldots a^+(\boldsymbol{p}_n\lambda_n, m_nJ_n)|0\rangle, \tag{122}$$

where A allows for the possible existence of identical bosons: for n identical bosons $A = (n!)^{-1/2}$.

Under Lorentz transformations the states (122) transform as the product of creation operators. The explicit transformation properties of the states (122) are easily established using (47). These states, consequently, transform according to a reducible representation of the Poincaré group.

To describe a multi-particle state from the point of view of the Poincaré group, we must find the basis vectors of the irreducible representations, depending in an invariant way on all specific degrees of freedom of the multi-particle state. Moreover, it is necessary to expand the state vector (122) in terms of these multi-particle basis vectors, i.e. to expand the product of representations in terms of irreducible representations.[(58–62)] The case of n particles may be studied in the simplest example $n = 2$.

Let us consider the two-particle state of non-identical particles

$$|1, 2\rangle \equiv |\boldsymbol{p}_1\lambda_1, m_1J_1; \boldsymbol{p}_2\lambda_2, m_2J_2\rangle = |1\rangle|2\rangle. \tag{123}$$

These states are normalized by the condition

$$\langle 1', 2'|1, 2\rangle = 2p_0 2p_0' \delta_{\lambda_1\lambda_1'}\delta_{\lambda_2\lambda_2'}\delta(\boldsymbol{p}_1-\boldsymbol{p}_1')\,\delta(\boldsymbol{p}_2-\boldsymbol{p}_2'), \tag{124}$$

arising from the normalization (11) of single-particle states.

The generators P_μ and $M_{\mu\nu}$ in the space of two-particle states (123) may be represented in the form of a sum:

$$P_\mu = P_\mu(1)+P_\mu(2), \quad M_{\mu\nu} = M_{\mu\nu}(1)+M_{\mu\nu}(2), \tag{125}$$

where $P_\mu(i)$ and $M_{\mu\nu}(i)$ $(i = 1, 2)$ act only on the single-particle states $|i\rangle$. The mass m_i and spin J_i of the state $|i\rangle$ are defined by the eigenvalues of the operators

$$P^2(i)|i\rangle = m_i^2|i\rangle, \quad w^2(i)|i\rangle = -m_i^2J_i(J_i+1)|i\rangle, \tag{126}$$

where

$$w(i) = \tfrac{1}{2}\varepsilon^{\mu\nu\lambda\sigma}M_{\nu\lambda}(i)\,P_\sigma(i). \tag{127}$$

The irreducible representations of the Poincaré group are characterized, as we saw in § 1.3, by the values of the mass m and spin J. In our case

$$P_\mu P^\mu = (P_\mu(1)+P_\mu(2))(P^\mu(1)+P^\mu(2)) = m^2 > 0 \tag{128}$$

and, consequently, the total spin J is equal to the total angular momentum in the center of mass system:

$$\boldsymbol{P} = \boldsymbol{p}_1+\boldsymbol{p}_2 = 0;$$

here

$$w^2 = w_\mu w^\mu = -m^2 J(J+1), \tag{129}$$

where

$$w_\mu = \tfrac{1}{2}\varepsilon_{\mu\nu\lambda\sigma}(M^{\nu\lambda}(1)+M^{\nu\lambda}(2))(P^\sigma(1)+P^\sigma(2)). \tag{130}$$

The canonical or helicity basis in the case of a single-particle state contains four variables: the momentum $\boldsymbol{p}_i$ and the projection of the spin j_{i3} or the helicity λ_i. In the case of a two-particle state the analogous variables—the total momentum $\boldsymbol{P}$ and the spin projection J_{i3} or the total helicity $\Lambda = \boldsymbol{J}\cdot\boldsymbol{P}/|\boldsymbol{P}|$—do not describe all eight degrees of freedom. These variables, together with $m^2 = s$ and J, correspond to only six degrees of freedom, so that we must construct two additional variables in order to characterize unambiguously each two-particle state in an irreducible basis of the Poincaré group.

Let us denote the two additional variables by γ_1 and γ_2. The two-particle state vector may then be written in the form

$$|\,\boldsymbol{P}, m; J, \Lambda; \gamma_1, \gamma_2\rangle, \tag{131}$$

where we have used the same symbol for the eigenvalue of total momentum as for the generator itself. The quantities γ_1 and γ_2 must be measured simultaneously with all remaining quantities in (131), so that the operators γ_1 and γ_2 commute with each other and with all operators characterizing the basis of the irreducible representation of the Poincaré group. Consequently, γ_1 and γ_2 must be invariant operators, i.e. commuting with all generators P_λ and $M_{\mu\nu}$ $(i = 1, 2)$:

$$[\gamma_i, P_\mu] = 0, \quad [\gamma_i, M_{\mu\nu}] = 0. \tag{132}$$

Let us construct from $P_\mu(i)$ and $M_{\mu\nu}(i)$ a set of two invariant operators which do not reduce to single-particle ones, i.e. which do not contain $P_\mu(i)$ and $M_{\mu\nu}(i)$ with the same index i. It is easily checked that the operators

$$\gamma_1 = w_\mu(1)P^\mu, \quad \gamma_2 = w_\mu(2)P^\mu \tag{133}$$

satisfy these demands. Instead of γ_1 and γ_2 it is convenient to choose other invariant operators:

$$\lambda_i = [(p_i^\mu P_\mu)^2 - m_i^2 m^2]^{-1/2}\gamma_i. \tag{134}$$

In the center of mass $\boldsymbol{P} = 0$ the quantity λ is the helicity of the ith particle.

Thus the two-particle basis state for an irreducible representation (m, J) of the Poincaré group is

$$|\,\boldsymbol{P}, m; J, \Lambda; \lambda_1, \lambda_2\rangle, \tag{135}$$

where λ_1 and λ_2 are the helicities of particles in the center of mass system, while the remaining quantities are given by (125), (128), and (129). In the c.m.s. $\lambda_1 - \lambda_2$ is the projection of the spin J on the direction

$$\boldsymbol{p}_{12} = \tfrac{1}{2}(\boldsymbol{p}_1 - \boldsymbol{p}_2),$$

so that λ_1 and λ_2 are bounded by the condition

$$|\lambda_1 - \lambda_2| \leqslant J. \tag{136}$$

Under Lorentz transformations and displacements the quantities λ_1 and λ_2 remain unchanged, so that for $J \geqslant J_1 + J_2$ [when (136) does not limit λ_1 and λ_2 in (123)] there are $(2J_1+1)(2J_2+1)$ different states (123) transforming independently.

The invariant normalization condition for the states (135) will be written in the form

$$\langle \boldsymbol{P}, m, J; \Lambda; \lambda_1, \lambda_2 | \boldsymbol{P}', \Lambda'; m', J'; \lambda_1', \lambda_2' \rangle = \frac{P_0}{m}\, \delta(\boldsymbol{P}-\boldsymbol{P}')\, \delta(m-m')\delta_{\Lambda\Lambda'}\delta_{JJ'}\delta_{\lambda_1\lambda_1'}\delta_{\lambda_2\lambda_2'}. \tag{137}$$

Let us obtain the relation between the product of single-particle states (123) in the helicity basis and the basis (two-particle) state (135) of an irreducible representation of the Poincaré group with mass m and spin J. We first pass to the two-particle variables $\boldsymbol{P}$ and $\boldsymbol{p}_{12}$. Instead of the relative momentum of $\boldsymbol{p}_{12}$ in the c.m.s., let us introduce the invariant mass $m = \sqrt{s}$ of the two-particle state

$$p_{12}{}^2 = \frac{1}{4s}(s+(m_1+m_2)^2)(s-(m_1-m_2)^2), \quad p_{12} = |\boldsymbol{p}_{12}|, \tag{138}$$

and two Euler angles φ and ϑ describing the rotation $\mathrm{R}(\varphi, \vartheta, -\varphi)$ from the z-axis to the direction $\boldsymbol{p}_{12}$. Then in the c.m.s. the product of single-particle states (123) may be written in the form

$$\begin{aligned} &|\boldsymbol{p}_{12}, \lambda_1; m_1, J_1\rangle | -\boldsymbol{p}_{12}, \lambda_2; m_2, J_2\rangle \\ &\quad = \mathrm{U}(R(\varphi, \vartheta, -\varphi))\, \{|p_{12}e_z, \lambda_1; m_1, J_1\rangle | -p_{12}e_z, \lambda_2; m_2, J_2\rangle\} \\ &\quad \equiv \sqrt{\frac{4m}{p_{12}}}\, |0, m; \varphi, \vartheta; \lambda_1, \lambda_2\rangle. \end{aligned} \tag{139}$$

Here we have used eqn. (61) for particle 1 and eqn. (63) for particle 2, so that any two-particle state in the c.m.s. may be obtained using the same rotation $R(\varphi, \vartheta, -\varphi)$ from a two-particle state having only momentum components along the z-axis. The choice of the normalization factor in (139) follows from

$$p_1^0 p_2^0 \delta^3(\boldsymbol{p}_1 - \boldsymbol{p}_1')\, \delta(\boldsymbol{p}_2 - \boldsymbol{p}_2') = \frac{m}{|\boldsymbol{p}_{12}|}\, \delta^4(p-p')\, \delta^2(\Omega - \Omega').$$

A Lorentz transformation of the state (139) from the c.m.s. to a system moving with velocity $\boldsymbol{P}/P_0$ defines the general two-particle helicity state

$$|\boldsymbol{P}, m; \varphi, \vartheta; \lambda_1, \lambda_2\rangle \equiv \mathrm{U}(h(\boldsymbol{P}))\, |0, m; \varphi, \vartheta; \lambda_1, \lambda_2\rangle, \tag{140}$$

where $h(\boldsymbol{P})$ is the Wigner operator (61) in the helicity basis. The states (140) are orthonormal in the sense that

$$\langle \boldsymbol{P}, m; \varphi, \vartheta; \lambda_1, \lambda_2 | \boldsymbol{P}', m'; \varphi', \vartheta'; \lambda_1', \lambda_2' \rangle = \delta^4(P-P')\, \delta^2(\Omega-\Omega')\delta_{\lambda_1\lambda_1'}\delta_{\lambda_2\lambda_2'}. \tag{141}$$

The desired connection between the bases (123) and (135) or (135) and (140) is expressed by the transformation function[58]

$$\langle \boldsymbol{P}', m'; J, \Lambda; \lambda_1, \lambda_2 | \boldsymbol{P}, m; \varphi, \vartheta; \lambda_1', \lambda_2' \rangle = \frac{P_0}{m} N(J)\, \delta(\boldsymbol{P}-\boldsymbol{P}')\, \delta(m-m') \delta_{\lambda_1 \lambda_1'} \delta_{\lambda_2 \lambda_2'}$$

$$\times \langle 0; J; \Lambda | \mathrm{U}(R(\varphi, \vartheta, -\varphi)) | 0; \varphi = \vartheta = 0; \lambda_1 - \lambda_2 = \lambda \rangle$$

$$= \frac{P_0}{m} N(J)\, \delta(\boldsymbol{P}-\boldsymbol{P}')\, \delta(m-m') \delta_{\lambda_1 \lambda_1'} \delta_{\lambda_2 \lambda_2'} \mathcal{D}^J_{\Lambda\lambda}(\varphi, \vartheta, -\varphi). \tag{142}$$

To obtain (142) it is necessary to use (139), (140), and also (when the dependence on J, Λ and φ is desired) the Lorentz invariance of all normalizations [$N(J)$ is a normalization factor]. We note that in the c.m.s., where $\boldsymbol{p}_{12} = \boldsymbol{e}_z |\boldsymbol{p}_{12}|$, the total helicity is

$$\Lambda = \frac{\boldsymbol{J}\cdot\boldsymbol{P}}{|\boldsymbol{P}|} = \frac{1}{|\boldsymbol{p}_{12}|} [\boldsymbol{J}_1 \cdot \boldsymbol{p}_{12} + \boldsymbol{J}_2 \cdot \boldsymbol{p}_{12}] = \lambda_1 - \lambda_2 .$$

The expansion of the state (140) (i.e. the product (123) of single-particle states) may be consequently written in the form ($\lambda = \lambda_1 - \lambda_2$)

$$|\boldsymbol{P}, m; \varphi, \vartheta; \lambda_1, \lambda_2\rangle = \sum_J \sum_{\Lambda=-J}^{J} N(J)\, \mathcal{D}^J_{\Lambda\lambda}(\varphi, \vartheta, -\varphi) |\boldsymbol{P}, m; J, \Lambda; \lambda_1, \lambda_2\rangle. \tag{143}$$

To preserve the usual normalization (3.95) of the functions $\mathcal{D}^J$ it is necessary to choose

$$N(J) = \left(\frac{2J+1}{4\pi}\right)^{1/2}, \tag{144}$$

where we use the normalization (124) and relation (137).

Inverting (143) using (3.96), we find an expression for the basis state of an irreducible representation (m, J) in terms of a two-particle state with diagonal relative momentum $p_{12}(m, \vartheta, \varphi)$:

$$|\boldsymbol{P}, m; J, \Lambda; \lambda_1, \lambda_2\rangle = N(J) \int D^{J*}_{\Lambda\lambda}(\varphi, \vartheta, -\varphi) |\boldsymbol{P}, m; \varphi, \vartheta; \lambda_1, \lambda_2\rangle\, d\Omega. \tag{145}$$

Let us now turn to the case in which particles 1 and 2 are identical. The rule for constructing states with given J for identical particles may be introduced using (139) and (145) and proceeding from known properties of single-particle states.

Let A_{12} be the operator interchanging particles 1 and 2:

$$A_{12}\{|\boldsymbol{p}_1, \lambda_1; 1\rangle |\boldsymbol{p}_2, \lambda_2; 2\rangle^-\} = |\boldsymbol{p}_1, \lambda_1; 2\rangle |\boldsymbol{p}_2, \lambda_2; 1\rangle^-.$$

Using (63) and (64) we find a phase factor arising from the interchange

$$A_{12}\{|p\boldsymbol{e}_z, \lambda_1; 1\rangle | -p\boldsymbol{e}_z, \lambda_2; 2\rangle^-\} = (-1)^{2j-(\lambda_1-\lambda_2)} e^{i\pi J_y} \{|p\boldsymbol{e}_z, \lambda_2; 1\rangle | -p\boldsymbol{e}_z, \lambda_1; 2\rangle^-\}, \tag{146}$$

where $j = J_1 = J_2$ and $\boldsymbol{J} = \boldsymbol{J}_1 + \boldsymbol{J}_2$. But because of (145) this result means that

$$A_{12} |\boldsymbol{P}, m; J, \Lambda; \lambda_1, \lambda_2\rangle = (-1)^{2j+J} |\boldsymbol{P}, m; J, \Lambda; \lambda_2, \lambda_1\rangle, \tag{147}$$

since $e^{i\pi J_y}|J,\lambda\rangle = (-1)^{J-\lambda}|J,-\lambda\rangle$. Consequently, the symmetric (antisymmetric) combination corresponds to the choice of the sign $+$ $(-)$ in

$$\frac{1}{\sqrt{2}}(|\boldsymbol{P}, m; J, \Lambda; \lambda_1, \lambda_2\rangle \pm (-1)^{2j+J}|\boldsymbol{P}, m; J, \Lambda; \lambda_2, \lambda_1\rangle)$$
$$= \frac{1}{\sqrt{2}}(|\boldsymbol{P}, m; J, \Lambda; \lambda_1, \lambda_2\rangle + (-1)^{J}|\boldsymbol{P}, m; J, \Lambda; \lambda_2, \lambda_1\rangle). \quad (148)$$

We have eliminated the signs $\pm$ on the right-hand side of (148) using the fact that symmetrization (antisymmetrization) is required for integral (half-integral) spins J, so that $(-1)^{2j}$ (j integral) $= -(-1)^{2j}$ (j half-integral) $= 1$.

From (148) it follows that in the case of odd J the helicities of the particles must be different ($\lambda_1 \neq \lambda_2$) both for bosons and fermions. For example, there does not exist a two-photon state with $J = 1$ and with the same polarizations $\lambda_1 = \lambda_2 = \pm 1$. For exactly the same reason there does not exist a two-nucleon state with $J = 1$ and the same nucleon helicities.

Chapter 5

WAVE FUNCTIONS AND EQUATIONS OF MOTION FOR PARTICLES WITH ARBITRARY SPIN

In the previous chapter it was shown that for zero-mass particles the equations of motion are conditions imposed on spinor functions so that they describe physical states transforming with respect to irreducible (one-dimensional) representations of the quantum mechanical Poincaré group. In other words, the equations of motion for $m = 0$ were obtained as a consequence of the relativistic properties of these quantum systems. Such a relation between equations of motion and physical representations of the quantum mechanical group is a general feature of the approach, and is characteristic also of states with rest mass.

The role of equations of motion consists in the description of the time evolution of a physical state. Equations of motion always contain a time translation operator P_0, and—by covariance—hence contain the spatial displacement operators P_k as well. But these operators are generators of the Poincaré group, as is the angular momentum $M_{\mu\nu}$ with which the condition of relativistic invariance is formulated.

Equations (4.47), (4.89), and (4.103) for irreducible representations of the Poincaré group also define the time evolution of the state of a free particle, i.e. they solve the problem which is usually posed after writing equations of motion. For this reason, equations of motion, as a source of information about non-interacting systems, are not of independent interest. Their construction proceeds backwards from a known solution—the field operator $\Psi(x)$ or the wave function

$$u(\boldsymbol{p}, \sigma) = (2\pi)^{3/2} \langle 0 | \Psi(0) | \boldsymbol{p}, \sigma \rangle, \tag{1}$$

i.e. in the final analysis, from the properties of the states $|\boldsymbol{p}, \sigma\rangle$. However, knowledge of various forms of wave functions and the equations corresponding to them is very useful for parametrizing matrix elements and introducing phenomenological interactions.

Equations of motion were first investigated as a basic source of information about the properties of relativistic particles. Beginning with the work of Dirac, Fierz, and Pauli,[63–65] many linear equations of various types describing particles with higher spins were proposed.

We shall consider only the Duffin–Kemmer,[66, 67] Rarita–Schwinger[68] and Bargmann–Wigner[69] equations, although equations of other types[70–74] also have their merits.

In the past decade the problem of describing particles with higher spins has taken on practical significance in connection with the discovery of resonances. It has also revived interest in equations of motion (see, for example, refs. 75–79).

§ 5.1. Wave functions, bilinear Hermitian forms, and equations of motion

Let us consider a general method for constructing the wave functions (1) and the derivation of equations of motion for them. Under the coordinate transformation $x \to x' = \Lambda(A)x+a$ the annihilation operators $a(\boldsymbol{p}, \sigma)$ and $b(\boldsymbol{p}, \sigma)$ of particles and antiparticles with mass m and spin J both transform according to the rule (4.85)

$$\mathrm{U}(\mathrm{a}, A)\, b(\boldsymbol{p}, \sigma)\, \mathrm{U}^{-1}(\mathrm{a}, A) = e^{-ip'a} \sum_{\sigma} \mathcal{D}^J_{\sigma\sigma'}(\alpha^{-1}(\boldsymbol{p})\, A^{-1} \alpha(\boldsymbol{p}'))\, b(\boldsymbol{p}', \sigma'), \tag{2}$$

$$p' = \Lambda(A)p.$$

The operators

$$\sum_{\sigma'} \mathcal{D}^J_{\sigma\sigma'}(C^{-1})\, b^{+}(\boldsymbol{p}, \sigma'), \quad \sum_{\sigma'} \mathcal{D}^J_{\sigma\sigma'}(C^{-1})\, a^{+}(\boldsymbol{p}, \sigma')$$

containing the creation operators b^+ and a^+ also transform analogously under homogeneous transformations.

Let us introduce the field $\Psi_r(x)$ for particles with mass m and spin J:

$$\Psi_r(x) = \frac{1}{(2\pi)^{3/2}} \sum_{\sigma} \int \frac{d^3p}{2p_0} \{u_r(\boldsymbol{p}, \sigma)\, a(\boldsymbol{p}, \sigma) e^{-ipx} + v_r(\boldsymbol{p}, \sigma)\, b^{+}(\boldsymbol{p}, \sigma) e^{ipx}\}, \tag{3}$$

which differs from the spinor fields considered earlier (see § 4.4) by the replacement of the matrices $D^J_{r\sigma}(\alpha)$ and $D^J_{r\sigma}(\alpha C^{-1})$ by the (still undefined) "wave functions" $u_r(\boldsymbol{p}, \sigma)$ and $v_r(\boldsymbol{p}, \sigma)$. These wave functions must be such that the field $\Psi(x)$ has the following properties:

1. Under a Lorentz transformation $x' = \Lambda(A)x$ the field $\Psi(x)$ must transform according to a finite dimensional (nonunitary) representation L of the homogeneous Lorentz group. Here

$$\mathrm{U}(0, A)\, \Psi_r(x)\, \mathrm{U}^{-1}(0, A) = \sum_{r'} L_{rr'}(A^{-1})\, \Psi_{r'}(\Lambda x) \tag{4}$$

(the representation L may be reducible, and the number of components of $\Psi_r(x)$ may exceed $2J+1$).

2. The field $\Psi_r(x)$ must admit the construction of an invariant bilinear Hermitian form. This condition is equivalent to the demand that there exist a Hermitian operator $\bar{g}$ such that

$$\Psi^{+}_r(x)\, \bar{g}_{rr'} \Psi_{r'}(x) = \Psi^{+}_r(\Lambda x)\, \bar{g}_{rr'} \Psi_{r'}(\Lambda x). \tag{5}$$

The operator $\bar{g}$ is assumed to be independent of momentum.

3. The field Ψ_r must satisfy a covariant equation of motion in momentum or coordinate space and supplementary conditions limiting the number of independent components of Ψ_r to $2J+1$.

Condition 1 allows for a broad class of fields. The use of representations of the homogeneous Lorentz group in constructing fields is motivated by the desire to have locally transforming quantities instead of the nonlocal transformation law (2). As was already discussed in § 4.4, to get rid of nonlocality in the transformation law it is necessary to subdivide the matrix $\mathcal{D}^J(\alpha^{-1}A^{-1}\alpha')$ entering into (2) into the product of three matrices of different types, depending respectively on α^{-1}, A^{-1}, and α'. These other matrices must refer to a representation of the Lorentz group, but not to the rotation group, since the rotation group does not contain the unitary transformations α and A. Thus the Lorentz group plays here the role of an auxiliary group essential for the convenient representation of transformation properties.

The most economical construction leads to spinor fields $\Phi^J(x)$ (see § 4.4) with $2(2J+1)$ components.

Spinor functions transform with respect to the representation $(0, J)$ or $(J, 0)$ of the Lorentz group. If one considers only irreducible representations of the Lorentz group, the wave functions $\Psi_r(x)$ for spin J may transform according to the representation (j_1, j_2) of the Lorentz group, as long as J lies between the numbers $|j_1-j_2|, |j_1-j_2|+1, \ldots, j_1+j_2$. The use of reducible representations L opens up broad possibilities for introducing various wave functions. A particle with spin J may, for example, be described in terms of a tensor of Jth rank or a spinor of $2J$th rank.

Condition 2 significantly limits the possible representations of the Lorentz group by which Ψ_r may transform. In accord with (5) the matrix L must be such that

$$L^+\bar{g}L = \bar{g}. \tag{6}$$

holds. Consequently, the matrices L and L^{-1+} must belong to the same representation of the Lorentz group used in constructing the field Ψ. As was shown in § 3.3, this representation must have the form (j, j) or $(j_1, j_2)+(j_2, j_1)$ or must be the direct product of representations of this type.

Let us obtain the relation between the matrices L in the transformation rule (4) of the field $\Psi(x)$ and the wave function $u_r(\boldsymbol{p}, \sigma)$. Let $e_r(L)$ be our choice for the basis of the representation L. The vectors of the rest state form a basis for a representation of the rotation group: $|0, \sigma, J\rangle \equiv e_\sigma(J)$. The matrix taking the basis $e_r(L)$ into the basis $e_\sigma(J)$ is

$$(e_r(L), e_\sigma(J)) = c_{r\sigma}(L, J).$$

If L refers to the representation $(J, 0)$, then

$$\big(e_\sigma(J), e_r((J, 0))\big) = \big(e_r((J, 0)), e_\sigma(J)\big) = \delta_{\sigma r}.$$

Thus the matrix $\mathcal{D}^J$ in (2), describing a spin rotation, may be written in the form

$$\mathcal{D}^J_{\sigma\sigma'}(\alpha^{-1}(\boldsymbol{p})\,A^{-1}\alpha(\boldsymbol{p}')) = \sum_{rr'} c_{\sigma r}(J, L)\,L_{rr'}(\alpha^{-1}(\boldsymbol{p})\,A^{-1}\alpha(\boldsymbol{p}'))\,c_{r'\sigma'}(L, J). \tag{7}$$

But the Lorentz matrices L are defined also for the nonunitary matrices α and A, and we may decompose $L(\alpha^{-1}A^{-1}\alpha')$ into three factors, as in the case of spinor fields (§ 4.4). The wave function $u_r(\boldsymbol{p}, \sigma)$ for a particle with spin J is thus

$$u_r(\boldsymbol{p}, \sigma) = \sum_{r'} L_{rr'}(\alpha(\boldsymbol{p}))\,c_{r'\sigma}(L, J) = \sum_{r'} L_{rr'}(\alpha(\boldsymbol{p}))\,u_{r'}(0, \sigma). \tag{8}$$

Here $\alpha(\boldsymbol{p})$ is the Wigner operator, so that $\Lambda(\alpha)$ is the Lorentz transformation from the rest state $\mathring{p}^\mu = (m, 0, 0, 0)$ to the momentum p^μ.

Using (2), (7), and (8) it is easily checked that the operator

$$\mathrm{u}_r(\boldsymbol{p}) = \sum_{\sigma} u_r(\boldsymbol{p}, \sigma)\, a(\boldsymbol{p}, \sigma) \tag{9}$$

transforms under $x \to \Lambda(A)x$ according to the rule

$$\mathrm{U}(0, A)\,\mathrm{u}_r(\boldsymbol{p})\,\mathrm{U}^{-1}(0, A) \equiv \mathrm{u}_r'(\boldsymbol{p}) = \sum_{r'} L_{rr'}(A^{-1})\, u_{r'}(\boldsymbol{p}'), \quad p' \equiv \Lambda(A)p. \tag{10}$$

The operator

$$\mathrm{v}_r(\boldsymbol{p}) = \sum_{\sigma'} v_r(\boldsymbol{p}, \sigma)\, b^+(\boldsymbol{p}, \sigma), \tag{11}$$

also transforms analogously, where

$$v_r(\boldsymbol{p}, \sigma) = \sum_{r'} L_{rr'}(\alpha(\boldsymbol{p})C^{-1})\, c_{r'\sigma}(L, J). \tag{12}$$

It is obvious that the properties of the field $\Psi(x)$ are fully defined by the wave functions u_r and v_r. According to (8) and (12) the construction of the wave functions then reduces to calculating the matrix $L(\alpha(\boldsymbol{p}))$ of the Lorentz transformation $\alpha(\boldsymbol{p})$ in the representation L and to establishing the relation $c_{r\sigma}(L, J) = u_r(0, \sigma)$ between the basis of the rotation group (spin J, component σ) and the basis of the Lorentz group (representation L, component r), i.e. to the choice of the wave function $u_r(0, \sigma)$ in the rest state.

The choice of the correspondence $u_r(0, \sigma) = c_{r\sigma}(L, J)$ defines the equation of motion and the subsidiary conditions which are satisfied by the wave function $u(\boldsymbol{p}, \sigma)$ for any momentum. These equations of motion and conditions are just a covariant expression for the relations defining the basis $u(0, \sigma)$.

As an example, let us derive the equation of motion for $u(\boldsymbol{p}, \sigma)$. Let the wave function in the rest state $u(0, \sigma)$ be the eigenstate of the operator $\bar{g}$, introduced in formulae (5) and (6):

$$\bar{g}u(0, \sigma) = \gamma u(0, \sigma), \tag{13}$$

where γ is a real number. Keeping in mind eqn. (8), i.e.

$$u(\boldsymbol{p}, \sigma) = L(\alpha(\boldsymbol{p}))\, u(0, \sigma),$$

we may rewrite the condition (13) using (6) in the form

$$L(\alpha(\boldsymbol{p})\,\alpha^+(\boldsymbol{p}))\,\bar{g}u(\boldsymbol{p}, \sigma) = \gamma u(\boldsymbol{p}, \sigma)$$

or

$$L\left(\frac{\mathrm{p}}{m}\right)\bar{g}u(\boldsymbol{p}, \sigma) = \gamma u(\boldsymbol{p}, \sigma). \tag{14}$$

This is the equation of motion for $u(\boldsymbol{p}, \sigma)$. For a suitable choice of the representation L this equation may be linear in p.

Equation (14) is covariant by construction, starting with (13). However, to verify the relativistic invariance of (14) in the usual way, it is necessary to know the transformation properties of the wave functions $u(\boldsymbol{p}, \sigma)$. So far we have considered the transformations $\mathrm{U}(a, A)$ in the Hilbert space of state vectors $|\,\boldsymbol{p}, \sigma\rangle$ and not in the space of functions $u(\boldsymbol{p}, \sigma)$

and $v(\mathbf{p}, \sigma)$. But it is the transformation properties of the wave functions that are used to introduce spinor amplitudes (see § 7.5).

The transformed wave function may be introduced by starting from eqns. (9) and (10) as that function which defines the transformed operator $\mathrm{u}'(\mathbf{p})$:

$$\mathrm{u}_r'(\mathbf{p}) = \sum_\sigma u_r'(\mathbf{p}, \sigma)\, a(\mathbf{p}', \sigma), \quad p' = \Lambda(A)p.$$

Consequently

$$u_r'(\mathbf{p}, \sigma) = L_r^{r'}(A^{-1})\, u_{r'}(\mathbf{p}', \sigma). \tag{15}$$

But the transformed wave function (15) does not in general have the same form as $u(\mathbf{p}, \sigma)$ since the relation $A\alpha(\mathbf{p}) = \alpha(\mathbf{p}')$ holds only for a special class of matrices A.

In the general case [see (4.19)]

$$A\alpha(\mathbf{p}) = \alpha(\mathbf{p}')\alpha^{-1}(\mathbf{p}')\, A\alpha(\mathbf{p}) = \alpha(\mathbf{p}')\, \tilde{A}(\mathring{p}, A),$$

so that $A\alpha(\mathbf{p})$ differs from $\alpha(\mathbf{p}')$ by a spin rotation $\tilde{A}(\mathring{p}, A)$, undergone by the functions $u_r(0, \sigma)$. Thus u may be expressed in terms of u' as follows:

$$u_r'(\mathbf{p}', \sigma) = \sum_{\sigma'} u_r(\mathbf{p}', \sigma')\, \mathcal{D}_{\sigma'\sigma}(\tilde{A}(\mathring{p}, A)), \tag{16}$$

where we have also used (7) and $p' = \Lambda p$.

One may construct the conjugate wave function of $u(\mathbf{p}, \sigma)$: $\bar{u}(\mathbf{p}, \sigma) = u^+(\mathbf{p}, \sigma)\bar{g}$. Because of (6) and (15) this function transforms according to the rule

$$\bar{u}'(\mathbf{p}', \sigma) = \bar{u}(\mathbf{p}, \sigma)\, L(A^{-1}),$$

so that there exists an invariant bilinear Hermitian form

$$\bar{u}'(\mathbf{p}', \sigma)\, u'(\mathbf{p}', \sigma) = \bar{u}(\mathbf{p}, \sigma)\, u(\mathbf{p}, \sigma).$$

Let us turn to the equation of motion (14). To verify its invariance, it is sufficient to learn how the matrix $L(\mathrm{p}/m)\bar{g}$ transforms. We have

$$L(A)\, L\left(\frac{\mathrm{p}}{m}\right) \bar{g} L(A^{-1}) = L\left(A \frac{\mathrm{p}}{m} A^+\right) \bar{g} = L\left(\frac{\mathrm{p}'}{m}\right) \bar{g},$$

where $p' = \Lambda(A)p$. From this equation and (15) it follows that after the substitution $p \to p'$, (14) will hold for the transformed function $u'(p', \sigma)$, which was to be proven.

In the next section we shall examine in detail the introduction and properties of wave functions using the example of Dirac wave functions.

§ 5.2. The Dirac equation

In the case of spin$-\frac{1}{2}$ particles, the simplest field $\psi(x)$ with the invariant density (5) will transform according to the reducible representation $(\frac{1}{2}, 0)+(0, \frac{1}{2})$ of the Lorentz group.

The two-component spinor field $\varphi(x)$ transforming via the representation $(\frac{1}{2}, 0)$ has already been studied in § 4.4:

$$\mathrm{U}(0, A)\, \varphi_\sigma(x)\, \mathrm{U}^{-1}(0, A) = \sum_{\sigma'} A_{\sigma\sigma'}^{-1} \varphi_{\sigma'}(\Lambda(A)x).$$

Using only the field $\varphi(x)$ it is impossible to form an invariant bilinear Hermitian form of the type (5) with momentum-independent coefficients, since there does not exist a 2 by 2 matrix $\bar{g}$ with the property $\bar{g}A\bar{g}^{-1} = A^{-1+}$. Thus the condition (6) may be satisfied only when the number of components of the field is doubled by passing to a reducible representation $(\frac{1}{2}, 0)+(0, \frac{1}{2})$, and $\bar{g}$ is described by a 4 by 4 matrix. Consequently we must consider, along with the field $\varphi(x)$, also the field $\chi(x)$ transforming via the matrix A^+:

$$\mathrm{U}(0, A)\, \chi_\sigma(x)\, \mathrm{U}^{-1}(0, A) = \sum_{\sigma'} A^+_{\sigma\sigma'} \chi_{\sigma'}(\Lambda(A)x).$$

If one now passes to a four-component (bispinor) field $\psi_\alpha(x)$ ($\alpha = 1, 2, 3, 4$):

$$\psi(x) = \begin{pmatrix} \varphi(x) \\ \chi(x) \end{pmatrix}, \tag{17}$$

then it will transform according to the representation $(\frac{1}{2}, 0)+(0, \frac{1}{2})$:

$$\mathrm{U}(0, A)\, \psi(x)\, \mathrm{U}^{-1}(0, A) = S(A^{-1})\, \psi(\Lambda(A)x), \tag{18}$$

where the matrix $S(A)$ is

$$S(A) = \begin{pmatrix} A & 0 \\ 0 & A^{-1+} \end{pmatrix}. \tag{19}$$

Let us introduce the 4 by 4 matrices:

$$\gamma_4 = \begin{pmatrix} 0 & 1 \\ 1 & 0 \end{pmatrix}, \qquad \gamma_5 = \begin{pmatrix} 1 & 0 \\ 0 & -1 \end{pmatrix}, \tag{20}$$

where, as in (19), each number denotes a 2 by 2 matrix. The matrix γ_5 distinguishes the representations $(\frac{1}{2}, 0)$ and $(0, \frac{1}{2})$, while the matrix γ_4 interchanges them. Consequently, the matrix γ_4 replaces the matrix $S(A)$ by the Hermitian conjugate of its inverse $S^{-1+}(A)$:

$$\gamma_4 S(A) \gamma_4 = S(A^{-1+}) = S^{-1+}(A). \tag{21}$$

Here the conjugate bispinor $\bar{\psi} = \psi^+(x)\gamma_4$ transforms by the rule

$$\mathrm{U}(0, A)\, \bar{\psi}(x)\, \mathrm{U}^{-1}(0, A) = \bar{\psi}(\Lambda(A)x)\, S(A) \tag{22}$$

and the bilinear Hermitian forms

$$\bar{\psi}(x)\, \psi(x) = \psi^+(x)\, \gamma_4 \psi(x), \quad i\bar{\psi}(x)\, \gamma_5 \psi(x) \tag{23}$$

are invariant in the sense (5) with $\bar{g} = \gamma_4, i\gamma_4\gamma_5$.

The fields $\psi(x)$ and $\bar{\psi}(x)$ may be expanded in terms of creation and annihilation operators for particles and antiparticles:

$$\left.\begin{aligned} \psi_\alpha(x) &= \frac{1}{(2\pi)^{3/2}} \sum_\sigma \int \frac{d^3p}{2p_0} \{u_\alpha(\boldsymbol{p}, \sigma)\, a(\boldsymbol{p}, \sigma) e^{-ipx} + v_\alpha(\boldsymbol{p}, \sigma)\, b^+(\boldsymbol{p}, \sigma) e^{ipx}\}, \\ \bar{\psi}^\alpha(x) &= \frac{1}{(2\pi)^{3/2}} \sum_\sigma \int \frac{d^3p}{2p_0} \{\bar{u}^\alpha(\boldsymbol{p}, \sigma)\, a^+(\boldsymbol{p}, \sigma) e^{ipx} + \bar{v}^\alpha(\boldsymbol{p}, \sigma)\, b(\boldsymbol{p}, \sigma) e^{-ipx}\}. \end{aligned}\right\} \tag{24}$$

In these expansions we have introduced new quantities $u_\alpha(\boldsymbol{p}, \sigma)$ and $v_\alpha(\boldsymbol{p}, \sigma)$—the four-component wave functions.

Let us first consider u_α. This wave function is defined by the expression (8), or, in our case, by

$$u_\beta(\boldsymbol{p}, \sigma) = S_\beta^{\beta'}(\alpha(\boldsymbol{p}))\, u_{\beta'}(0, \sigma). \tag{25}$$

The wave function in the rest state $u_\beta(0, \sigma)$ is characterized by the spin projection σ. It has twice the number of components needed to describe spin $\frac{1}{2}$. The upper and lower components of $u(0, \sigma)$ (for diagonal γ_5) transform identically under rotations R since $R = R^{-1+}$. Thus, choosing $\xi(\sigma)$ as an independent two-component function, we may write $u(0, \sigma)$ in the form

$$u(0, \sigma) = \begin{pmatrix} \xi(\sigma) \\ \pm\xi(\sigma) \end{pmatrix}.$$

The usual Dirac function is obtained for the positive choice of sign if one subjects $u_\beta(0, \sigma)$ to the supplementary condition

$$\gamma_4 u(0, \sigma) = u(0, \sigma), \tag{26}$$

eliminating the two superfluous components of $u(0, \sigma)$. As we shall see below, the choice of the eigenvalue $\gamma_4 = -1$ leads to the function $v(\boldsymbol{p}, \sigma)$.

Let us find an equation of the type (14) whose solution is u. For this we insert into (26) the expression $u(0, \sigma) = S^{-1}(\alpha)\, u(\boldsymbol{p}, \sigma)$ from (25):

$$S(\alpha)\,(\gamma_4 - 1)\, S^{-1}(\alpha)\, u(\boldsymbol{p}, \sigma) = \left\{\tfrac{1}{2}(1+\gamma_5)\gamma_4\alpha\alpha^+ + \tfrac{1}{2}(1-\gamma_5)\,\gamma_4(\alpha\alpha^+)^{-1} - 1\right\} u(\boldsymbol{p}, \sigma) = 0.$$

But the 2 by 2 matrix $\alpha(\boldsymbol{p})\,\alpha^+(\boldsymbol{p})$ is just the momentum matrix $\mathrm{p}/m = (1/m)(\sigma_0 p^0 + \boldsymbol{\sigma}\cdot\boldsymbol{p})$. Thus if we pass to γ-matrix notation ($\gamma^0 = \gamma_0 \equiv \gamma_4$):

$$\gamma^k = \gamma_4\gamma_5\sigma_k = \begin{pmatrix} 0 & -\sigma_k \\ \sigma_k & 0 \end{pmatrix}, \qquad \{\gamma^\mu, \gamma^\nu\} = 2g^{\mu\nu}, \tag{27}$$

this relation takes the form of the Dirac equation in momentum space:

$$(\not{p} - m)u \equiv (\gamma_4 p^0 - \boldsymbol{\gamma}\cdot\boldsymbol{p} - m)u = 0. \tag{28}$$

Thus the condition (26) which isolates the physical two-component function from a four-component quantity is equivalent to the Dirac equation.

The conjugate bispinor $\bar{u} = u^*\gamma_4$ satisfies the equation

$$\bar{u}(\not{p} - m) = 0.$$

If we had limited ourselves to the two-component field $\varphi(x)$ (or $\chi(x)$), a covariant equation would have been impossible to write. Indeed, a 2 by 2 momentum matrix p, with which one would try to construct an equation for $\varphi(p) = (\frac{1}{2})\,(1+\gamma_5)\, u(\boldsymbol{p})$, transforms as $\mathrm{p}' = A\mathrm{p}A^+$, so that the transformation properties of $\mathrm{p}\varphi$ and $m\varphi$ do not coincide with one another. A covariant equation is possible only when $m = 0$: in this case the Dirac equation breaks up into separate equations [cf. the Weyl equation (4.119)] for the components

$\nu = (\frac{1}{2})(1+\gamma_5)\psi$ and $\tilde{\nu} = (\frac{1}{2})(1-\gamma_5)\psi$. The functions ν and $\tilde{\nu}$, of course, are two-component only in a representation of the γ matrices in which γ_5 is diagonal.

An explicit expression for the Dirac function $u(\boldsymbol{p}, \sigma)$ may be found by substituting a concrete expression for $\alpha(\boldsymbol{p})$ into (25) [under the condition (26)]. For example, in the canonical basis, in which α is given by (4.24),

$$u(\boldsymbol{p}, \boldsymbol{\sigma}) = \begin{pmatrix} \alpha(\boldsymbol{p})\,\xi(\sigma) \\ \alpha^{-1+}(\boldsymbol{p})\,\xi(\sigma) \end{pmatrix} \sqrt{m} = \frac{1}{\sqrt{2}}(p^0+m)^{-1/2} \begin{pmatrix} (m+p^0+(\boldsymbol{\sigma}\cdot\boldsymbol{p}))\,\xi(\sigma) \\ (m+p^0-(\boldsymbol{\sigma}\cdot\boldsymbol{p}))\,\xi(\sigma) \end{pmatrix}. \tag{29}$$

Here $\xi(\sigma)$ is the normalized spin wave function in the rest state, whose components $\sigma = \pm\frac{1}{2}$ differ by the eigenvalues of the spin projection $\frac{1}{2}\sigma_3$, so that

$$\xi_{\sigma'}(\sigma) = \delta_{\sigma'\sigma}, \qquad \sigma, \sigma' = \pm\tfrac{1}{2}.$$

The wave functions (25) and (29) are orthonormal:

$$\left.\begin{aligned} \bar{u}(\boldsymbol{p}, \sigma')\,u(\boldsymbol{p}, \sigma) \equiv \bar{u}^{\beta}(\boldsymbol{p}, \sigma')\,u_{\beta}(\boldsymbol{p}, \sigma) &= 2m\delta_{\sigma\sigma'}, \\ u^{*}(\boldsymbol{p}, \sigma')\,u(\boldsymbol{p}, \sigma) &= 2p_0\delta_{\sigma\sigma'}. \end{aligned}\right\} \tag{30}$$

Another expression for u, which corresponds to a representation of the γ matrices with diagonal γ_4, is frequently used. Diagonalizing γ_4 using the unitary transformation $e^{(\gamma_5\gamma_4\pi/4)}$,

$$u(\boldsymbol{p}, \sigma) = (m+p_0)^{1/2} \begin{pmatrix} \xi(\sigma) \\ \dfrac{\boldsymbol{\sigma}\cdot\boldsymbol{p}}{m+p_0}\,\xi(\sigma) \end{pmatrix}. \tag{31}$$

Equation (28) for the wave function $u(\boldsymbol{p}, \sigma)$ allows one to write [using (24)] an equation for the part of the field $\psi(x)$ containing this function. The equation for $\psi(x)$ must be the same for the part with creation operators as for the part with annihilation operators. Consequently, $\psi(x)$ satisfies the Dirac equation

$$(i\gamma^{\mu}\,\partial_{\mu} - m)\,\psi(x) = 0, \tag{32}$$

and the conjugate field $\overline{\psi}(x)$ must satisfy the equation

$$(i\,\partial_{\mu}\overline{\psi}(x)\gamma^{\mu} + m\overline{\psi}(x)) = 0.$$

Let us now turn to the study of the wave function $v(\boldsymbol{p}, \sigma)$ contained in the field $\psi(x)$. Its explicit form may be found from the same expression (8) since the operators

$$\mathrm{u}_{\beta}(\boldsymbol{p}) = \sum_{\sigma} u_{\beta}(\boldsymbol{p}, \sigma)\,a(\boldsymbol{p}, \sigma) \quad \text{and} \quad \mathrm{v}_{\beta}(\boldsymbol{p}) = \sum_{\sigma} v_{\beta}(\boldsymbol{p}, \sigma)\,b^{+}(\boldsymbol{p}, \sigma) \tag{33}$$

have the same properties with respect to homogeneous Lorentz transformations. Then

$$v_{\beta}(\boldsymbol{p}, \sigma) = S_{\beta}^{\ \beta'}(\alpha(\boldsymbol{p}))\,v_{\beta'}(0, \sigma). \tag{34}$$

From expression (24) for the field $\psi(x)$ and from the Dirac equation (32) it is clear that $v(\boldsymbol{p}, \sigma)$ satisfies the equation

$$(\not{p}+m)\,v(\boldsymbol{p}, \sigma) = 0. \tag{35}$$

If one follows the derivation of (28) for $u(\boldsymbol{p}, \sigma)$ in reverse, one may easily find the condition on $v(0, \sigma)$ (picking out the physical components), equivalent to (35). This condition has the form

$$\gamma_4 v(0, \sigma) = -v(0, \sigma), \tag{36}$$

i.e. $v(0, \sigma)$ must belong to a different eigenvalue of γ_4 than $u(0, \sigma)$. If one does not refer to the Dirac equation (32) but demands that the functions $u(0, \sigma)$ and $v(0, \sigma)$ be independent for the rest state, (36) will follow uniquely from (26). It is obvious that the functions u and v are orthogonal:

$$\bar{u}^a(\boldsymbol{p}, \sigma)\, v_a(\boldsymbol{p}, \sigma') = \bar{u}^a(0, \sigma)\, v_a(0, \sigma') = 0. \tag{37}$$

Expression (34) for $v(\boldsymbol{p}, \sigma)$ may now be written in the form

$$v(\boldsymbol{p}, \sigma) = \begin{pmatrix} \alpha(\boldsymbol{p})\, C\xi^*(\sigma) \\ -\alpha^{-1+}(\boldsymbol{p})\, C\xi^*(\sigma) \end{pmatrix} \sqrt{m}, \tag{38}$$

where $C = -i\sigma_2$, and the spin functions $\xi(\sigma)$ are the same as in (29). In (38) it is assumed that the function $v(\boldsymbol{p}, \sigma)$ is normalized according to

$$\left.\begin{aligned} \bar{v}(\boldsymbol{p}, \sigma)\, v(\boldsymbol{p}, \sigma') &= -2m\delta_{\sigma\sigma'}, \\ v^+(\boldsymbol{p}, \sigma)\, v(\boldsymbol{p}, \sigma') &= 2p_0\delta_{\sigma\sigma'}. \end{aligned}\right\} \tag{39}$$

The functions u and v are sometimes called the Dirac wave functions for states with positive and negative energy.

Transformation properties of wave functions

So far we have been considering the transformation rules of state vectors or field operators constructed linearly from creation and annihilation operators. Let us turn to the transformations of wave functions.

The operator $\mathrm{u}(\boldsymbol{p})$, which involves the wave function $u(\boldsymbol{p}, \sigma)$ [formula (33)] transforms via the general formula (10), or, in the case of a bispinor,

$$\mathrm{U}(0, A)\, \mathrm{u}(\boldsymbol{p})\, \mathrm{U}^{-1}(0, A) = S(A^{-1})\, \mathrm{u}(p'). \tag{40}$$

Let us define the transformed function $u'(\boldsymbol{p}, \sigma)$ by

$$\mathrm{U}(0, A)\, \mathrm{u}(\boldsymbol{p})\, \mathrm{U}^{-1}(0, A) = \sum_{\sigma'} u'(\boldsymbol{p}, \sigma')\, a(\boldsymbol{p}', \sigma') \tag{41}$$

and let us calculate the matrix element of both sides of formula (40) between the vacuum $|0\rangle$ and the single-particle state $|\boldsymbol{p}', \sigma; a\rangle = a^+(\boldsymbol{p}', \sigma)|0\rangle$. We have

$$u'_\beta(p, \sigma)\, 2p_0\delta(\boldsymbol{p}-\boldsymbol{p}'') = S_\beta^\gamma(A^{-1})\, u_\gamma(\boldsymbol{p}', \sigma)\, 2p'_0\delta(\boldsymbol{p}'-\boldsymbol{p}'').$$

Since $2p_0\delta(\boldsymbol{p}-\boldsymbol{p}')$ is invariant with respect to Lorentz transformations, we find the following expression for $u'_\beta(\boldsymbol{p}, \sigma)$:

$$u'_\beta(\boldsymbol{p}, \sigma) \equiv (\mathrm{U}(0, A)u)_\beta\, (\boldsymbol{p}, \sigma) = S_\beta^\gamma(A^{-1})\, u_\gamma(\boldsymbol{p}', \sigma). \tag{42}$$

This is the transformation rule for the wave function $u(\boldsymbol{p}, \sigma)$. Another convenient form of (42) is

$$u'_\beta(\boldsymbol{p}', \sigma) = S_\beta^{\ \gamma}(A)\, u_\gamma(\boldsymbol{p}, \sigma).$$

With this definition of u' the bilinear Hermitian forms

$$\bar{u}'(\boldsymbol{p}', \sigma)\, u'(\boldsymbol{p}', \sigma) = \bar{u}(\boldsymbol{p}, \sigma)\, u(\boldsymbol{p}, \sigma),$$
$$\bar{u}'(\boldsymbol{p}', \sigma)\, \gamma_5 u'(\boldsymbol{p}', \sigma) = u(\boldsymbol{p}, \sigma)\, \gamma_5 u(\boldsymbol{p}, \sigma).$$

will be invariant.

Let us consider the infinitesimal transformation $x^\mu \to x'^\mu = x^\mu + \omega^\mu{}_\nu x^\nu$ and find the generators $M_{\mu\nu}$ for Dirac wave functions from (42). This transformation corresponds to the matrix

$$A = 1 - i\tfrac{1}{2}\varepsilon_{ikl}\,\tfrac{1}{2}\sigma_l\omega^{ik} - \tfrac{1}{2}\sigma_i\omega^{0i}. \tag{43}$$

Calculating $S(A)$ by (19) and (27),

$$S(A) = 1 - i\tfrac{1}{4}\sigma^{\mu\nu}\omega_{\mu\nu}, \tag{44}$$

$$\sigma^{\mu\nu} = i\tfrac{1}{2}[\gamma^\mu, \gamma^\nu]. \tag{45}$$

In the case of the infinitesimal transformation (43), the left-hand side of (42) may be expressed in terms of the generators $M_{\mu\nu}$ as follows:

$$u'_\beta(\boldsymbol{p}, \sigma) = (1 + i\tfrac{1}{2}M_{\mu\nu}\omega^{\mu\nu})_\beta^\gamma\, u_\gamma(\boldsymbol{p}, \sigma).$$

Consequently, the generators $M_{\mu\nu}$ in the case of Dirac functions $u(\boldsymbol{p}, \sigma)$ have the form

$$M_{\mu\nu} = \frac{1}{2}\sigma_{\mu\nu} + \frac{1}{i}\left(p_\mu \frac{\partial}{\partial p^\nu} - p_\nu \frac{\partial}{\partial p^\mu}\right), \tag{46}$$

i.e. $\frac{1}{2}\sigma_{\mu\nu}$ may be called the spin part of $M_{\mu\nu}$.

The commutation relations between the matrices $\frac{1}{2}\sigma_{\mu\nu}$ and γ_λ are the same as between the generators $M_{\mu\nu}$ and P_λ:

$$\left.\begin{aligned}[\gamma_\lambda, \sigma_{\mu\nu}] &= 2i(g_{\lambda\mu}\gamma_\nu - g_{\lambda\nu}\gamma_\mu),\\ [\sigma_{\varrho\lambda}, \sigma_{\mu\nu}] &= 2i(g_{\varrho\nu}\sigma_{\lambda\mu} + g_{\lambda\mu}\sigma_{\varrho\nu} - g_{\varrho\mu}\sigma_{\lambda\nu} - g_{\lambda\nu}\sigma_{\varrho\mu}).\end{aligned}\right\} \tag{47}$$

Consequently, in a linear transformation of the matrices γ^μ

$$\gamma'^\mu = S(A)\,\gamma^\mu S^{-1}(A) = \Lambda^\mu{}_\nu(A)\gamma^\nu, \tag{48}$$

the coefficients coincide with those of the Lorentz transformation of a vector $\Lambda^\mu{}_\nu(A)$. This means that the matrix $\not{p}$ does not change if, along with a Lorentz transformation of momenta $p^\mu \to p'^\mu = \Lambda^\mu{}_\nu(A)p^\nu$, one also replaces the matrices γ^μ by γ'^μ according to (48):

$$\gamma'^\mu p'_\mu = \gamma^\mu p_\mu = \not{p}.$$

We may conclude from this that the Dirac equation is invariant with respect to transformations of the Poincaré group

$$\begin{aligned}(\gamma^\mu p'_\mu - m)\, u'(\boldsymbol{p}', \sigma) &= (\gamma^\mu p'_\mu - m)\, S^{-1}u(\boldsymbol{p}, \sigma)\\ &= S^{-1}S(\gamma'^\mu p_\mu - m)\, S^{-1}u(\boldsymbol{p}, \sigma) = S^{-1}(\gamma^\mu p_\mu - m)\, u(\boldsymbol{p}, \sigma) = 0.\end{aligned}$$

Since [by (48)] the index μ of the matrices γ^μ may be considered as vectorial, transformations of products of matrices $\gamma^\mu\gamma^\nu \ldots$ are easily defined if these matrices stand between Dirac wave functions. The quantities $\bar{u}u$, $\bar{u}\gamma_\mu u$, and $\bar{u}\sigma_{\mu\nu}u$ are respectively scalar, vector, and tensor, and, as we shall see below, the quantities $\bar{u}\gamma_5 u$, $\bar{u}\gamma_5\gamma_\mu u$ are pseudo-scalar and pseudo-vector.

Using the generators $M_{\mu\nu}$ one may write an expression for the covariant spin operator. According to (1.50) it is defined as

$$J_l = -\frac{1}{m} w_\mu n^\mu_{(l)}, \qquad w_\mu = \frac{1}{2}\varepsilon_{\mu\nu\lambda\sigma} M^{\nu\lambda} p^\sigma,$$

where

$$n^\mu_{(l)} n^\nu_{(k)} g_{\mu\nu} = -\delta_{lk}, \quad p_\mu n^\mu_{(k)} = 0.$$

Substituting (46) into the expression for J_l we find

$$J_l = -\frac{1}{2m}\varepsilon_{\lambda\mu\nu\varrho}\sigma^{\mu\nu} p^\varrho n^\lambda_{(l)} = -\frac{i}{2m}\gamma_5\sigma_{\mu\nu} n^\mu_{(l)} p^\nu. \tag{49}$$

In the rest state $\boldsymbol{p} = 0$, $n^0_{(k)} = 0$, and the spin projection J_l is

$$J_l = -i\tfrac{1}{2}\gamma_5\sigma_{k0} = -\tfrac{1}{2}\gamma_5\gamma_4\gamma_k n^k_{(l)} = \tfrac{1}{2}\boldsymbol{\Sigma}\cdot\boldsymbol{n}_{(l)}.$$

Here the matrix Σ_k is to be taken as the matrix $\sigma_k \times \mathbf{1}$, which in the representation (20) and (27) of the γ-matrices has the 2 by 2 matrix σ_k in both diagonal blocks.

Completeness of the system of Dirac wave functions and projection operators

The Dirac wave functions $u(\boldsymbol{p}, \sigma)$ and $v(\boldsymbol{p}, \sigma)$ form a complete set of four-dimensional functions for a given momentum $\boldsymbol{p}$. This follows from the fact that the four independent functions $u(0, \sigma)$ and $v(0, \sigma)$, $\sigma = \pm\frac{1}{2}$, are complete in the rest state ($\boldsymbol{p} = 0$). The functions $u(0, \sigma)$ and $v(0, \sigma)$ differ by the eigenvalues of the matrix γ_4 and may be distinguished by the projection operators

$$\Lambda_\pm(0) = \tfrac{1}{2}(1 \pm \gamma_4),$$
$$\Lambda_+(0)\, u(0, \sigma) = u(0, \sigma).$$

States with spin projections $\sigma = \pm\frac{1}{2}$ on the direction $\boldsymbol{n}$ may be distinguished by the projection operator

$$\Lambda(0, \pm\tfrac{1}{2}) = \tfrac{1}{2}(1 + \boldsymbol{\Sigma}\cdot\boldsymbol{n}) = \tfrac{1}{2}(1 \pm \gamma_5\gamma_4\boldsymbol{\gamma}\cdot\boldsymbol{n}).$$

Let us choose a representation of the γ-matrices in which γ_4 and $\boldsymbol{\Sigma}\cdot\boldsymbol{n}$ are diagonal:

$$\gamma_4 = \begin{pmatrix} 1 & 0 \\ 0 & -1 \end{pmatrix}, \quad \boldsymbol{\Sigma}\cdot\boldsymbol{n} = \begin{pmatrix} \sigma_3 & 0 \\ 0 & \sigma_3 \end{pmatrix}.$$

Then the functions u and v take the form

$$\left.\begin{aligned} u_\beta(0, \tfrac{1}{2}) &= \sqrt{2m}\,\delta_{\beta 1}, & v_\beta(0, \tfrac{1}{2}) &= \sqrt{2m}\,\delta_{\beta 3}, \\ u_\beta(0, -\tfrac{1}{2}) &= \sqrt{2m}\,\delta_{\beta 2}, & v_\beta(0, -\tfrac{1}{2}) &= \sqrt{2m}\,\delta_{\beta 4}. \end{aligned}\right\} \tag{50}$$

The normalization and orthogonality conditions (30), (37), and (39) hold automatically, while the projection operators $\Lambda_{\pm}(0, \sigma)$ are equal to

$$\Lambda_{+\beta}{}^{\gamma}\left(0, \frac{1}{2}\right) = \left\{\frac{1}{2}(1+\boldsymbol{\Sigma}\cdot\boldsymbol{n})\,\frac{1}{2}(1+\gamma_4)\right\}_{\beta}^{\gamma} = \frac{1}{2m}\,u_\beta\left(0, \frac{1}{2}\right)\bar{u}^{\gamma}\left(0, \frac{1}{2}\right),$$

$$\Lambda_{+\beta}{}^{\gamma}(0) = \left(\Lambda_{+}\left(0, \frac{1}{2}\right)+\Lambda_{+}\left(0, -\frac{1}{2}\right)\right)_{\beta}^{\gamma} = \frac{1}{2m}\sum_{\sigma} u_\beta(0, \sigma)\,\bar{u}^{\gamma}(0, \sigma),$$

$$\Lambda_{-\beta}{}^{\gamma}(0) = -\frac{1}{2m}\sum_{\sigma} v_\beta(0, \sigma)\,\bar{v}^{\gamma}(0, \sigma).$$

The completeness of the system of functions $u(0, \sigma)$ and $v(0, \sigma)$ follows directly from its explicit form. The unit 4 by 4 matrix is

$$\delta_\beta^\gamma = \frac{1}{2m}\sum_{\sigma}\{u_\beta(0, \sigma)\,\bar{u}^{\gamma}(0, \sigma)-v_\beta(0, \sigma)\,\bar{v}^{\gamma}(0, \sigma)\} = \Big[\Lambda_{+}(0)+\Lambda_{-}(0)\Big]_{\beta}^{\gamma}.$$

To pass to the case of arbitrary momentum, one may use relations (25) and (34) by which all expressions for $\boldsymbol{p} = 0$ are subjected to the transformation $S(\alpha)$:

$$\Lambda_{\pm}(\boldsymbol{p}) = S(\alpha(\boldsymbol{p}))\,\Lambda_{\pm}(0)S^{-1}(\alpha(\boldsymbol{p})) = \frac{m \pm \not{p}}{2m}. \tag{51}$$

The completeness condition now takes the form

$$\delta_\beta^\gamma = \frac{1}{2m}\sum_{\sigma}\{u_\beta(\boldsymbol{p}, \sigma)\,\bar{u}^{\gamma}(\boldsymbol{p}, \sigma)-v_\beta(\boldsymbol{p}, \sigma)\,\bar{v}^{\gamma}(\boldsymbol{p}, \sigma)\}. \tag{52}$$

Transforming the projection operator $\Lambda_{+}(0, \frac{1}{2})$ to a state with momentum $\boldsymbol{p}$, we find

$$\begin{aligned}\Lambda_{+}\left(\boldsymbol{p}, \frac{1}{2}\right) &= S(\alpha(\boldsymbol{p}))\,\Lambda_{+}\left(0, \frac{1}{2}\right)S^{-1}(\alpha(\boldsymbol{p}))\\ &= \frac{1}{4m}\left(1+\gamma_5\not{n}\,\frac{\not{p}}{m}\right)(\not{p}+m) = \frac{1}{4m}(1+\gamma_5\not{n})(\not{p}+m).\end{aligned} \tag{53}$$

Here the vector n^μ satisfies the same conditions as in (49). Since n and p are orthogonal, the matrices $\not{n}$ and $\not{p}$ commute with one another. Thus for $\boldsymbol{p} \neq 0$ each of the functions $u(p, \sigma)$ and $v(p, \sigma)$ is an eigenfunction of the operators $\not{p}/m$ and $\gamma_5\not{n}$, with eigenvalues ± 1.

The normalization conditions for the wave functions (30) and (39) are correlated with (24) for the field $\psi(x)$ and the commutation relations between creation and annihilation operators

$$\{a(\boldsymbol{p}, \sigma), a^{+}(\boldsymbol{p}', \sigma'\} = 2p_0\delta_{\sigma\sigma'}\delta(\boldsymbol{p}-\boldsymbol{p}').$$

In short, the equal-time commutator of the fields $\psi(x)$ and $\psi^{+}(y)$ may be written in the usual form:

$$\delta(x_0-y_0)\,\{\psi_\alpha(x), \psi^{+\beta}(y)\} = \delta_\alpha^\beta\delta^4(x-y). \tag{54}$$

When writing (54) one must use expressions (51) for the projection operators.

Properties of Dirac matrices

The matrices γ^μ and $\sigma^{\lambda\nu}$ are examples of 4 by 4 matrices acting on the Dirac functions u and v. Let us study the properties of such matrices. There are sixteen independent 4 by 4 matrices γ^R which may be chosen so that the quantities $\bar{u}\gamma^R u$ are real:

$$\gamma^R = \{1, \gamma^\mu, \sigma^{\mu\nu}, \sigma_5^\mu = \gamma^\mu\gamma_5, i\gamma_5\}. \tag{55}$$

The matrices γ^R are orthonormal in the following sense:

$$\left.\begin{aligned} \operatorname{Tr}(\gamma^R\gamma^{R'}) &= 0, \quad R \neq R', \\ \operatorname{Tr}(\gamma^\mu\gamma^\nu) &= -\operatorname{Tr}(\sigma_5^\mu\sigma_5^\nu) = 4g^{\mu\nu}, \\ \operatorname{Tr}(\sigma^{\mu\nu}\sigma^{\lambda\sigma}) &= 4(g^{\nu\sigma}g^{\mu\lambda}-g^{\nu\lambda}g^{\mu\sigma}), \\ \operatorname{Tr}\gamma_5^2 &= 4. \end{aligned}\right\} \tag{56}$$

Any 4 by 4 matrix O may be expanded in the complete set γ^R:

$$\mathrm{O} = \sum_{R=1}^{16} \mathrm{O}_R\gamma^R, \quad \mathrm{O}_R = \frac{\operatorname{Tr}(\mathrm{O}\gamma^R)}{\operatorname{Tr}\gamma_R^2}. \tag{57}$$

The Hermiticity condition for O has the form

$$\mathrm{O}^+ = \gamma_4\mathrm{O}\gamma_4, \tag{58}$$

which by virtue of (55) reduces to the reality of O_R.

The matrices γ^R in (55) are just the matrices γ^μ, and all of their products up to four factors, since

$$\gamma_5 = -i\gamma^1\gamma^2\gamma^3\gamma_4.$$

The matrices γ^R in the set (55) may be classified according to their transformation properties. The concrete form of the matrices γ^R depends on which of them are chosen to be diagonal.

It is convenient to have a set of fixed basis (Hermitian) matrices Γ_R. These matrices may be expressed as a direct product of two systems of Pauli matrices σ_k and ϱ_l, as well as the unit matrices σ_0 and ϱ_0:

$$\Gamma_{00} = \varrho_0\times\sigma_0, \quad \Gamma_{0k} = \varrho_0\times\sigma_k, \quad \Gamma_{l0} = \varrho_l\times\sigma_0, \quad \Gamma_{lk} = \varrho_l\times\sigma_k. \tag{59}$$

Of the sixteen matrices Γ_R, six are antisymmetric: $\Gamma_{02}, \Gamma_{20}, \Gamma_{12}, \Gamma_{21}, \Gamma_{23}, \Gamma_{32}$; the remaining matrices are symmetric.

A Lorentz transformation $S(A)$ decomposes into two blocks (19) in the representation in which

$$\gamma_5 = \varrho_3\times\sigma_0, \quad \gamma_4 = \varrho_1\times\sigma_0, \quad \gamma^k = -i\varrho_2\times\sigma_k. \tag{60}$$

It is just the matrices (60) that were used earlier [see (20) and (27)] in introducing the four-component field $\psi(x)$. The wave functions $u(\boldsymbol{p}, \sigma)$ and $v(\boldsymbol{p}, \sigma)$ are given by (29) and (38).

In the standard Dirac representation of the gamma matrices,

$$\gamma_5 = \varrho_1\times\sigma_0, \quad \gamma_4 = \varrho_3\times\sigma_0, \quad \gamma^k = -i\varrho_2\times\sigma_k. \tag{61}$$

Here the wave function has the form (31).

In the Majorana representation

$$\left.\begin{aligned} \gamma_5 &= -\varrho_2\times\sigma_3, \quad \gamma^1 = \varrho_3\times\sigma_0, \quad \gamma^2 = \varrho_2\times\sigma_2, \\ \gamma^3 &= -\varrho_1\times\sigma_0, \quad \gamma_4 = \varrho_2\times\sigma_1. \end{aligned}\right\} \tag{62}$$

Any two sets of 4 by 4 matrices γ^μ obeying the condition $\{\gamma^\mu, \gamma^\nu\} = 2g^{\mu\nu}$ are connected by a similarity relation

$$\gamma'^\mu = X\gamma^\mu X^{-1}, \tag{63}$$

where X is some 4 by 4 matrix. If, moreover, the condition $\gamma^0 = \gamma_4 = \gamma_4^+$, $\gamma^j = -\gamma^{j+}$ [conserved under transformations (63)] is imposed, then X will be a unitary matrix. We shall consider the conditions $\gamma_4 = \gamma_4^+$, $\gamma^j = -\gamma^{j+}$ always to hold.

Let us turn to symmetry properties of the matrices γ^R. The symmetry of the matrices γ^R depends on the representation since, like the Lorentz matrix $S_\beta^{\ \gamma}(A)$, they have mixed indices.

Let us introduce the special matrices $\mathcal{C}^{-1\beta\gamma}$ and $\mathcal{C}_{\beta\gamma}$, for raising or lowering spinor indices:

$$u^\beta = \mathcal{C}^{-1\beta\gamma}u_\gamma, \quad v_\beta = \mathcal{C}_{\beta\gamma}v^\gamma.$$

This notation implies that the bilinear combinations $\mathcal{C}^{-1\beta\gamma}u_\beta v_\gamma$ are invariant with respect to transformations $u'_\alpha = S_\alpha^{\ \beta}u_\beta$, $v'_\gamma = S_\gamma^{\ \varrho}v_\varrho$. This imposes on $\mathcal{C}$ the condition

$$\mathcal{C}^{-1}S = (\mathcal{C}S^{\mathrm{T}})^{-1},$$

which holds if

$$\mathcal{C}^{-1}\gamma_\mu\mathcal{C} = \pm\gamma_\mu^{\mathrm{T}}. \tag{64}$$

Let $S(A)$ be written in the quasi-diagonal form (19). Then there exist two operators

$$\mathcal{C}_1 = \begin{pmatrix} C & 0 \\ 0 & C \end{pmatrix}, \quad \mathcal{C}_2 = \gamma_5\mathcal{C}_1,$$

where C is the 2 by 2 matrix (3.5) of the spinor group, with which one may satisfy (64). The matrix C has the property that $A = C^{-1}A^{-1\mathrm{T}}C$.

Both operators $\mathcal{C}_1$ and $\mathcal{C}_2$ are antisymmetric and unitary:

$$\mathcal{C}^+\mathcal{C} = \mathcal{C}\mathcal{C}^+ = 1, \quad \mathcal{C}^{\mathrm{T}} = -\mathcal{C}. \tag{65}$$

This symmetry property of the matrices $\mathcal{C}$ is preserved when passing to another set of matrices γ^μ by means of the unitary transformation (63); here $\mathcal{C}$ transforms into

$$\mathcal{C}' = X\mathcal{C}X^{\mathrm{T}},$$

while the condition (64) holds for the new γ'^μ and $\mathcal{C}'$.

The matrix $\mathcal{C}_1$ does not change the eigenvalues of γ_4 and $\mathfrak{p}$ while the matrix $\mathcal{C}_2$ changes their sign. Consequently, $\mathcal{C}_2$ can connect the functions $u(\boldsymbol{p}, \sigma)$ and $v(\boldsymbol{p}, \sigma')$: $\mathcal{C}_2^{-1}\mathfrak{p}\mathcal{C}_2 = -\mathfrak{p}^{\mathrm{T}}$. Using operators $\mathrm{u}_\alpha(\boldsymbol{p})$ and $v_\beta(\boldsymbol{p})$ of the type (33) with identical Lorentz transformation properties let us construct the bilinear form $\mathcal{C}^{-1\alpha\beta}\mathrm{u}_\alpha(\boldsymbol{p})\,\mathrm{v}_\beta(\boldsymbol{p})$. This form will be invariant with respect to translations only for $\mathcal{C} = \mathcal{C}_2$. It then makes sense to choose $\mathcal{C}_2$ as the metric matrix $\mathcal{C}$. Thus

$$\mathcal{C} = \begin{pmatrix} C & 0 \\ 0 & -C \end{pmatrix} \quad \text{for} \quad \gamma_5 = \begin{pmatrix} 1 & 0 \\ 0 & -1 \end{pmatrix}. \tag{66}$$

One may now formulate the symmetry properties of the matrices γ^R. As a consequence of (64)–(66), the matrices

$$\gamma^\mu\mathcal{C}, \quad \sigma^{\mu\nu}\mathcal{C} \tag{66a}$$

are symmetric, while the matrices

$$\mathcal{C}, \quad \gamma_5\gamma_\mu\mathcal{C}, \quad \gamma_5\mathcal{C} \tag{66b}$$

are antisymmetric.

The connection between the functions u and v related to solutions with positive and negative energy has the form

$$\bar{u}^\beta = \mathcal{C}^{-1\beta\gamma}v_\gamma. \tag{67}$$

Relation (67) may be checked by resorting to the explicit form of u, v and the matrix $\mathcal{C}$.

Let us also give an expression for the matrix $\mathcal{C}$ in the case (61):

$$\mathcal{C} = i\varrho_1\times\sigma_2. \tag{68}$$

In the Majorana representation (62) the matrix $\mathcal{C}$ is

$$\mathcal{C} = i\varrho_2\times\sigma_1 = i\gamma_4 \tag{69}$$

We also list some commutators and anticommutators containing the matrices γ^R:

$$\left.\begin{aligned} [\gamma_\lambda, \sigma_{\mu\nu}]_+ &= -2\varepsilon_{\lambda\mu\nu\varrho}\sigma_5^\varrho, \\ [\sigma_{\varrho\lambda}, \sigma_{\mu\nu}]_+ &= 2(g_{\varrho\mu}g_{\lambda\nu}-g_{\lambda\mu}g_{\varrho\nu})-2i\varepsilon_{\varrho\lambda\mu\nu}\gamma_5, \\ [i\gamma_5, \sigma_{\mu\nu}]_+ &= i\varepsilon_{\mu\nu\varrho\lambda}\sigma^{\varrho\lambda}, \\ [\sigma_{\varrho5}, \sigma_{\mu\nu}]_- &= i(g_{\varrho\mu}\sigma_{\nu5}-g_{\varrho\nu}\sigma_{\mu5}). \end{aligned}\right\} \tag{70}$$

Together with relations (27) and (47), eqns. (70) give the rules for multiplying any two matrices γ^R.

Case m = 0 and Hermitian fields

In concluding this section let us dwell briefly on the description of particles of zero mass using the Hermitian bispinor field $\eta(x) = \eta^+(x)$.

Let $\psi(x)$ be a non-Hermitian field with $m = 0$. In the representation (27) of the matrices γ^μ with diagonal γ_5 this field, by (19), may be split up into the parts

$$\nu = \tfrac{1}{2}(1+\gamma_5)\psi, \quad \tilde{\nu} = \tfrac{1}{2}(1-\gamma_5)\psi, \tag{71}$$

which transform separately (and differently) under Lorentz transformations. The Dirac eqn. (32) thus takes the form

$$\begin{pmatrix} 0 & \partial_0+\boldsymbol{\sigma}\cdot\boldsymbol{\partial} \\ \partial_0-\boldsymbol{\sigma}\cdot\boldsymbol{\partial} & 0 \end{pmatrix}\begin{pmatrix} \nu(x) \\ \tilde{\nu}(x) \end{pmatrix} = 0$$

and consequently decomposes into two separate Weyl equations for the neutrino field $\nu(x)$ and the antineutrino field $\tilde{\nu}(x)$. These equations were obtained earlier in § 4.5.

Instead of the non-Hermitian fields ψ, $\bar{\psi}$ one sometimes considers the Hermitian fields

$$\eta_1 = \psi + \psi^+, \quad \eta_2 = \frac{1}{i}(\psi - \psi^+), \quad \bar{\eta} = \eta\gamma_4.$$

For the Hermiticity to be maintained under Lorentz transformation, one must have

$$\gamma_4 S^{-1}(A)\gamma_4 = S^{\mathrm{T}}(A), \quad \text{or} \quad (\gamma_4\sigma_{\mu\nu}\gamma_4)^{\mathrm{T}} = \sigma_{\mu\nu}. \tag{72}$$

This is equivalent to the condition that the γ_k be symmetric matrices, and γ_4 antisymmetric. Here γ_5 is also antisymmetric. The Majorana representation (62) is such a representation of γ-matrices. Since γ_5 is not a diagonal matrix in the Majorana representation, $S(A)$ does not decompose into 2 by 2 blocks, and relativistic invariance demands that the Hermitian field $\eta(x)$ have at least four components even in the case of particles with zero mass.

The operator $\not{p}$ or $\not{\partial}$ becomes real in this representation and, consequently, the Dirac equation indeed allows real functions or Hermitian fields as solutions.

In the case of particles of zero mass, one can make a relativistically invariant distinction between non-Hermitian fields describing, for example, the neutrino $\nu(x)$ and antineutrino $\tilde{\nu}(x)$ and a single overall Hermitian field of the type $\eta_1(x)$ or $\eta_2(x)$. In fact, by virtue of (71) the fields $\nu(x)$ and $\tilde{\nu}(x)$ satisfy the relativistically invariant conditions

$$\gamma_5\nu(x) = \nu(x), \quad \gamma_5\tilde{\nu}(x) = -\tilde{\nu}(x).$$

These conditions, of course, do not depend on the choice of representation of the matrices γ^μ. Thus if $\eta(x)$ $(=\eta_1(x))$ is a Hermitian neutrino field, the respective neutrino field $\nu(x)$ and the field of the antiparticle $\tilde{\nu}(x)$ may be obtained from $\eta(x)$ by their projections with respect to γ_5:

$$\left.\begin{aligned} \nu(x) &= \tfrac{1}{2}(1+\gamma_5)\,\eta(x), \\ \tilde{\nu}(x) &= \tfrac{1}{2}(1-\gamma_5)\,\eta(x) = \nu^+(x). \end{aligned}\right\} \tag{73}$$

One may also verify directly that $\tilde{\nu}(x) = \nu^+(x)$ in the Majorana representation. In this representation, the matrix γ_5 is antisymmetric and Hermitian. Consequently, if we consider the condition Hermitian conjugate to the condition $\gamma_5\nu(x) = \nu(x)$, we obtain $\gamma_5\nu^+(x) = -\nu^+(x)$, defining the antineutrino field $\tilde{\nu}(x)$.

§ 5.3. $2(2J+1)$-component functions for spin J

In § 3.3 spinor functions were constructed which transformed according to the $(J, 0)$ representation of the Lorentz group. To be able to construct bilinear Hermitian invariants, we must, according to § 5.2, double the space of functions, including those quantities which transform via the representation $(0, J)$. In the previous section, the steps in proceeding from a spinor field to a bispinor field were followed using the example of the Dirac equation and its solutions. $2(2J+1)$-component fields are introduced in an analogous manner.[54, 55]

Let us rewrite formula (4.94) for the transformation of the spinor field $\varphi(x)$ of particles with spin J:

$$\mathrm{U}(a, A)\,\varphi(x)\,\mathrm{U}^{-1}(a, A) = D^J(A^{-1})\,\varphi(\Lambda(A)x + a), \tag{74}$$

where D^J is the $(J, 0)$ representation of the spinor group (see § 3.3). Along with $\varphi(x)$ we shall consider also the field $\chi(x)$ transforming via the representation $(0, J)$:

$$\mathrm{U}(a, A)\,\chi(x)\,\mathrm{U}^{-1}(a, A) = D^{J+}(A)\,\chi(\Lambda(A)x+a), \tag{75}$$

where D^J is the same matrix as in (74).

Let us now form the $2(2J+1)$-component field

$$\Psi(x) = \begin{pmatrix}\varphi(x)\\ \chi(x)\end{pmatrix}, \tag{76}$$

which by (74) and (75) will transform as

$$\left.\begin{aligned} \mathrm{U}(a, A)\,\Psi(x)\,\mathrm{U}^{-1}(a, A) &= S(A^{-1})\,\Psi(\Lambda x+a),\\ S(A) &= \begin{pmatrix}D^J(A) & 0\\ 0 & D^{J+}(A^{-1})\end{pmatrix}. \end{aligned}\right\} \tag{77}$$

If we now introduce the matrices

$$\gamma_4 = \begin{pmatrix}0 & 1\\ 1 & 0\end{pmatrix}, \quad \gamma_5 = \begin{pmatrix}1 & 0\\ 0 & -1\end{pmatrix}, \tag{78}$$

where each number denotes a $(2J+1)$-row matrix, then

$$\gamma_4 S^{+}(A^{-1})\gamma_4 = S(A). \tag{79}$$

The matrices S^{-1+} and S are equivalent. Thus, as in the case of the Dirac field, one may define the conjugate field $\overline{\Psi} = \Psi^{+}\gamma_4$, so that

$$\mathrm{U}(0, A)\,\overline{\Psi}(x)\,\Psi(x)\,\mathrm{U}^{-1}(0, A) = \overline{\Psi}(\Lambda x)\,\Psi(\Lambda x).$$

An explicit expression for the field $\Psi(x)$ may be written at once if one notes that passing from spin $\frac{1}{2}$ to arbitrary J entails replacing the Wigner operators $\alpha(\boldsymbol{p})$ by $D^J(\alpha)$ and $\alpha(\boldsymbol{p})C^{-1}$ by $D^J(\alpha(\boldsymbol{p})C^{-1})$. Then we may write $\chi_\sigma(x)$ in the form

$$\begin{aligned}\chi_\sigma(x) = \frac{1}{(2\pi)^{3/2}}\int\frac{d^3p}{2p_0}\sum_{\sigma'}\{&D^J_{\sigma\sigma'}(\alpha^{-1+}(\boldsymbol{p}))\,a(\boldsymbol{p}, \sigma')e^{-ipx}\\ &+(-1)^{2J}\eta D^J_{\sigma\sigma'}(\alpha^{-1+}(\boldsymbol{p})C^{-1})\,b^{+}(\boldsymbol{p}, \sigma')e^{ipx}\},\end{aligned} \tag{80}$$

at the same time writing the former expression (4.93) for $\varphi(x)$:

$$\varphi_\sigma(x) = \frac{1}{(2\pi)^{3/2}}\int\frac{d^3p}{2p_0}\sum_{\sigma'}\{D^J_{\sigma\sigma'}(\alpha(\boldsymbol{p}))\,a(\boldsymbol{p}, \sigma')e^{-ipx}+D^J_{\sigma\sigma'}(\alpha(\boldsymbol{p})C^{-1})\,b^{+}(\boldsymbol{p}, \sigma')e^{ipx}\}. \tag{81}$$

Expression (80) is defined up to a phase factor η by the condition that the commutator $[\chi(x), \chi^{+}(y)]_{\pm}$ vanishes for spacelike intervals. If one imposes the further demand $[\chi(x), \varphi^{+}(y)]_{\pm} = 0$ in the space-like region $(x-y)^2 < 0$, one obtains

$$[\chi(x), \varphi^{+}(y)]_{\pm} = \frac{1}{(2\pi)^3}\int\frac{d^3p}{2p_0}\{e^{-ip(x-y)}\pm(-1)^{2J}\eta e^{ip(x-y)}\},$$

or $\eta = 1$.

The field $\Psi(x)$ obeys the wave equation

$$(\Box - m^2)\,\Psi(x) = 0. \tag{82}$$

To obtain the equation of motion which takes the place of the Dirac equation in this case, let us repeat the derivation of the latter, replacing $\alpha(\boldsymbol{p})$ by $D^J(\alpha)$. The wave functions of the positive-energy states

$$u(\boldsymbol{p}) = \begin{pmatrix} D^J(\alpha(\boldsymbol{p})) & \varphi^J(\sigma) \\ D^J(\alpha^{-1+}(\boldsymbol{p})) & \varphi^J(\sigma) \end{pmatrix} \tag{82a}$$

satisfy the condition $(\gamma_4 - 1)\,u(\boldsymbol{p}) = 0$ when $\boldsymbol{p} = 0$. Here $\varphi(\sigma)$ is a $(2J+1)$-component spin function. Since $u(\boldsymbol{p}) = S(\alpha)\,u(0)$, in analogy with (22) we find $S(\alpha)\,(\gamma_4 - 1)\,S^{-1}(\alpha)\,u(\boldsymbol{p}) = 0$. This means that $u(\boldsymbol{p})$ is the solution of the equation

$$\{\tfrac{1}{2}(1+\gamma_5)\,\gamma_4 D^J(p^0\sigma_0 + \boldsymbol{\sigma}\cdot\boldsymbol{p}) + \tfrac{1}{2}(1-\gamma_5)\,\gamma_4 D^J(p^0\sigma_0 - \boldsymbol{\sigma}\cdot\boldsymbol{p}) - m^{2J}\}\,u(\boldsymbol{p}) = 0, \tag{83}$$

where we have used $m\alpha\alpha^+ = p^0\sigma_0 + \boldsymbol{\sigma}\cdot\boldsymbol{p}$, and $D^J(m) = m^{2J}$.

The wave functions of the negative energy states

$$v(\boldsymbol{p}) = \begin{pmatrix} D^J(\alpha(\boldsymbol{p})C^{-1}) & \varphi^{J*}(\sigma) \\ D^J(\alpha^{-1+}(\boldsymbol{p})C^{-1})\,(-1)^{2J} & \varphi^{J*}(\sigma) \end{pmatrix}$$

satisfy the condition $(\gamma_4 - (-1)^{2J})\,v(0) = 0$. From this it follows that $v(\boldsymbol{p})$ is the solution of the equation

$$\{\tfrac{1}{2}(1+\gamma_5)\,\gamma_4 D^J(p^0\sigma_0 + \boldsymbol{\sigma}\cdot\boldsymbol{p}) + \tfrac{1}{2}(1-\gamma_5)\,\gamma_4 D^J(p^0\sigma_0 - \boldsymbol{\sigma}\cdot\boldsymbol{p}) - (-1)^{2J}\,m^{2J}\}\,v(\boldsymbol{p}) = 0. \tag{84}$$

Both eqns. (83) and (84) correspond to the same equation in coordinate space for the field $\Psi(x)$:

$$\begin{pmatrix} -m^{2J} & D^J(i\partial) \\ D^J(i\tilde{\partial}) & -m^{2J} \end{pmatrix} \begin{pmatrix} \varphi \\ \chi \end{pmatrix} = 0, \quad [\partial = \sigma_0\,\partial^0 + \boldsymbol{\sigma}\cdot\boldsymbol{\partial}], \tag{85}$$

since the factor $(-1)^{2J}$ is absorbed when replacing $D^J(i\partial)$ by $D^J(-i\partial)$ by virtue of the fact that $D^J(-1) = (-1)^{2J}$.

Thus the Dirac equation is the only linear equation of the type (85) containing derivatives in first order. In the general case, (85) contains $(i\partial)^{2J}$. Moreover, only in the case of spin $\frac{1}{2}$ is the wave equation a consequence of the Dirac equation. For $J > \frac{1}{2}$ one cannot obtain the wave equation (82) from (85), and the former must be postulated in addition.

§ 5.4. Particles with spin $J = 1$

As the next example let us describe particles with spin 1 using the vector representation $(\frac{1}{2}, \frac{1}{2})$ of the Lorentz group. The wave function of a vector particle is a four-vector which we shall denote by $\varepsilon_\mu(\boldsymbol{p})$, i.e. the index r in (8) has the sense of a vector index μ, while $L_{rr'}$ is the usual Lorentz transformation $\Lambda_\mu^{\ \nu'}$. The transformation function from the basis of the little group for $\boldsymbol{p} = 0$ to the four-vector basis is $\varepsilon_\mu(0, \sigma)$:

$$\left.\begin{aligned} c_{r\sigma}(\tfrac{1}{2}, \tfrac{1}{2}, J) &= \varepsilon_\mu(0, \sigma), \\ \Lambda_\mu^{\ \nu}(\alpha)\,\varepsilon_\nu(0, \sigma) &= \varepsilon_\mu(\boldsymbol{p}, \sigma), \end{aligned}\right\} \tag{86}$$

where, as earlier, $\Lambda_\mu^{\ \nu}(\alpha)$ is a Lorentz transformation from the rest system to the system moving with velocity $\boldsymbol{v} = \boldsymbol{p}/p_0$.

The quantities $\varepsilon_\mu(0, \sigma)$ forming the basis of the vector representation of the three-dimensional rotation group are easily found from the relation between spinors and vectors of the rotation group [see (3.49) and (3.62)]:

$$\left.\begin{aligned} \varepsilon_\mu(0, 1) &= \frac{1}{\sqrt{2}}(0, 1, i, 0), \\ \varepsilon_\mu(0, 0) &= (0, 0, 0, -1), \\ \varepsilon_\mu(0, -1) &= \frac{1}{\sqrt{2}}(0, -1, i, 0). \end{aligned}\right\} \tag{87}$$

It is obvious that the zero-th component ε_0 of all spin projections is equal to 0, or $p^0\varepsilon_0(0, \sigma) = 0$, in the rest system. The covariant form of this equation may be written as

$$p^\mu \varepsilon_\mu(\boldsymbol{p}, \sigma) = 0. \tag{88}$$

Here it is assumed that the momenta p_μ obey $p_\mu p^\mu = m^2$.

The normalization condition for ε_μ may be obtained from (86) and (87), since it does not depend on the particle momentum

$$g^{\mu\nu}\varepsilon_\mu^+(\boldsymbol{p}, \sigma)\, \varepsilon_\nu(\boldsymbol{p}, \sigma') = -\delta_{\sigma\sigma'}. \tag{89}$$

The explicit expression for $\varepsilon_\mu(\boldsymbol{p})$ as a function of momentum, by (86), has the form

$$\varepsilon_\mu(\boldsymbol{p}) = \varepsilon_\mu(0) - \frac{p_\mu + g_{0\mu}m}{p_0 + m}\,\frac{\varepsilon(0)p}{m}. \tag{90}$$

We give the expression for projection operator on the state with spin 1:

$$X_{\mu\nu} = -\sum_\sigma \varepsilon_\mu(\boldsymbol{p}, \sigma)\, \varepsilon_\nu^+(\boldsymbol{p}, \sigma) = g_{\mu\nu} - \frac{p_\mu p_\nu}{p^2}. \tag{91}$$

The tensor $X_{\mu\nu}$ has the property

$$X_{\mu\nu}X^\nu_{\ \sigma} = X_{\mu\sigma}, \tag{92}$$

characterizing it as a projection operator; moreover, of course,

$$p^\mu X_{\mu\nu} = p^\nu X_{\mu\nu} = 0. \tag{93}$$

The functions $\varepsilon_\mu(\boldsymbol{p}, \sigma)$ transform according to

$$\varepsilon'_\mu(\boldsymbol{p}) = \Lambda_\mu^{\ \nu}\varepsilon_\nu(\Lambda^{-1}\boldsymbol{p}), \tag{94}$$

under Lorentz transformations, from which one may find the angular momentum operator

$$(M_{\varrho\sigma})_\mu^{\ \nu} = -i\left(p_\varrho \frac{\partial}{\partial p_\sigma} - p_\sigma \frac{\partial}{\partial p_\varrho}\right)\delta_\mu^\nu - i(g_{\mu\varrho}\delta_\sigma^\nu - g_{\mu\sigma}\delta_\varrho^\nu). \tag{95}$$

The second term in (95) gives the spin part of the angular momentum; in the rest system, the spin angular momentum is equal to the spatial part of this quantity:

$$(J_k)_{il} = -ie_{kil}. \tag{96}$$

In the general case, the spin projection on the direction n_μ, $n_\mu p^\mu = 0$, $n_\mu n^\mu = -1$, is equal to $J_n = (1/n)w_\mu n^\mu$; its matrix elements are equal to

$$(J_n)^{\sigma\varrho} = i\tfrac{1}{2}\varepsilon^{\lambda\nu\sigma\varrho} n_\lambda p_\nu. \tag{97}$$

Thus a particle with spin 1 may be described both within the framework of a formalism with three-component wave functions (§ 4.3) and using the four-vector function ε_μ. In the latter case, the function satisfies the supplementary condition (88) guaranteeing the absence of particles with spin 0.

Introducing creation and annihilation operators for particles $a^+(\boldsymbol{p}, \sigma)$, $a(p, \sigma)$ and antiparticles and bearing in mind that the wave function for antiparticles is also $\varepsilon_\mu(\boldsymbol{p})$, one may write the field operator in the usual way:

$$B_\mu(x) = \frac{1}{(2\pi)^{3/2}} \sum_\sigma \int \frac{d^3p}{2p_0} \{\varepsilon_\mu(\boldsymbol{p}, \sigma)\, a(\boldsymbol{p}, \sigma) e^{-ipx} + \varepsilon_\mu^*(\boldsymbol{p}, \sigma)\, b^+(\boldsymbol{p}, \sigma) e^{ipx}\}, \tag{98}$$

where the operators a, a^+ and b, b^+ obey the standard commutation relations

$$[a(\boldsymbol{p}, \sigma), a^+(\boldsymbol{p}', \sigma')] = [b(\boldsymbol{p}, \sigma), b^+(\boldsymbol{p}', \sigma')] = 2p_0 \delta(\boldsymbol{p}-\boldsymbol{p}')\delta_{\sigma\sigma'}. \tag{99}$$

In place of the operators $a(\boldsymbol{p}, \sigma)$ one sometimes uses their linear combinations

$$\left.\begin{aligned} a_1 &= \frac{1}{\sqrt{2}}(a(-1)-a(1)), & a_2 &= \frac{1}{i\sqrt{2}}(a(1)+a(-1)), \\ a_3 &= a(0), & a(\boldsymbol{p}, \sigma) &\equiv a(\sigma). \end{aligned}\right\} \tag{100}$$

The expression for the field then takes the form

$$B_\mu(x) = \frac{1}{(2\pi)^{3/2}} \sum_{\lambda=1}^{3} \int \frac{d^3p}{2p_0} \varepsilon_\mu(\boldsymbol{p}, \lambda)\, \{a_\lambda(\boldsymbol{p}) e^{-ipx} + b_\lambda^+(\boldsymbol{p}) e^{ipx}\}, \tag{101}$$

where the vectors $\varepsilon_\mu(0, \lambda)$ have the components $\varepsilon_\mu(0, \lambda) = -\delta_{\mu\lambda} (\lambda = 1, 2, 3)$.

Under the transformation $\bar{g} = (\mathrm{a}, A)$ of the Poincaré group, the field $B_\mu(x)$ transforms according to

$$\mathrm{U}(\mathrm{a}, A)\, B_\mu(x)\, \mathrm{U}^{-1}(\mathrm{a}, A) = \Lambda_\mu^{\ \nu}(A^{-1})\, B_\nu(\Lambda x + a), \tag{102}$$

which is easily checked using the formulae written earlier.

§ 5.5. Rarita–Schwinger wave functions

In the Rarita–Schwinger formalism the wave functions of particles with half-integral spin $J = j + \frac{1}{2}$ are described using functions transforming as a product of a Dirac spinor and a tensor of rank j. In the case of integral spin J it is assumed that the wave functions have the transformation properties of a tensor of rank J.[68]

Spin $\frac{3}{2}$

Let us first discuss the wave functions of particles with spin $\frac{3}{2}$. The wave function $\psi_{\alpha\mu}(\boldsymbol{p}, \sigma)$ has one Dirac index α and one four-vector index μ. Consequently, in the transformation matrix $c_{r\sigma}(L, J)$ the index L denotes the direct product of the representations $L = \big((\frac{1}{2}, 0)\times(0, \frac{1}{2})\big)\times(\frac{1}{2}, \frac{1}{2})$, while $r = (\alpha, \mu)$. Let us use formula (8). In constructing $\psi_{\alpha\mu}(0, \sigma) = c_{\alpha\mu,\sigma}(L, \frac{3}{2})$ we shall use the wave function of a particle with spin 1, $\varepsilon_\mu(0, \gamma)$, and a Dirac spinor in the rest system, $u_\alpha(0, \lambda)$. By the rule for addition of angular momenta

$$c_{\alpha\mu,\sigma}(L, \tfrac{3}{2}) = \sum_{\gamma,\lambda} \langle \tfrac{1}{2} 1 \lambda\gamma \,|\, \tfrac{1}{2} 1 \tfrac{3}{2}\sigma\rangle\, \varepsilon_\mu(0, \gamma)\, u_\alpha(0, \lambda) \equiv \psi_{\alpha\mu}(0, \sigma). \tag{103}$$

Thus, performing a Lorentz transformation from the momentum $\mathring{p}^\mu = (m, 0, 0, 0)$ to the momentum p^μ, we find

$$\psi_{\alpha\mu}(\boldsymbol{p}, \sigma) = \Lambda_\mu^{\ \nu}(\alpha)\, S_\alpha^{\ \beta}(\alpha)\, \psi_{\beta\nu}(0, \sigma) = \sum_{\gamma,\lambda} \langle \tfrac{1}{2} 1 \lambda\gamma \,|\, \tfrac{1}{2} 1 \tfrac{3}{2}\sigma\rangle\, \varepsilon_\mu(\boldsymbol{p}, \gamma)\, u_\alpha(\boldsymbol{p}, \lambda). \tag{104}$$

The form of the matrix $\alpha(\boldsymbol{p})$ depends on whether one is using the canonical or helicity basis (see § 4.3).

By construction $\psi_{\alpha\mu}$ does not contain spin $J = \frac{1}{2}$. This means that it is impossible to form a spinor from $\psi_{\alpha\mu}$ by multiplying by the vectors γ^μ and p^μ. It is easily checked directly that $\gamma^\mu\psi_\mu(0) = 0$. Passing to the function $\psi_\mu(\boldsymbol{p}, \sigma)$ using (104) and keeping in mind that by (35) γ^μ transforms as a vector, we find

$$\gamma^\mu \psi_\mu(\boldsymbol{p}, \sigma) = 0. \tag{105}$$

From formula (104) it is clear that by virtue of $p^\mu\varepsilon_\mu(\boldsymbol{p}, \gamma) = 0$ one has

$$p^\mu \psi_{\alpha\mu}(\boldsymbol{p}, \sigma) = 0. \tag{106}$$

Since the Dirac spinor $u(\boldsymbol{p})$ satisfies the equation $(\not{p} - m)\, u(\boldsymbol{p}) = 0$, $\psi_{\alpha\mu}$ is also a solution of the equation

$$(\not{p} - m)_\alpha^{\ \beta}\, \psi_{\beta\mu}(\boldsymbol{p}, \sigma) = 0. \tag{107}$$

Equation (106) is not independent; it follows from (105) and (107).

Let us define the conjugate function

$$\bar{\psi}_\mu(\boldsymbol{p}, \sigma) = \psi_\mu^\dagger(\boldsymbol{p}, \sigma)\gamma_4. \tag{108}$$

Then the normalization condition may be written in the form

$$\bar{\psi}_\mu^\alpha(\boldsymbol{p}, \sigma)\, \psi_\alpha^\mu(\boldsymbol{p}, \sigma') = -2m\delta_{\sigma\sigma'}; \tag{109}$$

it holds because of the properties of Clebsch–Gordan coefficients.

The projection operator on the state with spin $\frac{3}{2}$ is equal to

$$X_{\mu\nu} = \sum_\sigma \psi_\mu(\boldsymbol{p}, \sigma)\, \bar{\psi}_\nu(\boldsymbol{p}, \sigma)\, (2m)^{-1} = -\frac{\not{p}+m}{2m}\left(g_{\mu\nu} - \frac{1}{3}\gamma_\mu\gamma_\nu - \frac{2p_\mu p_\nu}{3m^2} + \frac{p_\mu\gamma_\nu - p_\nu\gamma_\mu}{3m}\right); \tag{110}$$

here $X_{\mu\nu}X^\nu{}_\lambda = -X_{\mu\lambda}$.

Thus the Rarita–Schwinger wave function for spin $\frac{3}{2}$ may be defined as a solution of the Dirac eqn. (107) with the subsidiary condition (105).

Spin 2

The wave function for a particle with spin 2 is contructed in the same way. The wave function in this case may be described by a tensor of second rank $\Phi_{\mu\nu}(\boldsymbol{p}, \gamma)$, where γ is the spin projection J_3 or the helicity. The wave function in the rest state $\Phi_{\mu\nu}(0, \gamma) = c_{\mu\nu,\gamma}(L, 2)$ is constructed from the spin-1 wave functions $\varepsilon_\mu(0, \lambda_1)$ and $\varepsilon_\nu(0, \lambda_2)$.

Coupling the two spin angular momenta to spin 2,

$$\Phi_{\mu\nu}(0, \nu) = \sum_{\lambda_1, \lambda_2} \langle 11\lambda_1\lambda_2 | 112\gamma\rangle \, \varepsilon_\mu(0, \lambda_1)\, \varepsilon_\nu(0, \lambda_2). \tag{111}$$

By construction this function is symmetric: $\Phi_{\mu\nu}(0, \gamma) = \Phi_{\nu\mu}(0, \gamma)$.

Passing to a moving reference frame, we find

$$\Phi_{\mu\nu}(\boldsymbol{p}, \gamma) = \Lambda_\mu^{\ \mu'}(\alpha)\, \Lambda_\nu^{\ \nu'}(\alpha) \Phi_{\mu'\nu'}(0, \gamma) = \sum_{\lambda_1, \lambda_2} \langle 11\lambda_1\lambda_2 | 112\gamma\rangle \, \varepsilon_\mu(\boldsymbol{p}, \lambda_1)\, \varepsilon_\nu(\boldsymbol{p}, \lambda_2). \tag{112}$$

The wave function does not contain lower spins ($J = 0$ and $J = 1$) if all vectors and scalars constructed from $\Phi_{\mu\nu}$, p_λ, and $g_{\mu\nu}$ vanish identically. Indeed, by construction,

$$p^\mu \Phi_{\mu\nu} = 0 \tag{113}$$

and (as is easily checked for $\boldsymbol{p} = 0$)

$$g^{\mu\nu} \Phi_{\mu\nu} = 0. \tag{114}$$

The equation of motion in this case is the wave equation

$$(p^2 - m^2) \Phi_{\mu\nu}(\boldsymbol{p}, \sigma) = 0, \tag{115}$$

since in (112) the momentum lies on the mass shell.

The normalization condition has the form

$$\Phi^+_{\mu\nu}(\boldsymbol{p}, \gamma) \Phi^{\mu\nu}(\boldsymbol{p}, \gamma') = \delta_{\gamma\gamma'}, \tag{116}$$

which follows from (112) and properties of Clebsch–Gordan coefficients.

The projection operator on the state with spin 2 is equal to

$$X_{\mu\nu,\lambda\varrho} = \sum_\gamma \Phi_{\mu\nu}(\boldsymbol{p}, \gamma) \Phi^+_{\lambda\varrho}(\boldsymbol{p}, \gamma) = \tfrac{1}{2} X_{\mu\lambda} X_{\nu\varrho} + \tfrac{1}{2} X_{\mu\varrho} X_{\nu\lambda} - \tfrac{1}{3} X_{\mu\nu} X_{\lambda\varrho}, \tag{117}$$

where $X_{\mu\nu}$ is the projection operator (93) on the state with spin 1.

Half-integral spins

The generalization of the method just presented for constructing Rarita–Schwinger wave functions in the case of arbitrary spin presents no problems. A state with half-integral spin $J = j \pm \frac{1}{2}$ is described by the function $\psi_{\alpha\mu_1 \ldots \mu_j}(\boldsymbol{p}, \gamma)$ $(-J \leqslant \gamma \leqslant J)$, where γ is the spin projection J_3 or the helicity. The function $\psi_{\alpha\mu_1 \ldots \mu_j}$ may be constructed inductively from $\psi_{\alpha\mu_1 \ldots \mu_{j-1}}$ and the wave function ε_{μ_j} for spin 1:

$$\psi_{\alpha\mu_1 \ldots \mu_j}(\boldsymbol{p}, \gamma) = \sum_{\lambda, \lambda'} \langle 1j-1\lambda\lambda' | 1j-1j\gamma\rangle \, \varepsilon_{\mu_j}(\boldsymbol{p}, \lambda')\, \psi_{\alpha\mu_1 \ldots \mu_{j-1}}(\boldsymbol{p}, \lambda). \tag{118}$$

The simplest function $\psi_{\alpha\mu}$, describing spin $\frac{3}{2}$, is already known to us and because of (118) is fully determined. For $\boldsymbol{p} = 0$ the expression (118) reduces to the $(2J+1)$-component basis of the rotation group with angular momentum J.

By construction $\psi_{\mu_1 \dots \mu_j}$ is symmetric with respect to $\mu_1 \dots \mu_j$, so that the condition

$$\gamma^{\mu_1}\psi_{\mu_1 \dots \mu_j} = 0 \tag{119}$$

holds if it holds for spin $J = \frac{3}{2}$. The symmetry of $\psi_{\mu_1 \dots \mu_j}$ also may be checked directly, if one uses the explicit expression for the coefficients $\langle 1j-1\lambda\lambda' \mid 1j-1j\gamma\rangle$ for adding the angular momenta $j-1$ and 1 to the maximum angular momentum j. Then, expression (118) for the wave function takes the form

$$\begin{aligned}\psi_{\alpha\mu_1 \dots \mu_j}(\boldsymbol{p}, \gamma) = &\left[\frac{2^j(j+\frac{1}{2}+\gamma)!\,(j+\frac{1}{2}-\gamma)!}{(2j+1)!}\right]^{1/2} \\ &\times \sum_{\lambda_1 \dots \lambda_j} [(1+\lambda_1)!\,(1-\lambda_1)! \dots (1+\lambda_j)!\,(1-\lambda_j)!]^{-1/2} \\ &\times \varepsilon_{\mu_1}(\boldsymbol{p}, \lambda_1) \dots \varepsilon_{\mu_j}(\boldsymbol{p}, \lambda_j)\, u_\alpha(\boldsymbol{p}, \gamma-\lambda_1- \dots -\lambda_j).\end{aligned} \tag{120}$$

The wave function (118) or (120) satisfies

$$(\not{p}-m)\,\psi_{\mu_1 \dots \mu_j}(\boldsymbol{p}) = 0. \tag{121}$$

As a consequence of (119) and (121) one has the relations

$$p^\mu \psi_{\mu\mu_2 \dots \mu_j}(\boldsymbol{p}) = 0, \tag{122}$$

$$g^{\mu\nu}\psi_{\mu\nu \dots \mu_j}(\boldsymbol{p}) = 0. \tag{123}$$

If the conjugate function is defined in the usual way: $\bar{\psi} = \psi^+\gamma_4$, the normalization condition reads:

$$\bar{\psi}^\alpha{}_{\mu_1 \dots \mu_j}(\boldsymbol{p}, \gamma)\,\psi_\alpha{}^{\mu_1 \dots \mu_j}(\boldsymbol{p}, \gamma') = (-1)^{J-\frac{1}{2}}\,\delta_{\gamma\gamma'}. \tag{124}$$

It follows from (118) and the properties of $\varepsilon_\mu(\boldsymbol{p})$.

Integral spins

In the case of integral spins J we choose as a basis the tensor of Jth rank $\Phi_{\mu_1 \dots \mu_J}(\boldsymbol{p}, \gamma)$ which is constructed inductively as in (118):

$$\Phi_{\mu_1 \dots \mu_J}(\boldsymbol{p}, \gamma) = \sum_{\lambda, \lambda'} \langle 1J\lambda\lambda' \mid 1J-1J\gamma\rangle\, \varepsilon_{\mu_J}(\boldsymbol{p}, \lambda')\Phi_{\mu_1 \dots \mu_{J-1}}(\boldsymbol{p}, \lambda). \tag{125}$$

In the rest state ($\boldsymbol{p} = 0$), $\Phi_{\mu_1 \dots \mu_J}$ reduces to the $(2J+1)$-component basis for the angular momentum J. As above, $\Phi_{\mu_1 \dots \mu_J}(\boldsymbol{p})$ may be obtained from $\Phi_{\mu_1 \dots \mu_J}(0)$ by a Lorentz transformation given by the operator α, whose form depends on the meaning of the quantum number γ (spin projection or helicity):

$$\Phi_{\mu_1 \dots \mu_J}(\boldsymbol{p}, \gamma) = \Lambda_{\mu_1}{}^{\mu'_1}(\alpha) \dots \Lambda_{\mu_J}{}^{\mu'_J}(\alpha)\Phi_{\mu'_1 \dots \mu'_J}(0, \gamma). \tag{126}$$

$\Phi_{\mu_1 \dots \mu_J}$ is symmetric in the indices $\mu_1 \dots \mu_J$, since the angular momenta $j-1$ and 1 are coupled up to the maximum angular momentum $j\,(j = 2, 3, \ldots, J)$. For this reason, the condition

$$g^{\mu\nu}\Phi_{\mu\nu\,\ldots\,\mu_J} = 0 \tag{127}$$

holds in general if it holds for $J = 2$.

Since each vector index originates with the wave function of a particle with spin 1, it is always true that

$$p^{\mu}\Phi_{\mu\mu_2\,\ldots\,\mu_J}(\boldsymbol{p}) = 0. \tag{128}$$

Equations (127) and (128) are a consequence of the fact that $\Phi_{\mu_1\ldots\mu_J}$ does not contain lower spins $J-1, J-2, \ldots, 1, 0$. The wave functions (126) are normalized according to

$$\Phi^{+}_{\mu_1\,\ldots\,\mu_J}(\boldsymbol{p}, \gamma)\Phi^{\mu_1\,\ldots\,\mu_J}(\boldsymbol{p}, \gamma') = (-1)^J\,\delta_{\gamma\gamma'}. \tag{129}$$

Both for integral and half-integral spins, the projection operator on a state with spin J may be written in general as

$$X^{J}_{(\mu)\,(\nu)} = \sum_{\gamma=-J}^{J} \psi_{(\mu)}(\boldsymbol{p}, \gamma)\,\bar{\psi}_{(\nu)}(\boldsymbol{p}, \gamma), \tag{130}$$

where $\bar{\psi} = \psi^{+}\gamma_4$ for half-integral spins and $\bar{\psi} = \psi^{+}$ for integral spins. It is obvious that as a consequence of the normalizations (124) and (129) we have

$$(X^J)^2 = \begin{cases} (-1)^J X^J & \text{for integral spins} \\ (-1)^{J-\frac{1}{2}} X^J & \text{for half-integral spins.} \end{cases}$$

The equations for wave functions of particles with higher spins were obtained above as a consequence of the fact that the initial function (for the rest state) describes particles with a given spin. The use of the wave functions (118) and (125) for parametrizing matrix elements is a significant convenience as a result of the simplicity of the equations and their tensor character.

§ 5.6. Bargmann–Wigner wave functions

In the rest state, when the spin angular momentum coincides with the total angular momentum, the wave function of a particle with spin $J = n/2$ is a symmetric nth rank spinor $\varphi_{\sigma_1\,\ldots\,\sigma_r}$, where σ_r may take on two values: $\sigma_r = 1, 2$ or $\sigma_r = \pm\frac{1}{2}$ (see § 3.2).

In the case of spin $\frac{1}{2}$ we associated the function $\varphi(\sigma)$ with the Dirac wave function $u(0, \sigma)$ restricted by the condition $(1-\gamma_4)u = 0$, specifically $\varphi = \frac{1}{2}(1+\gamma_4)\,u(0)$, if γ_4 is chosen as a diagonal matrix. Analogously, the spinor $\varphi_{\sigma_1\,\ldots\,\sigma_n}$ may be associated with the Bargmann–Wigner wave function in the rest state, or the symmetric Dirac spinor of nth rank $u_{\alpha_1\,\ldots\,\alpha_n}(\boldsymbol{p})$ which satisfies the condition $(\gamma_4-1)_{\beta}^{\beta'}\,u_{\beta'\beta_2\,\ldots\,\beta_n}(0) = 0$ in each index $\beta_1 \ldots \beta_n$ and for a diagonal matrix γ_4 coincides with $\varphi_{\sigma_1\,\ldots\,\sigma_n}$.[(69)]

The wave function of a particle with momentum $\boldsymbol{p}$ may be obtained from $u_{\beta_1\,\ldots\,\beta_n}(0)$ using the standard Lorentz transformation

$$u_{\beta_1\,\ldots\,\beta_n}(\boldsymbol{p}) = S_{\beta_1}^{\ \beta'_1}(\alpha(\boldsymbol{p}))\,\ldots\,S_{\beta_n}^{\ \beta'_n}(\alpha(\boldsymbol{p}))\,u_{\beta'_1\,\ldots\,\beta'_n}(0), \tag{131}$$

where $S(A)$ is the matrix for the transformation of the Dirac wave function.

By construction, the function (131) satisfies the Dirac equation in each index (see § 5.2):

$$(\not{p}-m)_{\beta}^{\beta'}\, u_{\beta'\beta_2 \ldots \beta_n}(\boldsymbol{p}) = 0. \tag{132}$$

Under Lorentz transformations and translations $g = (a, A)$ the wave function $u_{\beta_1 \ldots \beta_n}(\boldsymbol{p})$ transforms into the function

$$u'_{\beta_1 \ldots \beta_n}(\boldsymbol{p}) = e^{-ipa} S_{\beta_1}^{\beta_1'}(A) \ldots S_{\beta_n}^{\beta_n'}(A)\, u_{\beta_1' \ldots \beta_n'}(\Lambda^{-1}\boldsymbol{p}). \tag{133}$$

Let us introduce the conjugate function

$$\bar{u}^{\beta_1 \ldots \beta_n} = u^{+\beta_1' \ldots \beta_n'}(\gamma_4)_{\beta_1'}^{\beta_1} \ldots (\gamma_4)_{\beta_n'}^{\beta_n}. \tag{134}$$

Then $\bar{u}^{\beta} u_{\beta}$ is invariant in the sense that

$$\bar{u}'(\boldsymbol{p})\, u'(\boldsymbol{p}) = \bar{u}(\Lambda^{-1}\boldsymbol{p})\, u(\Lambda^{-1}\boldsymbol{p}).$$

Since the Dirac spinor $u(\boldsymbol{p})$ obeys $u^{+}(\boldsymbol{p})\, u(\boldsymbol{p}) = (p_0/m)\, \bar{u}(\boldsymbol{p})\, u(\boldsymbol{p})$, the function (131) is normalized as

$$u^{+\beta_1 \ldots \beta_n}(\boldsymbol{p})\, u_{\beta_1 \ldots \beta_n}(\boldsymbol{p}) = \left(\frac{p_0}{m}\right)^n \bar{u}^{\beta_1 \ldots \beta_n}(\boldsymbol{p})\, u_{\beta_1 \ldots \beta_n}(\boldsymbol{p}). \tag{135}$$

Consequently, the scalar product must be defined by the formula

$$(u_1, u_2) = \int \frac{d^3p}{2p_0^{n+1}}\, \bar{u}^{\beta_1 \ldots \beta_n}(\boldsymbol{p})\, u_{\beta_1 \ldots \beta_n}(\boldsymbol{p}), \qquad p_0 > 0. \tag{136}$$

With this definition of the scalar product, the transformations (133) are unitary.

The angular momentum tensor $M_{\mu\nu}$ may be found from (133) as a direct generalization of expression (46) for $M_{\mu\nu}$ in the case of spin $\frac{1}{2}$. The spin part of $M_{\mu\nu}$ is equal to

$$J_{\mu\nu} = \tfrac{1}{2} \sum_{r=1}^{n} \sigma_{\mu\nu}^{(r)}, \tag{137}$$

where the matrix $\sigma_{\nu\mu}^{(r)}$ acts only on the rth index:

$$(\sigma_{\mu\nu}^{(r)})_{\alpha_1 \ldots \alpha_n}^{\beta_1 \ldots \beta_n} = \delta_{\alpha_1}^{\beta_1} \ldots \delta_{\alpha_{r-1}}^{\beta_{r-1}} (\sigma_{\mu\nu})_{\alpha_r}^{\beta_r} \delta_{\alpha_{r+1}}^{\beta_{r+1}} \ldots \delta_{\alpha_n}^{\beta_n}.$$

To find an expression for the case of negative energies, one may use relation (69) connecting solutions with positive and negative energies in the case of the Dirac equation, or formula (38) for $v(\boldsymbol{p})$. When $\boldsymbol{p} = 0$ the function $v(\boldsymbol{p})$ satisfies the condition $(\gamma_4+1)_{\beta}^{\beta'} \times v_{\beta'\beta_2 \ldots \beta_n}(0) = 0$ in each index, while in the general case $v_{\beta_1 \ldots \beta_n}(\boldsymbol{p})$ obeys

$$(\not{p}+m)_{\beta}^{\beta'}\, v_{\beta'\beta_2 \ldots \beta_n} = 0, \tag{138}$$

in each index. $v_{\beta_1 \ldots \beta_n}$ is symmetric in $\beta_1 \ldots \beta_n$.

Knowing $v_{\beta_1 \ldots \beta_n}$ and $u_{\beta_1 \ldots \beta_n}$, we may construct a quantum field for particles with spin $J = n/2$ in the standard way:

$$\psi_{\beta_1 \ldots \beta_n}(x) = \frac{1}{(2\pi)^{3/2}} \int \frac{d^3p}{2p^0} \sum_{\gamma} \{u_{(\beta)}(\boldsymbol{p}, \gamma)\, a(\boldsymbol{p}, \gamma)\, e^{-ipx} + \ldots\}, \tag{139}$$

where the creation and annihilation operators satisfy the usual commutation relations (4.84).

The parametrization of the symmetric spinors $u_{\beta_1 \ldots \beta_n}$ is much easier if one uses the symmetry properties (66a) and (66b) of the matrices γ_μ. For example, the general form of the symmetric spinor of second rank is

$$u_{\alpha\beta} = [\gamma^\mu \mathcal{C} \varphi_\mu + \tfrac{1}{2}(\sigma^{\mu\nu}\mathcal{C})\varphi_{\mu\nu}]_{\alpha\beta}, \tag{140}$$

since $\gamma_\mu \mathcal{C}$ and $\sigma_{\mu\nu}\mathcal{C}$ are the only symmetric 4 by 4 matrices with lower indices. The Bargmann–Wigner equations

$$(\not{p}-m)_\alpha^{\alpha'} u_{\alpha'\beta} = 0, \quad (\not{p}-m)_\beta^{\beta'} u_{\alpha\beta'} = 0 \tag{141}$$

after multiplication by $(\mathcal{C}^{-1}\sigma_{\mu\nu})^{\alpha\beta}$ and $(\mathcal{C}^{-1}\gamma_\lambda)^{\alpha\beta}$ and contraction of indices, are equivalent to the following equations for a spin-1 particle:

$$p^\mu \varphi_{\mu\nu} = -im\varphi_\nu, \quad p_\mu\varphi_\nu - p_\nu\varphi_\mu = im\varphi_{\mu\nu}. \tag{142}$$

Thus while if a particle with spin 1 is described by a tensor $\varphi_{\mu\nu}$ in the $2(2J+1)$-component formalism, and by a four-vector in the usual formalism (corresponding in spirit to the Rarita–Schwinger formalism), according to the Bargmann–Wigner equation these particles must be described *simultaneously* by a vector φ_ν and a tensor $\varphi_{\mu\nu}$.

In the case of particles with spin $\frac{3}{2}$, the Bargmann–Wigner wave function is a symmetric third-rank spinor, which by virtue of the symmetry properties of the matrices γ^μ may be written in the form

$$u_{\alpha\beta\gamma} = \psi_{\alpha\mu}(\gamma^\mu \mathcal{C})_{\beta\gamma} + \tfrac{1}{2}\psi_{\alpha\mu\nu}(\sigma^{\mu\nu}\mathcal{C})_{\beta\gamma}. \tag{143}$$

In (143) only symmetry in β and γ has been taken into account. The antisymmetric matrices $\mathcal{C}^{-1}$, $\mathcal{C}^{-1}\gamma_5$ and $\mathcal{C}^{-1}\gamma^\mu\gamma_5$ must give zero when contracted with the symmetric spinor $u_{\alpha\beta\gamma}$. From this it follows that $u_{\alpha\beta\gamma}$ is symmetric in its remaining indices:

$$\gamma^\mu\psi_\mu(p) = 0, \quad \gamma^\mu\psi_{\mu\nu}(p) + i\psi_\nu(p) = 0. \tag{144}$$

Analogously the equation

$$(\not{p}-m)_\alpha^{\alpha'} u_{\alpha'\beta\gamma}(p) = 0 \tag{145}$$

may be written as a system of equations for the spinor-vector ψ_μ and the spinor-tensor $\psi_{\mu\nu}$:

$$(\not{p}-m)\psi_\mu = 0, \quad p^\nu\psi_{\nu\mu} = -im\psi_\mu, \quad p_\mu\psi_\nu - p_\nu\psi_\mu = im\psi_{\mu\nu}. \tag{146}$$

Comparison of the system (146) and the conditions (144) with the Rarita–Schwinger equations (105), (106), and (107) shows the equivalence of the two descriptions.

In the rest state the wave function must become an irreducible (i.e. symmetric) spinor of the rotation group describing a given spin. If we wish to describe several spins using one higher-rank spinor we must, in general, use reducible spinors. Cases are also possible for which spinors of higher rank n must be used in describing lower spins $J < n/2$. In particular, in the quark model, baryons with spin $\frac{1}{2}$ must be described using a third-rank spinor of mixed symmetry, and particles with spin 0 by an antisymmetric second-rank spinor.

Let $\Phi_{\alpha}{}^{\beta}$ be a second-rank spinor. Let us expand the 4 by 4 matrix $\Phi_{\alpha}{}^{\beta}$ in a complete set of matrices $(\gamma^R)_{\alpha}{}^{\beta}$:

$$\Phi_{\alpha}{}^{\beta} = (\varphi + i\gamma_5\varphi_5 + \gamma^{\mu}\gamma_5\varphi_{\mu 5} + \gamma^{\mu}\varphi_{\mu} + \tfrac{1}{2}\sigma_{\mu\nu}\varphi^{\mu\nu})_{\alpha}{}^{\beta}, \tag{147}$$

and use the Bargmann–Wigner equations (132) whose validity does not depend on symmetry properties of the spinor:

$$[\not{p}, \Phi]_{-} = 2m\Phi, \quad [\not{p}, \Phi]_{+} = 0. \tag{148}$$

The first of eqns. (148) leads to (142) for the functions φ_{μ} and $\varphi_{\mu\nu}$ and also to the following equations for φ_5 and $\varphi_{\mu 5}$:

$$p_{\mu}\varphi_5 = im\varphi_{\mu 5}, \quad p_{\mu}\varphi_{\mu 5} = -im\varphi_5. \tag{149}$$

The second eqn. (148) gives only one new relation $\varphi = 0$, while the remaining equations are satisfied identically as a result of (142) and (149). Thus, in the general case, the second-rank spinor (147) describes both particles with spin 1 (the symmetric part of $(\Phi\mathscr{C})_{\alpha\beta}$), and particles with spin 0 (the antisymmetric part of $(\Phi\mathscr{C})_{\alpha\beta}$).

§ 5.7. The Duffin–Kemmer equation

In the case of particles with spin 0 and 1 the equation of motion may be written in the form of a single equation reminiscent of the Dirac equation but with different matrices β_{μ} instead of γ_{μ}. Let us return to the second-rank spinor (147) and write it with lower indices $\tilde{\Phi} = \Phi\mathscr{C}$, assuming that $\varphi = 0$. For the spinor $\tilde{\Phi}_{\alpha\beta}$ the first of equations (148) may be re-written in the form

$$(\not{p}_{\alpha}{}^{\alpha'} + \not{p}_{\beta}{}^{\beta'})\tilde{\Phi}_{\alpha'\beta'} = 2m\tilde{\Phi}_{\alpha\beta}, \tag{150}$$

while the second equation (148) may be discarded, since it is a consequence of the first when $\varphi = 0$. Let us introduce the 16 by 16 matrices:

$$\beta_{\mu} = \tfrac{1}{2}(\gamma_{\mu}1' + 1\gamma'_{\mu}), \tag{151}$$

where the unprimed matrices 1 and γ_{μ} act on the first index in $\tilde{\Phi}_{\alpha\beta}$, and the primed on the second. With this notation, eqn. (150) takes the form

$$(\beta^{\mu}p_{\mu} - m)\tilde{\Phi} = 0. \tag{152}$$

It is called the Duffin–Kemmer equation.[66, 67]

The matrices β^{μ} satisfy the commutation relations

$$\beta^{\mu}\beta^{\nu}\beta^{\lambda} + \beta^{\lambda}\beta^{\nu}\beta^{\mu} = g^{\mu\nu}\beta^{\lambda} + g^{\nu\lambda}\beta^{\mu}, \tag{153}$$

which follow directly from the properties of the matrices γ^{μ} and (151). In particular (not summing over μ)

$$(\beta^{\mu})^3 = g_{\mu\mu}\beta^{\mu}, \quad \beta^{\mu}\beta^{\nu}\beta^{\mu} = 0, \quad (\nu \neq \mu). \tag{154}$$

From the second of eqns. (154) one may conclude that the matrices β^{μ} do not have inverses.

Besides the matrices β_μ it is convenient to introduce the Hermitian matrices

$$\eta_\mu = 2\beta_\mu^2 - g_{\mu\mu} = \gamma_\mu \gamma'_\mu \tag{155}$$

with the properties

$$\left.\begin{aligned} \eta_\mu^2 &= 1, & \eta_\mu \eta_\nu &= \eta_\nu \eta_\mu, \\ \beta_\mu \eta_\mu &= \eta_\mu \beta_\mu, & \beta_\mu \eta_\nu &= -\eta_\nu \beta_\mu \end{aligned}\right\} \tag{156}$$

(not summed over μ, $\nu \neq \mu$).

Using the matrix η_4 one may define the conjugate function $\bar{\Phi} = \tilde{\Phi}\eta_4$, which satisfies

$$\bar{\Phi}(\beta^\mu p_\mu - m) = 0. \tag{157}$$

From the properties of the matrices γ^μ and (151) it is clear that the matrices β^μ transform as a vector:

$$S^{-1}(A)\, S'^{-1}(A)\, \beta^\mu S(A)\, S'(A) = \Lambda^\mu{}_\nu(A)\beta^\nu. \tag{158}$$

Here $S(A)$ and $S'(A)$ act respectively on the unprimed and primed matrices γ_μ in (151). Let us denote $SS' = \tilde{S}$. Then, when $\tilde{\Phi}' = \tilde{S}\tilde{\Phi}$ we will have $\bar{\Phi}' = \bar{\Phi}\tilde{S}^{-1}$, so that $\bar{\Phi}\Phi$ is invariant.

The wave equation

$$(p^2 - m^2)\Phi = 0 \tag{159}$$

is a consequence of the Duffin–Kemmer equation (152). This may be seen directly from the fact that eqns. (152) and (150) or (148) are equivalent.

The 16 by 16 matrices β_μ may be decomposed into five-dimensional and ten-dimensional irreducible representations as well as the trivial one-dimensional representation $\beta_\mu = 0$. The five-dimensional representation corresponds to spin 0 and the ten-dimensional one to spin 1. In terms of Bargmann–Wigner functions, the five-dimensional matrices β_μ act on an antisymmetric second-rank spinor, and the ten-dimensional matrices β_μ on a symmetric spinor $\Phi\mathcal{C}$ with the trace of Φ equal to zero: $\mathrm{Tr}\,\Phi = 0$.

CHAPTER 6

REFLECTIONS

SPACE and time reflections are included in the transformations that leave the square of the interval $s^2 = (x-y)^2$ invariant. Together with transformations (a, Λ) of the proper orthochronous Poincaré group $\mathcal{P}_+^\uparrow$, they form the full Poincaré group $\mathcal{P}$.

In contrast to the group $\mathcal{P}_+^\uparrow$, the full group $\mathcal{P}$ is not a symmetry group of the laws of nature. In 1956 the breaking of symmetry with respect to spatial reflections was discovered[80] (anticipated by Lee and Yang,[81]), and in 1964 a reaction was found[82] in which invariance with respect to time reversal was broken. However, the degree of breaking of both symmetries is small. In phenomena due to the strong and electromagnetic interactions, the group $\mathcal{P}$ may be considered a symmetry group, and there exist quantum objects whose properties are determined by this group.

In § 1.3 and Chapter 4 we considered the transition from the classical proper Poincaré group $\mathcal{P}_+^\uparrow$ to the quantum mechanical Poincaré group $\overline{\mathcal{P}}_+^\uparrow$ and found the unitary representations of the latter. Starting from these results, we will discuss in this chapter the full quantum mechanical Poincaré group $\overline{\mathcal{P}}$ (including reflections), and will find its irreducible representations.

There is a profound difference between transformations of the group $\overline{\mathcal{P}}_+^\uparrow$ and reflections. Lorentz transformations and displacements are continuous, and we can ensure that an object be the same in different reference frames by applying consecutive infinitesimal transformations. This procedure is impossible for spatial reflections, where we cannot compare the same state in different reference frames (original and reflected) but always compare different states (in different reference frames) with one another. In the case of time reversal the corresponding inertial reference frame does not even exist in the literal sense. Time reversal is usually understood in the sense of reversal of motion. The identity of an object under direct and reversed motion also cannot be guaranteed.

In quantum theory the state of a system is described by a unit ray in Hilbert space. Transformations of vectors connected continuously with the identity transformation are always unitary (see § 1.3). Discrete transformations may be either unitary or antiunitary. As we shall see below, transformations must be antiunitary when the initial state transforms into the final one, i.e. under time reversal. Continuous transformations of the group $\mathcal{P}_+^\uparrow$ connect states of the same coherent subspace; discrete operations may connect different coherent subspaces, but the square of a reflection cannot take a state vector out of a given coherent subspace.

In the presence of internal symmetries, there are generally ambiguities in defining reflection operations, since if one performs an internal symmetry transformation the state thus obtained will be indistinguishable from the initial one.

The operation θ, related to total space–time reflection $x_\mu \to -x_\mu$, occupies a special place among the discrete symmetry operations. In theories symmetric with respect to spatial reflection P, time reversal T, and charge conjugation C (the operation which substitutes particles by antiparticles) the operation θ is equal to the product CPT. By the CPT theorem[29, 83, 84] a relativistically invariant local theory is automatically invariant with respect to the discrete transformation θ. This theorem is well supported by experimental data. It has been proved in axiomatic quantum field theory[85, 86] and is one of the basic postulates of axiomatic S-matrix theory. It has also been shown in axiomatic field theory that the transformation θ commutes with internal symmetry transformations G:

$$\theta G = G\theta. \tag{1}$$

We shall consider θ-invariance as a basic postulate. One should note that so far no self-consistent theory has been constructed in which θ-invariance is violated. The degeneracy of states in quantum mechanics associated with the complex nature of the wave function was first noted by Kramers[87] and formulated by Wigner[88] in the language of antiunitary time-reversal operators.

§ 6.1. Total reflection θ, or CPT

In classical physics, space–time reflection I changes the sign of the coordinates

$$Ix_\mu = -x_\mu, \quad I^2 = 1, \tag{2}$$

but does not change the Lorentz transformation matrix

$$I\Lambda_\mu^{\ \nu} I^{-1} = \Lambda_\mu^{\ \nu},$$

so that the effect of I on an element $g = (a, \Lambda)$ of the Poincaré group is

$$I(a, \Lambda)I^{-1} = (-a, \Lambda). \tag{3}$$

Adding the reflection I to the proper orthochronous group $\mathcal{P}_+^\uparrow$ we arrive at an enlarged group

$$\mathcal{P}_+ = \{\mathcal{P}_+^\uparrow, I\mathcal{P}_+^\uparrow\}.$$

Time reversal must be understood not as a transformation to a reversed sense of time but as a reversal of motion. Instead of motion with momentum $\boldsymbol{p}$ from the point $\boldsymbol{q}$ to the point $\boldsymbol{q}'$, where the particle has the momentum $\boldsymbol{p}'$, after time reversal one must consider the reversed motion from the point $\boldsymbol{q}'$ with momentum $-\boldsymbol{p}'$ to the point $\boldsymbol{q}$ with momentum $-\boldsymbol{p}$. Adding spatial reflection, the inversion I finally leads to motion from the point $-\boldsymbol{q}'$ with initial momentum $\boldsymbol{p}'$ to the point $-\boldsymbol{q}$, where the final momentum is equal to $\boldsymbol{p}$.

In quantum theory the probability of the transition from the state a to the state b is characterized by the matrix element

$$\langle b, \text{out} \,|\, a, \text{in}\rangle = \langle b \,|S|\, a\rangle.$$

The reversed process, obtained by replacing the initial state by the final one, is described by matrix elements of the type

$$\langle a_\theta, \text{out} \,|\, b_\theta, \text{in}\rangle = \langle a_\theta \,|\, S_\theta \,|\, b_\theta\rangle.$$

Here a_θ and b_θ refer to states arising as a result of the inversion I, while S_θ is defined with respect to the states $|\, a_\theta\rangle$:

$$S_\theta \,|\, a_\theta, \text{out}\rangle = |\, a_\theta, \text{in}\rangle. \tag{4}$$

If the probability of the transition $a \to b$ is invariant with respect to the inversion I, then

$$|\langle b, \text{out} \,|\, a, \text{in}\rangle|^2 = |\langle a_\theta, \text{out} \,|\, b_\theta, \text{in}\rangle|^2. \tag{5}$$

The replacement of the initial state by the final one is a nonunitary operation, for which

$$\langle b, \text{out} \,|\, a, \text{in}\rangle = \langle a_\theta, \text{out} \,|\, b_\theta, \text{in}\rangle. \tag{6}$$

For the scattering matrix we obtain from (4) and (6) the relation

$$\langle b \,|\, S \,|\, a\rangle = \langle a_\theta \,|\, S_\theta \,|\, b_\theta\rangle, \tag{7}$$

defining the transformed operator S_θ. A formula of the type (7) will also be used to give a meaning to any operator H_θ acting on the states $|\, a_\theta\rangle$:

$$\langle b \,|\, H \,|\, a\rangle = \langle a_\theta \,|\, H_\theta \,|\, b_\theta\rangle. \tag{8}$$

The transformation from the states $|\, a\rangle$ and the operators H to the time-reversed states $|\, a_\theta\rangle$ and operators H_θ [respecting relations (6) and (8)] may be carried out in two ways: using an antiunitary operator (Wigner),[18, 89] or using a transposition operator (Schwinger).[83] We shall use the Wigner treatment of the inversion I; its relation to the Schwinger treatment will be shown at the end of this section.

An antiunitary operator A is an operator with the following properties:

$$\left.\begin{aligned} A(c_1 \,|\, a_1\rangle + c_2 \,|\, a_2\rangle) &= c_1^* A \,|\, a_1\rangle + c_2^* A \,|\, a_2\rangle, \\ \langle Aa \,|\, Ab\rangle &= \langle b \,|\, a\rangle, \end{aligned}\right\} \tag{9}$$

i.e. A includes complex conjugation, which may be symbolically written as $Ac = c^*A$ or $Ai = -iA$. An antiunitary transformation A leaves the transition probability $|\langle b \,|\, a\rangle|^2$ invariant.

In contrast to the square of a unitary operator, the square of an antiunitary operator A^2 is invariant with respect to a phase transformation: when $A' = e^{i\omega}A$ one will have $A'^2 = A^2$. According to (9) the operator A^2 is unitary.

An antiunitary operator A may be represented in the form

$$A = \beta K, \quad K^2 = 1, \quad A^2 = \beta^* \beta, \tag{10}$$

where K is the complex conjugation operator, while β is unitary.

The space–time reflection (2) in the Hilbert space of states will be described by an antiunitary operator θ:

$$|\, a_\theta, \text{out}\rangle = \theta \,|\, a, \text{in}\rangle, \quad |\, b_\theta, \text{in}\rangle = \theta \,|\, b, \text{out}\rangle,$$

assuming here that the vacuum $|0\rangle$ is invariant with respect to total reflection θ:

$$\theta|0\rangle \equiv |0_\theta\rangle = |0\rangle.$$

The transition amplitude in this case satisfies the condition (6):

$$\langle a_\theta, \text{out}\,|\, b_\theta, \text{in}\rangle = \langle \theta a, \text{in}\,|\,\theta b, \text{out}\rangle = \langle a, \text{in}\,|\, b, \text{out}\rangle^*.$$

Let us now find explicitly the transformation of the operators $H \to H_\theta$ determined by (8). According to (6) and (11)

$$\begin{aligned}\langle a_\theta, \text{out}\,|\, H_\theta |\, b_\theta\ \text{in}\rangle &= \langle \theta^{-1} H_\theta b_\theta, \text{in}\,|\,\theta^{-1} a_\theta, \text{out}\rangle = \langle \theta^{-1}\theta a, \text{in}\,|\,\theta^{-1} H_\theta \theta\,|\, b, \text{out}\rangle^* \\ &= \langle b, \text{out}\,|\,(\theta^{-1} H_\theta \theta)^+\,|\, a, \text{in}\rangle,\end{aligned}$$

which, along with the definition (8), leads to the formula

$$H_\theta = \theta H^+ \theta^{-1}. \tag{12}$$

Consequently, the condition of invariance of the operator H with respect to reflections θ in the Wigner treatment has the form

$$H_\theta = H = \theta H^+ \theta^{-1}. \tag{13}$$

To find the transformation of the generators of the Poincaré group, we shall use the group multiplication law (3), rewriting it first for the quantum mechanical group $\bar{\mathcal{P}}_+^\uparrow$:

$$I(\text{a}, A) I^{-1} = (-\text{a}, A). \tag{14}$$

Passing to the operators θ and U(a, A) acting on the state vectors, we obtain

$$\theta \text{U}(\text{a}, A)\theta^{-1} = \text{U}(-\text{a}, A); \tag{15}$$

so that

$$\theta e^{iP_\mu a^\mu} \theta^{-1} = e^{-iP_\mu a^\mu}, \tag{16}$$

$$\theta e^{iM_{\mu\nu}\omega^{\mu\nu}} \theta^{-1} = e^{iM_{\mu\nu}\omega^{\mu\nu}}. \tag{17}$$

Since an antilinear operator θ "anticommutes" with i, relations (16) and (17) are equivalent to the following rules transforming momentum and angular momentum under reflections θ:

$$P_{\theta\mu} = \theta P_\mu \theta^{-1} = P_\mu, \quad M_{\theta\mu\nu} = \theta M_{\mu\nu}\theta^{-1} = -M_{\mu\nu}. \tag{18}$$

Thus the reflection θ does not change the invariants of the Poincaré group—the mass m and the spin J, as well as the sign of the energy. Under the reflection θ the momentum does not change, while the spin vector J changes sign. If the reflection θ were not represented by an antiunitary operator, the sign of the energy would have changed, in contradiction with the observed positivity of the energy. In fact, if θ had been unitary, from (16) it would have followed that $\theta P_\mu \theta^{-1} = -P_\mu$.

The transformation rules (18) do not contradict the commutation relations. Let us apply the reflection θ to both sides of the commutation relations of the Hermitian operators M, M', and M'':

$$[M, M'] = iM''.$$

Then the commutator of the transformed operators is

$$[M_\theta, M'_\theta] = -iM''_\theta,$$

so that the operators $-M_\theta$ when applied to $|a_\theta\rangle$ and the vacuum $|0\rangle$ have the same properties as the operators M acting on $|a\rangle$ and $|0\rangle$. It is easy to see that the operators (18) satisfy the transformed commutation relations (1.41).

The condition (1) that the reflection θ and internal symmetry transformations be independent allows one to introduce the transformation law for self-adjoint generators F_r of an internal symmetry group $\mathfrak{G}$. Inserting into (1) the expression $u(G) = \exp[iF_r\alpha_r]$ for a unitary representation of the group $\mathfrak{G}$,

$$\theta e^{iF_r\alpha_r}\theta^{-1} = e^{iF_r\alpha_r}, \quad \theta F_r\theta^{-1} = -F_r \tag{19}$$

(the parameters α_r are real). In particular, all additive quantum numbers, i.e. all charges Q change sign under the reflection θ. This leads to an important conclusion: under the reflection θ a particle goes into its antiparticle.

From the group theoretic point of view our problem consists in finding the irreducible representations or, more precisely, "co-representations" of the quantum mechanical Poincaré group with the reflection θ. Co-representations take the place of representations when the group contains both linear and antilinear transformations. The unitary equivalence of the representations R and R' means that

$$R' = \mathrm{U}R\mathrm{U}^+, \quad \mathrm{UU}^+ = 1. \tag{20}$$

In the case of co-representations, the transition to another unitarily equivalent basis entails (for antiunitary A) a transformation

$$\mathrm{O}(A) = \mathrm{O}'(A) = \mathrm{UO}(A)\mathrm{U}^\mathrm{T}, \tag{21}$$

reducing to a similarity transformation for real U. A co-representation is called irreducible if it cannot be broken down into diagonal blocks using the transformations (20) and (21).

If we fix the operator θ in the space of states, we may pass from the classical group $\mathcal{P}_+$ to the quantum mechanical group with reflection

$$\overline{\mathcal{P}}_+ = \{\overline{\mathcal{P}}_+, \theta\overline{\mathcal{P}}^\dagger_+\}.$$

However, although there is only one classical group $\mathcal{P}_+$, in the quantum mechanical case there exist several groups $\overline{\mathcal{P}}_+$ differing from one another by the values of the operator θ^2.

The square of the classical inversion I^2 is unity; it takes into itself each physical state and consequently, also each unit ray in the Hilbert space of state vectors. This means that the square of the quantum mechanical inversion operator $\theta^2 = \varepsilon_\theta$ must be a phase factor. Because of the unitarity of θ^2 the phase factors must be the same in each coherent subspace. Because of the relations $\theta^3 = \varepsilon_\theta\theta = \theta\varepsilon_\theta = \varepsilon^*_\theta\theta$, the quantity ε_θ is real, and, consequently,

$$\theta^2 = \varepsilon_\theta = \pm 1. \tag{22}$$

The sign of ε_θ may be different in different coherent subspaces.

Let us temporarily assume that the system we are considering does not have an internal symmetry. The form of the operator θ is then determined only by the Poincaré group. Let us apply the reflection θ to the Lorentz-transformed state and recall that as a consequence of the antiunitarity of θ the transformation matrices must be replaced by their complex conjugates:

$$\theta e^{-i\frac{1}{2}M_{\mu\nu}\omega^{\mu\nu}}\,|\boldsymbol{p},\lambda;Q_i\rangle = \theta\sum_{\lambda'}\mathcal{D}^J_{\lambda'\lambda}(\tilde{A})\,|\boldsymbol{p}',\lambda';Q_i\rangle = \sum_{\lambda'}\mathcal{D}^{J*}_{\lambda'\lambda}(\tilde{A})\theta\,|\boldsymbol{p}',\lambda';Q_i\rangle, \tag{23}$$

where $\mathcal{D}^J(\tilde{A})$ is a unitary matrix of the rotation group depending on the Wigner rotation $\tilde{A}$ [we have used (4.47)].

Thus when we include the reflection θ we must, along with the basis states $|a\rangle$ of an irreducible representation of the group $\bar{\mathcal{P}}^{\uparrow}_{+}$, consider the states $\theta\,|a\rangle$ which transform via the complex conjugate matrices.

The matrices $\mathcal{D}^J$ and $\mathcal{D}^{J*}$ are unitarily equivalent:

$$\mathcal{D}^{J*}(\tilde{A}) = \mathcal{D}^J(C^{-1})\,\mathcal{D}^J(\tilde{A})\,\mathcal{D}^J(C), \quad C = -i\sigma_2, \tag{24}$$

Thus, applying the operation θ once more to (23), we find

$$\theta^2\mathcal{D}^J(\tilde{A})\theta^{-2} = (\mathcal{D}^J(CC^*))^{-1}\,\mathcal{D}^J(\tilde{A})\,\mathcal{D}^J(CC^*),$$

from which we conclude that

$$\varepsilon_\theta = \theta^2 = \pm\mathcal{D}^J(CC^*) = \pm\mathcal{D}^J(-1) = \pm(-1)^{2J}. \tag{25}$$

Depending on the choice of sign in (25), we obtain two characteristic types of operators θ

1. For the choice of sign $+$ in (25) (type 1) the transformation θ^2 belongs to the Poincaré group

$$\theta^2 = \mathrm{U}(0, -E) = (-1)^{2J}. \tag{26}$$

Here E is the unit matrix. The number of states does not change; the states $\theta\,|a\rangle$ refer to the same representation of $\bar{\mathcal{P}}^{\uparrow}_{+}$ as the states $|a\rangle$. In fact, according to (10) and (25) one must set (for spin J)

$$\theta\,|\boldsymbol{p},\lambda;Q_r=0\rangle = \sum_{\lambda'}(K\,|\boldsymbol{p},\lambda';Q_r=0\rangle)\,\mathcal{D}^J_{\lambda'\lambda}(C^*), \tag{27a}$$

which may be written symbolically in the form

$$\theta = \mathcal{D}^J(C^*)K. \tag{27b}$$

Here K is an antiunitary operator satisfying the condition

$$K\,|\boldsymbol{p},\lambda;Q_r=0,\mathrm{in}\rangle = |\boldsymbol{p},\lambda;Q_r=0,\mathrm{out}\rangle, \tag{28}$$

which defines K uniquely (since it is applied to states of neutral particles).

2. For the sign $-$ in (25) (type 2) the unitary transformation $\theta^2 = -\mathrm{U}(0, -E)$ does not belong to the Poincaré group, and the number of states is doubled although $|a\rangle$ and $\theta\,|a\rangle$ transform in the same way under Lorentz transformations. We shall distinguish these states from one another by a new quantum number $r_{1,2} = \pm 1$, so that the basis states are now

$|a, r_1 = +1\rangle$ and $|a, r_2 = -1\rangle$. In this basis a Lorentz transformation $0(\mathrm{a}, A)$ has diagonal form

$$\mathrm{O}(\mathrm{a}, A) = \begin{pmatrix} \mathrm{U}(\mathrm{a}, A) & 0 \\ 0 & \mathrm{U}(\mathrm{a}, A) \end{pmatrix}, \tag{29}$$

while the reflection θ (in the case of spin J) is

$$\theta = \begin{pmatrix} 0 & \mathcal{D}^J(C^*) \\ -\mathcal{D}^J(C^*) & 0 \end{pmatrix}, \quad \theta^2 = -(-1)^{2J}. \tag{30}$$

Thus for type 2 a neutral particle (in the sense $Q_i = 0$) does not transform into itself under the operation θ but transforms into another particle differing by the sign of the quantum number r. In other words, if one includes the quantum number r in the characterization of particles ($r_1 = 1$) and antiparticles ($r_2 = -1$), then in the case $\theta^2 = -(-1)^{2J}$ no neutral particles exist (in the sense $Q_r = 0$, $r_1 = r_2$).

The case of physical interest is type 1. It is these reflections that are observed in experiment and are the only possible ones in axiomatic quantum field theory.[86] The CPT theorem applies only to this case. We shall thus choose operators θ only of type 1, where $\theta^2 = (-1)^{2J}$ does not depend on whether or not an internal symmetry exists.

We have thus obtained the irreducible co-representations of the quantum mechanical Poincaré group $\bar{\mathcal{P}}_+$ with reflection θ. Under the reflection θ a neutral particle with spin J, momentum $\boldsymbol{p}$, and helicity λ changes helicity to $-\lambda$; here the initial and final states are interchanged and [by virtue of (27)] acquire a phase factor:

$$\theta\,|\boldsymbol{p}, \lambda; J, m; Q_i = 0, \mathrm{in}\rangle = (-1)^{J-\lambda}|\boldsymbol{p}, -\lambda; J, m; Q_i = 0, \mathrm{out}\rangle. \tag{31}$$

For stable particles the single-particle states in (31) do not require the symbols in, out, since they are stationary and do not depend on boundary conditions. Nonetheless, we shall continue to retain these symbols in all formulae for reflections θ (and later for time-reversal T. This allows us to pass automatically from formulae of the type (31) for single-particle states to analogous formulae for multi-particle states.

We note that because of (26) the group $\bar{\mathcal{P}}_+$ will be only the semi-direct (but not the direct) product of the quantum mechanical group $\bar{\mathcal{P}}^{\dagger}_+$ and the finite inversion group (in contrast to the classical group $\mathcal{P}_+$).[90]

Internal symmetries

The most general invariance group includes internal symmetry as well as θ and $\bar{\mathcal{P}}^{\dagger}_+$.

Let us return to (23) and take account of internal symmetry,[91] setting $\theta^2 = (-1)^{2J}$. For simplicity, we shall consider a particle at rest. To eliminate double-valuedness, we introduce the operator

$$\theta_0 = \theta e^{i\pi M_{31}}, \quad \theta_0^2 = 1, \tag{32}$$

describing the reflection θ with a rotation by π about the second axis. In spin space, θ_0 is equivalent to $\mathcal{D}^J(C)\theta$ and, as a result of (24), commutes with rotations

$$\mathcal{D}^J(C)\,\theta\mathcal{D}^J(V) = \mathcal{D}^J(C)\,\mathcal{D}^J(V^*)\theta = \mathcal{D}^J(V)\,\mathcal{D}^J(C)\theta.$$

Consequently, the operator θ_0 is a scalar in spin space; when it is applied to a multiplet of the rest state the result depends only on the internal symmetry properties connected with the group $\mathfrak{G}$. Omitting the spin variables and the momentum, we obtain analogously to (23),

$$\theta_0 e^{iF_i a^i} |\alpha\rangle = e^{iF_i a^i \theta_0} |\alpha\rangle = \theta_0 \sum_\beta d_{\beta\alpha}(G) |\beta\rangle = \sum_\beta d^*_{\beta\alpha}(G) \theta_0 |\beta\rangle, \tag{33}$$

where α denotes the component of the multiplet transforming via the irreducible representation d of the group $\mathfrak{G}$. Thus the states $\theta_0 |\alpha\rangle$ form the adjoint multiplet the basis of the adjoint representation d^*.

Let us find the representations $\vec{d}$ of the internal symmetry group $\mathfrak{G}$ with the reflection θ_0 or the group

$$\mathfrak{G}_\theta = \{G, \theta_0 G\},$$

containing the transformations G and $\theta_0 G$ as elements. According to (33) the multiplicity of the co-representation $\vec{d}$ will be different from d if the set $|\alpha\rangle$ does not contain the states $\theta_0 |\alpha\rangle$.

Depending on the properties of d and d^*, three types of co-representations of the group $\mathfrak{G}_\theta$ are possible.

If the representations d and d^* are unitarily equivalent,

$$d^*(G) = \beta^{-1} d(G) \beta, \quad \beta\beta^+ = 1, \tag{34}$$

then applying θ_0 once more to (33),

$$d(G) = (\beta\beta^*)^{-1} d(G) \beta\beta^*,$$

which, because of the irreducibility of d and the unitarity of β, means that

$$\beta\beta^* = \pm 1. \tag{35}$$

Here three cases are possible.

1. If $\beta\beta^* = +1 (\beta \equiv \beta_+)$, then $\theta_0 |\alpha\rangle$ is contained in the set $|\alpha\rangle$ and, consequently, the operator $\vec{d}(\theta_0) \equiv \theta_0$ acts inside the multiplet. The irreducible co-representations of d have the form

$$\vec{d}(G) = d(G), \quad \vec{d}(\theta_0) = \beta_+. \tag{36}$$

Examples are the regular and other real irreducible representations of finite-dimensiona compact groups, in particular of the groups SU_2 and SU_3 (see Chapters 8 and 9).

2. If $\beta\beta^* = -1$, then in analogy with the case $\theta^2 = -1$ for the Poincaré group [formulae (29) and (30)], the number of states must be doubled, since otherwise it would be impossible to construct θ_0. Let us perform a unitary transformation of states $\theta_0 |\alpha\rangle \to |\alpha_2\rangle = -\theta_0 \beta |\alpha\rangle$. Then, with respect to the column vector formed from $|\alpha_1\rangle \equiv |\alpha\rangle$ and $|\alpha_2\rangle$, the irreducible representations $\vec{d}$ will have the form

$$\vec{d}(G) = d(G) \times \begin{pmatrix} 1 & 0 \\ 0 & 1 \end{pmatrix}, \quad \vec{d}(\theta_0) = \beta \times \begin{pmatrix} 0 & -1 \\ 1 & 0 \end{pmatrix}. \tag{37}$$

where, of course, $\vec{d}(G\theta_0) \equiv \vec{d}(G) \vec{d}(\theta_0)$. Here the following group relations will hold:

$$\vec{d}(\theta_0 G) = \vec{d}(\theta_0) \vec{d}^*(G), \quad \vec{d}(\theta_0^2) = \vec{d}(\theta_0) \vec{d}^*(\theta_0). \tag{38}$$

Illustrations of this case are the representations of the isospin group SU_2 with half-integral isospin (see Chapter 8).

3. The representations d and d^* are not equivalent, so that it is impossible to find a unitary operator β with the properties $d^* = \beta d \beta^{-1}$. This case has no analogy for reflections θ in the Poincaré group. The irreducible representations $\tilde{d}$ then have doubled dimension:

$$\tilde{d}(G) = \begin{pmatrix} d(G) & 0 \\ 0 & d^*(G) \end{pmatrix}, \quad \tilde{d}(\theta_0) = \begin{pmatrix} 0 & 1 \\ 1 & 0 \end{pmatrix}, \tag{39}$$

while the basis of the co-representation is the column vector

$$|\tilde{\alpha}\rangle = \begin{pmatrix} |\alpha\rangle \\ \theta_0 |\alpha\rangle \end{pmatrix}. \tag{40}$$

In cases 2 and 3 the second row of the set of states describes antiparticles. Consequently particles and antiparticles may be placed in the same multiplet only for representations of internal symmetries satisfying condition (34) in case 1.

Case of exact symmetry

The simplest example of an internal symmetry leading to type 3 is the internal symmetry group $\mathfrak{G}^0$ whose transformations have the form

$$d(\gamma, \delta, \varrho_e, \varrho_\mu) = \exp\{i(Q\gamma + B\delta + L_e\varrho_e + L_\mu\varrho_\mu)\}, \tag{41}$$

where $Q_i = Q, B, L_e$ and L_μ are, respectively, the electric, baryonic, and leptonic charges, while γ, δ, ϱ_e, and ϱ_μ are parameters of the transformation.

The irreducible representations of each of the four one-dimensional unitary groups in (41) are one-dimensional; for example, the values Q and $-Q$ characterize different irreducible representations of the group $U_Q(1)$. Thus, limiting ourselves to irreducible representations, we cannot connect the representations d and d^* via a unitary transformation (34); the operator β does not exist. The state $|\alpha\rangle$ in the case of an exact symmetry group is determined by the charges $Q_i = Q, B, L_e$, and L_μ; if $|\alpha\rangle \equiv |Q_i', \text{in}\rangle$, the state $\theta_0|\alpha\rangle$ is then $\theta_0|\alpha\rangle \equiv |-Q_i', \text{out}\rangle$.

Formulae (39) and (40) give the matrices of the irreducible co-representation of the exact symmetry group with reflection θ_0:

$$\left.\begin{aligned} \tilde{d}(\gamma, \delta, \varrho_e, \varrho_\mu) &= \begin{pmatrix} \exp\{i(Q\gamma + B\delta + L_e\varrho_e + L_\mu\varrho_\mu)\} & 0 \\ 0 & \exp\{-i(Q\gamma + B\delta + L_e\varrho_e + L_\mu\varrho_\mu)\} \end{pmatrix}, \\ \tilde{d}(\theta_0) \equiv \theta_0 &= \begin{pmatrix} 0 & 1 \\ 1 & 0 \end{pmatrix}, \quad |\tilde{Q}_i'\rangle = \begin{pmatrix} |Q_i', \text{in}\rangle \\ |-Q_i', \text{out}\rangle \end{pmatrix}. \end{aligned}\right\} \tag{42}$$

All internal symmetry states $|\alpha\rangle$ were defined for particles at rest. Let us now turn to the full operator $\theta = \theta_0 \exp\{i\pi M_{31}\}$ [see (32)] and pass to a reference frame in which the particle has momentum $\boldsymbol{p}$ using the Lorentz transformation. Then the column vector $|\tilde{Q}_i\rangle$ in (42) is replaced by the column vector

$$\begin{pmatrix} |\,\boldsymbol{p}, \quad \lambda;\, J,\, m;\, Q_i',\, \text{in}\rangle \\ |\,\boldsymbol{p}, \,-\lambda;\, J,\, m;\, Q_i',\, \text{out}\rangle \end{pmatrix}; \tag{43}$$

the operator θ takes the form (for spin J)

$$\theta = \begin{pmatrix} 0 & \mathcal{D}^J(C) \\ \mathcal{D}^J(C^*) & 0 \end{pmatrix} \equiv \begin{pmatrix} 0 & \theta_{12} \\ \theta_{21} & 0 \end{pmatrix}. \tag{44}$$

Usually the operator θ refers to both θ_{12} and θ_{21} in (44) [which does not lead to confusion since θ_{12} and θ_{21} act on different lines in (43)]. In the usual notation (helicity basis)

$$\theta\,|\,\boldsymbol{p},\, \lambda;\, J,\, m;\, Q_i',\, \text{in}\rangle = (-1)^{J-\lambda}\,|\,\boldsymbol{p},\, -\lambda;\, J,\, m;\; -Q_i',\, \text{out}\rangle. \tag{45}$$

Formula (45) defines the transformation θ for any particle–antiparticle pair independent of the presence of other internal symmetry properties [which will be approximate, since all exact symmetries are exhausted by the transformation group (41)].

If one considers an approximate internal symmetry (the group $\mathfrak{G}'$), the particles will be connected with the irreducible representations f of the group $\mathfrak{G}'$. These representations may lead to an operator θ_0 of any type. Denoting by α the component of the multiplet f of the approximate symmetry group and writing separately the exactly conserved charges Q_i, we obtain, by virtue of (36) and (37), for types 1 and 2,

$$\theta\,|\,\boldsymbol{p},\, \lambda;\, J,\, m; f,\, \alpha,\, Q_i',\, \text{in}\rangle = (-1)^{J-\lambda}\sum_{\alpha'}|\,\boldsymbol{p},\, -\lambda;\, J,\, m;\, \bar{f},\, \alpha',\, -Q_i',\, \text{out}\rangle\beta^f_{\alpha'\alpha}, \tag{46}$$

where $\bar{f}$ refers to the antiparticle multiplet, while $\beta^f = \beta^{\bar{f}} = \beta_+$ for type 1; $\beta^f = -\beta^{\bar{f}} = \beta$ for type 2. On the right-hand side of (46) the multiplet $\bar{f}$ of antiparticles is summed over the states α'. By (36) and (37) the multiplets f and $\bar{f}$ transform in the same way.

Comparing (45) and (46) one may find the antiparticle state for any particle. If the quantum numbers (f, α) denote the particle α in the multiplet f, the antiparticle will have the quantum numbers $\overline{(f, \alpha)}$ and will be described by the state

$$|\,\boldsymbol{p},\, -\lambda;\, J,\, m;\, \overline{(f, \alpha)},\, -Q_i',\, \text{out}\rangle = \sum_{\alpha'}|\,\boldsymbol{p},\, -\lambda;\, J,\, m;\, \bar{f},\, \alpha';\; -Q_i',\, \text{out}\rangle\beta^f_{\alpha'\alpha}. \tag{47}$$

As a consequence of the unitarity of the matrices β these matrices vanish when calculating probability amplitudes.

Let us denote the set of invariant quantum numbers of particles and antiparticles by $a, b, \ldots$, and $\bar{a}, \bar{b}$. ... The many-particle in- and out-states

$$|\,\boldsymbol{p}_a,\, \lambda_a,\, a;\, \boldsymbol{p}_b,\, \lambda_b,\, b;\, \ldots\,;\, \text{in (out)}\rangle$$

transform under the reflection θ as the product of single-particle states. For example, for the two-particle state

$$\theta\,|\,\boldsymbol{p}_a,\, \lambda_a,\, a;\, \boldsymbol{p}_b,\, \lambda_b,\, b;\, \text{in (out)}\rangle = (-1)^{J_a+J_b-\lambda_a-\lambda_b}\,|\,\boldsymbol{p}_a,\, -\lambda_a,\, \bar{a};\, \boldsymbol{p}_b,\, -\lambda_b,\, \bar{b};\, \text{out (in)}\rangle. \tag{48}$$

This composition rule for transformations θ may either be postulated, relying on the properties of asymptotic fields, or introduced from quantum field theory, based on the explicit representation of multi-particle states in terms of creation operators.

The invariance of the S-matrix with respect to the reflection θ means that $S_\theta = S$, or

$$S = \theta S^+ \theta^{-1}.$$

The reflection θ connects the amplitude for the process $a+b+\ \dots \to c+d+\ \dots$ with the amplitude for the process $\bar{c}+\bar{d}+\ \dots \to \bar{a}+\bar{b}+\ \dots$, with opposite helicities:

$$\begin{aligned}&\langle \boldsymbol{p}_c, \lambda_c, c; \boldsymbol{p}_d, \lambda_d, d; \dots | S | \boldsymbol{p}_a, \lambda_a, a; \boldsymbol{p}_b, \lambda_b, b; \dots \rangle \\ &\quad = \eta_\theta \langle \boldsymbol{p}_a, -\lambda_a, \bar{a}; \boldsymbol{p}_b, -\lambda_b, \bar{b}; \dots | S | \boldsymbol{p}'_c, -\lambda_c, \bar{c}; \boldsymbol{p}_d, -\lambda_d, \bar{d}; \dots \rangle,\end{aligned} \tag{49}$$

where η_θ is equal to the product of the factors $(-1)^{J-\lambda}$ for all particles in the process. According to the CPT theorem, invariance in the sense of (49) is an automatic consequence of relativistic invariance and locality (for positivity of the energy spectrum).

The transformation law of creation and annihilation operators which follows from (46) has the form

$$\theta a^+_{\text{in}}(\boldsymbol{p}, \lambda)\theta^{-1} = \beta_a(-1)^{J_a-\lambda}\, b^+_{\text{out}}(\boldsymbol{p}, -\lambda), \tag{50}$$

where a^+ refers to a particle and b^+ to an antiparticle. Consequently the (non-Hermitian) $2(2J+1)$-component fields $\Phi^J(x)$ (see § 5.3) transform according to the rule

$$\left.\begin{aligned}\theta \Phi^J_{\text{in}}(x)\theta^{-1} &= \Phi^{+J}_{\text{out}}(-x), && J = 0, 1, 2, \dots, \\ \theta \Phi^J_{\text{in}}(x)\theta^{-1} &= \gamma_5 \Phi^{+J}_{\text{out}}(-x), && J = \tfrac{1}{2}, \tfrac{3}{2}, \dots .\end{aligned}\right\} \tag{51}$$

We note that according to (51) the $(2J+1)$-component fields $\frac{1}{2}(1\pm\gamma_5)\Phi^J$ transform separately. In the case of the vector field B_μ, reflection entails

$$\theta B_\mu(x)\theta^{-1} = -B^+_\mu(-x). \tag{52}$$

Knowing the reflection properties of the vector and spinor field, one may easily write the reflection rule for the Rarita–Schwinger field:

$$\theta \psi_{\mu_1 \dots \mu_n}(x)\theta^{-1} = \gamma_5(-1)^{J-\frac{1}{2}}\psi^+_{\mu_1 \dots \mu_n}(-x), \quad J = n+\tfrac{1}{2}, \tag{53}$$

($\mu = 0, 1, 2, 3$; n an integer).

For the Bargmann–Wigner field

$$\theta \psi_{\beta_1 \dots \beta_n}(x)\theta^{-1} = (\gamma_5^{(1)})_{\beta_1}^{\beta'_1} \dots (\gamma_5^{(n)})_{\beta_n}^{\beta'_n} \psi^{+'}_{\beta'_1 \dots \beta'_n}(-x), \tag{54}$$

which is easily derived from the properties of the spinor field (n is an integer).

The Schwinger treatment of reflections follows from the same relations (4), (6)–(8), but the transformed state here is defined using the operator θ^S with the properties

$$\left.\begin{aligned}\theta^S |a, \text{in}\rangle &= \langle a_\theta, \text{out}|, \\ \theta^S |a, \text{out}\rangle &= \langle a_\theta, \text{in}| \\ \langle \theta^S a, \text{out} | \theta^S b, \text{in}\rangle &= \langle b_\theta, \text{out} | a_\theta, \text{in}\rangle = \langle a_{\text{out}} | b_{\text{in}}\rangle.\end{aligned}\right\} \tag{55}$$

The operators θ^S may be represented in the form of a product

$$\theta^S = UT, \tag{56}$$

where U is unitary and T is the transposition operator: $T|a\rangle = \langle a|$. A transformed operator H_θ, by (8) and (49), is defined by the formula

$$H_\theta = \theta^S H(\theta^S)^{-1}, \tag{57}$$

and instead of the invariance condition (13) in the Wigner treatment, we now have $H = \theta^S H(\theta^S)^{-1}$ for a θ-invariant operator. The order of operator multiplication is reversed through the operation θ^S:

$$(MN)_\theta = N_\theta M_\theta. \tag{58}$$

Comparison of (13) and (57) shows that Hermitian operators transform in the same way in both treatments, and the formulae for transformation of specific operators via θ^S may be obtained from the respective formulae with the operator θ by removing or adding the Hermitian conjugation operation. For example, instead of (50) we have

$$\theta^S a_{\text{in}}^{+}(\boldsymbol{p}, \lambda)\,(\theta^S)^{-1} = \beta_a(-1)^{J_a-\lambda}\, b_{\text{out}}(\boldsymbol{p}, -\lambda), \tag{59}$$

while the transformation of the boson field Φ^J now has the form

$$\theta^S \Phi^J_{\text{in}}(x)\,(\theta^S)^{-1} = \Phi^J_{\text{out}}(-x). \tag{60}$$

The transformation properties of fields are summarized in Table 6.1 (p. 144).

§ 6.2. The operations P, C, and T

While the classical inversion $Ix = -x$ does not change internal symmetry variables, the quantum mechanical reflection θ, by virtue of its antiunitarity, exchanges particles and antiparticles. Let us discuss separately the transformation C which exchanges particles and antiparticles and the "geometrical" reflections P and T, describing the reflection θ as the product of three discrete operations:

$$\theta = \text{CPT}, \tag{61}$$

where C and P are unitary and T is antiunitary.

Charge conjugation C transforms a particle into an antiparticle but does not act on the variables of the Poincaré group:

$$[\text{C}, M_{\mu\nu}] = 0, \quad [\text{C}, P_\mu] = 0. \tag{62}$$

The operator P describes "geometrical" spatial reflection, i.e. that transformation of states or operators (induced by the coordinate reflection $\boldsymbol{x} \to -\boldsymbol{x}$), which commutes with C:

$$\text{CPC}^{-1} = \text{P}. \tag{63}$$

The operation T when combined with CP gives the total reflection θ. From the definition of C, P, and T it is clear that the effect of the geometrical reflections P and T does not depend on the charges of the particles. The operations P and T connect state vectors in the same coherent subspace (characterized by the set of charges Q_i). Charge conjugation C takes a state vector from one coherent subspace into another differing from the first by the sign of all charges Q_i.

Parity

The commutation relations of the parity operation P with momentum and angular momentum in classical physics are easily derived from

$$\mathrm{P}(\mathrm{a}, A)\mathrm{P}^{-1} = (\mathrm{a}', \sigma_2 A^* \sigma_2), \quad a' = (a^0, -a), \tag{64}$$

where (a, A) is a transformation of the Poincaré group. The parity operation P anticommutes with spatial components of vectors and commutes with time components, e.g.

$$\left.\begin{aligned} \mathrm{P}P_k\mathrm{P}^{-1} &= -P_k, & \mathrm{P}P_0\mathrm{P}^{-1} &= P_0 \\ \mathrm{P}M_{ik}\mathrm{P}^{-1} &= M_{ik}, & \mathrm{P}M_{0k}\mathrm{P}^{-1} &= -M_{0k}, \end{aligned}\right\} \tag{65}$$

These commutation relations also hold in quantum theory, where P is a unitary operator.

According to (63) and (65) the operator P takes a state of a particle with momentum $\boldsymbol{p}$ into a state of the same particle with momentum $-\boldsymbol{p}$, without changing the three-dimensional angular momentum. The rest state vector in the canonical basis thus acquires at most a phase factor η_{P}^* under the action of P:

$$\mathrm{P}\,|0, \sigma; a\rangle = \eta_{\mathrm{P}}^*(a)\,|0, \sigma; a\rangle, \tag{66}$$

where σ is the projection of the spin on the z-axis and a is the type of the particle (a set of invariant quantum numbers); $\eta_{\mathrm{P}}^*(a)$ in the case of bosons is called the (intrinsic) parity of the particle a.

Passing to a moving reference frame using the Lorentz transformation $\alpha(\boldsymbol{p})$, we find the effect of P on a state vector in the canonical basis:

$$\begin{aligned} \mathrm{P}\,|\boldsymbol{p}, \sigma; a\rangle &= \mathrm{PU}(0, \alpha(\boldsymbol{p}))\mathrm{P}^{-1}\mathrm{P}\,|0, \sigma; a\rangle \\ &= \eta_{\mathrm{P}}^*(a)\,\mathrm{U}(0, \alpha(-p))\,|0, \sigma; a\rangle = \eta_{\mathrm{P}}^*(a)\,|-\boldsymbol{p}, \sigma; a\rangle, \end{aligned} \tag{67}$$

since $\mathrm{PU}(0, \alpha(\boldsymbol{p}))\mathrm{P}^{-1} = \mathrm{U}(0, \alpha(-\boldsymbol{p}))$ by virtue of (65). For the antiparticle $\bar{a}$ we shall write

$$\mathrm{P}\,|\boldsymbol{p}, \sigma; \bar{a}\rangle = \bar{\eta}_{\mathrm{P}}^*(a)\,|-\boldsymbol{p}, \sigma; \bar{a}\rangle. \tag{68}$$

In the helicity basis, the application of P changes the helicity state of a particle for $\boldsymbol{p} \neq 0$. Since in this basis the momentum is introduced along the z-axis to start with, it is convenient to use instead of P the operator

$$Y = e^{-i\pi M_{31}}\mathrm{P}, \tag{69}$$

which commutes with a Lorentz transformation along the z-axis. Since in the case of spin $J(\boldsymbol{p} = 0)$

$$e^{-i\pi M_{31}}\,|0, \lambda; J\rangle = (-1)^{J-\lambda}\,|0, -\lambda; J\rangle.$$

for nonzero momentum $\boldsymbol{p}$ along the z-axis

$$Y\,|\boldsymbol{p}, \lambda; J\rangle = \eta_{\mathrm{P}}^*(-1)^{J-\lambda}\,|\boldsymbol{p}, -\lambda; J\rangle. \tag{70}$$

Performing the rotation $R(\varphi, \vartheta, -\varphi)$, we find the effect of P on a state of a particle in the helicity basis with momentum $\boldsymbol{p} = (p, \vartheta, \varphi)$:

$$\mathrm{P}\,|p, \vartheta, \varphi, \lambda; J\rangle = \eta_{\mathrm{P}}(-1)^{J-\lambda}\, e^{i\pi M_{31}}\,|p, \vartheta, -\varphi, -\lambda; J\rangle. \tag{71}$$

Assuming that the vacuum is invariant with respect to P,

$$\mathrm{P}\,|0\rangle = |0\rangle, \tag{72}$$

we find from (67) the transformation of the creation and annihilation operators, e.g.:

$$\mathrm{P}a(\boldsymbol{p}, \sigma)\mathrm{P}^{-1} = \eta_{\mathrm{P}}a(-\boldsymbol{p}, \sigma). \tag{73}$$

Let us find the transformation of the $(2J+1)$-component local fields φ and χ under P. According to (5.80) and (5.81)

$$\left.\begin{aligned}\varphi(x) &= \frac{1}{(2\pi)^{3/2}}\int \frac{d^3p}{2p_0}\{D^J(\alpha(\boldsymbol{p}))\, e^{-ipx}a(\boldsymbol{p}) + D^J(\alpha(\boldsymbol{p})C^{-1})\, e^{ipx}b^+(\boldsymbol{p})\},\\ \chi(x) &= \frac{1}{(2\pi)^{3/2}}\int \frac{d^3p}{2p_0}\{D^J(\alpha^{-1+}(\boldsymbol{p}))\, e^{-ipx}a(\boldsymbol{p}) + (-1)^{2J}\, D^J(\alpha^{-1+}(\boldsymbol{p})C^{-1})\, e^{ipx}b^+(\boldsymbol{p})\}.\end{aligned}\right\} \tag{74}$$

Let us apply P to the field φ; by virtue of (73)

$$\mathrm{P}\varphi(x)\mathrm{P}^{-1} = \eta_{\mathrm{P}}\chi^{(a)}(x_0, -\boldsymbol{x}) + (-1)^{2J}\,\bar{\eta}_{\mathrm{P}}^{*}\chi^{(b)}(x_0, -\boldsymbol{x}),$$

where $\chi^{(a)}$ and $\chi^{(b)}$ are the parts of the field χ containing respectively the operators a and b^+. Let us demand that the transformation of the field $\varphi(x)$ be local, i.e. that the field $\mathrm{P}\varphi\mathrm{P}^{-1}$ be expressible in terms of the field χ:

$$\mathrm{P}\varphi(x)\mathrm{P}^{-1} = \eta_{\mathrm{P}}(a)\,\chi(x_0, -\boldsymbol{x}). \tag{75}$$

This is possible when

$$\bar{\eta}_{\mathrm{P}}\eta_{\mathrm{P}} = (-1)^{2J}. \tag{76}$$

Consequently, for fermions the product of the parities of particles and antiparticles is negative.

Analogously we find

$$\mathrm{P}\chi(x)\mathrm{P}^{-1} = \eta_{\mathrm{P}}(a)\,\varphi(x_0, -\boldsymbol{x}), \tag{77}$$

so that P connects upper and lower components of the $2(2J+1)$-component field

$$\Phi = \begin{pmatrix}\varphi\\ \chi\end{pmatrix}, \quad \mathrm{P}\Phi(x)\mathrm{P}^{-1} = \eta_{\mathrm{P}}\gamma_4\Phi(x_0, -\boldsymbol{x}), \tag{78}$$

where the matrix γ_5 is assumed diagonal (see § 5.3).

Consequently, when taking into account the operation P the irreducible object is a $2(2J+1)$-component field Φ. Earlier, in § 5.3, such a field was introduced to enable the construction of a bilinear Hermitian form. The two conditions (the condition of the existence of P and the condition of the existence of a bilinear Hermitian form) are related to one another: according to (78), under the parity operation the same matrix γ_4 appears which enters into the bilinear Hermitian form and changes the sign of the generators of the Lorentz transformation M_{0k}.

The parity η_P was required to satisfy $|\eta_P| = 1$, but since P does not depend on the charges of the particles, the parity η_P is equal to the parity of a neutral particle. Consequently, one must have $\bar{\eta}_P = \eta_P$, so that by (76)

$$\bar{\eta}_P = \eta_P = \pm i \quad \text{for fermions} \quad \bar{\eta}_P = \eta_P = \pm 1 \quad \text{for bosons.} \tag{79}$$

The parity of bosons and the spin J are usually combined in the combination J^P (where the superscript denotes the eigenvalue η_P of P in the rest state). Pseudo-scalar mesons (pions, kaons, etc.) have $J^P = 0^-$, vector mesons (ϱ and ω mesons, etc.) have $J^P = 1^-$, pseudo-vector (A_1, K^*, etc.) have $J^P = 1^+$, and so on.

In the case of fermions, where the choices $\pm i$ are equivalent, since the state vectors of all fermions are defined up to a sign (see § 1.3), one may discuss only the relative parity of particles. For example, the notation $J^P = \frac{1}{2}^+$ for a Λ particle reflects the fact that its parity is the same as that of the neutron, which is taken as a standard for baryons.

Performing the parity operation P twice, we must obtain a state differing from the initial one only by a phase factor. Since the operation P is the same as for a neutral particle, this phase must be related only to the Poincaré group. In fact, according to (79)

$$P^2 = U(0, -E) = (-1)^{2J} \quad \text{for spin} \quad J, \tag{80}$$

where $U(0, -E) = z$ is a transformation of the spinor group corresponding to a rotation $z = e^{iM_{12}2\pi}$ by an angle 2π.

The imaginary factor $\pm i$, as a phase factor η_P for fermions, entails the anticommutativity of P and θ for fermions, since θ is antiunitary; for bosons P and θ commute, so that, in general,

$$P\theta = (-1)^{2J}\,\theta P. \tag{81}$$

The parity $\eta_P^*(a, b, \ldots)$ of multi-particle states is equal to the product of the parities $\eta_P^*(a)\,\eta_P^*(b)\ldots$ of single-particle states. For example, in the canonical basis, we have, for a two-particle state,

$$P\,|\boldsymbol{p}_a, \sigma_a; \boldsymbol{p}_b, \sigma_b\rangle = \eta_P^*(a)\,\eta_P^*(b)\,|-\boldsymbol{p}_a, \sigma_a; -\boldsymbol{p}_b, \sigma_b\rangle. \tag{82}$$

In the helicity basis when constructing a two-particle state (see § 4.6) two types of helicity states were used: (4.61) and (4.63). In one of them the momentum was first introduced along the positive z-axis, while in the other the momentum was first directed along the negative z-axis. It is obvious that these states may be obtained from one another by a reflection. For the momentum $|\boldsymbol{p}|\,\boldsymbol{e}_z$ we find from (4.63) and (71), using (4.64),

$$P\,|p, 0, 0, \lambda\rangle^- = \overset{*}{\eta}_P\,|p, 0, 0, -\lambda\rangle = (-1)^{J+\lambda}\,e^{i\pi M_{31}}\,\overset{*}{\eta}_P\,|p, 0, 0, -\lambda\rangle^-.$$

For the two-particle helicity states (4.139) in the center of mass system and with relative momentum along the z-axis (i.e. $\vartheta = \varphi = 0$), the parity transformation thus has the form

$$\begin{aligned} P\,|0; p, 0, 0; \lambda_a\,\lambda_b\rangle &\equiv P\{|p, 0, 0; \lambda_a\rangle\,|p, 0, 0; \lambda_b\rangle^-\} \\ &= \overset{*}{\eta}_P(1)\,\overset{*}{\eta}_P(2)\,(-1)^{J_a+J_b-\lambda_a+\lambda_b}\,e^{iM_{31}\pi}\,|0; p, 0, 0; -\lambda_a, -\lambda_b\rangle. \end{aligned}$$

Since the operation P commutes with rotations, we may rotate the relative momentum from the z-axis to the final direction $\boldsymbol{p} = (p, \vartheta, \varphi)$:

$$\begin{aligned} \mathrm{P}\,|0; p, \vartheta, \varphi; \lambda_a, \lambda_b\rangle &\equiv \mathrm{P}R(\varphi, \vartheta, -\varphi)\,|0; p, 0, 0; \lambda_a, \lambda_b\rangle \\ &= \overset{*}{\eta}_{\mathrm{P}}(1)\,\overset{*}{\eta}_{\mathrm{P}}(2)\,(-1)^{J_a+J_b-\lambda_a+\lambda_b}\,R(\varphi, \vartheta, -\varphi) \\ &\quad\times R(0, -\pi, 0)\,|0; p, 0, 0; -\lambda_a, -\lambda_b\rangle. \end{aligned} \tag{83}$$

The effect of the parity operation on the state $|0, \Lambda; J, m; \lambda_a, \lambda_b\rangle$ with definite (two-particle) spin J, mass m, and spin projection Λ (in the c.m.s. $\boldsymbol{P} = 0$) we find, inserting (83) into the general formula (4.145) to be:

$$\mathrm{P}\,|0, \Lambda; J, m; \lambda_a, \lambda_b\rangle = \overset{*}{\eta}_{\mathrm{P}}(a)\,\overset{*}{\eta}_{\mathrm{P}}(b)\,(-1)^{J-J_a-J_b}\,|0, \Lambda; J, m; -\lambda_a, -\lambda_b\rangle. \tag{84}$$

Thus the spatial reflection of a two-particle state with spin J differs from the initial one by the sign of the helicities of the particles.

The superposition of states with opposite helicities (and the same total spin J)

$$|\ldots; \lambda_a, \lambda_b\rangle \pm \overset{*}{\eta}_{\mathrm{P}}(a)\,\overset{*}{\eta}_{\mathrm{P}}(b)\,(-1)^{2J-J_a-J_b}\,|\ldots; -\lambda_a, -\lambda_b;\rangle$$

will have definite parity.

If the S-matrix is invariant with respect to spatial reflections:

$$\mathrm{P}S\mathrm{P}^{-1} = S,$$

then according to (82) the following relation will hold between its matrix elements

$$\begin{aligned} \langle \boldsymbol{p}_a\sigma_a; \boldsymbol{p}_b\sigma_b; \ldots |S| \boldsymbol{p}_{a'}\sigma_{a'}; \boldsymbol{p}_{b'}\sigma_{b'}; \ldots\rangle &= \eta_{\mathrm{P}}(a, b, \ldots)\,\overset{*}{\eta}_{\mathrm{P}}(a', b', \ldots) \\ &\quad\times \langle -\boldsymbol{p}_a\sigma_a; -\boldsymbol{p}_b\sigma_b; \ldots |S| -\boldsymbol{p}_{a'}\sigma_{a'}; -\boldsymbol{p}_{b'}\sigma_{b'}; \ldots\rangle. \end{aligned} \tag{85}$$

Here $\eta_{\mathrm{P}}(a', b', \ldots)$ and $\eta_{\mathrm{P}}(a, b, \ldots)$ are the products of the parities of the particles in the initial and final states. The right- and left-hand sides of these relations refer to the same reaction occurring under different conditions. Equation (85) thus constitutes a restriction on transition amplitudes.

Charge conjugation

Charge conjugation C is defined as a unitary operation replacing particles by antiparticles:

$$\mathrm{C}\,|p, \sigma; a\rangle = \overset{*}{\eta}_{\mathrm{C}}\,|p, \sigma; \bar{a}\rangle, \quad \mathrm{CC}^{+} = 1, \tag{86}$$

which by virtue of the charge invariance of the vacuum ($\mathrm{C}\,|0\rangle = |0\rangle$) may be formulated via creation and annihilation operators for particles (a^{+}, a) and antiparticles (b^{+}, b) as follows:

$$\left.\begin{aligned} \mathrm{C}a(\boldsymbol{p}, \sigma)\mathrm{C}^{-1} &= \eta_{\mathrm{C}} b(\boldsymbol{p}, \sigma), & \overset{*}{\eta}_{\mathrm{C}}\eta_{\mathrm{C}} &= 1, \\ \mathrm{C}b(\boldsymbol{p}, \sigma)\mathrm{C}^{-1} &= \bar{\eta}_{\mathrm{C}} a(\boldsymbol{p}, \sigma), & \overset{*}{\bar{\eta}}_{\mathrm{C}}\bar{\eta}_{\mathrm{C}} &= 1. \end{aligned}\right\} \tag{87}$$

The operator C commutes with the generators P_μ and $M_{\mu\nu}$, and also with the parity operation P. By virtue of (86), C anticommutes with all charges Q_i:

$$\mathrm{C}Q_i = -Q_i\mathrm{C}, \tag{88}$$

and, consequently, connects different coherent subspaces with one another. Only when all charges are equal to zero does the operator C act within one coherent subspace.

We note that charge conjugation C, being a unitary operator, cannot change the sign of all generators F_r of an internal symmetry group if they are noncommuting generators with commutation relations of the type

$$[F_i, F_j] = ic_{ijk}F_k$$

(for real antisymmetric c_{ijk}). Indeed, substituting $F_i \to -F_i$ leads to a different algebra of generators.

If a particle is neutral ($a = \bar{a}$), i.e. all its charges $Q_i = 0$, then according to (86) and (88), it is characterized by an additional quantum number—the charge parity $\eta_C^* = \eta_C = \pm 1$, showing how the state of a neutral particle behaves under charge conjugation.

The charge parity C of neutral particles is indicated along with their spin and parity in the combination J^{PC}. For example, the particle π^0 has even charge parity, $J^{PC} = 0^{-+}$, the meson ω has negative charge parity, $\eta_C(\omega) = -1$, $J^{PC} = 1^{--}$. Any neutral state may have a charge parity. Such a state must contain the same number of particles and their antiparticles.

If the scattering matrix is invariant with respect to charge conjugation, $CSC^{-1} = S$, and the initial state has a definite charge parity $\eta_C(\text{in})$, then the interaction will not change $\eta_C(\text{in})$, so that $\eta_C(\text{out}) = \eta_C(\text{in})$. If the initial state is not neutral, then the C-invariance of the scattering matrix does not entail a conservation law but allows one to express the amplitude for the process with particles $a+b+\ldots \to c+d+\ldots$ in terms of the amplitude for the process with antiparticles $\bar{a}+\bar{b}+\ldots \to \bar{c}+\bar{d}+\ldots$:

$$\langle c, d, \ldots | S | a, b, \ldots \rangle = \eta_C(\text{in})\, \bar{\eta}_C(\text{out})\, \langle \bar{c}, \bar{d}, \ldots | S | \bar{a}, \bar{b}, \ldots \rangle, \tag{89}$$

where $\eta_C(\text{in})$ is the product of the phases η_C in the in-state. The momenta and helicities of all particles and antiparticles in (89) are equal in pairs.

Let us perform the C-transformation on the $2(2J+1)$-component field Φ of the form (78). According to (87) the field $\Phi_C = C\Phi C^{-1}$ will contain annihilation operators with the factor η_C and creation operators with the factor $\bar{\eta}_C^*$. In order that the field Φ transform as a whole, it is necessary that

$$\eta_C = \bar{\eta}_C^*. \tag{90}$$

Bearing in mind that $C\alpha^{-1T}C^{-1} = \alpha$ and $CC^* = -1$,

$$\left.\begin{aligned} \Phi_C(x) &= \eta_C \mathscr{C}_J \gamma_5 \gamma_4 \Phi^+(x) \quad \text{(bosons)}, \\ \Phi_C(x) &= \eta_C \mathscr{C}_J \Phi^+(x) \gamma_4 \quad \text{(fermions)}, \end{aligned}\right\} \tag{91}$$

where the matrix $\mathscr{C}_J$ is

$$\mathscr{C}_J = \begin{pmatrix} D^J(C) & 0 \\ 0 & -D^J(C) \end{pmatrix}. \tag{92}$$

In the case of the Dirac bi-spinor, $D^J(C) = C$, so that

$$\psi_C = \eta_C \mathscr{C} \bar{\psi},$$

where the matrix $\mathscr{C}$ was introduced in (5.66) and has the properties (5.64) (with the choice of the lower sign) and (5.65).

It was shown earlier (78) that the parity operation P interchanges the upper and lower components of the field $\Phi : \varphi \leftrightarrow \chi$. Charge conjugation expresses φ in terms of χ^+ and χ in terms of φ^+. For this reason, the combined reflection CP separately connects the upper and lower components Φ and Φ^+:

$$\left.\begin{aligned} \mathrm{CP}\Phi(x)(\mathrm{CP})^{-1} &= \eta_{\mathrm{C}}\eta_{\mathrm{P}}\mathscr{C}_J\gamma_5\Phi^+(x^0, -\boldsymbol{x}) \quad \text{(bosons)}, \\ \mathrm{CP}\Phi(x)(\mathrm{CP})^{-1} &= -\eta_{\mathrm{C}}\eta_{\mathrm{P}}\mathscr{C}_J\Phi^+(x^0, -\boldsymbol{x}) \quad \text{(fermions)}. \end{aligned}\right\} \tag{93}$$

According to the definition (63), P commutes with charge conjugation: PC = CP. The choice of the phase $\eta_{\mathrm{P}} = \pm i$ for fermions [see (79)] guarantees the independence of the operations P and C.

The commutation relations of C and CP with total reflection θ may be introduced via (45), (71), and (86):

$$\theta \mathrm{C} = \mathrm{C}\theta, \quad \theta(\mathrm{CP}) = (-1)^{2J}(\mathrm{CP})\theta. \tag{94}$$

Time reversal

The antiunitary time-reversal transformation T may be found by comparing formulae (45), (71), and (86) for θ = CPT and the operations P and C:

$$\mathrm{T} = (\mathrm{CP})^{-1}\theta, \quad \mathrm{T}^2 = \mathrm{P}^2 = (-1)^{2J}. \tag{95}$$

Bearing in mind that in the canonical basis the reflection θ acts according to (45), or

$$\theta\,|\,\boldsymbol{p}, \sigma; a, \mathrm{in}\rangle = (-1)^{J_a-\sigma}\,|\,\boldsymbol{p}, -\sigma; \bar{a}, \mathrm{out}\rangle,$$

we find

$$\mathrm{T}\,|\,\boldsymbol{p}, \sigma; a, \mathrm{in}\rangle = \eta_{\mathrm{T}}(-1)^{J_a-\sigma}\,|\,-\boldsymbol{p}, -\sigma; a, \mathrm{out}\rangle, \tag{96}$$

where the phase factor η_{T} does not depend on p or σ. As in the case of reflections θ, we have retained the symbols in and out in the formulae for single-particle states to facilitate the transition to multi-particle states and quantum fields.

Time reversal changes the sign of momentum and spin projection of particles, and also takes an in-state into an out-state and vice versa. The particle type remains unchanged here: all charges Q_i and other internal symmetry quantum numbers are invariant with respect to time-reversal T, which justifies the treatment of T as a "geometrical" time reflection.

The commutation relations for T may be found by comparing the commutation relations for θ, P, and C:

$$\left.\begin{aligned} \mathrm{T}P_k\mathrm{T}^{-1} &= -P_k, & \mathrm{T}P_0\mathrm{T}^{-1} &= P_0, \\ \mathrm{T}M_{ij}\mathrm{T}^{-1} &= -M_{ij}, & \mathrm{T}M_{0j}\mathrm{T}^{-1} &= M_{0j}, \\ \mathrm{T}Q_i\mathrm{T}^{-1} &= Q_i. \end{aligned}\right\} \tag{97}$$

From these relations it is clear that the operation T does not change a helicity state. Thus in the helicity basis with momentum $\boldsymbol{p}$ along the z-axis,

$$\mathrm{T}\,|\,p, 0, 0, \lambda; a, \mathrm{in}\rangle = \eta_{\mathrm{T}}^*\,|\,p, \pi, 0, \lambda; a, \mathrm{out}\rangle = \eta_{\mathrm{T}}^* e^{-i\pi M_{31}}\,|\,p, 0, 0, \lambda; a, \mathrm{out}\rangle. \tag{98}$$

Rotations commute with T:

$$\mathrm{T}R(\varphi, \vartheta, \psi)\mathrm{T}^{-1} = R(\varphi, \vartheta, \psi),$$

so that (98) leads to the general transformation:

$$\begin{aligned}\mathrm{T}\,|\,p, \vartheta, \varphi, \lambda; a, \text{in}\rangle &= \mathrm{T}R(\varphi, \vartheta, -\varphi)\,|\,p, 0, 0, \lambda; a, \text{in}\rangle \\ &= \eta_{\mathrm{T}}^{*} R(\varphi, \vartheta, -\varphi)\, R(0, \pi, 0)\,|\,p, 0, 0, \lambda; a, \text{out}\rangle. \end{aligned} \tag{99}$$

This formula remains valid for the other single-particle helicity state, $|\boldsymbol{p}, \lambda\rangle^{-}$ [see (4.63) and (4.64)].

The transformation of multi-particle states under time-reversal T may be found as a product of the transformations (96) or (99) of single-particle states. The effect of T on a two-particle state with definite total spin J in the helicity basis may now be found in analogy with the way in which the effect (84) of P was found. Using (99) we obtain instead of (83) the expression

$$\mathrm{T}\,|\,0; p, \vartheta, \varphi; \lambda_a, \lambda_b\rangle = \eta_{\mathrm{T}}^{*}(a)\,\eta_{\mathrm{T}}^{*}(b)\, R(\varphi, \vartheta, -\varphi)\; R(0, \pi, 0)\,|\,0; p, 0, 0; \lambda_a\, \lambda_b\rangle,$$

from which, by (4.145), we may introduce the desired transformation law

$$\mathrm{T}\,|\,0, \Lambda; J, m; \lambda_a, \lambda_b\rangle = \eta_{\mathrm{T}}^{*}(a)\,\eta_{\mathrm{T}}^{*}(b)\,(-1)^{J-\lambda}\,|\,0, -\Lambda; J, m; \lambda_a, \lambda_b\rangle, \tag{100}$$

where Λ is the projection of the total spin J on the z-axis.

The transformation of operators for antiunitary T is defined by a formula of the same type (12) as in the case of total reflection θ. If the S-matrix is invariant with respect to time reversal, then the condition

$$S = S_{\mathrm{T}} = \mathrm{T}S^{+}\mathrm{T}^{-1}.$$

holds. This condition is known as the reciprocity theorem, and in detailed notation reads

$$\begin{aligned}&\langle \boldsymbol{p}_c, \lambda_c; \boldsymbol{p}_d, \lambda_d; \ldots\,|S|\,\boldsymbol{p}_a, \lambda_a; \boldsymbol{p}_b, \lambda_b; \ldots\rangle \\ &\quad = \pm\langle -\boldsymbol{p}_a, \lambda_a; -\boldsymbol{p}_b, \lambda_b; \ldots\,|S|\,-\boldsymbol{p}_c, \lambda_c; -\boldsymbol{p}_d, \lambda_d; \ldots\rangle. \end{aligned} \tag{101}$$

The reciprocity theorem connects the amplitudes for the processes $a+b+\ \ldots \rightarrow c+d+\ \ldots$ and $c+d+\ \ldots \rightarrow a+b+\ \ldots$, where in the second process the momenta of the particles differ in sign from the momenta of particles in the first process. For elastic processes when the initial and final particles are identical, the condition (101) limits the possible form of the transition amplitude.

§ 6.3. Reflections and interactions. Decays

Writing the total reflection operator in the form $\theta = \mathrm{CPT}$ implies that each of the operations P, C, and T may be defined separately. If the S-matrix were invariant with respect to the operation C, P, and T, i.e. if the conditions

$$\mathrm{C}S\mathrm{C}^{-1} = S, \quad \mathrm{P}S\mathrm{P}^{-1} = S, \quad \mathrm{T}S^{+}\mathrm{T}^{-1} = S \tag{102}$$

held, then the formulae of § 6.2 would have defined these operations. However, in reality, all the symmetries P, C, and T are only approximate. In the case of broken symmetry, the definitions of the operations P, C and T presented in § 6.2 are, strictly speaking, unsound or useless.

The operations P, C, and T were defined in § 6.2 both in terms of state vectors and by commutation relations. Two cases are possible in which these operators lose their meaning.

1. The definitions (67), (82), and others assume that the application of the operation P to the physical state $|\alpha\rangle$ leads to another physical state. But if symmetry with respect to spatial reflections is broken, the vector $P|\alpha\rangle$ may not correspond to a physical state it may be missing in the Hilbert space $\mathcal{H}$. The space $\mathcal{H}$ will then be noninvariant with respect to reflections P. The operation P thus loses its meaning. Analogous considerations hold for the operations C and T.

An example is the neutrino. The operations P and C when applied to the state vector of a neutrino have no meaning since both P and C take a neutrino state into a nonexistent one and only the operation CP connects physical states—the neutrino ν and the antineutrino $\bar{\nu}$:[92, 93]

$$\mathrm{CP}\,|\,\boldsymbol{p},\lambda_\nu,\nu\rangle = \eta_\nu|-\boldsymbol{p},-\lambda_\nu,\bar{\nu}\rangle. \tag{103}$$

The helicity of the neutrino is equal to $\lambda_\nu = -\frac{1}{2}$, while the helicity of the antineutrino differs in sign: $\lambda_{\bar{\nu}} = -\lambda_\nu$. As was shown in § 4.5, from the point of view of the Poincaré group, the neutrino and antineutrino are associated with different irreducible representations.

2. The vectors $P|\alpha, \text{in}\rangle$ and $P|\alpha, \text{out}\rangle$ exist among the set of physical in- or out-vectors and, consequently, a definition of the type (67) has meaning, but the condition (102) for P does not hold (the S-matrix is noninvariant with respect to spatial reflections). In this case, one can introduce two different operators P_{in} and P_{out} in the overall Hilbert space $\mathcal{H} = \mathcal{H}_{\text{in}} = \mathcal{H}_{\text{out}}$. These operators will be defined correspondingly with respect to in- and out-states. The connection between the operators P_{in} and P_{out} is established via the S-matrix:

$$P_{\text{in}} = SP_{\text{out}}S^{-1}.$$

Such operations P_{in} and P_{out} will be characterized by the properties of the initial and final configurations of the system but not by the symmetry of the laws of motion or interactions. For this reason the operators P_{in} and P_{out} are useless in classifying states and introducing selection rules for transitions.

Analogously, for a C-noninvariant S-matrix one may introduce the operators C_{in} and C_{out}, and for a T-noninvariant S-matrix, the operators T_{in} and T_{out}. Since the theory is always invariant with respect to total reflection θ,

$$\theta = C_{\text{in}}P_{\text{in}}T_{\text{in}} = C_{\text{out}}P_{\text{out}}T_{\text{out}}.$$

When the S-matrix is symmetric with respect to P, the operators P_{in} and P_{out} coincide ($P_{\text{in}} = P_{\text{out}}$) and are spatial reflection operators. If this symmetry is lacking, the operators P_{in} and P_{out} do not satisfy the commutation relations for a spatial reflection operator, according to which the energy operator should remain unchanged under spatial reflections. In other words, when the S-matrix is not symmetric with respect to P and T these operators do not correspond to the geometrical meaning ascribed to them.

These general theoretical considerations do not diminish the importance of C, P, and T as approximate symmetry operations. The possibility of introducing approximate discrete symmetries is based on definite properties of interactions of various types with respect to reflections and on the sharp subdivision of interactions with respect to strength.

All present experiments indicate that the strong interactions are invariant with respect to P, C, and T separately. In processes with only electromagnetic interactions P, C, and T may also be considered as symmetry operations. The weak interactions break C- and P-invariance separately, but in nearly all processes induced by the weak interactions CP is conserved, which by virtue of the CPT theorem also implies T-symmetry. CP is broken in decays of K-mesons, in which the effective constant of the CP-violating interaction is approximately 1000 times smaller than the weak interaction constant. Thus if one neglects the effects of the weak and the CP-violating (superweak) interactions, the scattering matrix will be invariant separately with respect to all three operations C, P, and T. In cases of the weak interactions, one may use those results of § 6.2 which refer to CP- and T-invariance. When taking account of the superweak interaction, the total reflection θ is the only exact symmetry operation.

An approximate treatment of the reflections C, P, and T is easily formulated in the Lagrangian approach. The full Lagrangian

$$L = L_0 + L_s + L_\gamma + L_w + L_{sw}$$

is invariant with respect to the reflection $\theta = \mathrm{CPT}$:

$$\theta L \theta^{-1} = L.$$

The terms L_0, L_s, and L_γ describing free fields and their strong and electromagnetic interactions are invariant separately with respect to C, P, and T, e.g.:

$$\mathrm{C} L_s \mathrm{C}^{-1} = L_s, \quad \mathrm{P} L_s \mathrm{P}^{-1} = L_s, \quad \mathrm{T} L_s \mathrm{T}^{-1} = L_s.$$

The Lagrangian for the weak interaction L_w has only CP- and, consequently, T-symmetry:

$$\mathrm{CP} L_w (\mathrm{CP})^{-1} = L_w, \quad \mathrm{T} L_w \mathrm{T}^{-1} = L_w.$$

The Lagrangian for the superweak interaction L_{sw} is invariant only with respect to the total reflection θ.

An approximate treatment of discrete symmetries—with respect to interactions—is an indispensable element of the theory of unstable particles. The results of §§ 6.1 and 6.2 refer, strictly speaking, only to absolutely stable particles whose lifetime is infinite, since the in- and out-states may be defined rigorously only for such particles. However, absolutely stable particles are a small fraction of the full set of particles and resonances. In the case of unstable particles, symmetry properties with respect to reflections may be defined aproximately, i.e. with respect to the symmetries of the interaction under which the particle is absolutely stable. Inclusion of the remaining part of the interaction induces the decays of these particles. Assuming that the decay interactions are much smaller than the interaction "forming" the particle, we may consider decay processes in the framework of perturbation theory. The symmetry properties of the decay amplitude are defined by properties of the corresponding parts of the Lagrangian or Hamiltonian.

Let particle a be absolutely stable if one neglects the interaction energy H' in the total Hamiltonian $H = H_a + H'$. The state vector of the particle is an eigenvector of H_a (see

Chapter 2). The transition probability for $a \to b$ is defined by the matrix element of the perturbation H'. To first order in H', the decay matrix is

$$2\pi\delta^4(p-p_b)\langle b|M|\boldsymbol{p}, \lambda, a\rangle = \int_{-\infty}^{\infty} \langle b, \text{out}\,|\,H'(x^0)\,|\,\boldsymbol{p}, \lambda, a\rangle\, dx^0. \tag{104}$$

The probability for the decay $w(a \to b)$ is expressed in terms of the formula (see § 2.3)

$$w(a \to b) = \frac{2\pi}{m_a}\int |\langle b, \text{out}\,|M|\,\boldsymbol{p}, \lambda, a\rangle|^2\, \delta^4(p-p_b)\, d^4p_b, \tag{105}$$

where $|\, b, \text{out}\rangle$ is the final (multi-particle) state of the decay.

Expressions (104) and (105) allow one to determine the behavior of the amplitude and probability of the decay $a \to b$ under reflections. The conditions

$$\text{P}H'\text{P}^{-1} = H', \quad \text{C}H'\text{C}^{-1} = H', \quad \text{T}H'(x^0)\text{T}^{-1} = H'(-x^0)$$

entail the invariance of the decay amplitude with respect to spatial reflection P, charge conjugation C, and time-reversal T, respectively.

The total decay probability, or the inverse lifetime $1/\tau(a)$, is obtained by summing (105) over all possible decay channels and helicities of the final particles and averaging over the helicities of the decaying particle a:

$$\frac{1}{\tau(a)} = \frac{2\pi}{m_a}\,\frac{1}{2J_a+1}\sum_\lambda\sum_b \int |\langle b, \text{out}\,|M|\,\boldsymbol{p}, \lambda, a\rangle|^2\, \delta^4(p-p_b)\, d^4p_b. \tag{106}$$

By virtue of the CPT theorem, the lifetimes of a particle $\tau(a)$ and antiparticle $\tau(\bar{a})$ are the same regardless of the interaction under which they are stable or the interaction causing their decays.[94] We shall prove $\tau(a) = \tau(\bar{a})$ to first order in the decay Hamiltonian H', using the expression (104) for the decay matrix in formula (106) for the mean lifetime.

Let us pass from particles to antiparticles using the reflection operator θ:

$$|\boldsymbol{p}, \lambda, a\rangle = (-1)^{J_a-\lambda}\,\theta^{-1}\,|\boldsymbol{p}, -\lambda, \bar{a}\rangle,$$
$$\langle b, \text{out}\,| = \langle b_\theta, \text{in}\,|\,\theta.$$

In comparison with $|\, b, \text{out}\rangle$, the state $|\, b_\theta, \text{in}\rangle$ has all particles replaced by antiparticles. The helicities have the opposite signs, and to each particle there corresponds the phase factor $(-1)^{J_i-\lambda_i}$ [see (48)]. Then the right-hand side of (104) takes the form

$$\int_{-\infty}^{\infty} \langle b, \text{out}\,|H'(x^0)|\,\boldsymbol{p}, \lambda, a\rangle\, dx^0 = (-1)^{J_a-\lambda}\int_{-\infty}^{\infty} \langle b_\theta, \text{in}\,|\,\theta H'(x^0)\theta^{-1}|\,\boldsymbol{p}, -\lambda, \bar{a}\rangle\, dx^0$$
$$= \int_{-\infty}^{\infty} (-1)^{J_a-\lambda}\langle b_\theta, \text{out}\,|\,S_0^{-1}H'(x^0)\,|\,\boldsymbol{p}, -\lambda, \bar{a}\rangle\, dx^0.$$

where S_0 is the S-matrix in the absence of the interaction H'. But the expression

$$\sum_\lambda\sum_b \left|\int \langle b_\theta, \text{out}\,|\,S_0^{-1}\,H'(x^0)\,|\,\boldsymbol{p}, -\lambda, \bar{a}\rangle\, dx^0\right|^2$$

does not depend on the operator S_0 (since S_0 is unitary) or on the signs of the helicities of all particles (since they are summed). Consequently, the particle and antiparticle lifetimes are equal:

$$\tau(a) = \tau(\bar{a}).$$

This result may be obtained to any order in perturbation theory.[95] From the proof it does not follow, however, that the relative probabilities of various decay channels and polarizations must also be identical for particles and antiparticles.

Let us turn to some consequences of reflection symmetries for decays. Rapid decays are due to the strong interactions; hence rapid and electromagnetic decays posses P-, C-, and T-symmetry.

The simplest consequences are those of C-invariance, by which charge parity is conserved. Recalling that the particles π^0, η, η', A_1^0, and A_2^0 are charge-even, while the particles γ, ϱ^0, ω, and φ are charge-odd, we may easily write reactions forbidden by C-invariance, e.g.:

$$\eta' \nrightarrow \varrho^0+\pi^0, \quad \omega^0 \nrightarrow n\pi^0 \,(\text{all } n).$$

In the case of charged particles the probabilities and angular distributions for charge-conjugate reactions will be identical. If, for example, a charge-even neutral particle decays to charged particles, the distribution of charged products of the decay must not depend on the sign of the charge. By studying the charge distribution of pions in the reactions

$$\eta \to \pi^+ + \pi^- + \pi^0, \quad \eta \to \pi^+ + \pi^- + \gamma, \quad \eta' \to \pi^+ + \pi^- + \gamma,$$

one may check the accuracy of C-symmetry. According to experiment the asymmetry of the pion distribution does not exceed $\sim \frac{1}{2}\%$.[96] The P-invariance of strong and electromagnetic interactions leads to a condition for the decay amplitude

$$\langle \boldsymbol{p}_1, \lambda_1, b_1; \boldsymbol{p}_2, \lambda_2, b_2; \ldots | M | \boldsymbol{p}, \lambda, a\rangle$$
$$= \langle -\boldsymbol{p}_1, -\lambda_1, b_1; -\boldsymbol{p}_2, -\lambda_2, b_2; \ldots | M | -\boldsymbol{p}, -\lambda, a\rangle.$$

Let the particle a at rest with spin $J_a = 0$ decay into two spinless particles b_1 and b_2. Then the total spin J of the final state is equal to zero. According to (84) the application of P to a two-particle state with zero total spin gives a factor $\eta_P(b_1)\,\eta_P(b_2)$. The P-invariance of the decay amplitude for a spinless particle into two spinless particles is thus equivalent to the conservation of intrinsic parity:

$$\eta_P(a) = \eta_P(b_1)\,\eta_P(b_2).$$

If the spin of the particle a is different from zero, then this formula is replaced by another:

$$\eta_P(a) = \eta_P(b_1)\,\eta_P(b_2)\,(-1)^{J_a}.$$

Both P- and T-invariance forbid a particle from having a nonzero electric dipole moment d. Under the spatial reflection P the electric field $\boldsymbol{E}$, being a polar vector, changes sign: $\boldsymbol{E} \to -\boldsymbol{E}$, while the particle spin remains unchanged. Hence the energy of a particle at rest undergoing dipole interaction with an electric field changes sign: $\mathrm{P} d\boldsymbol{J}\cdot\boldsymbol{E}\mathrm{P}^{-1} = -d\boldsymbol{J}\cdot\boldsymbol{E}$. This contradicts the assumption of P-invariance and, consequently, d must vanish.

Under time-reversal T the electric field $\boldsymbol{E}$ does not change sign (we write, for example, $\boldsymbol{E} = (e/r^3)\boldsymbol{r}$ and recall that $\boldsymbol{r}$ and the charge are not affected by the transformation T). However, $\mathrm{T}J_k\mathrm{T}^{-1} = -J_k$, so that the dipole interaction energy changes sign in this case as well: $\mathrm{T}\,d\boldsymbol{J}\cdot\boldsymbol{E}\mathrm{T}^{-1} = -d\boldsymbol{J}\cdot\boldsymbol{E}$. Consequently, under T-invariance one must also have $d = 0$. The measurement of the dipole moment of the neutron gives $d \leqslant 5\times 10^{-24}\,e$[97, 98, 244]. While the decays of resonances generally are due to the strong and electromagnetic interactions, the decays of "stable" particles (see Table A.1, p. 359) are mostly due to the weak interactions (slow decays). For this reason, slow decays have only CP- and, consequently, T-invariance. As an interesting illustration of processes of this type we shall now consider decays of the neutral K-mesons K^0 and $\overline{\mathrm{K}}^0$.[99, 100]

K-mesons, or kaons, are pseudo-scalar particles, i.e. have spin-parity $J^{\mathrm{P}} = 0^-$. Kaons are produced in strong interactions, e.g. in the reaction $\pi^- + \mathrm{p} \to \mathrm{K}^0 + \Lambda$, and decay vi-weak interactions (into pions and leptons). The properties of the neutral kaons K^0 and $\overline{\mathrm{K}}^0$ are defined by the strong interactions; K^0 and $\overline{\mathrm{K}}^0$ are defined as the particle and antia particle, respectively: one may set

$$\mathrm{C}\,|\,\mathrm{K}_0\rangle = |\,\overline{\mathrm{K}}_0\rangle, \quad \mathrm{C}\,|\,\overline{\mathrm{K}}_0\rangle = |\,\mathrm{K}_0\rangle.$$

Here and below, $|\,\mathrm{K}^0\rangle$ and $|\,\overline{\mathrm{K}}^0\rangle$ denote the state vectors of kaons in the c.m.s.

The neutral kaons K^0 and $\overline{\mathrm{K}}^0$ are neutral only in the electrical sense—they differ with respect to hypercharge and isospin projection:

$$Y(\mathrm{K}^0) = 1, \quad Y(\overline{\mathrm{K}}^0) = -1, \quad Q = I_3 + \tfrac{1}{2}Y = 0.$$

The weak interaction does not conserve hypercharge, and thus, with respect to the weak interaction alone, the particles K^0 and $\overline{\mathrm{K}}^0$ are indistinguishable. But the weak interaction is CP-invariant and can distinguish the CP-symmetric combination of K^0 and $\overline{\mathrm{K}}^0$ from the CP-antisymmetric one. Let us form the two combinations:

$$|\,\mathrm{K}_1^0\rangle = -\frac{i}{\sqrt{2}}\left(|\,\mathrm{K}^0\rangle - |\,\overline{\mathrm{K}}^0\rangle\right),$$

$$|\,\mathrm{K}_2^0\rangle = \frac{1}{\sqrt{2}}\left(|\,\mathrm{K}^0\rangle + |\,\overline{\mathrm{K}}^0\rangle\right).$$

Since $\mathrm{P}\,|\,\mathrm{K}\rangle = -|\,\mathrm{K}\rangle$, then $\mathrm{CP}\,|\,\mathrm{K}_1^0\rangle = |\,\mathrm{K}_1^0\rangle$. The states $|\,\mathrm{K}_1^0\rangle$ and $|\,\mathrm{K}_2^0\rangle$ are orthogonal: $\langle \mathrm{K}_1^0\,|\,\mathrm{K}_2^0\rangle = 0$.

Because of the conservation of hypercharge in the strong interactions the transitions $\mathrm{K}^0 \leftrightarrow \overline{\mathrm{K}}^0$ are impossible; such transitions are admitted by the weak interactions, which, however, do not have matrix elements connecting the states K_1^0 and K_2^0. For the strong and electromagnetic interactions these states are also different—there are no transitions between K_1^0 and K_2^0. Hence, when taking account of the weak interactions, K_1^0, K_2^0 must be considered as particles; specifically, K_1^0 and K_2^0 will be characterized by definite values of mass and lifetime.

Since K_1^0 and K_2^0 belong to different eigenvalues of the operator CP, the decay modes of K_1^0 and K_2^0 are different. States of two pions ($\pi^+\pi^-$ or $\pi^0\pi^0$) with total angular momentum

$J = 0$ are even with respect to spatial reflections P [see (84)] and even with respect to charge conjugation C [since, according to (90), $\eta_C(\pi^+)\eta_C(\pi^-) = 1$]. Consequently, this state has the same CP eigenvalue as K_1^0, and the two-pion decay channel is possible only for K_1^0:

$$K_1^0 \to \pi^+ + \pi^-, \quad 2\pi^0; \qquad K_2^0 \nrightarrow \pi^+ + \pi^-, \quad 2\pi^0,$$

The three-pion state $\pi^+\pi^-\pi^0$ is even with respect to C. By virtue of the pseudo-scalar nature of the pions ($\eta_P(\pi) = -1$) it will be odd with respect to P if the pions are in a state with zero relative angular momentum, and will be even if the relative angular momentum of one of the pairs is different from zero. Hence the decay $K_1^0 \to 3\pi$ is suppressed by a centrifugal barrier, while for K_2^0 this is the main pionic decay channel.

The lifetimes of K_1^0 and K_2^0 differ significantly from one another:

$$\tau(K_1^0) = \tau_S = (0.8947 \pm 0.0033) \times 10^{-10} \text{ sec}, \text{ (244)}$$
$$\tau(K_2^0) = \tau_L = (5.18 \pm 0.04) \times 10^{-8} \text{ sec}, \text{ (8)}$$

which is largely related to the difference between two- and three-particle phase space.

In reality the decays $K_2^0 \to 2\pi$, in which CP invariance is broken, do exist. The amplitude for these decays is small in comparison with the CP-symmetric amplitude. The violation of CP invariance in kaon decays will be examined in § 15.6.

§ 6.4. Summary of formulae for reflection transformations

1. Total reflection θ (Wigner treatment) or θ^S (Schwinger treatment):

$$\theta^2 = (\theta^S)^2 = (-1)^{2J}.$$

The operator H is invariant:

$$H = \theta H^+ \theta^{-1}, \quad H = \theta^S H (\theta^S)^{-1}.$$

If one introduces separately the operation P and C, then

$$\theta = \text{CPT}, \quad \theta^S = \text{CPT}^S.$$

θ and T are antiunitary, while P and C are unitary:

$$\begin{aligned} \text{CP} &= \text{PC}, & \text{P}^2 &= \text{T}^2 = (\text{T}^S)^2 = (-1)^{2J}, \\ \text{T} &= \theta(\text{CP})^{-1}, & \text{P}\theta &= (-1)^{2J}\theta\text{P}, \\ \text{T}^S &= \theta^S(\text{CP})^{-1}, & \text{C}\theta &= \theta\text{C}, \quad \text{C}^2 = 1. \end{aligned}$$

2. Transformation of single-particle states:

$$\theta\,|\boldsymbol{p}, \lambda; a, \text{in}\rangle = (-1)^{J-\lambda}|\boldsymbol{p}, -\lambda; \bar{a}, \text{out}\rangle,$$
$$\theta^S\,|\boldsymbol{p}, \lambda; a, \text{in}\rangle = (-1)^{J-\lambda}\langle\boldsymbol{p}, -\lambda; \bar{a}, \text{out}|,$$
$$\text{C}\,|\boldsymbol{p}, \lambda; a\rangle = \eta_C^*(a)\,|\boldsymbol{p}, \lambda; \bar{a}\rangle.$$

The above formulae do not change when one replaces the helicity λ by the spin projection on a fixed axis $\sigma = (J_3)'$.

$$\begin{aligned} \mathrm{P}\,|\,\boldsymbol{p}, \sigma; a\rangle &= \eta_{\mathrm{P}}^{*}(a)\,|\,-\boldsymbol{p}, \sigma; a\rangle, \\ \mathrm{P}\,|\,\boldsymbol{p}, \lambda; a\rangle &= \eta_{\mathrm{P}}^{*}(a)\,(-1)^{J-\lambda}\,e^{i\pi M_{31}}\,|\,p, \vartheta, -\varphi, -\lambda; a\rangle, \\ \mathrm{T}\,|\,\boldsymbol{p}, \sigma; a, \mathrm{in}\rangle &= \eta_{\mathrm{T}}^{*}(a)\,(-1)^{J-\sigma}\,|\,-\boldsymbol{p}, -\sigma; a, \mathrm{out}\rangle, \\ \mathrm{T}\,|\,\boldsymbol{p}, \lambda; a, \mathrm{in}\rangle &= \eta_{\mathrm{T}}^{*}e^{-i\pi M_{31}}\,|\,p, \vartheta, -\varphi, \lambda; a, \mathrm{out}\rangle, \\ \mathrm{T^S}\,|\,\boldsymbol{p}, \sigma; a, \mathrm{in}\rangle &= \eta_{\mathrm{T}}^{*}(a)\,(-1)^{J-\sigma}\langle -\boldsymbol{p}, -\sigma; a, \mathrm{out}\,|. \end{aligned}$$

The transformation of two-particle states:

$$P\,|\,0, \Lambda; J, m; \lambda_1, \lambda_2\rangle = \eta_{\mathrm{T}}(1)\,\eta_{\mathrm{T}}(2)\,(-1)^{J-J_1-J_2}\,|\,0, \Lambda; J, m; -\lambda_1, -\lambda_2\rangle.$$

The phase factors satisfy the relations

$$\begin{aligned} \eta_{\mathrm{P}}(a)\,\eta_{\mathrm{P}}(\bar{a}) &= (-1)^{2J}, & \eta_{\mathrm{P}}(a) &= \eta_{\mathrm{P}}(\bar{a}), \\ \eta_{\mathrm{C}}(a)\,\eta_{\mathrm{C}}(\bar{a}) &= 1, & \eta_{\mathrm{C}}(a)\,\eta_{\mathrm{P}}(a)\,\eta_{\mathrm{T}}(a) &= 1. \end{aligned}$$

For a neutral particle $\eta_{\mathrm{C}}(a_0) = \eta_{\mathrm{C}}^{*}(a_0) = \pm 1$. In formulae for helicity states the momentum is parametrized using the angles ϑ and φ: $\boldsymbol{p} = (p, \vartheta, \varphi)$.

3. Scattering matrix. The initial process is $a+b+\ldots \to c+d+\ldots$

θ-invariance:

$$\begin{aligned} &\langle \boldsymbol{p}_c, \lambda_c, c; \boldsymbol{p}_d, \lambda_d, d; \ldots\,|S|\,\boldsymbol{p}_a, \lambda_a, a; \boldsymbol{p}_b, \lambda_b, b; \ldots\rangle \\ &\quad = \eta_\theta \langle \boldsymbol{p}_a, -\lambda_a, \bar{a}; \boldsymbol{p}_b, -\lambda_b, \bar{b}; \ldots\,|S|\,\boldsymbol{p}_c, -\lambda_c, \bar{c}; \boldsymbol{p}_d, -\lambda_d, \bar{d}; \ldots\rangle. \end{aligned}$$

P-invariance:

$$\begin{aligned} &\langle \boldsymbol{p}_c, \lambda_c, c; \boldsymbol{p}_d, \lambda_d, d; \ldots\,|S|\,\boldsymbol{p}_a, \lambda_a, a; \boldsymbol{p}_b, \lambda_b, b; \ldots\rangle = \eta_{\mathrm{P}}(\mathrm{in})\,\eta_{\mathrm{P}}^{*}(\mathrm{out}) \\ &\quad \times \langle -\boldsymbol{p}_c, -\lambda_c, c; -\boldsymbol{p}_d, -\lambda_d, d; \ldots\,|S|\,-\boldsymbol{p}_a, -\lambda_a, a; -\boldsymbol{p}_b, -\lambda_b, b; \ldots\rangle. \end{aligned}$$

C-invariance:

$$\begin{aligned} &\langle \boldsymbol{p}_c, \lambda_c, c; \boldsymbol{p}_d, \lambda_d, d; \ldots\,|S|\,\boldsymbol{p}_a, \lambda_a, a; p_b, \lambda_b, b; \ldots\rangle \\ &\quad = \eta_{\mathrm{C}}(\mathrm{in})\,\eta_{\mathrm{C}}^{*}(\mathrm{out})\,\langle \boldsymbol{p}_c, \lambda_c, \bar{c}; \boldsymbol{p}_d, \lambda_d, \bar{d}; \ldots\,|S|\,\boldsymbol{p}_a, \lambda_a, \bar{a}; -\boldsymbol{p}_b, \lambda_b, \bar{b}; \ldots\rangle. \end{aligned}$$

T-invariance:

$$\begin{aligned} &\langle \boldsymbol{p}_c, \lambda_c, c; \boldsymbol{p}_d, \lambda_d, d; \ldots\,|S|\,\boldsymbol{p}_a, \lambda_a, a; \boldsymbol{p}_b, \lambda_b, b; \ldots\rangle \\ &\quad = \pm\langle -\boldsymbol{p}_a, \lambda_a, a; -\boldsymbol{p}_b, \lambda_b, b; \ldots\,|S|\,-\boldsymbol{p}_c, \lambda_c, c; -\boldsymbol{p}_d, \lambda_d, d; \ldots\rangle \end{aligned}$$

Here η_θ is the product of the factors $(-1)^{J-\lambda}$ for all particles; $\eta_{\mathrm{P}}(\mathrm{in})$ and $\eta_{\mathrm{P}}(\mathrm{out})$ are the products of parities for initial and final states, and $\eta_{\mathrm{C}}(\mathrm{in})$ and $\eta_{\mathrm{C}}(\mathrm{out})$ are the analogous products of the phase factors η_{C}. All formulae are shown for the helicity basis.

4. Transformation of fields. $\Phi^J(x)$ are $2(2J+1)$-component fields with spin J; the field $\Phi^{1/2}(x)$ coincides with the Dirac field $\psi(x)$; $\Psi_{\alpha\mu_1\ldots\mu_n}(x)$ is the Rarita–Schwinger field, and $\Psi_{\alpha_1\alpha_2\ldots\alpha_n}(x)$ is the Bargman–Wigner field.

One of the possible representations for the matrices γ_4, γ_5, and $\mathscr{C}_J$ has the form

$$\gamma_4 = \begin{pmatrix} 0 & 1 \\ 1 & 0 \end{pmatrix}, \quad \gamma_5 = \begin{pmatrix} 1 & 0 \\ 0 & -1 \end{pmatrix}, \quad \mathscr{C}_J = \begin{pmatrix} D^J(C) & 0 \\ 0 & -D^J(C) \end{pmatrix}.$$

The results are shown in Table 6.1.

TABLE 6.1

$\zeta(x)$	J	$\theta\zeta\theta^{-1}$	$\theta^s\zeta(\theta^s)^{-1}$
$\Phi^J(x)$	$0, 1, 2, \ldots$	$\Phi^{J+}(-x)\gamma_5$	$\gamma_5\Phi^J$
$\Phi^J(x)$	$\frac{1}{2}, \frac{3}{2}, \ldots$	$\Phi^{J+}(-x)\gamma_5$	$\gamma_5\Phi^J$
$\Psi_{\mu_1 \ldots \mu_n}(x)$	$n+\frac{1}{2},\ n = 0, 1, 2, \ldots$	$(-1)^{J-\frac{1}{2}}\Psi^+_{\mu_1 \ldots \mu_n}(-x)\gamma_5$	$\gamma_5(-1)^{J-\frac{1}{2}}\Psi_{\mu_1 \ldots \mu_n}(-x)$
$\Psi_{\mu_1 \ldots \mu_n}(x) \equiv \Phi_{\mu_1 \ldots \mu_n}(x)$	$n = 0, 1, 2, \ldots$	$(-1)^J\Phi^+_{\mu_1 \ldots \mu_n}(-x)$	$(-1)^J\Phi_{\mu_1 \ldots \mu_n}(-x)$
$\Psi_{\alpha_1 \ldots \alpha_n}(x)$	$\frac{1}{2}n$	$\Psi^+(-x)\gamma_5^{(1)} \ldots \gamma_5^{(n)}$	$\gamma_5^{(1)} \ldots \gamma_5^{(n)}\Psi(-x)$

$\zeta(x)$	$C\zeta C^{-1}$	$\mathrm{P}\zeta\mathrm{P}^{-1}$	$\mathrm{T}\zeta\mathrm{T}^{-1}$
$\Phi^J(x)$	$\eta_C\mathscr{C}_J\gamma_5\gamma_4\Phi^{J+}(x)$	$\eta_P\gamma_4\Phi^J(x^0, -x)$	$\eta_T\mathscr{C}_J^{-1}\Phi^J(-x^0, x)$
$\Phi^J(x)$	$\eta_C\mathscr{C}_J\gamma_4\Phi^{J+}(x)$	$\eta_P\gamma_4\Phi^J(x^0, -x)$	$\eta_T\mathscr{C}_J^{-1}\gamma_5\Phi^J(-x^0, x)$
$\Psi_{\mu_1 \ldots \mu_n}(x)$	$\eta_C\mathscr{C}\gamma_4\Psi_{\mu_1 \ldots \mu_n}(x)$	$\eta_P(-1)^n\gamma_4\Psi_{\mu_1 \ldots \mu_n}(x^0, -x)$, $\mu_i \neq 0$	$\eta_T\mathscr{C}^{-1}\gamma_5\Psi_{\mu_1 \ldots \mu_n}(-x^0, x)$
$\Psi_{\mu_1 \ldots \mu_n}(x) \equiv \Phi_{\mu_1 \ldots \mu_n}(x)$	$\eta_C\Phi^+_{\mu_1 \ldots \mu_n}(x)$	$\eta_P(-1)^J\Phi_{\mu_1 \ldots \mu_n}(x^0, -x)$, $\mu_i \neq 0$	$\eta_T\Phi_{\mu_1 \ldots \mu_n}(-x^0, x)$
$\Psi_{\alpha_1 \ldots \alpha_n}(x)$	$\eta_C(\mathscr{C}\gamma_4)^{(1)} \ldots (\mathscr{C}\gamma_4)^{(n)}\Psi^+(x)$	$\eta_P\gamma^{(1)} \ldots \gamma_4^{(n)}\Psi(x^0, -x)$	$\eta_T(\mathscr{C}^{-1}\gamma_5)^{(1)} \ldots \Psi(-x^0, x)$

5. Transformation of bilinear Hermitian quantities ("currents") in the case of the Dirac field ($J = \frac{1}{2}$). The bilinear Hermitian "currents" j^R are defined by the formula

$$j^R(x) = \tfrac{1}{2}[\overline{\psi}^\alpha(x), \psi_\beta(x)]\,(\gamma^R)_\alpha{}^\beta, \quad j^{R+} = j^R,$$

where γ^R is one of the Dirac matrices (5.55) satisfying the condition $\gamma^R = \gamma_4\gamma^{R+}\gamma_4$.

Under reflections the "currents" j^R acquire the factor $\varepsilon^R = \pm 1$, e.g.

$$\theta j^R(x)\theta^{-1} = \varepsilon^R_\theta j^R(x_\theta), \quad \varepsilon^R_\theta = \pm 1,$$

where x_θ denotes the transformed coordinate. The quantities j_R transform in the same way under the Wigner and Schwinger treatments of the reflections θ and T. The factors ε^R_σ for the reflections $\sigma = \theta$, P, C, T, and CP are shown in Table 6.2.

TABLE 6.2

γ^R	θ	P	C	T	CP
I	+	+	+	+	+
γ^k	−	−	−	−	+
γ_4	−	+	−	+	−
σ^{ij}	+	+	−	−	−
σ^{k4}	+	−	−	+	+
$i\gamma^k\gamma_5$	−	+	+	−	−
$i\gamma_4\gamma_5$	−	−	+	−	−
$i\gamma_5$	+	−	+	−	−

CHAPTER 7

THE SCATTERING MATRIX. KINEMATICS

THE properties of the scattering matrix may be divided into kinematical and dynamical ones. Kinematical properties are based on space–time symmetry or the invariance of the theory with respect to transformations of the quantum mechanical Poincaré group. Dynamical properties of the S-matrix are determined by particular features of interactions. In this chapter we shall examine kinematical properties of the S-matrix.

§ 7.1. The problem of kinematics

In Chapter 4 we studied the description of asymptotic particle states from the point of view of the Poincaré group. For free particles, and hence for asymptotic states in the theory of interacting particles, this approach permits a full description. The element of the S-matrix corresponding to the process $1+2+\ldots \to 1'+2'+\ldots$ is equal (in the helicity basis) to

$$\langle f|S|i\rangle = \langle \boldsymbol{p}_1', \lambda_1', 1'; \boldsymbol{p}_2', \lambda_2', 2'; \ldots | S | \boldsymbol{p}_1, \lambda_1, 1; \boldsymbol{p}_2, \lambda_2, 2; \ldots\rangle, \tag{1}$$

where for brevity the particle number ($k = 1, 2, \ldots$) denotes its relativistic properties—the mass m_i, spin J_i, as well as other quantum numbers.

In experiment one measures directly noninvariant quantities (such as energies, angles, etc.) which depend on the reference frame. Consequently the initial and final asymptotic states change under translations and rotations:

$$|f\rangle \to |f_g\rangle, \quad |i\rangle \to |i_g\rangle.$$

But by relativistic invariance the matrix element for the transition (1) cannot depend on the reference frame:

$$\langle f_g|S|i_g\rangle = \langle f|S|i\rangle,$$

i.e. the elements of the S-matrix must be functions of the invariants (expressed in terms of measurable noninvariant quantities). This condition of relativistic invariance of the S-matrix allows one to display part of the dependence (the kinematic dependence) of the matrix elements (1) on invariant quantities: specifically, that dependence which may be obtained from the transformation properties of asymptotic states.

The transformation of any state vector induced by translations and Lorentz rotations,

$g = (a, A)$, is described by the unitary operator $U(a, A)$, e.g.:

$$|f_g\rangle = U(a, A)\,|f\rangle.$$

Hence, the condition of relativistic invariance of the S-matrix is

$$S = U^{-1}(a, A)\, S U(a, A).$$

If one writes $U(a, A)$ in terms of the generators P_μ and $M_{\mu\nu}$,

$$U(a, A) = e^{iP_\mu a^\mu} e^{-i\frac{1}{2}M_{\lambda\nu}\omega^{\lambda\nu}},$$

we obtain

$$[P_\mu, S] = 0, \quad [M_{\mu\nu}, S] = 0,$$

from which the conservation laws for P_μ and $M_{\mu\nu}$ follow.

The asymptotic states transform as the product of single-particle states (see §§ 1.4 and 4.6). The transformation properties of single-particle states were established in § 4.3 [formula (4.47)].

Under the coordinate transformation $x^\mu \to x'^\mu = \Lambda^\mu{}_\nu(A)x^\nu + a^\mu$ the particle state vector undergoes a unitary transformation $U(a, A)$:

$$U(a, A)\,|\boldsymbol{p}, \xi\rangle = e^{ip'a} \sum_{\xi'=-J}^{J} |\boldsymbol{p}', \xi'\rangle\, \mathcal{D}^J_{\xi'\xi}(\tilde{A}(p, A)),$$

where $p' = \Lambda(A)p$, $\mathcal{D}^J_{\xi'\xi}$ is a matrix of the rotation group for spin J (see § 3.5), and $\tilde{A}(p, A)$ is the 2 by 2 matrix (4.45) of the Wigner spin rotation, which accompanies Lorentz transformations; this rotation depends on momentum via the matrix $\alpha(\boldsymbol{p})$, corresponding to the Lorentz transformation from the rest state ($\boldsymbol{p} = 0$) to the momentum $\boldsymbol{p}$:

$$\tilde{A}(p, A) = \alpha(\boldsymbol{p}')\, A\alpha^{-1}(\boldsymbol{p}).$$

In the helicity basis ξ is the helicity λ, while the operator $\alpha(\boldsymbol{p})$ is chosen in the form (4.60) or (4.63): $\alpha(\boldsymbol{p}) = h(\boldsymbol{p})$ or $\alpha(\boldsymbol{p}) = h^-(\boldsymbol{p})$. In the canonical basis ξ is σ—the projection of the spin on a given direction n_μ orthogonal to the momentum: $p^\mu n_\mu = 0$. In this basis the operator $\alpha(\boldsymbol{p})$ is usually chosen self-adjoint: $\alpha(\boldsymbol{p}) = \alpha^+(\boldsymbol{p})$ [see (4.22) and (4.24)]. Thus the condition for relativistic invariance of the S-matrix

$$\langle f_g\,|\,S\,|\,i_g\rangle = \langle f\,|\,U^{-1}(g)\, SU(g)\,|\,i\rangle = \langle f\,|\,S\,|\,i\rangle$$

means that

$$\begin{aligned}
&\langle \boldsymbol{p}_1', \lambda_1', 1'; \ldots |\,S\,|\,\boldsymbol{p}_1, \lambda_1, 1; \ldots\rangle \\
&\quad = \sum_{(\mu,\mu')} \mathcal{D}^{J_1'*}_{\lambda_1'\mu_1'}(\tilde{A}(p_1', A)) \ldots \langle \Lambda p_1', \mu_1', 1'; \ldots |\,S\,|\,\Lambda p_1, \mu_1, 1; \ldots\rangle \\
&\quad \times \mathcal{D}^{J_1}_{\mu_1\nu_1}(\tilde{A}(p_1, A)) \ldots \exp\,[ia\Lambda(p_1 + \ldots - p_1' - \ldots)]. \qquad (2)
\end{aligned}$$

If all particles in (1) are spinless, then according to (2) the matrix element (1) can depend only on the relativistically invariant variables $p_r p_j$ formed from the momenta of particles. If the particles have spin, the problem of subdividing the matrix element (1) into independent relativistically invariant "scalar" amplitudes F_l and kinematic factors Z_l, connected with spin, arises:

$$\langle f \mid S \mid i \rangle = \sum_{l} Z_l(f, i)\, F_l(p_r p_j), \tag{3}$$

where F_l depend only on the invariant momentum variables $p_r p_j$.

An expansion of the type (3) allows one to separate the kinematic and dynamical aspects of the theory: the kinematics determine Z_l, while the dynamics establish the dependence of the scalar amplitudes F_l on the scalar momentum variables. The kinematic factors Z_l are found relatively easily; they may be considered as known. On the other hand, the calculation of the functions F_l relies as a rule on unsolved theoretical questions.

The expansion (3) of the transition matrix element in terms of scalar amplitudes is not the only means of kinematical analysis. The multi-particle states $\mid i\rangle$ and $\mid f\rangle$ may be expanded in terms of irreducible representations of the Poincaré group (see § 4.6) characterized by the total mass m and total spin J. As a consequence of relativistic invariance the operator S is diagonal in m and J, so that the matrix element (1) also may be written (symbolically) in diagonal form:

$$\langle f \mid S \mid i \rangle = \sum_{J} S_{if}(J, m)\, Y_{Jm}(i, f), \tag{4}$$

where $S(J, m)$ describes the scattering with a given J, m, and the known functions Y_{Jm} are determined by the Poincaré group. The multi-particle states $\mid J, m\rangle$ may be given in various bases, and for this reason there are various expansions of the type (4).

To find expansions (3) and (4), i.e. to find a set of quantities Z_l and Y_{Jm}, is the basic problem of the kinematics of the scattering matrix.

§ 7.2. The variables s, t, u

The scalar amplitudes F_l in expansion (3) for the transition matrix element depend on invariants formed from particle momenta. We shall examine in this section the construction of such variables for processes with a total of four particles in the initial and final states. Then there are four momenta p_r ($r = 1, 2, 3, 4$) on the mass shell $p_r^2 = m_r^2$ and related by the conservation law

$$p_1 + p_2 = p_3 + p_4. \tag{5}$$

Depending on the sign of the time component p_r^0 these momenta are involved in the kinematics of one of the three following reactions:

$$1+2 \to 3+4 \qquad (s\text{-channel}), \tag{6}$$

$$1+\bar{3} \to \bar{2}+4 \qquad (t\text{-channel}), \tag{7}$$

$$1+\bar{4} \to 3+\bar{2} \qquad (u\text{-channel}), \tag{8}$$

where the symbol $\bar{r}$ denotes the antiparticle of the particle r.

Reactions (6)–(8) are called crossed reactions of one another. Each one of them may be obtained from any other by replacing a particle in the initial (final) state by an antiparticle in the final (initial) state. The fact that such processes are related to one another is connected with properties of amplitudes under crossing symmetry (see below, § 11.2). In a local theory, amplitudes for all three processes (6)–(8) are described by formulae of the same type.

All initial and final states in (6)–(8) have positive energy. In accord with (5) we shall set all $p_{r0} > 0$ for reactions (6). Then in reaction (7) $p_{10}, p_{40} > 0$ and $p_{20}, p_{30} < 0$, while in reaction (8) $p_{10}, p_{30} > 0$ and $p_{20}, p_{40} < 0$. When passing to the crossed reaction the momentum p of a particle in the initial state is changed to the momentum $-p$ of an antiparticle in the final state, so that, for example, the conservation law (5) for reaction (7) becomes

$$p_1+(-p_3) = (-p_2)+p_4 \qquad (p_{30} < 0,\ p_{20} < 0).$$

The physical four-momenta of particles in reaction (7) are equal to p_1, $-p_2$, $-p_3$, and p_4, so that the condition $p_{20} < 0$, $p_{30} < 0$ is also a condition for positivity of particle energies. Crossed processes are depicted graphically in Fig. 3.

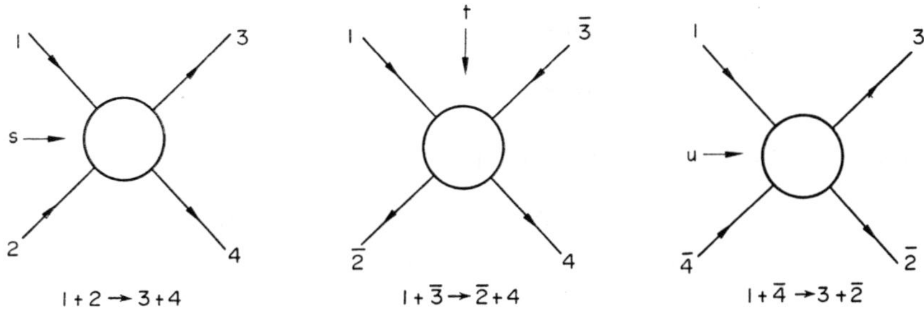

FIG. 3. s-, t-, and u-channel processes related to one another by crossing.

By CPT invariance, along with reactions (6)–(8) one must consider simultaneously the three CPT-conjugate reactions obtained from (6)–(8) by replacing all particles by antiparticles and interchanging the initial and final states. When the theory is invariant with respect to charge conjugation C (which always holds when only strong and electromagnetic interactions are taken into account), the six reactions are supplemented by six more C-conjugate reactions, in which all particles are replaced by antiparticles. Thus, in strong interaction theory it is convenient to consider simultaneously 12 matrix elements [the amplitudes for the processes (6)–(8) and nine others, obtained from (6)–(8) both by interchanging the initial and final states and by changing all particles to antiparticles]).

With four momenta p_r satisfying $p_r^2 = m_r^2$ and (5), one may construct two independent invariants. To preserve the symmetry among the reactions (6)–(8) one usually introduces the three invariants:

$$\left.\begin{aligned} s &= (p_1+p_2)^2 = (p_3+p_4)^2, \\ t &= (p_1-p_3)^2 = (p_2-p_4)^2, \\ u &= (p_1-p_4)^2 = (p_2-p_3)^2, \end{aligned}\right\} \tag{9}$$

connected by

$$s+t+u = \sum_r m_r^2 = h.$$

The quantity s is the square of the energy in the c.m.s. for reaction (6); analogously the quantities t and u are squares of the energies in the c.m.s. respectively for reactions (7) and (8). For this reason, the reactions (6), (7) and (8) are referred to as the s-, t-, and u-channels of the system of the three crossed reactions. It is obvious that in the s-channel $s > 0$, in the

t-channel $t > 0$, and in the u-channel $u > 0$. Let us define reaction parameters in the s-channel in the c.m.s.:

$$\left.\begin{aligned} \boldsymbol{q}_{12}(s) &= \boldsymbol{p}_1 = -\boldsymbol{p}_2, & \boldsymbol{q}_{34}(s) &= \boldsymbol{p}_3 = -\boldsymbol{p}_4, \\ s &= (p_{10}+p_{20})^2, & \boldsymbol{p}_1\cdot\boldsymbol{p}_3 &= q_{12}\,q_{34}\,z_s, \end{aligned}\right\}, \tag{10}$$

where $z_s = \cos\vartheta_s$ is the cosine of the scattering angle (in the s-channel), while $q_{ij} = |\boldsymbol{q}_{ij}|$. Forward scattering corresponds to $z_s = 1$, or $\boldsymbol{p}_1 = \boldsymbol{p}_3$ (but not $\boldsymbol{p}_1 = \boldsymbol{p}_4$).

Analogous parameters $q_{13}(t)$, $q_{24}(t)$, and $z_t = \cos\vartheta_t$ in the t-channel may be obtained from (10) in correspondence with (6) and (7) by the substitution

$$p_1 \to p_1, \quad p_2 \to -p_3, \quad p_3 \to p_4, \quad p_4 \to -p_2.$$

To find the parameters of the u-channel one must perform the substitution

$$p_1 \to p_1, \quad p_2 \to -p_4, \quad p_3 \to -p_2, \quad p_4 \to p_3.$$

Consequently, in the c.m.s. of the t-channel one will have

$$\left.\begin{aligned} \boldsymbol{q}_{13}(t) &= \boldsymbol{p}_1 = \boldsymbol{p}_3, & \boldsymbol{q}_{24} &= \boldsymbol{p}_2 = \boldsymbol{p}_4, \\ t &= (p_{10}+|p_{30}|)^2, & \boldsymbol{p}_1\cdot\boldsymbol{p}_4 &= q_{13}\,q_{24}\,z_t, \end{aligned}\right\} \tag{11}$$

while in the c.m.s. of the u-channel the sign of z_u is fixed by

$$\boldsymbol{p}_1\cdot\boldsymbol{p}_2 = -q_{14}\,q_{23}\,z_u.$$

Let us denote

$$\Delta_{ij}^2(x) = [x-(m_i+m_j)^2]\,[x-(m_i-m_j)^2]. \tag{12}$$

Then in all scattering channels

$$4xq_{ij}^2(x) = \Delta_{ij}^2(x). \tag{13}$$

The cosines of the scattering angles z_s, z_t, and z_u respectively in the s-, t-, and u-channels have the form

$$z_s = \frac{s(t-u)+(m_1^2-m_2^2)\,(m_3^2-m_4^2)}{\Delta_{12}(s)\,\Delta_{34}(s)}, \tag{14}$$

$$z_t = \frac{t(u-s)+(m_1^2-m_3^2)\,(m_2^2-m_4^2)}{\Delta_{13}(t)\,\Delta_{24}(t)}, \tag{15}$$

$$z_u = \frac{u(s-t)+(m_1^2-m_4^2)\,(m_2^2-m_3^2)}{\Delta_{14}(u)\,\Delta_{23}(u)}. \tag{16}$$

In the case of elastic scattering, when $m_1 = m_3$, $m_2 = m_4$, we find that $q_{12} = q_{34}$ and

$$\left.\begin{aligned} t &= -(\boldsymbol{q}_{12}-\boldsymbol{q}_{34})^2 = -2q_{12}^2(s)\,(1-z_s) = -\frac{\Delta_{12}^2(s)}{2s}\,(1-z_s), \\ u &= -2q_{12}^2(s)\,(1+z_s)+\frac{(m_1^2-m_2^2)^2}{s}. \end{aligned}\right\} \tag{17}$$

All relations are especially simple when the four masses are the same ($m_1 = m_2 = m_3 = m_4 = m$):

$$\left.\begin{aligned} \Delta_{12}^2(s) &= s(s-4m^2), & q_{12}^2 \equiv q^2 &= \frac{s-4m^2}{4}, \\ t &= -\frac{s-4m^2}{2}(1-z_s), & u &= -\frac{s-4m^2}{2}(1+z_s). \end{aligned}\right\} \tag{18}$$

If $m_1 > m_2+m_3+m_4$, there is also the decay channel $1 \to \bar{2}+3+4$. In this case, obviously, all three invariants s, t, and u may be interpreted as momentum transfers.

In each channel the variables s, t, and u have values belonging to a specific region—the physical region of the given channel. The physical region for the s-channel includes only those s, t, and u for which

$$|z_s| \leqslant 1, \quad q_{12}(s) > 0, \quad q_{34}(s) > 0. \tag{19}$$

For example, for elastic scattering ($m_1 = m_3$, $m_2 = m_4$) the physical region of the s-channel consists of

$$s \geqslant (m_1+m_2)^2, \quad t \leqslant 0, \quad u \leqslant (m_1-m_2)^2. \tag{20}$$

In the general case, the boundaries of the physical region for each channel may be obtained by setting $z = \pm 1$ in (14)–(16).

On the boundary of the physical region the independent momenta p_1, p_2, and p_3 become linearly dependent, or $L^2 = 0$, where $L_\mu = \varepsilon_{\mu\nu\lambda\sigma} p_1^\nu p_2^\lambda p_3^\sigma$.

From this,[101]

$$\Delta = \begin{vmatrix} p_1^2 & p_1p_2 & p_1p_3 \\ p_2p_1 & p_2^2 & p_2p_3 \\ p_3p_1 & p_3p_2 & p_3^2 \end{vmatrix} = 0, \tag{21}$$

which gives a relation defining the boundaries of the physical regions in all three channels:

$$stu = as+bt+cu, \tag{22}$$

$$a = \frac{1}{h}(m_1^2m_2^2-m_3^2m_4^2)(m_1^2+m_2^2-m_3^2-m_4^2), \tag{23}$$

$$b = \frac{1}{h}(m_1^2 m_3^2-m_2^2m_4^2)(m_1^2+m_3^2-m_2^2-m_4^2), \tag{24}$$

$$c = \frac{1}{h}(m_1^2m_4^2-m_2^2m_3^2)(m_1^2+m_4^2-m_2^2-m_3^2). \tag{25}$$

Inside the physical region, $L^2 < 0$ or $\Delta > 0$, since L_μ is a space-like vector. Formula (22) defines a curve of third order with asymptotes $s = 0$, $t = 0$, and $u = 0$. The physical regions of the variables s, t, and u may be depicted graphically in the stu plane, or the Mandelstam[102] plane. To preserve the symmetry with respect to s, t, and u while taking account of $s+t+u = h$, it is convenient to introduce triangular coordinates for which the sum of the distances from the point $P(s, t, u)$ to the coordinate axes forming an equilateral triangle is always equal to h (Fig. 4).

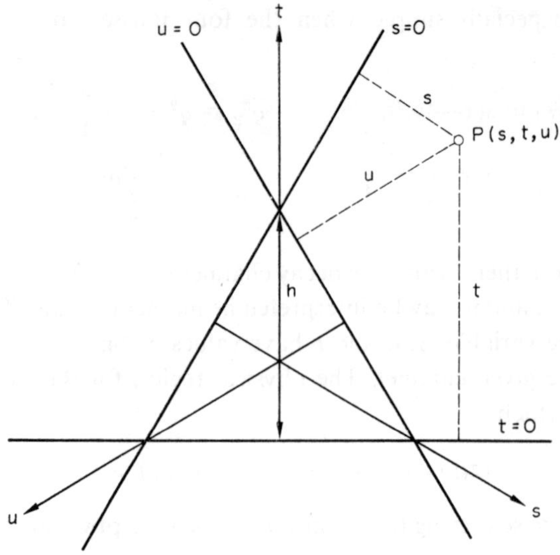

FIG. 4. Mandelstam *stu* plane, showing constraint $s+t+u = h$ obeyed by each point.

For elastic scattering of identical particles ($m_i = m$) the physical regions of the s-, t-, and u-channels occupy the triangular sectors:

$$\begin{aligned} &s \geqslant 4m^2, \quad t \leqslant 0, \quad u \leqslant 0 \quad (s\text{-channel}); \\ &s < 0, \quad t \geqslant 4m^2, \quad u \leqslant 0 \quad (t\text{-channel}); \\ &s \leqslant 0, \quad t \leqslant 0, \quad u \geqslant 4m^2 \quad (u\text{-channel}) \end{aligned}$$

(Fig. 5); in this case the picture in the *stu* plane is invariant with respect to rotations by 120°

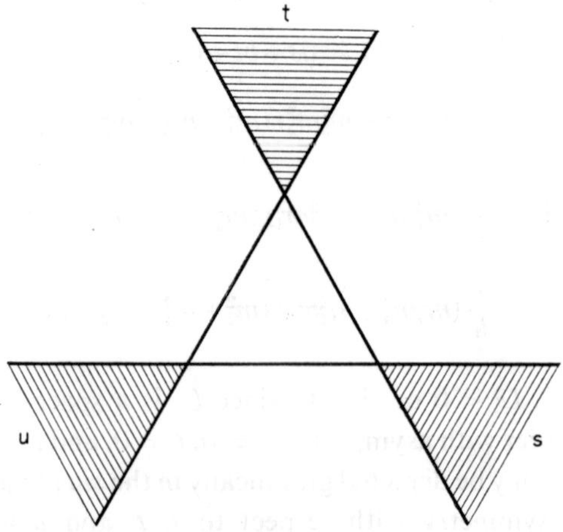

FIG. 5. Physical regions in the Mandelstam *stu* plane for equal-mass scattering.

§ 7.3. Cross-sections for processes. Unitarity and optical theorem

In scattering experiments one measures cross-sections, i.e. the probabilities for processes (per unit volume and unit time) for a unit flux of colliding particles. While the matrix element (1) and the probabilities for processes (1.70) depend on the normalization of single-particle states (and, for normalizations different from ours, may be noninvariant quantities), the cross-sections do not depend on this normalization.

The derivation of an expression for the cross-section in terms of the scattering matrix may be found in nearly any book on scattering theory or elementary particle theory. For this reason we shall give only the basic steps of the derivation, explaining the notation and particular aspects of the formulae in connection with our normalization conventions.

Let us consider a process initiated by the collision of two particles: $a+b \rightarrow 1+ \ldots +n$, and let us assume for simplicity that the initial and final states contain no identical particles. If the momenta of the particles in the final state are in the region $\delta = \Delta \boldsymbol{p}_1 \ldots \Delta \boldsymbol{p}_n$ with average momenta $\boldsymbol{p}_1 \ldots \boldsymbol{p}_n$, then the final state vector $|f\rangle$ is equal to

$$|f\rangle = \int_\delta d^3p_1 \ldots d^3p_n | \boldsymbol{p}_1, \lambda_1, 1\rangle \ldots | \boldsymbol{p}_n, \lambda_n, n\rangle. \tag{26}$$

The norm of the state (26) is equal to

$$\langle f | f \rangle = N_1 \ldots N_n \, \Delta \boldsymbol{p}_1 \ldots \Delta \boldsymbol{p}_n \, 2p_{10} \ldots 2p_{n0}, \tag{27}$$

where N_k appears in the norm of the single-particle state for particle k:

$$\langle \boldsymbol{p}, \lambda, k | \boldsymbol{p}', \lambda', k \rangle = N_k 2p_{k0} \delta_{\lambda\lambda'} \delta(\boldsymbol{p}-\boldsymbol{p}'). \tag{28}$$

Our choice

$$N_k = 1 \tag{29}$$

will be the same for bosons and fermions.

In the final state (26) the matrix element for the transition from the initial state

$$| i \rangle = | \boldsymbol{p}_a, \lambda_a, a; \boldsymbol{p}_b, \lambda_b, b\rangle \tag{30}$$

is equal, according to (1.67) and (26), to

$$\langle f | S-1 | i \rangle = i(2\pi)^4 \, \delta^4(p_a+p_b-p_1- \ldots -p_n) \times \langle 1 \ldots n | T | a, b\rangle \, \Delta \boldsymbol{p}_1 \ldots \Delta \boldsymbol{p}_n,$$

where for brevity we have denoted

$$| \boldsymbol{p}_1, \lambda_1, 1; \ldots; \boldsymbol{p}_l, \lambda_l, l\rangle = | 1 \ldots l\rangle.$$

The formula (1.70), refers to the probability for the transition (per unit time in unit volume) from the state (30) to a state normalized to unity. With the normalization (27) we must replace (1.70) by the formula

$$w(i \rightarrow f) = (2\pi)^4 \, \delta^4(p(i)-p(f)) \frac{|\langle f | T | i \rangle|^2}{\langle f | f \rangle}.$$

Inserting (27) and

$$\langle f | T | i \rangle = \langle 1 \ldots n | T | a, b \rangle \, \Delta \boldsymbol{p}_1 \ldots \Delta \boldsymbol{p}_n,$$

we find the probability for the process $a+b \to 1+ \dots +n$:

$$\begin{aligned} w(a+b \to 1+ \dots +n) &= (2\pi)^4 |\langle 1 \dots n|T|a, b\rangle|^2 \\ &\quad \times \delta^4(p_a+p_b-p_1- \dots -p_n)(2p_{10} \dots 2p_{n0}N_1 \dots N_n)^{-1} d\boldsymbol{p}_1 \dots d\boldsymbol{p}_n \\ &= (2\pi)^4 |\langle 1 \dots n|T|a, b\rangle|^2 (N_1 \dots N_n)^{-1} dR_n(p_a+p_b), \end{aligned} \tag{31}$$

where we have introduced the invariant volume element in momentum space for particles $1 \dots n$:

$$dR_n(p) = \delta^4(p-p_1- \dots -p_n)\, \delta(p_1^2-m_1^2) \dots \delta(p_n^2-m_n^2)\, d^4p_1 \dots d^4p_n. \tag{32}$$

The probability (31) may be represented as the product of the flux of particles in the initial state $j(a, b)$ by the differential cross-section for the process $d\sigma(a+b \to 1+ \dots +n)$:

$$w(a+b \to 1+ \dots +n) = j(a, b)\, d\sigma(a+b \to 1+ \dots +n). \tag{33}$$

The flux $j(a, b)$ is equal to the product of the density of the number of particles a and b by the relative velocity v_{ab} in the c.m.s.

$$j(a, b) = \varrho_a \varrho_b v_{ab} = \varrho_a \varrho_b q_{ab}\left(\frac{1}{p_{a0}} + \frac{1}{p_{b0}}\right) = \varrho_a \varrho_b \frac{q_{ab}\sqrt{s}}{p_{a0}\, p_{b0}} = \varrho_a \varrho_b \frac{F}{p_{a0}\, p_{b0}}, \tag{34}$$

where

$$F = \frac{\Delta_{ab}}{2} = \frac{\sqrt{[s-(m_a+m_b)^2]\,[s-(m_a-m_b)^2]}}{2} \tag{35}$$

is called the invariant Møller flux.

The single-particle states $|\boldsymbol{p}, \lambda, r\rangle$ are normalized in (28) so that the particle density ϱ_r (with given λ) is equal to

$$\varrho_r = \frac{N_r 2p_0}{(2\pi)^3}. \tag{36}$$

Here the spinors u and v must be normalized according to

$$u^+(\boldsymbol{p}, \lambda)\, u(\boldsymbol{p}, \lambda') = N_r 2p_0 \delta_{\lambda\lambda'}, \tag{37}$$

so that, for example, the wave function of a particle with spin $\frac{1}{2}$ is

$$\langle 0|\psi_\alpha(x)|\boldsymbol{p}, \lambda\rangle = \frac{u_\alpha(\boldsymbol{p})}{(2\pi)^{3/2}} e^{ipx}. \tag{38}$$

In the case of a particle with spin 0

$$\langle 0|\varphi(x)|\boldsymbol{p}, J=0\rangle = \frac{N^{1/2}}{(2\pi)^{3/2}} e^{ipx}. \tag{39}$$

Thus the flux density $j(a, b)$ is equal to

$$j(a, b) = \begin{cases} \dfrac{2\Delta_{ab}}{(2\pi)^6} N_a N_b, \\[2ex] \dfrac{2\Delta_{ab}}{(2\pi)^6} \quad \text{in our normalization.} \end{cases} \tag{40}$$

From formulae (31), (33), and (40) it is clear that for the normalization (29) the flux density $j(a, b)$ is relativistically invariant, while the cross-section $d\sigma$ may be expressed only in terms of invariant quantities. In what follows we shall set $N_k = 1$. In fact, $d\sigma$ does not depend on the normalization constants N_k, since in $d\sigma$ the matrix element in (31) always appears in combination with $(N_1 \dots N_n)^{-1}$.

Finally, the differential cross-section for the process $a+b \to 1+ \dots +n$ is equal to

$$d\sigma = \frac{32}{B^2 \Delta_{ab}} |\langle 1 \dots n | T | a, b \rangle|^2 \, dR_n(p_a + p_b), \tag{41}$$

where the invariant quantities Δ_{ab} and the phase element dR_n are given by formulae (12) and (32). We have introduced in (41) the notation

$$B = \frac{8}{(2\pi)^5} \tag{42}$$

for a numerical coefficient which will be encountered often in subsequent formulae.

The total cross-section for the process $a+b \to 1+ \dots +n$ may be found by integrating with respect to momenta of the produced particles and summing over their spin states $(\lambda) = (\lambda_1 \dots \lambda_n)$:

$$\sigma(a+b \to 1+ \dots +n) = \frac{32}{B^2 \Delta_{ab}} \sum_{(\lambda)} \int |\langle 1 \dots n | T | a, b \rangle|^2 \, dR_n(p_a + p_b). \tag{43}$$

When the amplitude is independent of the momenta of the final state, this cross-section will be proportional to the invariant phase space

$$R_n = \int dR_n. \tag{44}$$

Finally, the total cross-section σ_{tot} for the scattering of particles a and b, i.e. the total cross-section for the process $a+b \to$ (anything), may be obtained by summing (43) over all possible final states $|f\rangle$ with various particles and various numbers of particles:

$$\sigma_{\text{tot}}(a, b) = \sum_f \sigma(a+b \to f). \tag{45}$$

If the initial beam is not polarized, then one must average the cross-section over the orientation of the spin in the initial state; this is equivalent to the operation

$$\frac{1}{(2J_a+1)(2J_b+1)} \sum_{\lambda_a, \lambda_b} . \tag{46}$$

Up to now it has been assumed that none of the particles $1 \dots n$ are identical. We shall now consider the case in which all particles $1 \dots n$ are identical. Let the particles be scalar. Set

$$|f\rangle = \int_\delta d^3p_1 \dots d^3p_n a^+(\mathbf{p}_1) \dots a^+(\mathbf{p}_n) |0\rangle,$$

where $\delta = \Delta \mathbf{p}_1 \dots \Delta \mathbf{p}_n$. If the regions Δp_i do not overlap, the normalization (27) continues to hold for all identical particles (we take $N_r = 1$):

$$\langle f | f \rangle = \Delta \mathbf{p}_1 \dots \Delta \mathbf{p}_n 2p_{10} \dots 2p_{n0}.$$

The factor $n!$ does not appear. Performing all the above calculations, we verify that for identical particles one obtains the same formula (41) for $d\sigma$ as in the case of distinguishable particles. In other words, the identity of the particles does not affect the differential cross-section if it is defined by formula (33):

$$d\sigma(a+b \to 1+\ \dots\ +n) = \frac{w(a+b \to 1+\dots+n)}{j(a, b)},$$

where w is the probability for a transition into the state $|f\rangle$ in which the identical particles lie in the regions $\Delta \boldsymbol{p}_1 \dots \Delta \boldsymbol{p}_n$.

However, in calculating the total cross-section and in the unitarity condition below [see (51)], the factor $(n!)^{-1}$ does appear. In fact in these formulae one must not integrate over all momentum space in the final state $(p_1 \dots p_n)$, but only over its $(1/n!)$th part. States differing by the interchange of identical particles must be considered as a single state, counted once. If there are ν_r particles of type r among the particles $1 \dots n$, the invariant phase space dR_n becomes $dR_n/\nu_r!$

The most interesting process is $a+b \to 1+2$. Passing to the c.m.s. with $\boldsymbol{p}_1 = -\boldsymbol{p}_2 = \boldsymbol{q}_{12}$ and integrating dR_2 over all variables except for the polar angles of the vector $\boldsymbol{q}_{12}$,

$$dR_2(\sqrt{s}) = \frac{1}{4}\frac{q_{12}}{\sqrt{s}}\, d\Omega, \quad d\Omega = -d\varphi\, dz_s. \tag{47}$$

Inserting (47) into (41) and recalling that according to (13) $\Delta_{ab}(s) = 2\sqrt{s}q_{ab}$, we find the differential cross-section for a binary process:

$$d\sigma(a+b \to 1+2) = \frac{4}{sB^2}|\langle 1, 2|T|a, b\rangle|^2 \frac{\Delta_{12}(s)}{\Delta_{ab}(s)}\, d\Omega. \tag{48}$$

In the case of elastic scattering $\Delta_{12} = \Delta_{ab}$ and, consequently,

$$d\sigma(a+b \to a+b) = \frac{4}{sB^2}|\langle a', b'|T|a, b\rangle|^2\, d\Omega,$$

where a' and b' denote particles a and b in the state after scattering. The usual definition of the normalized scattering amplitudes $f_{ab}(s, \vartheta, \varphi)$, originating in nonrelativistic theory, is based on the formula

$$d\sigma(a+b \to a+b) = |f_{ab}(s, \vartheta, \varphi)|^2\, d\Omega.$$

Consequently, the relation between the amplitude $f_{ab}(s, \vartheta, \varphi)$ and the matrix element

$$\langle a', b'|T|a, b\rangle \equiv \langle p'_a, \lambda'_a;\, p'_b, \lambda'_b|T|p_a, \lambda_a;\, p_b, \lambda_b\rangle,$$

which is connected with the S-matrix element by relation (30), may be written in the form

$$f_{ab}(s, \vartheta, \varphi) = \frac{2}{\sqrt{s}B}\langle p'_a, \lambda'_a;\, p'_b, \lambda'_b|T|p_a, \lambda_a;\, p_b, \lambda_b\rangle. \tag{49}$$

To avoid confusion we shall perform all calculations using the matrix elements $\langle a', b'|T|a, b\rangle$.

Let us give an expression for the differential cross-section (48) as a function of the invariant variable $t = (p_a - p_1)^2$, $dt = 2q_{ab}q_{12}\,dz_s$ for the case of spinless particles. In this case, the amplitude depends only on the invariant variables s, t, u related by the conditions $s+t+u = m_a^2+m_b^2+m_1^2+m_2^2$, i.e. $\langle 1, 2|T|a, b\rangle = F(s, t, u)$. By (48)

$$d\sigma(a+b \to 1+2) = \frac{16\pi}{B^2 \Delta_{ab}^2(s)} |F(s, t, u)|^2\, dt. \tag{50}$$

The unitarity condition (1.68) may now be written in explicit form.

The completeness of the system of states $|m\rangle$ means that

$$1 = \sum_m |m\rangle\langle m| = \sum_{(n)} \sum_{(\lambda)} \int \frac{d^3p_1}{2p_{10}} \ldots |\boldsymbol{p}_1, \lambda_1; \ldots\rangle\langle \boldsymbol{p}_1, \lambda_1; \ldots| \frac{1}{\nu_1! \ldots}$$

$$= \sum_{(n)} \sum_{(\lambda)} \int d^4p_1 \delta(p_1^2 - m_1^2) \ldots |\boldsymbol{p}_1, \lambda_1; \ldots\rangle\langle \boldsymbol{p}_1, \lambda_1; \ldots| \frac{1}{\nu_1! \ldots}, \tag{51}$$

where (λ) denotes the set of spin states $(\lambda_1 \ldots \lambda_n)$, of particles $1 \ldots n$ in the state $|\boldsymbol{p}_1, \lambda_1; \ldots; \boldsymbol{p}_n, \lambda_n\rangle$; ν_1 is the number of identical particles of type 1 in the state $\boldsymbol{p}_1, \lambda_1$; the summation on (n) means a summation over various types of particles for a given particle number n and the sum over all possible numbers of particles n. Using expression (32) for the phase space and replacing the states $|\alpha\rangle$, $|\beta\rangle$ in (1.68) by $|a, b\rangle$ and $|1 \ldots n\rangle$, we obtain the unitarity condition in the following form:

$$\frac{1}{2i}\langle 1, 2|T - T^+|a, b\rangle = \frac{1}{2}(2\pi)^4 \sum_{(n)} \sum_{(\lambda)} \int \langle 1, 2|T^+|1 \ldots n\rangle \frac{dR_n(p_a+p_b)}{\nu_1! \ldots} \langle 1 \ldots n|T|a, b\rangle. \tag{52}$$

In the particular case of forward elastic scattering (i.e. for $z_s = 1$ and $\varphi = 0$, for which $|a, b\rangle = |\boldsymbol{p}_a, \lambda_a; \boldsymbol{p}_b, \lambda_b\rangle$ and $|1, 2\rangle = |a, b\rangle$, the left-hand side of the unitarity condition (52) reduces to the imaginary part of the forward elastic scattering amplitude:

$$\frac{1}{2i}\langle a, b|T - T^+|a, b\rangle = \mathrm{Im}\,\langle a, b|T|a, b\rangle.$$

The right-hand side of (52) in this case, by (43) and (45), will be proportional to the total cross-section $\sigma_{tot}(a, b)$ for $a+b \to$ (anything).

We thus obtain the relation

$$\mathrm{Im}\,\langle a, b|T|a, b\rangle = \frac{\Delta_{ab}(s)B}{16\pi}\sigma_{tot}(a, b), \tag{53}$$

known as the optical theorem. If a and b are spinless particles of the same mass m, then for forward scattering $t = 0$, and the optical theorem takes the form

$$\mathrm{Im}\, F(s, 0) = \frac{[s(s-4m^2)]^{1/2}\, B}{16\pi}\sigma_{tot}(a, b),$$

where the scattering amplitude $F(s, t) = F(s, t, u(s, t))$ was introduced in connection with (50).

§ 7.4. Helicity amplitudes

Let us consider the amplitude for the process $1+2 \to 3+4$. Since the S-matrix commutes with the generators P_μ and $M_{\mu\nu}$ of the Poincaré group, it will also commute with the invariants of the group—the total mass $m^2 = P^2$ and the square of the total spin, $-w^2/m^2 = J^2$, whose eigenvalues denote the irreducible representations of this group. Consequently, the S-matrix is diagonal in m and J, and the amplitude may be expanded in a series of partial wave amplitudes $T(m, J)$. We shall find such an expansion for the amplitude (an expansion of the type (4)) when the initial and final configurations are described using two-particle helicity states.[58]

Two-particle helicity states were constructed in § 4.6. Let the momenta of particles a and b be $\boldsymbol{p}_a = \boldsymbol{q}_{ab}$ and $\boldsymbol{p}_b = -\boldsymbol{q}_{ab}$ in the c.m.s. In the helicity basis the momentum $\boldsymbol{q}_{ab}$ is defined by the polar angles φ and ϑ.

The two-particle state depending on the angles φ and ϑ was defined in the c.m.s. in terms of the product of single-particle states by formula (4.139) or

$$|0, m; \varphi, \vartheta; \lambda_a, \lambda_b\rangle = \sqrt{\frac{q_{ab}}{4m}}\,|\boldsymbol{q}_{ab}, \lambda_a\rangle\,|-\boldsymbol{q}_{ab}, \lambda_b\rangle^-,$$

where $m^2 = s = (p_{a0}+p_{b0})^2$, while λ is the helicity.

A two-particle state with momentum $\boldsymbol{P}$ may be obtained by passing to a moving reference frame [see (4.140)]:

$$|\boldsymbol{P}, m; \varphi, \vartheta; \lambda_a, \lambda_b\rangle = \mathrm{U}(0, h(\boldsymbol{P}))\,|0, m; \varphi, \vartheta; \lambda_a, \lambda_b\rangle.$$

Here $h(\boldsymbol{P})$ is the Wigner operator in the helicity basis. According to (4.143) this state may be expanded in states with various spins J (and the same mass m):

$$|\boldsymbol{P}, m; \varphi, \vartheta; \lambda_a, \lambda_b\rangle = \sum_J \sum_{\Lambda=-J}^{J} \left(\frac{2J+1}{4\pi}\right)^{1/2} \mathcal{D}^J_{\Lambda\lambda}(\varphi, \vartheta, -\varphi)\,|\boldsymbol{P}, m; J, \Lambda; \lambda_a, \lambda_b\rangle \tag{54}$$

$(\lambda = \lambda_a - \lambda_b)$. Let the momenta of particles 1 and 2 (in the c.m.s.) have components only along the z-axis or $\varphi_{12} = \vartheta_{12} = 0$. Then, writing the final state in the form (54),

$$\left.\begin{aligned} &\langle \boldsymbol{P}', m'; \varphi, \vartheta; \lambda_3, \lambda_4|S-1|\boldsymbol{P}, m; 0, 0; \lambda_1, \lambda_2\rangle| \\ &\quad = i(2\pi)^4\,\delta^4(P-P')\sum_J \frac{2J+1}{4\pi}\,\mathcal{D}^{J*}_{\lambda\lambda'}(\varphi, \vartheta, -\varphi)\,\langle\lambda_3, \lambda_4|T(m^2, J)|\lambda_1, \lambda_2\rangle, \\ &\qquad P_0 = (m^2+\boldsymbol{P}^2)^{1/2} = (s+\boldsymbol{P}^2)^{1/2}, \end{aligned}\right\} \tag{55}$$

where $\lambda = \lambda_1 - \lambda_2$, $\lambda' = \lambda_3 - \lambda_4$, ϑ is the angle between the momentum $\boldsymbol{q}_{34}$ and the z-axis in the c.m.s., and φ is the second polar angle of the vector $\boldsymbol{q}_{34}$.

We have introduced the spin matrix T in (55) using the relation

$$\begin{aligned} &\langle \boldsymbol{P}, m; J, \Lambda; \lambda_1, \lambda_2|S-1|\boldsymbol{P}', m'; J', \Lambda'; \lambda_3, \lambda_4\rangle \\ &\quad = i(2\pi)^4\,\delta^4(P-P')\,\delta_{JJ'}\delta_{\Lambda\Lambda'}\langle\lambda_1, \lambda_2|T(m^2, J)|\lambda_3, \lambda_4\rangle. \end{aligned} \tag{56}$$

Equation (55) is an expansion of the matrix element in terms of partial wave amplitudes $T(m^2, J)$.

We are interested in an expansion of the type (55) for the helicity amplitude

$$F_{\lambda_1\lambda_2\lambda_3\lambda_4}(s, \vartheta, \varphi) \equiv \langle \boldsymbol{p}_3, \lambda_3; \boldsymbol{p}_4, \lambda_4 | T | \boldsymbol{p}_1, \lambda_1; \boldsymbol{p}_2, \lambda_2 \rangle,$$

for which the product of single-particle helicity states is defined according to (4.139), where the mass of the two-particle state is $m = \sqrt{s}$. This amplitude is connected with the usual (normalized) helicity amplitude of Jacob and Wick, $f_{\lambda_1\lambda_2\lambda_3\lambda_4}$, by (49):

$$F_{\lambda_1\lambda_2\lambda_3\lambda_4} = \frac{B\sqrt{s}}{2} f_{\lambda_1\lambda_2\lambda_3\lambda_4}, \quad B = \frac{8}{(2\pi)^5}.$$

Let us write the expansion of the helicity amplitude $F_{\lambda_1\lambda_2\lambda_3\lambda_4}$ in terms of partial wave amplitudes in the form

$$F_{\lambda_1\lambda_2\lambda_3\lambda_4} = B \sum_J (2J+1)\, \mathcal{D}^{J*}_{\lambda\lambda'}(\varphi, \vartheta, -\varphi)\, a_{J\lambda_1\lambda_2\lambda_3\lambda_4}(s). \tag{57}$$

Then, comparing (55) and (57) and recalling (4.139) gives

$$a_{J\lambda_1\lambda_2\lambda_3\lambda_4}(s) = \left(\frac{s}{\pi q_{12} q_{34}}\right)^{1/2} \langle \lambda_3, \lambda_4 | (s, J) | \lambda_1, \lambda_2 \rangle.$$

The dependence of the helicity amplitude on the angle φ is easily isolated:

$$F_{\lambda_1\lambda_2\lambda_3\lambda_4}(s, \vartheta, \varphi) = e^{i(\lambda-\lambda')\varphi} \tilde{F}_{\lambda_1\lambda_2\lambda_3\lambda_4}(s, \vartheta),$$

since

$$\mathcal{D}^{J*}_{\lambda\lambda'}(\varphi, \vartheta, -\varphi) = e^{i(\lambda-\lambda')\varphi}\, d^J_{\lambda\lambda'}(\vartheta).$$

The differential cross-section is equal to [see (48)]

$$d\sigma(1+2 \to 3+4) = \frac{4}{sB^2} |\tilde{F}_{\lambda_1\lambda_2\lambda_3\lambda_4}(s, \vartheta)|^2 \frac{\Delta_{12}(s)}{\Delta_{34}(s)}\, d\Omega = |\tilde{f}_{\lambda_1\lambda_2\lambda_3\lambda_4}(s, \vartheta)|^2 \frac{\Delta_{12}(s)}{\Delta_{34}(s)}\, d\Omega.$$

The expansions (55) and (57) have a simple group theoretic meaning: In the c.m.s. the amplitude $F_{\lambda_1\lambda_2\lambda_3\lambda_4}(s, \vartheta, \varphi)$ is proportional to the average of the product of the invariant scattering operator T by the finite rotation $R(\varphi, \vartheta, -\varphi)$. This average is calculated over all pairs of quantum numbers J, Λ—the total spin and its projection. Since $T(m^2, J)$ does not depend on Λ, the presence of the quantum number Λ leads only to the factor $2J+1$.

If particles 1, 2, 3, and 4 are spinless, then inserting $\mathcal{D}^J_{00}(\varphi, \vartheta, -\varphi) = P_J(\cos\vartheta)$ into (57) we obtain an expansion of the amplitude in terms of Legendre polynomials:

$$f(s, z_s) = \sum_J (2J+1)\, P_J(z_s)\, f_J(s). \tag{58}$$

The number of independent helicity amplitudes is defined by the number of independent helicity states (§ 4.6) and, consequently, is equal to the number of different combinations of helicities λ_1, λ_2, λ_3, λ_4, subject to

$$|\lambda_1 - \lambda_2| \leqslant J, \quad |\lambda_3 - \lambda_4| \leqslant J. \tag{59}$$

Equation (59) holds for all values of λ_r whenever the total spin J satisfies $J \geqslant J_1+J_2$ and $J \geqslant J_3+J_4$, where J_r is the spin of the rth particle. The maximum number of independent amplitudes is equal to $(2J_1+1)(2J_2+1)(2J_3+1)(2J_4+1)$.

If particles 1 and 2 (or 3 and 4) are identical, the helicity amplitudes are restricted by an additional condition (see § 4.6):

$$\langle \lambda_1, \lambda_2 | T(0, J) | \lambda_3, \lambda_4 \rangle = 0 \quad \text{when} \quad \lambda_1 = \lambda_2 \tag{60}$$

for odd J. The partial wave amplitude is symmetric or antisymmetric with respect to $\lambda_1 \leftrightarrow \lambda_2$ respectively for even or odd J.

The invariance of the theory with respect to spatial reflection P and time-reversal T also imposes a limitation on the helicity amplitudes. According to (6.84) spatial reflection P takes a two-particle helicity state with spin J (in the c.m.s.) into a state with opposite helicities:

$$\mathrm{P}|0, m; J, \Lambda; \lambda_1, \lambda_2\rangle = \eta_1^*\eta_2^*(-1)^{J-J_1-J_2}|0, m; J, \Lambda; -\lambda_1, -\lambda_2\rangle,$$

where $\eta_{1,2}$ are the parities of particles 1 and 2. Consequently, if the S-matrix is invariant with respect to parity P ($\mathrm{P}^{-1}S\mathrm{P} = S$), then

$$\langle \lambda_3, \lambda_4 | T(0, J) | \lambda_1, \lambda_2 \rangle = \frac{\eta_3\eta_4}{\eta_1\eta_2}(-1)^{J_3+J_4-J_1-J_2}\langle -\lambda_3, -\lambda_4 | T(0, J) | -\lambda_1, -\lambda_2 \rangle. \tag{61}$$

The factor in front of the matrix element on the right-hand side of (61) depends only on the properties of individual particles, and the relation (61) holds for all partial wave amplitudes. This factor may equal ± 1; for elastic scattering it is equal to $(\eta_3\eta_4/\eta_1\eta_2)\times(-1)^{J_3+J_4-J_1-J_2} = 1$. Parity conservation thus divides the number of independent helicity amplitudes by two.

For time-reversal T, according to (6.100), the helicities do not change, but the total helicity (the projection of the total spin) changes sign:

$$\mathrm{T}|0, m; J, \Lambda; \lambda_1, \lambda_2\rangle = (-1)^{J-\Lambda}|0, m; J, -\Lambda; \lambda_1, \lambda_2\rangle.$$

Since in a theory invariant with respect to time-reversal T,

$$\langle a | S | b \rangle = \langle \mathrm{T}b | S | \mathrm{T}a \rangle,$$

we obtain

$$\langle \lambda_1, \lambda_2 | T(s, J) | \lambda_3, \lambda_4 \rangle = \langle \lambda_3, \lambda_4 | T(s, J) | \lambda_1, \lambda_2 \rangle \tag{62}$$

—a relation between partial wave amplitudes for direct and reversed processes. In the case of elastic scattering (62) is a condition for symmetry of the amplitude.

For baryon–baryon scattering $1+2 \to 3+4$, in the case of spins $J_i = \frac{1}{2}$, $i = 1 \ldots 4$, there are four independent partial wave amplitudes $\langle \lambda_3, \lambda_4 | T(s, 0) | \lambda_1, \lambda_2 \rangle$ or $a_{0\lambda_1\lambda_2\lambda_3\lambda_4}(s)$ with angular momentum $J = 0$, specifically:

$$a_{0++++}, \quad a_{0++--}, \quad a_{0--++}, \quad a_{0----}.$$

Here we have introduced the notation

$$a_{J+\frac{1}{2}+\frac{1}{2}+\frac{1}{2}+\frac{1}{2}} = a_{J++++} \quad \text{etc.}$$

In the case $J \geqslant 1$ there will be 16 partial wave amplitudes for each J, since according to (59) the helicities of the particles $\lambda_i = \pm\frac{1}{2}$ are not correlated for $J \geqslant 1$.

When parity conservation is taken into account the number of amplitudes is divided by two; according to (61) amplitudes of the type a_{J++--} and a_{J--++} may differ only by a sign. For elastic scattering subject to T-invariance, the symmetry condition (62) implies

$$a_{J+++-} = a_{J+-++}, \quad a_{J++-+} = a_{J-+++}. \tag{63}$$

Moreover, if particles 1 and 2 are identical, by (60)

$$\left.\begin{array}{ll}\left.\begin{array}{l} a_{J++\lambda_3\lambda_4} = a_{J--\lambda_3\lambda_4} = 0, \\ a_{J+-+-} = -a_{J-++-} \end{array}\right\} & \text{for odd } J, \\ a_{J+-\lambda_3\lambda_4} = a_{J-+\lambda_3\lambda_4} & \text{for even } J. \end{array}\right\} \tag{64}$$

From this it follows that the total number of independent elastic amplitudes $F_{\lambda_1\lambda_2\lambda_3\lambda_4}(s, \vartheta, \varphi)$ when all particles with $J = \frac{1}{2}$ are identical, will be equal to 5:

$$F_{++++}, \quad F_{+++-}, \quad F_{++--}, \quad F_{+-+-}, \quad F_{+--+}. \tag{65}$$

Two-particle unitarity condition

To pass from the general unitarity condition of (52) to the unitarity condition for partial wave amplitudes $\langle\lambda_3, \lambda_4 | T(s, J) | \lambda_1, \lambda_2\rangle$ or $a_{J\lambda_1\lambda_2\lambda_3\lambda_4}(s)$, we may insert the expansion (55) or (57) into (52). Since the total angular momentum J is conserved, the states $|a, b\rangle$ and $|1 \ldots n\rangle$ appearing in the unitarity condition (52) must all have the same J, i.e. the unitarity condition may be written separately for each partial wave amplitude, connecting the partial wave amplitudes for two-particle scattering with the infinite series (in n) of amplitudes $\langle a, b | T(s, J) | 1 \ldots n\rangle$.

If we neglect all amplitudes for higher processes with $n \geqslant 3$ in the unitarity condition (52), then in this approximation ("two-particle") the unitarity condition will contain only the amplitude $\langle\lambda_3, \lambda_4 | T(s, J) | \lambda_1, \lambda_2\rangle$. For sufficiently small s (so that s does not exceed the threshold for production of states with an additional number of particles) this approximation will be exact, while for somewhat larger s it will still be fairly good. Using the orthogonality condition for the functions $\mathcal{D}^J$, we obtain the two-particle unitarity condition for the process $1+2 \to 3+4$ in the form

$$a_{J\lambda_1\lambda_2\lambda_3\lambda_4}(s) - a^*_{J\lambda_3\lambda_4\lambda_1\lambda_2}(s) = 4i\left(\frac{q_{12}q_{34}}{s}\right)^{1/2} \sum_{\lambda_a, \lambda_b} a^*_{J\lambda_3\lambda_4\lambda_a\lambda_b}(s)\, a_{J\lambda_1\lambda_2\lambda_a\lambda_b}(s). \tag{66}$$

If the theory is T-invariant, and, consequently, the symmetry condition (62) holds, (66) is more conveniently written in terms of $T(s, J)$:

$$\text{Im}\,\langle\lambda_3, \lambda_4 | T - T^+ | \lambda_1, \lambda_2\rangle = \frac{2}{\sqrt{\pi}} \langle\lambda_3, \lambda_4 | T(s, J)\, T^+(s, J) | \lambda_1, \lambda_2\rangle. \tag{67}$$

In the spinless case, for which the partial wave amplitudes $a_J(s)$ are defined by (58), we find from (52)

$$a_J(s) - a_J^*(s) = 4i\left(\frac{q_{12}q_{34}}{s}\right)^{1/2} |a_J(s)|^2.$$

For elastic scattering of identical spinless particles $q_{12}^2 = q_{34}^2 = (s-4m^2)/4$ and, consequently, the unitarity condition reads

$$a_J(s)-a_J^*(s) = 2i\left(\frac{s-4m^2}{s}\right)^{1/2} |a_J(s)|^2. \tag{68}$$

in the elastic approximation. This relation will be used frequently in Chapters 12 and 13.

§ 7.5. Spinor amplitudes ($\mathcal{M}$-functions) and invariant amplitudes

In § 4.4 spinor states were introduced whose transformation law under Lorentz rotations was simple and did not depend on momentum. The transformation properties of the operators u($\boldsymbol{p}$) and wave functions (§ 5.1) also do not depend on momentum.

Using spinor states or wave functions we may define transition amplitudes of a new type—spinor amplitudes, or $\mathcal{M}$-functions, which transform in a simple way under Lorentz rotations. We shall rely to a large extent on the material of § 5.1 and consider the two-particle process $1+2 \to 3+4$ as an example.

Let us return to eqn. (2) for the transformation of the S-matrix. The transition amplitude, which differs from the S-matrix only by the invariant factor $i(2\pi)^4\,\delta^4(p_i-p_f)$, transforms by the same rule. The relativistic properties of the transition amplitude are characterized by

$$\langle \boldsymbol{p}_3, \sigma_3; \boldsymbol{p}_4, \sigma_4 | T | \boldsymbol{p}_1, \sigma_1; \boldsymbol{p}_2, \sigma_2\rangle = \sum_{\sigma'} \mathcal{D}^{J_3^*}_{\sigma_3\sigma_3'}(\tilde{A}_3)\, \mathcal{D}^{J_4^*}_{\sigma_4\sigma_4'}(\tilde{A}_4)\, \langle \boldsymbol{p}_3', \sigma_3'; \boldsymbol{p}_4', \sigma_4' | T | \boldsymbol{p}_1', \sigma_1'; \boldsymbol{p}_2', \sigma_2'\rangle$$
$$\times \mathcal{D}^{J_1}_{\sigma_1'\sigma_1}(\tilde{A}_1)\, \mathcal{D}^{J_2}_{\sigma_2'\sigma_2}(\tilde{A}_2), \quad p' = \Lambda(A)p, \tag{69}$$

where $\tilde{A}_n = \alpha^{-1}(\boldsymbol{p}_n')\, A\alpha(\boldsymbol{p}_n)$ describes a rotation of the spin basis. Equation (69) is written in the canonical basis since many formulae in § 5.2 are written explicitly in this basis. One passes to the helicity basis in (69) and below, as usual, by substituting $\sigma \to \lambda$ and by other choices of the operator $\alpha(\boldsymbol{p})$ in the definition of $\tilde{A}(\mathring{p}, A)$ (see § 4.2).

The transition amplitude depends on the momenta and spin variables of the particles, while the transformation of the spin basis depends on momentum. Let us attempt to isolate that part of the dependence of the amplitude related to properties of noninteracting individual particles. At the same time we shall be able to separate the transformation of spin variables from that of momenta.

We define the spinor function $\mathcal{M}^{r_1 r_2}_{r_3 r_4}(p_1 \dots p_4)$, writing the transition amplitude in the form[52, 53, 103]

$$\langle \boldsymbol{p}_3, \sigma_3; \boldsymbol{p}_4, \sigma_4 | T | \boldsymbol{p}_1, \sigma_1; \boldsymbol{p}_2, \sigma_2\rangle$$
$$= \bar{u}^{r_3}(\boldsymbol{p}_3, \sigma_3)\, \bar{u}^{r_4}(\boldsymbol{p}_4, \sigma_4)\, \mathcal{M}^{r_1 r_2}_{r_3 r_4}(p_1 \dots p_4)\, u_{r_1}(\boldsymbol{p}_1, \sigma_1)\, u_{r_2}(\boldsymbol{p}_2, \sigma_2), \tag{70}$$

where $u(\boldsymbol{p}_n, \sigma_n)$ is the wave function for the nth particle introduced in § 5.1. In (70) one sums over identical upper and lower indices as usual.

The introduction of the wave functions $u(\boldsymbol{p}, \sigma)$ in (70) eliminates the matrices $\mathcal{D}^J(\tilde{A})$ in the formula characterizing the relativistic properties of spinor functions.

If we insert formula (70) into eqn. (69), we obtain on the right-hand side an expression containing the wave functions $u(\boldsymbol{p}', \sigma)$ for momentum $p' = \Lambda(A)p$.

The relation

$$L(A)\,u(\boldsymbol{p},\sigma)=\sum_{\sigma'}u(\boldsymbol{p}',\sigma')\,\mathcal{D}^{J}_{\sigma'\sigma}(\alpha^{-1}(\boldsymbol{p}')\,A\alpha(\boldsymbol{p})),\tag{71}$$

holds between the functions $u(\boldsymbol{p},\sigma)$ and $u(\boldsymbol{p}',\sigma')$, as may be easily checked if one uses the explicit form of $u(\boldsymbol{p},\sigma)$:

$$u(\boldsymbol{p},\sigma)-L(\alpha(\boldsymbol{p}'))\,u(0,\sigma)\tag{72}$$

In fact,

$$L(A)\,L(\alpha(\boldsymbol{p}))\,u(0,\sigma)=L(\alpha(\boldsymbol{p}'))\,L(\tilde{A}(\mathring{p},A))\,u(0,\sigma)=\sum_{\sigma'}L(\alpha(\boldsymbol{p}'))\,u(0,\sigma')\,\mathcal{D}^{J}_{\sigma'\sigma}(\tilde{A}(\mathring{p},A)),$$

since, by virtue of (4.45) and (5.7), the matrix $L(\tilde{A})$, which depends on the transformation $\tilde{A}(\mathring{p},A)$ of the little group of the standard momentum $\mathring{p}^{\mu}=(m,0,0,0)$, must perform a rotation of the spin basis $u(0,\sigma)$.

Thus the condition of relativistic invariance for the spinor function has the form

$$\mathcal{M}^{r_1r_2}_{r_3r_4}(p_1\ldots p_4)=L^{r_3'}_{r_3}(A)\,L^{r_4'}_{r_4}(A)\,\mathcal{M}^{r_1'r_2'}_{r_3'r_4'}(p_1'\ldots p_4')\,L^{r_1}_{r_1'}(A^{-1})\,L^{r_2}_{r_2'}(A^{-1}),\tag{73}$$

or

$$\mathcal{M}(p_1\ldots p_4)=L_3(A)\,L_4(A)\,\mathcal{M}(p_1'\ldots p_4')\,L_1(A^{-1})\,L_2(A^{-1}),\tag{74}$$

where L_n is the matrix of the representation of the Lorentz group for the nth particle (with matrix elements $L^{r_n'}_{r_n}$).

To clarify the meaning of the conditions (73) and (74), let us consider two simple examples.

1. Let particles 1 and 3 have spin $\frac{1}{2}$, and particles 2 and 4 be spinless. We shall describe particles 1 and 3 using Dirac wave functions. Then the matrices $L_1(A)$ and $L_3(A)$ coincide with the matrix $S(A)$ [formula (5.19)], while the matrices $L_2(A)$ and $L_4(A)$ are one-dimensional, do not depend on A, and are equal to unity. The spinor amplitude in this case must satisfy the relation

$$\mathcal{M}^{\beta}(p_1\ldots p_4)=S_{\alpha}{}^{\alpha'}(A)\,\mathcal{M}_{\alpha}{}^{\beta'}(p_1'\ldots p_4')\,S_{\beta}{}^{\beta}(A^{-1}),\tag{75}$$

where α and β are Dirac indices, and $p'=\Lambda(A)p$. This invariant condition has the same form as the invariant condition for the Dirac momentum matrix:

$$S(A)\,\gamma^{\mu}p_{\mu}'S(A^{-1})=\gamma'^{\mu}p_{\mu}'=\gamma^{\mu}p_{\mu}.$$

2. Let particles 2 and 4 be spinless, and the spin of particles 1 and 3 be equal to $\frac{3}{2}$. Then there arises the question of which of the equivalent wave functions (spinor functions with four or eight components, Bargmann–Wigner functions or Rarita–Schwinger functions) will be used to describe the particles. The choice of a wave function determines the matrices $L_1(A)$ and $L_3(A)$ in (73) and (74).

Let us choose as wave functions $u(\boldsymbol{p},\sigma)$ of particles with spin $\frac{3}{2}$ the Rarita–Schwinger function $\psi_{\alpha\mu}(\boldsymbol{p},\sigma)$ (see § 5.5) having one Dirac index α and one vector index μ, so that the index r in (73) is $r=(\alpha,\mu)$. Then the matrices $L_1(A)$ and $L_3(A)$ are the direct product of the Dirac matrix $S_{\alpha}{}^{\beta}(A)$ and the Lorentz transformation matrix for a vector $\Lambda_{\mu}{}^{\nu}$:

$$L_r{}^{r'}(A)=S_{\alpha}{}^{\alpha'}(A)\,\Lambda_{\mu}{}^{\mu'}(A).$$

The condition (73) may be written in this case in the form

$$\mathcal{M}^{\beta\nu}_{\alpha\mu}(p_1, \dots p_4) = S_\alpha^{\ \alpha'}(A)\, \Lambda_\mu^{\ \mu'}(A)\, \mathcal{M}^{\beta'\nu'}_{\alpha'\mu'}(p'_1 \dots p'_4)\, S_{\beta'}^{\ \beta}(A^{-1})\, \Lambda_{\nu'}^{\ \nu}(A^{-1}). \tag{76}$$

Equation (76) describes the transformation of a tensor in the indices μ, ν. In particular, a product of momenta of the type $p_\mu p^\nu(\not{p})_\alpha^{\ \beta}$ will satisfy (76).

Invariant amplitudes and kinematic singularities

After a choice of wave functions has established the relativistic properties of the spinor amplitude $\mathcal{M}$, one may find its expansion in terms of invariant amplitudes F_l which depend only on invariant momentum variables. In the case of two-particle processes

$$\mathcal{M}(p_1 \dots p_4) = \sum_l F_l(s, t, u)\, X_l(p_1 \dots p_4). \tag{77}$$

The invariant variables s, t, and u (77) are related by $s+t+u = h = \Sigma m_n^2$.

The independent kinematic covariants X_l have the same transformation properties as the spinor amplitude $\mathcal{M}$. The covariants X_l are constructed from particle momenta and matrices associated with the representations L_n. The matrix elements of the covariants X_l calculated with wave functions for free particles in the initial and final states u_1, u_2, $\bar{u}_3$, $\bar{u}_4$ are equal to the kinematic factors Z_l in the expansion of the amplitude (3):

$$Z_l = (\bar{u}_3\bar{u}_4 X_l u_1 u_2).$$

The number of covariants X_l is equal to the number of independent amplitudes F_l. Since this number cannot depend on the way in which the expansion is performed, it is equal to the number of helicity amplitudes. Finding the independent covariants X_l is the basic problem of kinematic analysis of spinor amplitudes.

The choice of covariants is not unique. There can be different sets of independent covariants and corresponding independent amplitudes. If there are two sets of linearly independent amplitudes, they are connected by the relation

$$F_l(s, t, u) = \sum_{l'} B_{ll'}(s, t, u)\, F'_{l'}(s, t, u).$$

If the sets F_l and F'_l are linearly independent, the matrix B is nonsingular. The singularities of the functions F_l and F'_l must thus coincide. It turns out, however, that many sets of independent amplitudes used in practice become linearly dependent for some momentum values p_n. As a rule, this holds only when [along with the usual relation (5) imposed by the conservation law] there is an additional linear dependence that arises between the momenta p_n. For those values of the variables s, t, and u for which there is a linear dependence between the amplitudes, the matrix $B(s, t, u)$ becomes singular. Consequently, for these values of s, t, and u the functions F_l may have singularities lacking in the functions F'_l, and vice versa. Singularities of this type are called kinematic ones. They always may be eliminated by passing to another set of amplitudes (sometimes with the result that other kinematic singularities appear). In contrast, singularities which are preserved for any choice of independent amplitudes are called dynamical, since they are related to the dynamics of the process.

Kinematic singularities may arise in any process involving particles with nonzero spin. For spinless particles there is a single invariant amplitude which (by definition) cannot have kinematic singularities.

Kinematic singularities arise when the invariants s, t, u lie on the boundary of the physical region (21) or take on threshold values. According to (22)–(25) the threshold values of the invariants depend on the masses of the particles in the initial and final states.

The threshold values of s may be defined by the condition that the momenta q_{ij} vanish in the c.m.s. According to (13) when all masses are different, there exist four thresholds:

$$\left.\begin{aligned} s_N &= (m_1+m_2)^2, \quad & s_A &= (m_1-m_2)^2, \\ s_N' &= (m_3+m_4)^2, \quad & s_A' &= (m_3-m_4)^2, \end{aligned}\right\} \tag{78}$$

of which s_N and s_N' correspond to the beginning of the physical s-channel region and are called normal thresholds, while s_A and s_A' characterize anomalous thresholds.

Thus the covariant factors X_l in the expansion (77) must satisfy the following conditions:

(a) X_l must transform in the same way as the spinor function $\mathcal{M}$;
(b) X_l must be linearly independent, and the number of covariants X_l must be equal to the number of independent helicity amplitudes;
(c) the expansion (77) in terms of the covariants X_l must not introduce kinematical singularities into the amplitudes F_l.†

The choice of a set of independent covariants X_l demands special care when the wave functions used for higher spin have a large number of extra components (e.g. the Rarita–Schwinger and Bargmann–Wigner wave functions). The total set of covariants which may be constructed in this case exceeds the set of independent covariants X_l. The reduction of these covariants to independent ones, carried out using the subsidiary conditions for wave functions, may introduce kinematic singularities.

Construction of independent covariants

Let us consider as simple examples the construction of sets of covariants X_l satisfying the conditions (a)–(c) above. In this problem the spin is the determining factor. Thus, along with the notation $1+2 \rightarrow 3+4$ for a two-particle reaction, we shall use the notation $J_1+J_2 \rightarrow J_3+J_4$, indicating the spins of the particles J_n.

1. $\frac{1}{2}+0 \rightarrow \frac{1}{2}+0$. In this case the spinor function $\mathcal{M}_\alpha^{\ \beta}(p_1 \dots p_4)$ is a 4 by 4 matrix with transformation properties (75). These properties are possessed by the independent Dirac matrices $\mathbf{1}$, γ_5, $\not{p}_n$, $\sigma_{\mu\nu}p_n^\mu p_{n'}^\nu$, and $\gamma_5\not{p}_n$. By counting the number of helicity amplitudes one finds that there are four independent covariants. The momenta are restricted by the momentum conservation law and by the conditions $\not{p}_1 - m_1 = 0$, if $\not{p}_1$ is the right-most factor, and $\not{p}_3 - m_3 = 0$ if $\not{p}_3$ stands on the left. Thus, only one of the quantities $\not{p}_n$ is independent, which we choose as $Q = (\frac{1}{2})(\not{p}_2+\not{p}_4)$. Analogously, only one of the matrices $\gamma_5\not{p}_n$ is independent—for example, $\gamma_5\not{Q}$.

† Expansions of spinor amplitudes have been studied for $(2J+1)$-component wave functions[42, 52, 103, 104] (including effects of reflection symmetry[105, 106]) and for Rarita–Schwinger functions.[107–109]

The matrices $\sigma_{\mu\nu}p_n^\mu p_{n'}^\nu = \frac{1}{2}[\not{p}_n, \not{p}_{n'}]$ may be expressed in terms of the remaining matrices. Consequently, the set of covariants has the form

$$X_l = \{\mathbf{1}, \gamma_5, \not{Q} = \tfrac{1}{2}(\not{p}_2+\not{p}_4), i\gamma_5\not{Q}\} \tag{79}$$

and the expansion of the spinor amplitude (77) is

$$\mathcal{M}_\alpha^\beta(p_1 \ldots p_4) = \{F_1(s, t, u)+\gamma_5 F_2(s, t, u)+\not{Q}F_3(s, t, u)+i\gamma_5\not{Q}F_4(s, t, u)\}_\alpha^\beta, \tag{80}$$

where F_l are the invariant amplitudes.

2. $\frac{1}{2}+0 \to \frac{3}{2}+0$. Choosing the Rarita–Schwinger wave functions $\bar{\psi}^{\alpha\mu}$ for spin $\frac{3}{2}$, we obtain the spinor amplitude $\mathcal{M}_{\alpha\mu}^\beta(p_1 \ldots p_4)$ with two Dirac indices and one vector index. The number of independent covariants is $(2J_1+1)(2J_3+1) = 8$. They are 4 by 4 matrices with a vector index. When constructing covariants one must take account of the subsidiary conditions and the equations of motion for the wave functions $\bar{\psi}^{\alpha\mu}$ and u_β:

$$(\not{p}_1-m_1)\,u(\boldsymbol{p}_1) = 0,$$

$$\bar{\psi}_\mu(\boldsymbol{p}_3)\,(\not{p}-m_3) = 0, \quad \bar{\psi}_\mu\gamma^\mu = 0,$$

so that $\bar{\psi}_\mu(\boldsymbol{p}_3)p_3^\mu = 0$.

As a set of linearly independent covariants, one may thus take the matrices

$$X_\mu = \{Q_\mu\cdot\mathbf{1} = \tfrac{1}{2}(p_2+p_4)_\mu\cdot\mathbf{1},\ P_\mu\cdot\mathbf{1} = \tfrac{1}{2}(p_1+p_3)_\mu\cdot\mathbf{1},$$
$$Q_\mu\not{Q},\ P_\mu\not{Q},\ \gamma_5 Q_\mu,\ \gamma_5 P_\mu,\ i\gamma_5 Q_\mu\not{Q},\ i\gamma_5 P_\mu\not{Q}\}. \tag{81}$$

Here P and Q are chosen symmetric with respect to initial and final momenta. We could have chosen other combinations of momenta instead and obtained a different expansion of the amplitude.

Let us show how a change in the set of covariants entails the appearance of kinematic singularities. One may construct a vector $N_\mu = -\varepsilon_{\mu\nu\lambda\sigma}p_1^\nu p_2^\lambda p_3^\sigma$ from the momenta p_1, p_2, and p_3. The vector N_μ becomes light-like ($N^2 = 0$) on the boundary of the physical region [see (21)]. Let us include the matrix $i\gamma_5\not{N}$ among the covariants. This can be done by virtue of the relation

$$i\gamma_5\not{N} = \tfrac{1}{2}[(m_1+m_3)^2-t]\not{N}+\tfrac{1}{4}(m_3-m_1)(m_2^2-m_4^2)-(m_1+m_3)(s-u), \tag{80'}$$

which may be derived using (5.70). According to this relation the matrix $i\gamma_5\not{N}$ may be substituted in the set of covariants for the matrix $\not{Q}$ or 1. In the second case, for example, instead of $Q_\mu\cdot\mathbf{1}$ one would write $Q_\mu i\gamma_5\not{N}$, etc.

However, if the invariant amplitudes F_l in the expansion with respect to covariants (80) have no singularities at $t = (m_1+m_3)^2$, then in the expansion with respect to the new set $X'_\mu = X_\mu(\not{Q} \to i\gamma_5\not{N})$ the new invariant amplitudes X'_l acquire a singularity at this point. Passing to the set $X''_\mu = X_\mu(1 \to i\gamma_5\not{N})$ also leads to the appearance of kinematic singularities.

3. $0+0 \to 0+2$. A wave function for a particle with spin 2 may be chosen in various ways. If one limits oneself to the wave functions considered in Chapter 5, there are three possibilities: either the wave function is (5.82) (a ten-component function transforming according to the representation $(2, 0)+(0, 2)$) or it is given by formula (5.111) (the symmetric tensor $\Phi_{\mu\nu}$), or it may be written in the form (5.131) (a symmetric spinor of fourth rank).

Let this wave function be the tensor $\Phi_{\mu\nu} = \Phi_{\nu\mu}$, so that the spinor amplitude is $\mathcal{M}_{\mu\nu}$ and, consequently, the covariants $X_{\mu\nu}$ are symmetric tensors of second rank. In this case, the set contains $(2J_4+1) = 5$ independent covariants $X_{\mu\nu}$, which may be constructed from the momenta of the particles p_n and their combinations, forming three independent vectors. The ten components of the symmetric tensor $\Phi_{\mu\nu}$ may be reduced to 5 linearly independent ones using the subsidiary conditions (5.113) and (5.114):

$$p_4^\mu \Phi_{\mu\nu} = p_4^\nu \Phi_{\mu\nu} = 0, \quad \Phi_\mu^\mu = 0.$$

It is thus convenient to choose as basis vectors

$$P = \frac{p_1+p_2}{2}, \qquad \Delta = \frac{p_1-p_3}{2}, \qquad \tilde{N}_\mu = \varepsilon_{\mu\nu\lambda\sigma} P^\nu \Delta^\lambda p_4^\sigma.$$

Then the set of covariants is

$$X_{\mu\nu} = \{P_\mu P_\nu,\ \Delta_\mu \Delta_\nu,\ P_\mu \Delta_\nu + P_\nu \Delta_\mu,\ \tilde{N}_\mu P_\nu + \tilde{N}_\nu P_\mu,\ \tilde{N}_\mu \Delta_\nu + \tilde{N}_\nu \Delta_\mu\}, \tag{82}$$

from which one obtains at once the expansion of $\mathcal{M}_{\mu\nu}$ in terms of invariant amplitudes.

The spinor amplitudes (70) were introduced using the wave functions § 5.1, for which there exists an invariant bilinear Hermitian form, and consequently there also exists a unitary matrix connecting the Lorentz matrices with their conjugates. This means that the relativistic transformation properties of the spinor amplitude (70) do not depend on which particles are in the initial or final states. Changing any particles to antiparticles also does not change the transformation properties of the spinor amplitude. Thus if one finds a general expansion of the spinor amplitude for the process $1+2 \to 3+4+ \ldots +n$, then in practice the expansion of the spinor amplitudes is known for all reactions with these particles or their antiparticles.

Two-particle decays $a \to b+c$ and interaction Lagrangian

Choosing wave functions for describing the free particles a, b, and c we can introduce a spinor amplitude for a decay of the type (70), and then find its expansion (77) in terms of covariants. The process $a \to b+c$ is characterized by two independent momenta of the three p_a, p_b, p_c, connected by the conservation law $p_a = p_b+p_c$ and the condition $p_a^2 = m_a^2$. ... The invariant momentum variables $(p_a-p_b)^2$ or $(p_a-p_c)^2$ in this case are uniquely fixed by the particle masses. Thus in the expansion (77) of the spinor function $\mathcal{M}^{r_a}_{r_b r_c}$ in terms of covariants

$$\mathcal{M}(p_a, p_b, p_c) = \sum_l g_l X_l(p_a, p_b, p_c)$$

the coefficients g_l are constants.

Let us assume that the X_l do not depend on momenta. The decay amplitude is then

$$\sum_l g_l \bar{u}^{r_b}(p_b, \sigma_b)\, \bar{u}^{r_c}(p_c, \sigma_c)\, (X_l)^{r_a}_{r_b r_c} u_{r_a}(p_a, \sigma_a). \tag{83}$$

The decay amplitude may be written in the form of a matrix element

$\langle \boldsymbol{p}_b, \sigma_b; \boldsymbol{p}_c, \sigma_c | \mathcal{L}_I(0) | \boldsymbol{p}_a, \sigma_a \rangle$ of the effective interaction Lagrangian

$$\mathcal{L}_I(x) = \sum_I g_I \bar{\Psi}^{r_b}(x) \bar{\Psi}^{r_c}(x) (X_I)^{r_a}_{r_b r_c} \Psi_{r_a}(x), \tag{84}$$

constructed from the fields describing particles a, b, and c. Here the g_I play the role of effective interaction constants.

In the general case, the covariants X_I are polynomials in the momenta. The corresponding Lagrangian will contain only derivatives of the fields. The relation

$$\langle \boldsymbol{p}_b, \sigma_b; \boldsymbol{p}_c, \sigma_c | T | \boldsymbol{p}_a, \sigma_a \rangle = \langle \boldsymbol{p}_b, \sigma_b; \boldsymbol{p}_c, \sigma_c | \mathcal{L}_I(0) | \boldsymbol{p}_a, \sigma_a \rangle, \tag{85}$$

on the one hand, allows one to find the decay amplitude to first order in perturbation theory, and, on the other, is a definition of the effective Lagrangian in terms of the decay amplitude. The construction of the covariants X_I in this case is thus equivalent to finding the independent trilinear interaction Lagrangians depending on free fields.

Spinor amplitudes and reflections

If the S-matrix is invariant with respect to spatial inversion P (or charge conjugation C, or time-reversal T), then relations between the scalar amplitudes $F_I(s, t, u)$ may arise.

The condition of P-invariance of the S-matrix (6.85) is equivalent, by virtue of (70), to

$$\begin{aligned} &\bar{u}(\boldsymbol{p}_3)\, \bar{u}(\boldsymbol{p}_4)\, \mathcal{M}(\boldsymbol{p}_1 \ldots \boldsymbol{p}_4)\, u(\boldsymbol{p}_1)\, u(\boldsymbol{p}_2) \\ &\qquad = \eta_{\mathrm{P}} \bar{u}(-\boldsymbol{p}_3)\, \bar{u}(-\boldsymbol{p}_4)\, \mathcal{M}(-\boldsymbol{p}_1 \ldots -\boldsymbol{p}_4)\, u(-\boldsymbol{p}_1)\, u(\boldsymbol{p}_2), \end{aligned} \tag{86}$$

where we have omitted the constant quantities p_{n0} and σ_n; η_{P} is the product of the "parities":

$$\eta_{\mathrm{P}} = \eta_{\mathrm{P}}^*(1)\, \eta_{\mathrm{P}}^*(2)\, \eta_{\mathrm{P}}(3)\, \eta_{\mathrm{P}}(4).$$

Explicit expressions for the wave functions $u(\boldsymbol{p}, \sigma)$ were obtained in Chapter 5, so that the connection between the functions $u(-\boldsymbol{p}, \sigma)$ and $u(\boldsymbol{p}, \sigma)$ is known. In particular, for the Dirac wave function $u(-\boldsymbol{p}) = \gamma_4 u(\boldsymbol{p})$, for the $2(2J+1)$-component function $\Phi(-\boldsymbol{p}) = \gamma_4 \Phi(\boldsymbol{p})$, and for spin 1 $\varepsilon_\mu(-\boldsymbol{p}) = g_{\mu\mu}\varepsilon_\mu(\boldsymbol{p})$. Using these formulae (or using equations introduced in deriving the reflection formulae in Chapter 6) one can easily find the relation between $u(-\boldsymbol{p})$ and $u(\boldsymbol{p})$ for Rarita–Schwinger wave functions with spin J:

$$\Psi^J_{\mu_1 \ldots \mu_n}(-\boldsymbol{p}) = \gamma_4 g_{\mu_1\mu_1} \ldots g_{\mu_n\mu_n} \Psi^J_{\mu_1 \ldots \mu_n}(\boldsymbol{p}), \quad n = J - \tfrac{1}{2},$$

and for Bargmann–Wigner wave functions

$$\Psi^J_{\alpha_1 \ldots \alpha_f}(-\boldsymbol{p}) = (\gamma_4)^{\alpha'_1}_{\alpha_1} \ldots (\gamma_4)^{\alpha'_f}_{\alpha_f} \Psi_{\alpha'_1 \ldots \alpha'_f}(\boldsymbol{p}), \quad (f = 2J).$$

The "parities" $\eta_{\mathrm{P}}(j)$ of fermions may have the values $\eta_{\mathrm{P}} = \pm i$ (see § 6.2), while the parities of bosons may be $\eta_{\mathrm{P}}(j) = \pm 1$. If, moreover, one takes into account the conservation of fermion number, the product of parities in (86) reduces to a sign factor $\eta_{\mathrm{P}} = \pm 1$.

Substituting for the functions $u(-\boldsymbol{p})$ and the total parity of the process η_{P} in (86) leads to an explicit form of the condition of P-invariance of the spinor amplitude. The relation between $u(-\boldsymbol{p})$ and $u(\boldsymbol{p})$ is of the form

$$u(-\boldsymbol{p}) = R_J u(\boldsymbol{p}),$$

where the Hermitian matrix R_J depends on the type of wave function. Consequently, the condition (86) leads to the relation

$$\mathcal{M}(p_1 \ldots p_4) = \eta_P R_{J_3} R_{J_4} \mathcal{M}(-p_1 \ldots -p_4) R_{J_1} R_{J_2}. \tag{87}$$

It is convenient to choose the covariants X_l so that they have definite properties under P, specifically:

$$R_{J_3} R_{J_4} X^{(\pm)}(-p_1 \ldots -p_4) R_{J_1} R_{J_2} = \pm X^{(\pm)}(p_1 \ldots p_4).$$

The invariant amplitudes entering into the expansion (77) as coefficients of the covariants $X_l^{(\pm)}$ will be called $F_l^{(\pm)}$. The condition (87) then demands that in the expansion (77) of the spinor amplitude some of the invariant amplitudes vanish. For $\eta_P = +1$ one must have $F_l^{(-)} = 0$, while for $\eta_P = -1$ $F_l^{(+)} = 0$.

Let us turn to some examples. To designate the reaction we shall now write the spin-parity J^P instead of the spin. These examples may be divided into two classes depending on the total parity η.

1. $\frac{1}{2}^+ + 0^- \to \frac{1}{2}^+ + 0^-$, $\eta = 1$. Relation (87) reduces to the condition $\mathcal{M}(p_1 \ldots p_4) = \gamma_4 \mathcal{M}(-p_1 \ldots -p_4)\gamma_4$. From (79) it follows that the invariant amplitudes for the covariants γ_5 and $i\gamma_5 Q$ must vanish. An example of this reaction is the elastic scattering of pions or kaons on nucleons.

$\frac{1}{2}^+ + 0^- \to \frac{1}{2}^- + 0^-$. Here $\eta = -1$ and, consequently, only the invariant amplitudes associated with the covariants γ_5 and $i\gamma_5 Q$ remain. This is a process in which a baryon resonance is produced with parity opposite to that of the neutron.

2. $\frac{1}{2}^+ + 0^- \to \frac{3}{2}^+ + 0^-$, $\eta = +1$. The expansion of the spinor amplitude contains only those covariants in (80) satisfying the condition $X_\mu(p_1 \ldots p_4) = g_{\mu\mu}\gamma_4 X_\mu(-p_1 \ldots -p_4)\gamma_4$, i.e. containing γ_5. These covariants will clearly determine the expansion of the spinor amplitude for all four processes $\frac{1}{2}^\pm + 0^\mp \to \frac{3}{2}^\pm + 0^\mp$ with the same total parity $\eta = 1$.

If the sign of one of the parities on the right- or left-hand side is changed (e.g. $\frac{1}{2}^+ + 0^- \to \frac{3}{2}^+ + 0^+$), the expansion of the spinor amplitude will include only those covariants of (80) which do not contain γ_5.

3. $0^- + 0^- \to 0^- + 2^+$, $\eta = -1$. The spinor amplitude is determined by the two pseudo-tensor covariants in (82).

The invariance of the S-matrix with respect to time reversal (6.101) connects amplitudes for direct and inverse processes. For elastic scattering this condition imposes a restriction on the form of the amplitude. In order to pass from (6.101) to the condition of T-invariance of the spinor amplitude, it is necessary, according to (70), to obtain the wave function of the particle in the transformed matrix element in terms of the initial one. Time reversal changes the sign of the momentum $\boldsymbol{p}$ and takes the initial state into the final one. Consequently, one must calculate (using the explicit form of the wave functions) the matrix $T^{rr'}$ in the relation

$$\bar{u}^r(-\boldsymbol{p}, \lambda) = T^{rr'} u_{r'}^*(\boldsymbol{p}, \lambda).$$

Using this relation one may get rid of the wave functions in the equation which results from (6.101) after substituting (70), and one then obtains a restriction on the spinor amplitude.

In the case of the Dirac wave function u_α and the wave function ε_μ for a particle with spin 1, we find, from the explicit expressions for these functions (see §§ 5.2 and 5.4),

$$\left.\begin{aligned} \bar{u}^\alpha(-\boldsymbol{p}, \lambda) &= (i\gamma_4\gamma_5\mathscr{C}^{-1})^{\alpha\beta}u_\beta(\boldsymbol{p}, \lambda) \equiv T^{\alpha\beta}u_\beta(\boldsymbol{p}, \lambda), \\ \varepsilon^*_\mu(-\boldsymbol{p}, \lambda) &= g_{\mu\mu}\varepsilon_\mu(\boldsymbol{p}, \lambda), \end{aligned}\right\} \tag{88}$$

The first of formulae (88) also holds for $2(2J+1)$-component wave functions if one replaces $\mathscr{C} \equiv \mathscr{C}^{1/2}$ by $\mathscr{C}^J$ [formula (6.92)]. Relation (88) allows one to construct analogous formulae for Rarita–Schwinger and Bargmann–Wigner wave functions.

The nature of the restriction on the spinor amplitude arising from its T-invariance is shown by the simple example of elastic scattering $\frac{1}{2}+\frac{1}{2} \to \frac{1}{2}+\frac{1}{2}$, for example: $N+\Sigma \to N+\Sigma$. The spinor amplitude for this process $\mathscr{M}^{\alpha\beta}_{\alpha'\beta'}$ has four Dirac indices, two upper and two lower, and, consequently, the covariants $X^{\alpha\beta}_{\alpha'\beta'}$ may be represented in the form of a direct product of two 4 by 4 matrices constructed from the Dirac γ matrices and the momenta. There are three ways of distributing the indices of such a covariant between two 4 by 4 matrices.

Let the indices of one 4 by 4 matrix belong to the initial state, and the indices of the other to the final state. Since the Dirac matrices have the indices $(\gamma^R)_\alpha{}^\beta$, in constructing $X^{\alpha\beta}_{\alpha'\beta'}$ we must use the combinations $(\gamma^R\mathscr{C})_{\alpha'\beta'}$ and $(\mathscr{C}^{-1}\gamma^R)^{\alpha\beta}$ (which have definite symmetry properties; see the end of § 5.2). The general expression for a covariant is then

$$X^{\alpha\beta}_{\alpha'\beta'} = (\mathrm{O}'\mathscr{C}')_{\alpha'\beta'}\times(\mathscr{C}^{-1}\mathrm{O})^{\alpha\beta} \equiv (\mathrm{O}'\times\mathrm{O})^{\alpha\beta}_{\alpha'\beta'}, \tag{89}$$

where O′ and O are constructed from the matrices γ^R and the momenta. The number of independent covariants is equal to $8 = (\frac{1}{2})(2\times\frac{1}{2}+1)^4$:

$$\begin{aligned} X_l = \{&\mathbf{1}'\cdot\mathbf{1},\ \gamma'^\mu\cdot\gamma_\mu,\ \gamma'_5\gamma'^\mu\cdot\gamma_5\gamma_\mu,\ \tfrac{1}{2}\sigma'_{\mu\nu}\cdot\sigma^{\mu\nu},\ \gamma'_5\cdot\gamma_5,\ \gamma'_5\gamma'_\mu(p_1-p_2)^\mu\cdot\gamma_5+\gamma'_5\cdot\gamma_5\gamma_\mu(p'_1-p'_2)^\mu, \\ &\gamma'_5\gamma'_\mu(p_1-p_2)^\mu\cdot\gamma_5-\gamma'_5\cdot\gamma_5\gamma_\mu(p'_1-p'_2)^\mu,\ (p_1-p_2)^\mu\gamma'_\mu\cdot\mathbf{1}-\mathbf{1}'\cdot(p'_1-p'_2)^\mu\cdot\gamma_\mu\}. \\ &(p_3 = p'_1, \quad p_4 = p'_2), \end{aligned} \tag{90}$$

The condition of T-invariance of the spinor amplitude, according to (6.101), (70), and (88), has the form

$$\mathscr{M}^{\alpha\beta}_{\alpha'\beta'}(\boldsymbol{p}_1 \ldots \boldsymbol{p}_4) = \eta_\mathrm{T}T^{-1}_{\alpha'\varrho'}T^{-1}_{\beta'\sigma'}\mathscr{M}^{\varrho'\sigma'}_{\varrho\sigma}(-\boldsymbol{p}'_1 \ldots -\boldsymbol{p}'_4)T^{\varrho\alpha}T^{\sigma\beta}, \tag{91}$$

in which $\eta_\mathrm{T} = \eta^*_\mathrm{T}(1)\,\eta^*_\mathrm{T}(2)\,\eta_\mathrm{T}(3)\,\eta_\mathrm{T}(4) = 1$ for elastic scattering. This means that the covariants X_l must be invariant with respect to the simultaneous substitution $p \leftrightarrow -p'$ and $(\gamma'\mathscr{C}) \leftrightarrow (\mathscr{C}^{-1}\gamma)^\mathrm{T}$ (in the notation of (89) and (90) for $\gamma' \leftrightarrow \gamma$). For this reason the expansion of a T-invariant spinor amplitude (91) does not contain X_7 and X_8.

Charge conjugation C relates amplitudes for different processes (see § 6.2). To convert the condition of C-invariance of the S-matrix (6.89) into a relation for spinor amplitudes, one must use the formulae (see § 5.2)

$$u_\mathrm{C}(\boldsymbol{p}, \lambda) = v(\boldsymbol{p}, \lambda) = \mathscr{C}\bar{u}(\boldsymbol{p}, \lambda), \quad \bar{v}(\boldsymbol{p}, \lambda) = \mathscr{C}^{-1}u(\boldsymbol{p}, \lambda).$$

In the case of $2(2J+1)$-component functions the wave functions of antiparticles may be obtained from (6.91) by omitting η_C.

The relation between the spinor amplitudes $\mathscr{M}$ and $\mathscr{M}_C$ for the process $\frac{1}{2}+\frac{1}{2}' \to \frac{1}{2}+\frac{1}{2}'$ and its charge conjugate $\overline{\frac{1}{2}}+\overline{\frac{1}{2}'} \to \overline{\frac{1}{2}}+\overline{\frac{1}{2}'}$ may be obtained from (6.89) using (70) and (5.67):

$$\mathscr{M}(p_1 \ldots p_4) = \eta_C \mathscr{C}\mathscr{C}\mathscr{M}_C^T(p_1 \ldots p_4)\mathscr{C}^{-1}\mathscr{C}^{-1}, \tag{92}$$

where each factor $\mathscr{C}$ on the left lowers an upper index of $\mathscr{M}_C^T$, while each factor $\mathscr{C}^{-1}$ on the right raises a lower index. The generalization of (92) to the case of higher spin particles described by Bargmann–Wigner or Rarita–Schwinger wave functions is obvious.

PART III

INTERNAL SYMMETRY

THE experimental data on particles and resonances[8] show clearly that the hadron spectrum is very complex. But it is just this spectrum that allows one to judge the properties of the dynamical system whose states it represents. This dynamical system differs from those known to us in nuclear and atomic physics primarily by the great strength of the interactions involved. It is possible that the hadron spectrum is some sort of "bootstrap" effect of the interaction of particles (i.e states of the spectrum) and of the self-consistency of the existence of such particles.

Another point of view, in which hadrons are bound states of quarks and antiquarks, is also quite possible. In either case, however, the "primary" classification of hadrons, i.e. the model on which the origin of the hadron spectrum is based, starts with the multiplet structure of the spectrum. Particle multiplets are described phenomenologically using approximate symmetries (see § 2.2). The detailed study of the approximate symmetries SU_2, SU_3, and SU_6 is the topic of the present part of this book.

We shall thus sidestep the question of the origin of internal symmetries (all the more so since the present status of the theory does not allow one to draw any conclusions in this respect). For the same reason we shall not dwell on interesting but so far unsuccessful attempts to unify space–time and internal symmetry groups in one non-trivial group.

The analysis of experimental data indicates the presence of an important correlation between the strength of an interaction and its symmetry, referred to in § 2.2 (the stronger the interaction and the more restricted its class, the broader its symmetry group). This correlation determines the order of presentation in this part of the book. We begin with the broad class of all hadronic interactions, possessing the isospin symmetry group SU_2, after which we isolate a stronger part of the interaction while enlarging the symmetry group—first to SU_3 and then to SU_6. The multiplets associated with the internal symmetry group are also enlarged, containing more and more particles.

CHAPTER 8

ISOSPIN SYMMETRY

THE concept of isotopic spin, or isospin, was introduced by Heisenberg to describe the properties of the neutron and proton and their interactions in connection with the charge independence of nuclear forces. Subsequently, it became clear that isospin symmetry was also a property of the pion–nucleon interaction. As new, strongly interacting particles were discovered—first the strange particles and then the resonances—the concept of isospin took on greater and greater generality since the (strong) interactions of the new particles also displayed isospin invariance. This led to the conviction that isospin symmetry was a universal property of strong interactions.

Isospin symmetry is broken by the electromagnetic and weak interactions. Effects due to these interactions are small, so that the isospin invariance of the S-matrix and the resulting conservation laws hold to the order of 1 per cent.

Isospin symmetry is based on two groups of facts: (a) the existence of multiplets of particles with similar masses (and identical spin and parity) but with different electrical charges; and (b) the presence of relations between decay constants and between cross-sections for different processes.

§ 8.1. Isospin multiplets, hypercharge, and the group SU_2

A survey of the properties of the elementary particles (see Tables A. 1–A.3 of pp. 359–364) shows that the strongly interacting particles exist in the form of charge, or isospin, multiplets. Individual particles of a multiplet differ with respect to electric charge. Particles of a multiplet have the same spin and almost the same mass. The mass splitting in a multiplet is related to the interaction which distinguishes the individual particles in a multiplet from one another, i.e. the electromagnetic interaction. Thus space–time properties of all particles of a multiplet are identical if one neglects the electromagnetic and weak interactions. For example, the stable particles (see Table A.1) fall into the multiplets N, Λ, Σ, Ξ, Ω (baryons) and K, π, η, $\overline{\mathrm{K}}$ (mesons) with one, two, or three particles in a multiplet. There is also a resonance multiplet Δ (see Table A.3) containing four particles: Δ^{++}, Δ^{+}, Δ^{0}, Δ^{-}. Some isospin multiplets are shown in Table 8.1.

The degeneracy of mass levels (when electromagnetism is neglected) indicates a symmetry of the strong interactions. Let us find the group $\mathfrak{G}$ of this symmetry, assuming that the degeneracy cannot be accidental. Then the number of particles in a multiplet n (or the

TABLE 8.1

Baryons	Mesons	Number of particles	Y	I
$N = (p, n)$	$K = (K^+, K^0)$	2	1	$\frac{1}{2}$
$\Sigma = (\Sigma^+, \Sigma^0, \Sigma^-)$	$\pi = (\pi^+, \pi^0, \pi^-)$	3	0	1
Λ	η	1	0	0
$\Xi = (\Xi^0, \Xi^-)$	$\bar{K} = (\bar{K}^0, K^-)$	2	-1	$\frac{1}{2}$
Ω^-		1	-2	0
$\Delta = (\Delta^{++}, \Delta^+, \Delta^0, \Delta^-)$		4	1	$\frac{3}{2}$

multiplicity) must coincide with the dimension of a basis of an irreducible representation of the group $\mathfrak{G}$. Thus, according to Table 8.1, the possible dimensionalities must include the values $n = 1, 2, 3, 4$.

Let us write a single-particle state vector in the form $|\ldots; \beta; I, t\rangle$, where the dots denote the variables of the Poincaré group, t and I are respectively the additive and nonadditive quantum numbers of isospin symmetry, and β are the remaining quantum numbers. Choosing the variables of the Poincaré group and the quantum numbers β in the same way for all particles of the multiplet I, we confine ourselves for the moment to the study of the state vector of the multiplet $|I, t\rangle$ in isospin space. In other words, we shall write the multiplet state vector in the form of a product of an isospin vector $|I, t\rangle$ by a state vector depending on the variables of the Poincaré group and the quantum numbers β; e.g.

$$|\boldsymbol{p}, \lambda; m, J; \beta; I, t\rangle = |\boldsymbol{p}, \lambda; m, J; \beta\rangle\, |I, t\rangle. \tag{1}$$

The multiplet is characterized by the value of I. Each particle in a multiplet is described uniquely by its electric charge Q or by a quantity connected with Q; e.g.

$$t = Q - \bar{Q} \equiv Q - \frac{Y}{2}, \tag{2}$$

where $\bar{Q}$ is the mean charge of the multiplet while Y is the hypercharge introduced by Gell-Mann and Nishijima. The quantity $Y = 2\bar{Q}$ obviously has the same value for all particles of a multiplet. The hypercharge Y characterizes the multiplet as a whole and may be included in the quantum numbers β in (1).

Let G be a transformation of the isospin group $\mathfrak{G}$. The existence of isospin symmetry means that the strong interactions do not distinguish a state $|I, t\rangle$ of the multiplet I from a linear combination of states of the same multiplet:

$$u(G)\,|I, t\rangle = \sum_{t'} |I, t'\rangle\, d^I_{t't}(G). \tag{3}$$

Here $d^I(G)$ is a unitary matrix for the multiplet I. The matrices $d^I(G)$ form a unitary group $\mathfrak{G}$, while I determines its irreducible representation.

Thus $\mathfrak{G}$ is a group of unitary matrices whose irreducible representations I may have dimension $n = 1, 2, 3, 4. \ldots$ We demand, moreover, that: (a) $\mathfrak{G}$ be the minimum possible group and (b) $\mathfrak{G}$ contain the transformations $e^{it\alpha}$, which along with a simultaneous hypercharge transformation $e^{i\alpha Y/2}$ lead, according to (2), to the transformation $U_Q = e^{iQ\alpha}$. As is well known (see § 2.2), U_Q is an exact symmetry transformation corresponding to the law of electric charge conservation.

All the above conditions are satisfied by the group SU_2 of unitary 2 by 2 matrices which was studied in Chapter 3 in connection with the group of three-dimensional rotations.

Let us briefly reproduce the results of Chapter 3 referring to the group SU_2. In contrast to the Pauli matrices σ_k of the rotation group, the isospin Pauli matrices will be denoted τ_k ($k = 1, 2, 3$). Then a transformation G in the fundamental representation is

$$G = \exp\{i\tfrac{1}{2}\boldsymbol{\tau}\cdot\boldsymbol{\omega}\}, \quad G^+ = G^{-1}. \tag{4}$$

For an irreducible representation with isospin I, the transformation (3) has the form

$$d^I(G) = \mathcal{D}^I(G) = \exp\{i\boldsymbol{I}\cdot\boldsymbol{\omega}\}. \tag{5}$$

The isospin components I_1, I_2, I_3 satisfy the commutation relations for angular momentum operators:

$$[I_i, I_j] = ie_{ijk}I_k\,. \tag{6}$$

The square of the isospin $I^2 = I_1^2+I_2^2+I_3^2$ commutes with all isospin generators I_k, and, consequently, is an invariant whose eigenvalues allow one to classify the irreducible representations $\mathcal{D}^I$. The eigenvalues I^2 may be written in the form $I(I+1)$, where I may be one of the numbers $0, \frac{1}{2}, 1, \frac{3}{2}, \ldots$, and the isospin projection I_3 varies within the limits $-I \leqslant I_3 \leqslant I$, taking on $2I+1$ values, so that the dimension is $n = 2I+1$.

Thus for isospin SU_2 symmetry, particles must fall into multiplets labeled by the isospin I, with $n = 1, 2, 3, 4, \ldots$, particles. Since the mean value of I_3 of a multiplet is equal to zero, I_3 must be identified with t:

$$Q = t+\tfrac{1}{2}Y = I_3+\tfrac{1}{2}Y.$$

The simplest nontrivial irreducible representation of the isospin group may be constructed using the unitary 2 by 2 matrices (4). The basis state $|\frac{1}{2}, t\rangle$, $t = \pm\frac{1}{2}$, in this case is an isospinor and characterizes an isodoublet. For example, the nucleon doublet N, consisting of the proton p and neutron n, is

$$|N\rangle = \begin{pmatrix} N_1 \\ N_2 \end{pmatrix} = N_1|\mathrm{p}\rangle + N_2|\mathrm{n}\rangle, \tag{7}$$

where

$$|\mathrm{p}\rangle = |\tfrac{1}{2}, \tfrac{1}{2}\rangle, \quad |\mathrm{n}\rangle = |\tfrac{1}{2}, -\tfrac{1}{2}\rangle$$

are the proton and neutron states. Here $|N_1|^2+|N_2|^2 = 1$. Any superposition of the type (7) describes the nucleon equally well, while two such superpositions are connected by a unitary transformation (4):

$$|N'\rangle = G|N\rangle. \tag{8}$$

The states $|I, t\rangle$ with isospin I may be described in the form of a $(2I+1)$-row column vector, each of whose lines is labeled by $t = I_3'$. We shall consider the states to form a symmetric spinor with lower indices and, consequently, to transform in the same way as Φ_t^I:

$$|I, t\rangle \sim \Phi_t^I = \frac{\sqrt{(2I)!}\,(N_1)^{I+t}(N_2)^{I-t}}{\sqrt{(I+t)!\,(I-t)!}}, \tag{9}$$

where N_1 and N_2 are the components of the isospinor (7). Under an isospin rotation (4)

$$|I, t'\rangle = \sum_t \mathcal{D}^I_{t't}(G)\,|I, t\rangle. \tag{10}$$

The operators

$$I_\pm = (I_1 \pm iI_2) \tag{11}$$

change the value of t by one unit:

$$[I_3, I_\pm] = \pm I_\pm .$$

Using (9) we find the isospin matrix elements

$$I_\pm\,|I, t\rangle = [(I \mp t)(I \pm t+1)]^{1/2}\,|I, t\pm 1\rangle. \tag{12}$$

The state vectors (1) correspond to the symmetry group $(SU_2)_I \times \mathrm{U}_Y(1)$—the direct product of the isospin group SU_2 and the group of one-dimensional unitary transformations $\mathrm{U}_Y(1) = e^{iY\lambda}$, associated with hypercharge. This agrees with the fact that there are independent conservation laws for hypercharge and isospin.

From (2) it is clear that, like the charge Q, the hypercharge Y is an additive quantum number. The total hypercharge is conserved in all interactions except the weak ones. The law of conservation of hypercharge is a generalization of the experimental facts reflecting the selective capacity of particles to undergo transformations. For example, in experiment one does not observe the processes

$$\pi + N \rightarrow \pi + \Lambda, \quad \mathrm{K} + \Lambda \rightarrow \pi + \Sigma$$

or rapid decays of the form

$$\Lambda \rightarrow \pi + \mathrm{N}, \quad \Sigma \rightarrow \pi + \mathrm{N} \quad \text{etc.}$$

contradicting the conservation of hypercharge Y.

The conservation of hypercharge means that the S-matrix commutes with Y, $[S, Y] = 0$, i.e. the S-matrix is invariant with respect to the hypercharge phase transformation

$$\mathrm{U}_Y^{-1} S \mathrm{U}_Y = S, \quad \mathrm{U}_Y = e^{iY\lambda}.$$

Since the electric charge is conserved in all interactions, the quantity I_3, like the hypercharge Y, is conserved in all but the weak interactions.

Now, if the internal symmetry were really the group $(SU_2)_I \times \mathrm{U}_Y(1)$, then any isomultiplet could have any value of Y. However, the values of I and Y are correlated. Both the baryon and meson isomultiplets exist only for definite combinations of Y and I; for example, multiplets with $Y = 0$ and $I = \frac{1}{2}$ or with $Y = 1$, $I = 0$ are unknown. We note that the relation

$$(-1)^Y = (-1)^{2I}$$

holds for all known isomultiplets.

The notation of Table 8.1 for baryonic and mesonic states is used for resonances as well as for stable particles.

Baryonic isomultiplets with quantum numbers of the nucleon ($I = \frac{1}{2}, Y = 1$) form the family of N states, including particles of various spins and parities. One speaks analogously of Λ-states ($I = Y = 0$), Σ-states ($I = 1$, $Y = 0$), etc.

The meson isomultiplets are labeled by the (I, Y) of the pseudo-scalars; for example, K-states have $I = \frac{1}{2}$, $Y = \pm 1$, but may have any spin and parity.

§ 8.2. Isospin and reflections. Antiparticle states. G-parity

The strong interactions are invariant with respect to the reflections C, P, and T separately. For this reason the symmetry group $\mathcal{P}_+^\uparrow \times \mathfrak{G}$, where $\mathfrak{G} = (SU_2)_I \times U_Y$, considered in the previous section, must be extended to the full invariance group containing reflections as well.

The reflection θ

First, we must extend the group $\mathcal{P}_+^\uparrow \times \mathfrak{G}$ by the total reflection $\theta =$ CPT, which is an exact symmetry operation. As was shown in § 6.1, the extension of an internal symmetry group by the reflection θ may be performed independently of the extension of the proper orthochronous Poincaré group $\mathcal{P}_+^\uparrow$ to the proper group $\mathcal{P}_+$. To do this, the operator θ must be replaced by another antiunitary operator θ_0 whose properties are defined only by the internal symmetry group.

The extended internal symmetry group $\mathfrak{G}_\theta$ then contains the transformations $\{G, \theta_0 G\}$, where G belongs to the original group. As the group SU_2, we consider the isospin group $(SU_2)_I$, since taking account of hypercharge transformations U_Y introduces nothing new (one-dimensional phase transformations were already studied in § 6.1 in the exact symmetry group example). Thus we must find the co-representations of the group $(SU_2)_{I\theta} = \{G, \theta_0 G\}$, where G has the form (5). The definitions (6.45) and (6.32) for the unitary operations θ and θ_0 may be written in the form

$$\theta \,|\, \boldsymbol{p}, \lambda; J, m; a\rangle = (-1)^{J-\lambda}\theta_0 \,|\, \boldsymbol{p}, -\lambda; J, m; a\rangle = (-1)^{J-\lambda} \,|\, \boldsymbol{p}, -\lambda; J, m; \bar{a}\rangle, \tag{13}$$

where a and $\bar{a}$ denote particle and antiparticle internal quantum numbers, and $\theta_0^2 = 1$.

Let $|\, a\rangle$ be that part of the state vector which depends on the internal symmetry quantum numbers, $|\, a\rangle = |\, Q_i'; I_a, t_a\rangle$, where the generalized charges Q_i include hypercharge as well. Under an isospin rotation G, the states $|\, I, t\rangle$ transform according to the rule (10)

$$u(G)\,|\, I, t\rangle = \sum_{t'} \mathcal{D}^I_{tt'}(G)\,|\, I, t'\rangle. \tag{14}$$

By virtue of (13) the state $\theta_0 | a\rangle = | \bar{a}\rangle$ describes an antiparticle; as a consequence of the antilinearity of θ_0 this state transforms via the complex conjugate matrix

$$u(G)\theta_0 | I, t\rangle = \sum_{t'} \mathcal{D}^{*I}_{tt'}(G)\, \theta_0 | I, t'\rangle, \tag{15}$$

while the isospin components I_k anticommute with $\theta_0 : \theta_0 I_k = -I_k \theta_0$.

According to the general theory (see § 6.1) the type of the co-representation determines the relation between the matrices d of the irreducible representation of the internal symmetry group $\mathfrak{G}$ and the complex conjugate matrices d^*. In the case of the group SU_2, $\mathcal{D}^I$ and $\mathcal{D}^{I*}$ are unitarily equivalent:

$$\mathcal{D}^{I*}(G) = \mathcal{D}^I(C)\, \mathcal{D}^I(G)\, \mathcal{D}^{I-1}(C), \quad C = -i\tau_2, \tag{16}$$

so that the unitary matrix β in formula (6.34) is equal to $\beta = \mathcal{D}^{I-1}(C)$. This means that co-representations of type 3 in the group $(SU_2)_{I\theta}$ are absent (see § 6.1). Co-representations of type 1 or 2 correspond to the values $\beta\beta^* = \pm 1$. In the case of the isospin group we have for a multiplet with isospin I:

$$\beta\beta^* = \mathcal{D}^I(C^{-1})\, \mathcal{D}^I(C^{-1*}) = (-1)^{2I}. \tag{17}$$

Consequently, the type of co-representation of the group $(SU_2)_{I\theta}$ depends on the value of isospin. For integral isospins, $I = 0, 1, 2, 3, \ldots$, type 1 [formula (6.36)] holds, while for half integral isospin type 2 is valid [formulae (6.37) and (6.38)]. This means that the states $| I, t\rangle$ and $\theta_0 | I, t\rangle$ for integral isospins may belong to the same multiplet of I, while for half-integral isospins they must belong to different multiplets with the same isospin I. In other words, only for integral isospins I may particles and antiparticles occur in the same multiplet, while the matrices $\mathcal{D}^I$ may be real ("self-adjoint multiplets").[110, 111]

The connection between the value of the isospin and the existence of multiplets containing both particles and antiparticles is a consequence of θ-invariance alone and does not depend on other particle quantum numbers. From Tables 8.1 and A.1 it is clear that this connection holds for all known particles.

Let us write formulae (6.36) and (6.37) explicitly for the co-representations of the extended isospin group $(SU_2)_{I\theta}$. First of all, we note that the antiparticle state $| \bar{a}\rangle$, by virtue of (14), (15), and (16), transforms in the same way as $| I, -t\rangle (-1)^{I-t}$.

$$| \bar{a}\rangle = \theta_0 | I, t\rangle = \varepsilon_I \sum_{t'} \mathcal{D}^I_{tt'}(C^{-1}) | I, t'\rangle = (-1)^{I-t}\, \varepsilon_I | I, -t\rangle, \tag{18}$$

where $| a\rangle = | I, t\rangle$ is a particle state and ε_I depends only on isospin.

In the case of integral isospin, when $| I, t\rangle$ and $\theta_0 | I, t\rangle$ are in the same multiplet, there must exist a unitary matrix $\mathcal{D}^I(\theta_0)$ such that

$$| \bar{a}\rangle = \theta_0 | I, t\rangle = \sum_{t'} \mathcal{D}^I_{tt'}(\theta_0) | I, t'\rangle. \tag{19}$$

From (14), (15), and (16) one then has

$$\mathcal{D}(\theta_0)\, \mathcal{D}(C)\, \mathcal{D}(G) = \mathcal{D}(G)\, \mathcal{D}(\theta_0)\, \mathcal{D}(C),$$

from which, by virtue of continuity and unitarity,

$$\mathcal{D}^I(\theta_0) = \varepsilon_I \mathcal{D}^I(C^{-1}), \quad \varepsilon_I = \pm 1, \tag{20}$$

where ε_I depends only on I. Formulae (14), (19), and (20) define the co-representations (6.36) for integral isospins.

Let us apply formula (19) to a neutral particle a^0 in a multiplet I, for which $|\bar{a}^0\rangle = |a^0\rangle$. Inserting (20) into (19) gives $|\bar{a}^0\rangle = \varepsilon_I(-1)^I|a^0\rangle$ or $\varepsilon_I = (-1)^I$. Formula (19) now allows one to find the phase factor connecting the antiparticle state $|\bar{a}\rangle = \theta_0|a\rangle = \theta_0|I, t\rangle$ with the isospin state $|I, t'\rangle$:

$$|\bar{a}\rangle = (-1)^{t_a}|I_a, -t_a\rangle, \tag{21}$$

which completes the definition of the phases in a self-adjoint multiplet.

Let us consider the pion triplet π^+, π^0, π^- as an example. We shall identify $|\pi^-\rangle$ with the isospin state $|I = 1, t = -1\rangle$. Then, according to (21), one must have $|\pi^+\rangle = -|1, 1\rangle$, so that the pion triplet is

$$|1, 1\rangle = -|\pi^+\rangle, \quad |1, 0\rangle = |\pi^0\rangle, \quad |1, -1\rangle = \pi^-\rangle \tag{22}$$

In the case of half-integral isospins the form of the irreducible co-representation (type 2) follows from (6.37) or from (14) and (15). If $|I, t\rangle$ is the original isomultiplet, then by adding the reflection θ_0, we obtain the double multiplet

$$|I, t, r\rangle = \begin{pmatrix} |I, t\rangle \\ -\mathcal{D}^I(C)\,\theta_0|I, t\rangle) \end{pmatrix} \quad (r = 1, 2), \tag{23}$$

with respect to which the matrices $d^I(G)$ and $d^I(\theta_0)$ are equal to

$$d^I(G) = \begin{pmatrix} \mathcal{D}^I(G) & 0 \\ 0 & \mathcal{D}^I(G) \end{pmatrix}, \quad d^I(\theta_0) = \begin{pmatrix} 0 & \mathcal{D}^I(C) \\ -\mathcal{D}^I(C) & 0 \end{pmatrix}, \tag{24}$$

which corresponds to the choice $\varepsilon_I = -1$ for half-integer spins in (8.18).

For example, let the original multiplet be the kaons K^+ and K^0, whose states are identified with isospin states as follows:

$$|K^+\rangle = |\tfrac{1}{2}, \tfrac{1}{2}\rangle, \quad |K^0\rangle = |\tfrac{1}{2}, -\tfrac{1}{2}\rangle.$$

Then the second line in (23) will contain the antiparticles $(-|\bar{K}^0\rangle, |K^-\rangle)$. The antiparticle multiplet has the same isospin content as the original one differing from it with respect to other additive quantum numbers (in this case with respect to hypercharge).

The reflection P *and charge conjugation* C

The relative parity of all particles in an isomultiplet is the same by definition:

$$P|\boldsymbol{p}, \sigma; I, t\rangle = \eta_P^*|-\boldsymbol{p}, \sigma; I, t\rangle;$$

the phase factor η_P does not depend on t; the isospin I_k commutes with the reflection P.

Charge conjugation takes particles into antiparticles (see § 6.2). In contrast to θ_0, charge conjugation C is described by a unitary operator which preserves commutation relations.

Hence C cannot change the sign of all components of isospin. The usual choice is

$$CI_1C^{-1} = -I_1, \quad CI_2C^{-1} = I_2, \quad CI_3C^{-1} = -I_3. \tag{25}$$

C takes states in a multiplet of particles into states of an antiparticle multiplet multiplied by an overall phase factor η_C:

$$C\,|\,Q_i'; I, t\rangle = \eta_C\,|\,-Q_i'; I, -t\rangle. \tag{26}$$

But the set of states $C\,|\,I, t\rangle$ is not an isospin multiplet since these states do not transform according to the rule (14).

For half-integral isospins there always exists an additional phase transformation outside the isospin group, since one may set $\eta_C = 1$. In the case of integral isospins and self-adjoint multiplets containing both particles and antiparticles, the value of η_C will be equal to the charge parity of the neutral member of the multiplet $a^0 = \bar{a}^0$:

$$C\,|\,a^0\rangle = \eta_C\,|\,\bar{a}^0\rangle = \eta_C\,|\,a^0\rangle. \tag{27}$$

For a neutral pion, $\eta_C = 1$, while for the particles ϱ^0 and ω^0 one will have $\eta_C = -1$.

G-*parity*

Charge conjugation C does not commute with the components of isospin [see (25)]. Hence the operator C is inconvenient for transforming a particle multiplet into an antiparticle multiplet since one must then change the isospin matrix as well.

Let us introduce the transformation G consisting of charge conjugation C and a rotation around the second isospin axis by 180°:[112]

$$G = e^{i\pi I_2}C. \tag{28}$$

Sometimes $e^{i\pi I_2}$ is called the charge-symmetry transformation since it changes I_3 into $-I_3$. Charge symmetry should not be confused with charge conjugation C, which changes the sign of all additive quantum numbers.

From (25) and (28) it follows that the operator G commutes with the components of isospin: $[G, I_k] = 0$.

Let us apply G to a state with a definite isospin and with other additive quantum numbers equal to 0. Since $e^{i\pi I_2}\,|\,I, t\rangle = (-1)^{I-t}\,|\,I, -t\rangle$,

$$G\,|\,I, t\rangle = (-1)^{I-t}\,\eta_C\,|\,I, t\rangle. \tag{29}$$

Thus the state $|\,I, t\rangle$ can be an eigenstate of the operator G, if all additive quantum numbers except I_3 are equal to 0.

The G-parity η_G of any self-adjoint multiplet is defined in terms of that of the neutral particle by

$$G\,|\,I, t = 0\rangle = \eta_G\,|\,I, t = 0\rangle, \tag{30}$$

from which, by comparison with (29), we find $\eta_G = (-1)^I\eta_C$. If one chooses the phase factors for antiparticles according to the rule (21) the factor η_G will be the same for all members of the self-adjoint multiplet. For example,

$$G\begin{pmatrix} -|\pi^+\rangle \\ |\pi^0\rangle \\ |\pi^-\rangle \end{pmatrix} = -\begin{pmatrix} -|\pi^+\rangle \\ |\pi^0\rangle \\ |\pi^-\rangle \end{pmatrix}, \quad (G(\pi) = -1).$$

The multiplets ϱ, η, ω, φ, A_2, ..., are also characterized by G-parity. The isospin and G-parity of a particle are usually combined in one notation I^G. For pions $I^G = 1^-$. Systems of particles and antiparticles with total $B = Y = 0$ also possess a definite G-parity. Here the particles involved in the system may not have definite G-parities themselves.

Since G commutes with isospin, in the case of an arbitrary isomultiplet the G transformation (28) takes a particle isomultiplet into an antiparticle isomultiplet; the same isospin matrices I_k act on these two isomultiplets.

G-parity is conserved in all processes which conserve isospin and C, i.e. in all strong interactions. Electromagnetic and weak interactions violate G-parity.

The G-parity of a system of particles, each of which has a definite G-parity, is equal to the product of the G-parities of the separate particles. This leads to a selection rule for strong decays. For example, a particle with $\eta_G = 1$ cannot decay into an odd number of G-odd particles, while a particle with $\eta_G = -1$ cannot decay into two pions. In particular, strong decays of the ϱ-meson into three pions and the φ and ω mesons into two pions cannot occur.

Choice of phase factors for particles and antiparticles

The phase factors ε_a determine a relation of the type $|a\rangle = \varepsilon_a |I_a, t_a\rangle$ between a particle state a and a basis isospin state transforming according to (14). These factors, are fixed by formula (21) for self-adjoint isomultiplets with $I = 0, 1, 2, \ldots$, and by formula (23) for half-integral isospins. Formula (22) is also used for the choice of phases in the case of baryon multiplets with isospin $I = 1$; e.g.:

$$|\Sigma^+\rangle = -|1, 1\rangle. \quad |\Sigma^0\rangle = 1, 0\rangle, \quad |\Sigma^-\rangle = |1, -1\rangle,$$

since the isospin properties of π and Σ must be the same. The antiparticle multiplet $\bar{\Sigma}$ has an analogous form:

$$|1, 1\rangle = -|\bar{\Sigma}^+\rangle, \quad |1, 0\rangle = |\bar{\Sigma}^0\rangle, \quad |1, -1\rangle = |\bar{\Sigma}^-\rangle.$$

Thus the states of particles and antiparticles may be expressed in terms of the basis isospin states $|I, t\rangle$ in the following manner. In the case of half-integral isospins ($I = \frac{1}{2}, \frac{3}{2}, \ldots$)

$$|a\rangle = |I_a, t_a\rangle, \quad |\bar{a}\rangle = (-1)^{I_a + t_a} |I_a, -t_a\rangle. \tag{31}$$

In the case of integral isospins ($I = 0, 1, 2, \ldots$)

$$\left.\begin{aligned} |a\rangle &= |I_a, t_a\rangle \quad (t_a \leqslant 0), \\ |a\rangle &= (-1)^{t_a} |I_a, t_a\rangle, \quad (t_a > 0), \\ |\bar{a}\rangle &= (-1)^{t_a} |a(I_a, -t_a)\rangle. \end{aligned}\right\} \tag{32}$$

For example, in the case $I = \frac{3}{2}$, the antiparticle multiplet has the form

$$|\bar{\Delta}^{+}\rangle, \quad -|\bar{\Delta}^{0}\rangle, \quad |\bar{\Delta}^{-}\rangle, \quad -|\bar{\Delta}^{--}\rangle,$$

while the antinucleon multiplet is

$$|\bar{\mathrm{n}}\rangle, \quad -|\bar{\mathrm{p}}\rangle.$$

§ 8.3. Multi-particle states and isospin amplitudes. Decays and relations between reactions

Multi-particle states

Multi-particle states transform according to the direct product of representations corresponding to single-particle states. The expansion of a direct product of representations of the group SU_2 into irreducible representations has been considered earlier, in Chapter 3, in connection with the rotation group. It contains Clebsch–Gordan coefficients, which are shown in Table A.4 (facing p. 364).

In the standard expansion of a product of two vectors (see § 3.4), it is assumed that both states transform with indices of the same type. In our case, the basis isospin states, whose set forms a multiplet I, are $|I, t\rangle$ and their transformation rule is given by formulae (10) and (14). We then find, by virtue of (3.82), that

$$|I_b, t_a, I_b, t_b\rangle \equiv |I_a, t_a\rangle\, |I_b, t_b\rangle = \sum_I \langle I_a I_b t_a t_b \,|\, I_a I_b I t\rangle\, |I, t\rangle, \tag{33}$$

where $t = t_a + t_b$, while $\langle I_a I_b t_a t_b \,|\, I_a I_b I t\rangle$ are Clebsch–Gordan coefficients.

Physical particle states may differ by phase factors from the basis isospin states $|I, t\rangle$. For example, if the two-particle state $|a, \bar{b}\rangle$ contains an antiparticle $\bar{b}$, then its states $|\bar{b}\rangle$ do not in general coincide with the basis states $|I_b, -t_b\rangle$, since the antiparticle transforms according to the complex conjugate representation. The choice of the factors ε_a in the relation $|a\rangle = \varepsilon_a |I_a, t_a\rangle$, connecting the particle state $|a\rangle$ with an isospin state, was made in § 8.2 [formulae (31) and (32)]. We obtain for the pion–nucleon system

$$\left.\begin{aligned}
|\mathrm{p}\pi^{+}\rangle &= -\,|\tfrac{3}{2}, \tfrac{3}{2}\rangle, \\
|\mathrm{n}\pi^{+}\rangle &= -\frac{1}{\sqrt{3}}\,|\tfrac{3}{2}, -\tfrac{1}{2}\rangle + \sqrt{\frac{2}{3}}\,|\tfrac{1}{2}, \tfrac{1}{2}\rangle, \\
|\mathrm{p}\pi^{0}\rangle &= \sqrt{\frac{2}{3}}\,|\tfrac{3}{2}, \tfrac{1}{2}\rangle + \frac{1}{\sqrt{3}}\,|\tfrac{1}{2}, \tfrac{1}{2}\rangle, \\
|\mathrm{n}\pi^{0}\rangle &= \sqrt{\frac{2}{3}}\,|\tfrac{3}{2}, -\tfrac{1}{2}\rangle - \frac{1}{\sqrt{3}}\,|\tfrac{1}{2}, -\tfrac{1}{2}\rangle, \\
|\mathrm{p}\pi^{-}\rangle &= \frac{1}{\sqrt{3}}\,|\tfrac{3}{2}, -\tfrac{1}{2}\rangle + \sqrt{\frac{2}{3}}\,|\tfrac{1}{2}, -\tfrac{1}{2}\rangle, \\
|\mathrm{n}\pi^{-}\rangle &= |\tfrac{3}{2}, -\tfrac{3}{2}\rangle.
\end{aligned}\right\} \tag{34}$$

Equation (33) may be illustrated explicitly (see § 3.4). In particular, for the nucleon–antinucleon system the states with definite isospin are

$$\left.\begin{aligned} |1, 1\rangle &= -|\mathrm{p\bar{n}}\rangle, \quad |1, 0\rangle = \frac{1}{\sqrt{2}}(|\mathrm{p\bar{p}}\rangle - |\mathrm{n\bar{n}}\rangle),\\ |1, -1\rangle &= |\mathrm{n\bar{p}}\rangle, \quad |0, 0\rangle = \frac{1}{\sqrt{2}}(|\mathrm{p\bar{p}}\rangle + |\mathrm{n\bar{n}}\rangle). \end{aligned}\right\} \tag{35}$$

The system of three nucleons N_1, N_2, N_3 in a state with isospin $\frac{3}{2}$ has the following isospin components:

$$\left.\begin{aligned} |\tfrac{3}{2}, \tfrac{3}{2}\rangle &= |\mathrm{p_1p_2p_3}\rangle, \quad |\tfrac{3}{2}, -\tfrac{3}{2}\rangle = |\mathrm{n_1n_2n_3}\rangle,\\ |\tfrac{3}{2}, \tfrac{1}{2}\rangle &= \frac{1}{\sqrt{3}}(|\mathrm{p_1p_2n_3}\rangle + |\mathrm{p_1n_2p_3}\rangle + |\mathrm{n_1p_2p_3}\rangle),\\ |\tfrac{3}{2}, -\tfrac{1}{2}\rangle &= \frac{1}{\sqrt{3}}(|\mathrm{p_1n_2n_3}\rangle + |\mathrm{n_1p_2n_3}\rangle + |\mathrm{n_1n_2p_3}\rangle). \end{aligned}\right\} \tag{36}$$

For identical particles, one must allow for the symmetry properties of the state vector with respect to interchange of the particle variables. An isospin multiplet may be treated as a single particle with an additional degree of freedom $I_3' = t$. The demand of total symmetry or antisymmetry hence introduces a correlation between the symmetry properties of the state vector with respect to interchanges (separately) of the variables of the Poincaré group and of the isospin variables. If, for example, the state of two identical bosons is symmetric or antisymmetric with respect to interchange of isospin variables of the particles, it must have the same symmetry with respect to interchange of the remaining variables. In particular, two pions in an $I = 1$ state (which is antisymmetric) may have only odd angular momenta, and only even angular momenta in an $I = 2$ state.

Isospin conservation and isospin amplitudes

In the approximation of exact isospin symmetry, in which electromagnetic and weak interactions are neglected, the S-matrix and therefore also the T-matrix are diagonal with respect to isospin I, i.e. the dynamics are insensitive to isospin rotations, and the total isospin I is conserved:

$$[S, I_k] = 0, \quad [T, I_k] = 0.$$

Consequently, the T-matrix may be written in the form of an expansion in terms of isospin amplitudes $T(I)$ describing scattering with isospin I. For the process

$$a(I_a, t_a) + b(I_b, t_b) \to c(I_c, t_c) + d(I_d, t_d)$$

this expansion is

$$\langle c, d | T | a, b\rangle = \sum_I T(I)\, \Lambda(I; a, b; c, d). \tag{37}$$

The amplitudes $T(I)$ are isospin invariant. The coefficient $\Lambda(I; a, b; c, d)$ is the matrix element of the operator for projection on a state with isospin I.

In order to obtain the transformation properties of the matrix (37), we relate it, using (31) and (32), to the T-matrix for isospin states:

$$\varepsilon_a\varepsilon_b\varepsilon_c\varepsilon_d\langle c, d|T|a, b\rangle = \langle I_c, t_c; I_d, t_d|T|I_a, t_a; I_b, t_b\rangle \equiv T^{t_c t_d}_{t_a t_b},$$

where the isospin indices are arranged in accord with the convention adopted in § 8.1. The invariance of the T-matrix with respect to an isospin transformation G is then expressed by

$$T^{t_c t_d}_{t_a t_b} = \mathcal{D}^{*t_c}_{t_c'}\mathcal{D}^{*t_d}_{t_d'}\mathcal{D}^{t_a'}_{t_a}\mathcal{D}^{t_b'}_{t_b}T^{t_c' t_d'}_{t_a' t_b'}, \tag{38}$$

where $\mathcal{D}^{t_a'}_{t_a} \equiv \left[\mathcal{D}^{I_a}(G)\right]^{t_a'}_{t_a}$ is the representation of the group SU_2 with isospin I_a. Equation (38) relates the initial amplitude to the amplitudes for scattering of other particles in t_a', t_b', t_c', t_d' of the same multiplets I_a, I_b, I_c, I_d.

Thus in the approximation of isospin symmetry, isomultiplets are treated as particles with an additional internal degree of freedom (the quantum number t) in which the strong interactions can distinguish only multiplets as a whole, but not separate particles in them. The isospin amplitudes $T(I)$ depend upon the isospins in the process $I_a+I_b \to I_c+I_d$, but not on the projections t_i. The set of the same amplitudes $T(I)$ describes the scattering of various particles in these multiplets. If the number of independent amplitudes $T(I)$ is not large then (37) allows one to obtain useful relations between cross-sections.

Let us find an explicit expression for the coefficients $\Lambda(I; a; b; c, d)$ in (37). For this according to (31) and (32), one must insert the isospin states $|I_a, t_a\rangle$, ... instead of the single-particle states $|a\rangle$, ..., into (37), and then use the reduction formula (33). Then

$$\langle c, d|T|a, b\rangle = \sum_{I, I', t, t'} \langle I_c I_d t_c t_d|I_c I_d I t\rangle\langle I, t|T|I', t'\rangle\langle I_a I_b I' t'|I_a I_b t_a t_b\rangle\varepsilon_a\varepsilon_b\varepsilon_c\varepsilon_d.$$

Since the isospin is conserved, T commutes with I^2 and I_3, so that

$$\langle I, t|T|I', t'\rangle = T(I, t)\delta_{II'}, \delta_{tt'}.$$

The conservation of isospin also means $[T, I_{1,2}] = 0$ by virtue of which $T(I, t)$ does not depend on t. This leads to formula (37) with coefficients

$$\Lambda(I; a, b; c, d\rangle = \sum_t \langle I_c I_d t_c t_d|I_c I_d I t\rangle\langle I_a I_b I t|I_a I_b t_a t_b\rangle\varepsilon_a\varepsilon_b\varepsilon_c\varepsilon_d. \tag{39}$$

In order that the isospin amplitude $T(I)$ be invariant, the coefficients (39) must transform in the same way as the T-matrix [rule (38)]. But the Clebsch–Gordan coefficients entering in (39) are components of an isospin tensor [see (3.86) and (3.89)];

$$\langle I_a I_b t_a t_b|I_a I_b I t\rangle = \left[I I_a I_b\right]^{t_a t_b}_{t}.$$

Hence the right-hand side of (39) is also an isotropic tensor

$$\left[I I_a I_b\right]^{t_a t_b}_{t}\left[I I_c I_d\right]^{t}_{t_c t_d}.$$

which, indeed, transforms according to (38).

Let us consider as an example a relation between the amplitudes for πN-scattering. The processes $\pi^- + p \to \pi^0 + n$ and $\pi^+ + n \to \pi^0 + p$ are connected by the charge symmetry transformation $e^{i\pi I_2}$, and, consequently, their cross-sections are equal;

$$\sigma(\pi^- p \to \pi^0 n) = \sigma(\pi^+ n \to \pi^0 p). \tag{40}$$

Among the remaining πN reactions there exist four more relations of the type (40). To take account of isospin conservation, let us consider the reactions

$$\left.\begin{array}{ll} \pi^+ + p \to \pi^+ + p, & \pi^- + p \to \pi^0 + n, \\ \pi^- + p \to \pi^- + p, & \pi^0 + p \to \pi^0 + p. \end{array}\right\} \tag{41}$$

In πN-scattering there are two independent amplitudes $T(\frac{1}{2})$ and $T(\frac{3}{2})$, corresponding to the possible isospins $I = \frac{1}{2}$ and $I = \frac{3}{2}$ of the πN system. Let us express the amplitudes for the processes (41) in terms of $T(\frac{1}{2})$ and $T(\frac{3}{2})$ using (34), (37), and (39)

$$\left.\begin{aligned} T(\pi^+ p \to \pi^+ p) &= T(\tfrac{3}{2}), \\ T(\pi^- p \to \pi^- p) &= \tfrac{1}{3}[T(\tfrac{3}{2}) + 2T(\tfrac{1}{2})], \\ T(\pi^- p \to \pi^0 n) &= \frac{\sqrt{2}}{3}[T(\tfrac{3}{2}) - T(\tfrac{1}{2})], \\ T(\pi^0 p \to \pi^0 p) &= \tfrac{1}{3}[2T(\tfrac{3}{2}) + T(\tfrac{1}{2})]. \end{aligned}\right\} \tag{42}$$

Consequently one has the relation

$$\sigma(\pi^+ p \to \pi^+ p) + \sigma(\pi^- p \to \pi^- p) = \sigma(\pi^- p \to \pi^0 n) + 2\sigma(\pi^0 p \to \pi^0 p). \tag{43}$$

The cross-section $\sigma(\pi^0 p \to \pi^0 p)$ is hard to measure experimentally, so that (43) cannot be tested directly. However, one can use eqns. (42) by noting that when $T(\frac{1}{2}) = 0$,

$$\sigma(\pi^+ p \to \pi^+ p) : \sigma(\pi^- p \to \pi^- p) : \sigma(\pi^- p \to \pi^0 n) = 9 : 1 : 2 \tag{44}$$

This, indeed, appears to be the case for the resonant contribution of the Δ (1236), indicating that the isospin of this resonance is $\frac{3}{2}$.

On the other hand, for large energies (and small scattering angles), the cross-sections for processes with charge exchange are small (see Chapter 13). Setting $\sigma(\pi - p \to \pi^0 n) \approx 0$, we find from (42) that for these energies one will have $\sigma(\pi^+ p \to \pi^+ p) \approx \sigma(\pi^- p \to \pi^- p)$.

Decays

Rapid decays of particles (i.e. those that conserve isospin) are analyzed in a similar way.

Let us write the final state in terms of isospin states using (31)–(33). Then, for example, the isospin part of the amplitude for the decay of the particle $a(I_a, t_a)$ into the particles $b(I_b, t_b)$ and $c(I_c, t_c)$ will have the form

$$\langle b, c | T | a \rangle = \langle I_b I_c t_b t_c | I_b I_c I_a t_a \rangle T, \tag{45}$$

where T does not depend on $t_a = t_b + t_c$, t_b, or t_c. Consequently, the relation between the decay amplitudes

$$a(I_a, t_a) \to b(I_b, t_b) + c(I_c, t_c),$$
$$a'(I_a, t'_a) \to b'(I_b, t'_b) + c'(I_c, t'_c)$$

is determined by Clebsch–Gordan coefficients:

$$\frac{\langle b, c \,|\, T \,|\, a\rangle}{\langle b', c' \,|\, T \,|\, a'\rangle} = \frac{\langle I_b I_c t_b t_c \,|\, I_b I_c I_a t_a\rangle}{\langle I_b I_c t'_b t'_c \,|\, I_b I_c I_a t'_a\rangle}. \tag{46}$$

In the case of two-particle decays, this form refers, for example, to the decays $\varrho \to 2\pi$, $f^0 \to 2\pi$, $A_2 \to K\bar{K}$, $\Delta \to N\pi$, etc. For $\varrho \to 2\pi$ we find the following expressions from (46) for the decay probabilities $w(\varrho \to 2\pi) \sim |T(\varrho \to 2\pi)|^2$:

$$\left.\begin{aligned} w(\varrho^+ \to \pi^+\pi^0) &= w(\varrho^- \to \pi^-\pi^0) = w(\varrho^0 \to \pi^-\pi^+), \\ w(\varrho^0 \to 2\pi^0) &= 0. \end{aligned}\right\} \tag{47}$$

Decays and effective Lagrangian

The effective Lagrangian $\mathcal{L}_I(x)$ was introduced earlier as a relativistically invariant local operator constructed from the fields involved in the decay $a \to b+c$ [see (7.84) and (7.85)]. The matrix element $\mathcal{L}_I(0)$ between the initial $|a\rangle$ and final $\langle b, c|$ states is equal to the decay amplitude. Knowledge of the effective isospin Lagrangian allows one easily to find the relation between the amplitudes for rapid decays of various particles of the isomultiplet; obviously, $\mathcal{L}_I$ is an isoscalar.

Under isospin transformations the basis states $|I_a, t_a\rangle$ transform according to the representation $\mathcal{D}^{I_a}$. The fields $a(x)$ transform in the same way as the annihilation operator, i.e. according to the conjugate representation $\mathcal{D}^{I_a*}$. Raising and lowering the isospin indices using the matrices $C_{\sigma\sigma'}$ and $C^{-1\,\sigma\sigma'}$, we can, of course, also use states of isospinors with indices of various types—only upper, only lower, or mixed—for describing fields. Here, however, additional phase factors appear connecting the components of the arbitrarily chosen isospinor with components of the basis states or fields. In particular, a field with isospin 1 is conveniently chosen as an isovector (i.e. with one upper and one lower isospinor index).

We shall now write the isospin part of several effective Lagrangians. Their construction reduces to forming "three-dimensional" invariants out of isovectors and isospinors. In the case of $\varrho \to 2\pi$ all three particles are isovector; consequently,

$$\mathcal{L}_I(\varrho \to 2\pi) = g_{\varrho\pi\pi}\boldsymbol{\varrho}\cdot(\boldsymbol{\pi}(1)\times\boldsymbol{\pi}(2)). \tag{48}$$

Similarly, one may construct the Lagrangian for the decay $Y^* \to \Sigma\pi$:

$$\mathcal{L}_I(Y^* \to \Sigma\pi) = g_{Y^*\Sigma\pi}\boldsymbol{\Sigma}\cdot(\boldsymbol{Y}^*\times\boldsymbol{\pi}).$$

In the case of the decay $f^0 \to 2\pi$,

$$\mathcal{L}_I(f^0 \to 2\pi) = g_{f^0\pi\pi}(\boldsymbol{\pi}(1)\cdot\boldsymbol{\pi}(2)), \tag{49}$$

since f^0 is an isoscalar ($I = 0$) with $J^P = 2^+$. Writing $\boldsymbol{\pi}(1)\cdot\boldsymbol{\pi}(2)$ explicitly, we see immediately that the probabilities for the processes $f^0 \to \pi^+\pi^-$ and $f^0 \to \pi^0\pi^0$, are connected by

$$w(f^0 \to \pi^+\pi^-) = 2w(f^0 \to 2\pi^0),$$

Furthermore, decays involving kaons (isospinors) may be described by the Lagrangians

$$\left.\begin{aligned} \mathcal{L}_I(f^0 \to K\bar{K}) &= g_{f^0K\bar{K}}\bar{K}K, \\ \mathcal{L}_I(A_2 \to K\bar{K}) &= g_{A_2K\bar{K}}\bar{K}(A_2\cdot\boldsymbol{\tau})K, \\ \mathcal{L}_I(K^* \to K\pi) &= g_{K^*K\pi}\bar{K}(\boldsymbol{\tau}\cdot\boldsymbol{\pi})K^*. \end{aligned}\right\} \quad (50)$$

In the Lagrangian for the decay $\Delta \to N\pi$ (with $I_\Delta = \frac{3}{2}$) it is most convenient to write the pion field with indices of one type:

$$\mathcal{L}_I(\Delta \to N\pi) = g_{\Delta N\pi}\bar{N}^\alpha\pi^{\{\beta\gamma\}}\Delta_{\{\alpha\beta\gamma\}}\,. \quad (51)$$

In formulae (48)–(51) the constants $g_{\varrho\pi\pi}$, ..., characterize the strength of the interaction giving rise to the decay.

CHAPTER 9

THE GROUP SU_3

THE fact that the observed multiplets have definite combinations of isospin and hypercharge (see § 8.1) indicates the presence of a broader internal structure than isospin. The small mass difference between the baryon isomultiplets N, Λ, Σ, and Ξ indicates that this internal structure may be interpreted as a broken symmetry, in which N, Λ, Σ, and Ξ are included in the same multiplet of the new symmetry group. The isospin group [together with the group of hypercharge transformations $U_Y(1)$] must be a subgroup of this new symmetry group. Such a symmetry is described by the group SU_3 suggested by Gell-Mann and Ne'eman[113, 114] It is sometimes called unitary symmetry (although from the mathematical point of view isospin symmetry is also unitary).

The choice of the group SU_3 as the next step in the construction of a generalized symmetry of the strong interaction was not at all obvious at the time. If one adheres to the minimum enlargement of the symmetry group $(SU_2)_I \times U_Y(1)$, the problem arises of finding the right symmetry group among the four existing semi-simple groups of second rank (i.e. with two diagonal generators corresponding to I_3 and Y). This group turned out to be exactly SU_3. In this chapter the basic facts of the group SU_3[116–117] are set forth.

§ 9.1. The matrices λ_a and structure constants

The group SU_3 is the group of unitary unimodular 3 by 3 matrices. Let us consider the transformations described by unitary 3 by 3 matrices $B(B^+ = B^{-1}, \det B = 1)$, acting on a three-component SU_3 spinor ξ†

$$\xi = \begin{pmatrix} \xi_1 \\ \xi_2 \\ \xi_3 \end{pmatrix}, \quad \xi' = B\xi. \tag{1}$$

As a consequence of the unitary of B the scalar product

$$(\xi^*\xi) = \xi_1^*\xi_1 + \xi_2^*\xi_2 + \xi_3^*\xi_3 \tag{2}$$

† One sometimes uses in place of ξ_1, ξ_2, ξ_3 the notation p, n, Λ originating in the Sakata model[118] in which all particles consisted of these three baryons. The Sakata model was a predecessor of SU_3.

remains unchanged; $(\xi'^*\xi') = (\xi^*\xi)$. The general form of a unitary operator B is

$$B = e^{i\hat{\omega}}, \quad \text{Tr}\,\hat{\omega} = 0, \tag{3}$$

where $\hat{\omega}$ is a Hermitian matrix. The condition det $B = 1$ holds if Tr $\hat{\omega} = 0$.

Let us choose as a basis the eight Hermitian 3 by 3 matrices λ_a ($a = 1 \dots 8$) normalized according to

$$\text{Tr}\,(\lambda_a\lambda_b) = 2\delta_{ab}, \quad \text{Tr}\,\lambda_a = 0, \quad \lambda_a = \lambda_a^+. \tag{4}$$

These matrices play the same role that the Pauli matrices τ_k did in the case of SU_2. One of the possible sets of matrices λ_a consistent with (4) has the form

$$\left.\begin{aligned}
&\lambda_1 = \begin{pmatrix} 0 & 1 & 0 \\ 1 & 0 & 0 \\ 0 & 0 & 0 \end{pmatrix}, \quad \lambda_2 = \begin{pmatrix} 0 & -i & 0 \\ i & 0 & 0 \\ 0 & 0 & 0 \end{pmatrix}, \quad \lambda_3 = \begin{pmatrix} 1 & 0 & 0 \\ 0 & -1 & 0 \\ 0 & 0 & 0 \end{pmatrix}, \\
&\lambda_4 = \begin{pmatrix} 0 & 0 & 1 \\ 0 & 0 & 0 \\ 1 & 0 & 0 \end{pmatrix}, \quad \lambda_5 = \begin{pmatrix} 0 & 0 & -i \\ 0 & 0 & 0 \\ i & 0 & 0 \end{pmatrix}, \quad \lambda_6 = \begin{pmatrix} 0 & 0 & 0 \\ 0 & 0 & 1 \\ 0 & 1 & 0 \end{pmatrix}, \\
&\lambda_7 = \begin{pmatrix} 0 & 0 & 0 \\ 0 & 0 & -i \\ 0 & i & 0 \end{pmatrix}, \quad \lambda_8 = \frac{1}{\sqrt{3}}\begin{pmatrix} 1 & 0 & 0 \\ 0 & 1 & 0 \\ 0 & 0 & -2 \end{pmatrix}.
\end{aligned}\right\} \tag{5}$$

The commutation relations for λ_a are

$$[\lambda_a, \lambda_b] = 2if_{abc}\lambda_c, \tag{6}$$

so that

$$f_{abc} = \frac{1}{4i}\,\text{Tr}\,([\lambda_a, \lambda_b]\lambda_c). \tag{7}$$

The structure constants f_{abc} are real and antisymmetric in all indices. The nonzero components of f_{abc} have the following values;

$$\left.\begin{aligned}
&f_{123} = 1, \\
&f_{147} = -f_{156} = f_{246} = f_{257} = f_{345} = -f_{367} = \tfrac{1}{2}, \\
&f_{458} = f_{678} = \frac{\sqrt{3}}{2}.
\end{aligned}\right\} \tag{8}$$

The product of two matrices λ_a may be expressed as a linear combination of these matrices and the unit matrix $\mathbf{1}$:

$$\lambda_a\lambda_b = if_{abc}\lambda_c + d_{abc}\lambda_c + \sqrt{\tfrac{2}{3}}\,\lambda_0\delta_{ab}, \tag{9}$$

where $\lambda_0 = \sqrt{\frac{2}{3}}\cdot\mathbf{1}$ and the structure constants d_{abc} are symmetric in all indices:

$$d_{abc} = \tfrac{1}{4}\,\text{Tr}\,(\{\lambda_a, \lambda_b\}\lambda_c). \tag{10}$$

A calculation of the components of d_{abc} via (10) gives

$$d_{146} = d_{157} = -d_{247} = d_{256} = d_{344} = d_{355} = -d_{366} = -d_{377} = \tfrac{1}{2},$$

$$d_{118} = d_{228} = d_{338} = -d_{888} = \frac{1}{\sqrt{3}},$$

$$d_{448} = d_{558} = d_{668} = d_{778} = -\frac{1}{2\sqrt{3}}.$$

The remaining components of d_{abc} are equal to zero.

The structure constants d_{abc} and f_{abc} satisfy the Jacobi identities

$$f_{abc}f_{cel}+f_{ceb}f_{acl}+f_{aec}f_{bcl} = 0, \tag{11}$$

$$d_{abc}f_{cel}+f_{ceb}d_{acl}-d_{cbl}f_{ace} = 0, \tag{12}$$

which may be easily derived with the help of (6), (7), and (10) from the identities

$$\left.\begin{aligned} [[A, B], C]+[[B, C], A]+[[C, A], B] = 0, \\ [A, \{B, C\}]-\{[A, B], C\}-\{[A, C], B\} = 0. \end{aligned}\right\} \tag{13}$$

From (11) and (12) it is clear that if the quantities f_{abc} and d_{abc} can be treated as the matrix elements of 8 by 8 matrices f_b and d_b:

$$(f_b)_{ac} = if_{abc}, \quad (d_b)_{ac} = d_{abc}, \tag{14}$$

then the matrices f_b and d_b will satisfy the commutation relations

$$[f_a, f_b] = if_{abc}f_c, \tag{15}$$

$$[d_a, f_b] = if_{abc}d_c. \tag{16}$$

By virtue of the linear independence of the matrices (5) and λ_0 any 3 by 3 matrix may be expanded in terms of the matrices λ_0 and $\lambda_a (a = 1 \dots 8)$. In particular, any unitary 3 by 3 matrix B [formula (3)] may be written in the form

$$B = e^{i\hat{\omega}}, \quad \hat{\omega} = \tfrac{1}{2}\lambda_a\omega_a \tag{17}$$

with real coefficients ω_a characterizing an "SU_3 rotation".

One sometimes uses a different (non-Hermitian) representation for the basis matrices. Let us introduce instead of the matrices λ_a the nine 3 by 3 matrices B_i^j with matrix elements

$$(B_i^j)_l^k = \delta_{il}\delta_{jk}-\tfrac{1}{3}\delta_{ij}\delta_{lk} \qquad (i, j, k, l = 1, 2, 3), \tag{18}$$

connected by the relation

$$B_1^1+B_2^2+B_3^3 = 0. \tag{19}$$

From (18) it follows that the commutation relations for B_i^j have the form

$$[B_i^k, B_j^l] = \delta_i^l B_j^k - \delta_j^k B_i^l. \tag{20}$$

§ 9.2. The fundamental representation and quarks. *U*- and *V*-spin

The matrices B give rise to an SU_3 transformation on the fundamental representation

$$\xi_i' = B_i^j \xi_j \qquad (i, j = 1, 2, 3). \tag{21}$$

The hypothetical particles corresponding to the SU_3 spinor ξ_i are called quarks. The generators of the group SU_3 in the case of quarks may be written in the form [see (17)]

$$F_a = \tfrac{1}{2}\lambda_a, \quad [F_a, F_b] = if_{abc}F_c. \tag{22}$$

In the case of the group SU_2, the matrices A and A^* are unitarily equivalent; for SU_3 the matrices B and B^* are not equivalent, and there are thus two basis spinors ξ_i and ξ^i:

$$\xi'^i = B^{*i}{}_j \xi^j, \quad \xi^i \sim \xi_i^*; \tag{23}$$

the spinor ξ^i with on upper index transforms as a conjugate spinor (or antiquark); in accord with this notation

$$\xi^i \xi_i = \xi'^i \xi_i' = \text{inv}. \tag{24}$$

The basis spinor ξ_i is denoted according to its dimension as $\underline{3}$, and ξ^i as $\underline{3}^*$.

The physical content of SU_3 multiplets will be examined in Chapter 10. However, in order to make our discussion less abstract we shall introduce the isospin I_k and hypercharge Y operators immediately.

Let us assume that the isospin and hypercharge symmetry group $(SU_2)_I \times U_Y(1)$ (see Chapter 8) forms a subgroup of SU_3. We can always choose the basis representations of the group SU_3 so as to identify the SU_3 generators F_k $(k = 1, 2, 3)$ with the components of isospin I_k. It is with this basis that formulae (5) for the matrices λ_a were written. (If $F_k \neq I_k$, one can ensure $F_k = I_k$ with the help of the unitary transformation $F_a \to uF_a u^{-1}$). The hypercharge Y commutes with the isospin $F_k = I_k$ and thus, according to (5) and (6), must be proportional to F_8. Let us set

$$Y = \frac{2}{\sqrt{3}} F_8, \quad I_k = F_k \quad (k = 1, 2, 3). \tag{25}$$

As we shall see below in § 10.1, this expression for Y leads to the correct values of the hypercharge of baryons and mesons.

From (25) it follows that the electric charge of the quarks is

$$Q = I_3 + \frac{1}{2} Y = \frac{1}{2}\left(\lambda_3 + \frac{1}{\sqrt{3}}\lambda_8\right) = B_1{}^1. \tag{26}$$

Let us write the quantum numbers of the quarks ξ_i and antiquarks ξ^i.

	I_3	Y	Q		I_3	Y	Q
ξ_1	$\frac{1}{2}$	$\frac{1}{3}$	$\frac{2}{3}$	ξ^1	$-\frac{1}{2}$	$-\frac{1}{3}$	$-\frac{2}{3}$
ξ_2	$-\frac{1}{2}$	$\frac{1}{3}$	$-\frac{1}{3}$	ξ^2	$\frac{1}{2}$	$-\frac{1}{3}$	$\frac{1}{3}$
ξ_3	0	$-\frac{2}{3}$	$-\frac{1}{3}$	ξ^3	0	$\frac{2}{3}$	$\frac{1}{3}$

(27)

Quarks are characterized by fractional charge and hypercharge. So far, particles with fractional charge have not been observed, at least for mass $m \lesssim 15$ GeV[118a]. Quarks may not even exist as particles, although, of course, it would be unfortunate if the simplest multiplet of SU_3 did not correspond to existing particles. But the main problem of SU_3 is the explanation of the properties of baryons and mesons, for which the existence of quarks is not required. We shall thus view quarks for the time being merely as a convenient mathematical concept.

Thus the quark SU_3 triplet contains the isodoublet ξ_1, ξ_2 and the isosinglet ξ_3. The content of the quarks ξ_i with respect to the subgroup consisting of the direct product of the isopin group $(SU_2)_I$ and the hypercharge phase transformation $U_Y(1)$ can be written as

$$\underline{3} = (\tfrac{1}{2}, \tfrac{1}{3})+(0, -\tfrac{2}{3}), \tag{28}$$

where (I, Y) denote the isospin and hypercharge of a representation of $(SU_2)_I \times U_Y(1)$. For the conjugate function ξ^i we have

$$\underline{3}^* = (\tfrac{1}{2}, -\tfrac{1}{3})+(0, -\tfrac{2}{3}), \tag{29}$$

It is convenient to depict the multiplets of SU_3 in the (I_3, Y) plane ("weight diagrams"). The weight diagrams of the states $\underline{3}$ and $\underline{3}^*$ are shown in Fig. 6a and b.

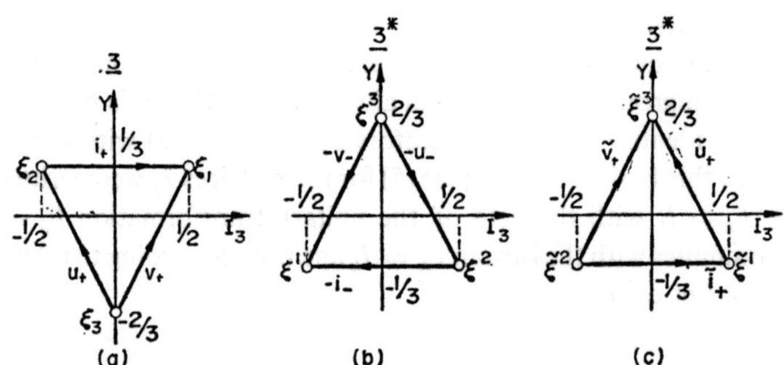

FIG. 6. Weight diagrams for (a) quarks belonging to $\underline{3}$, (b) antiquarks belonging to $\underline{3}^*$, and (c) antiquarks belonging to $\tilde{\underline{3}}^*$. (See text.)

The state ξ_1 (Fig. 6a) may be obtained from ξ_2 by applying the operator $i_+ = \frac{1}{2}(\lambda_1+i\lambda_2)$; the state ξ_2 may be obtained from ξ_3 by applying $u_+ = \frac{1}{2}(\lambda_6+i\lambda_7)$. The effect of the operators i_+, u_+, and v_+ is shown on Fig. 6 by arrows. Reversal of the arrows is equivalent to replacing these operators by i_-, u_-, v_-. For normalized states the matrix element of each such transition in $\underline{3}$ is equal to 1; for example, $\langle 2|i_+|1\rangle = 1$.

In the representation $\underline{3}^*$ (Fig. 6b) the operators i_+, u_+, v_+ are replaced by $i_+ = -i_-$, $u_+ = -u_-$, $v_+ = -v_-$, so that the states ξ^1, ξ^2, and ξ^1 are obtained respectively from the states ξ^3, ξ^3, and ξ^2 by applying $-v_-$, $-u_-$, and $-i_-$.

From the representation $\underline{3}^*$ it is convenient to pass to a unitarily equivalent representation $\tilde{\underline{3}}^*$. In the representation $\tilde{\underline{3}}^*$, the isospin components of the antiquarks $\tilde{\xi}^{1,2}$ transform under isospin rotations as $\xi_{1,2}$, i.e. they have the properties of the isospinor N_α (which is used to describe antiparticles in the isospin group). Let us introduce the unitary operator

$$W = \begin{pmatrix} 0 & -1 & 0 \\ 1 & 0 & 0 \\ 0 & 0 & 1 \end{pmatrix}, \qquad WW^{+} = 1, \tag{30}$$

whose 2 by 2 isospin part is equal to $C = -i\tau_2$. Then

$$\tilde{\xi}^i = W_k{}^i \xi^k = \begin{pmatrix} -\xi^2 \\ \xi^1 \\ \xi^3 \end{pmatrix} \equiv \begin{pmatrix} \tilde{\xi}^1 \\ \tilde{\xi}^2 \\ \tilde{\xi}^3 \end{pmatrix}. \tag{31}$$

Under SU_3 transformations B, the antiquark $\tilde{\xi}^i$ transforms according to the rule

$$\tilde{\xi}' = WB^*W^{-1}\tilde{\xi} \equiv \tilde{B}^*\tilde{\xi}. \tag{32}$$

The generators of the transformation (32) are thus equal to

$$\tilde{F}_a \equiv \tfrac{1}{2}\tilde{\lambda}_a = -\tfrac{1}{2}W\lambda_a^* W^{-1}. \tag{33}$$

The isospin matrices do not change; $\tilde{\lambda}_k = \lambda_k$, while

$$\left.\begin{aligned} \tilde{v}_+ &= \tfrac{1}{2}(\tilde{\lambda}_4 + i\tilde{\lambda}_5) = \tfrac{1}{2}(\lambda_6 + i\lambda_7) = u_+, \\ \tilde{u}_+ &= \tfrac{1}{2}(\tilde{\lambda}_6 + i\tilde{\lambda}_7) = -\tfrac{1}{2}(\lambda_4 + i\lambda_5) = -v_+. \end{aligned}\right\} \tag{34}$$

The transition from ξ^k to $\tilde{\xi}^k$ implies a special choice of basis states in the case of antiparticles and a special phase convention. The weight diagram of the representation $\underline{\tilde{3}^*}$ with the states denoted in (31) and with the new generators is shown in Fig. 6c. In $\underline{\tilde{3}^*}$ the matrix elements of $\tilde{v}_+$ and $\tilde{i}_+$ are positive while the matrix element of $\tilde{u}_+$ is negative.

U- and V-spin

To clarify the structure of the group SU_3 it is useful to introduce the concepts of U- and V-spin along with isospin. The isospin subgroup with generators F_1, F_2, and F_3 is one of the SU_2 subgroups of the group SU_3. Another SU_2 subgroup (the U-spin subgroup) is generated by

$$U_\pm = F_6 \pm iF_7, \quad U_3 = -\tfrac{1}{2}I_3 + \tfrac{3}{4}Y, \tag{35}$$

and a third SU_2 subgroup (the V-spin subgroup) by the generators

$$V_\pm = F_4 \pm iF_5, \quad V_3 = \tfrac{1}{2}I_3 + \tfrac{3}{4}Y. \tag{36}$$

The role of the operator Y in the case of U-spin is played by the electric charge with reversed sign: $Y_U = -Q$, since

$$[Q, U_\pm] = [Q, U_3] = 0.$$

In the case of V-spin, the operator $Y_V = I_3 - \tfrac{1}{2}Y$ plays an analogous role.

The U- and V-spin matrices for quarks have the form

$$u_+ = \tfrac{1}{2}(\lambda_6+i\lambda_7) = \begin{pmatrix} 0 & 0 & 0 \\ 0 & 0 & 1 \\ 0 & 0 & 0 \end{pmatrix}, \quad u_3 = \tfrac{1}{2}\begin{pmatrix} 0 & 0 & 0 \\ 0 & 1 & 0 \\ 0 & 0 & -1 \end{pmatrix}, \tag{37}$$

$$v_+ = \tfrac{1}{2}(\lambda_4+i\lambda_5) = \begin{pmatrix} 0 & 0 & 1 \\ 0 & 0 & 0 \\ 0 & 0 & 0 \end{pmatrix}, \quad v_3 = \tfrac{1}{2}\begin{pmatrix} 1 & 0 & 0 \\ 0 & 0 & 0 \\ 0 & 0 & -1 \end{pmatrix}. \tag{38}$$

Let us also introduce the matrices Y_V and Y_U:

$$Y_V = \begin{pmatrix} \frac{1}{3} & 0 & 0 \\ 0 & -\frac{2}{3} & 0 \\ 0 & 0 & \frac{1}{3} \end{pmatrix}, \quad Y_U = \begin{pmatrix} -\frac{2}{3} & 0 & 0 \\ 0 & \frac{1}{3} & 0 \\ 0 & 0 & \frac{1}{3} \end{pmatrix} = -Q. \tag{39}$$

From the properties of U- and V-spin it follows that the classification of particles in an SU_3 multiplet may be performed not only with respect to isospin I and hypercharge Y, but in two other ways, specifically: by U-spin U and Y_U or by V-spin V and Y_V. Here

$$\left.\begin{aligned} U^2 &= \tfrac{1}{2}(U_+U_-+U_-U_+)+U_3^2, \\ V^2 &= \tfrac{1}{2}(V_+V_-+V_-V_+)+V_3^2. \end{aligned}\right\} \tag{40}$$

From the explicit form of the matrices u_k and Y_U, it follows that in the U-spin group the quarks ξ_2 and ξ_3 form a U-spinor with U-hypercharge $Y_U = \frac{1}{3}$, while the quark ξ_1 forms a U-singlet with $Y_U = -\frac{2}{3}$. In the case of V-spin, the V-spinor is formed by the quarks ξ_1 and ξ_3 with $Y_V = \frac{1}{3}$, while quark ξ_2 is a V-singlet with $Y_V = -\frac{2}{3}$.

§ 9.3. Representations of the group SU_3

The commutation relations among the generators F_a $(a = 1 \ldots 8)$ do not depend on representation. Hence we can derive then using $F_a = \frac{1}{2}\lambda_a$ in the fundamental representation $\underline{3}$ from which, according to (6),

$$[F_a, F_b] = if_{abc}F_c. \tag{41}$$

The generators F_a are Hermitian: $F_a = F_a^+$.

Several equivalent ways are known for constructing irreducible representations of SU_3. We shall use the spinor method. Let us examine the mixed SU_3 spinor of higher rank transforming as the product of p quark functions with lower indices ξ_k and q quark functions with upper indices ξ^i:

$$\Phi_{k_1 \ldots k_p}^{i_1 \ldots i_q} \sim \xi_{k_1} \ldots \xi_{k_p}\xi^{i_1} \ldots \xi^{i_q}. \tag{42}$$

The function $\Phi_{k_1 \ldots k_p}^{i_1 \ldots i_q}$ in general describes a reducible representation of SU_3. To guarantee the absence in (42) of representations of lower dimension one must obtain zero when contracting the indices in $\Phi_{k_1 \ldots k_p}^{i_1 \ldots i_q}$ with constant SU_3 tensors.

In the group SU_3 there are three invariant tensors (i.e. tensors not changing under SU_3 transformations): δ_i^k, ε_{ijk}, and ε^{ijk} $(i, j, k = 1, 2, 3)$. The constancy of the Kronecker symbol δ_i^k is a consequence of the unitarity of the transformation B:

$$\delta_i'^k = B_i^{i'} B^{*k}{}_{k'}\, \delta_{i'}^{k'} = \delta_i^k .$$

The antisymmetric tensors ε_{ijk} and ε^{ijk} have constant components by virtue of the unimodularity of B:

$$\varepsilon'_{ijk} = B_i^{i'} B_j^{j'} B_k^{k'} \varepsilon_{i'j'k'} = \varepsilon_{ijk} \det B = \varepsilon_{ijk} . \tag{43}$$

Thus we arrive at the following two conditions for the irreducibility of $\Phi^{i_1 \dots i_q}_{k_1 \dots k_p}$:

$$\Phi^{\dots j \dots}_{\dots j \dots} = 0, \tag{44}$$

$$\Phi^{\dots i_1 i_2 \dots}_{\dots} = \Phi^{\dots i_2 i_1 \dots}_{\dots}, \qquad \Phi^{\dots}_{\dots j_1 j_2 \dots} = \Phi^{\dots}_{\dots j_2 j_1 \dots}, \tag{45}$$

i.e., firstly, contraction of an upper index with a lower one must give zero, and, secondly, the function $\Phi^{i_1 \dots i_q}_{k_1 \dots k_p}$ must be symmetric separately with respect to upper and lower indices. When these two conditions hold, the function transforms according to the irreducible representation $D(p, q)$ of the group SU_3.

Since the function with lower index ξ_k transforms via the matrix B [formula (1)], while ξ^i transforms via the complex conjugate matrix B^*, the representations $D(p, q)$ and $D(q, p)$ are the complex conjugate of one another:

$$D(p, q) = D^*(q, p). \tag{46}$$

A representation $D(p, p)$ with the same number of upper and lower indices is self-adjoint.

The quark ξ_k transforms according to the representation $D(1, 0)$ while the antiquark ξ^i (or $\tilde{\xi}^i$) transforms according to the representation $D(0, 1)$. The arbitrary SU_3 spinor of second rank is:

$$\left.\begin{aligned} \Phi_{ij} &= \tfrac{1}{2}\varphi_{\{i,j\}} + \tfrac{1}{2}\varphi_{[i,j]} = \tfrac{1}{2}\varphi_{\{i,j\}} + \varepsilon_{ijk}\varphi'^k, \\ \varphi_{\{i,j\}} &= \tfrac{1}{2}(\Phi_{ij} + \Phi_{ji}), \quad \varphi_{[i,j]} = \tfrac{1}{2}(\Phi_{ij} - \Phi_{ji}), \end{aligned}\right\} \tag{47}$$

and contains the parts $\varphi_{\{i,j\}}$ and φ'^k, transforming via the representations $D(2, 0)$ and $D(0, 1)$. The mixed SU_3 spinor of second rank Φ_i^j contains an SU_3 scalar Φ_i^i (the representation $D(0, 0)$) and an octet $\varphi_i^j = \Phi_i^j - (\frac{1}{3})\delta_i^j \Phi_k^k$, transforming according to the representation $D(1, 1)$:

$$\Phi_i^j = \tfrac{1}{3}\delta_i^j \Phi_k^k + (\Phi_i^j - \tfrac{1}{3}\delta_i^j \Phi_k^k) \equiv \tfrac{1}{3}\delta_i^j \Phi_k^k + \varphi_i^j. \tag{48}$$

Let us dwell in more detail on the octet state φ_i^j, since it is the most important for physical applications. The generators F_a form an octet; octets are the most widespread SU_3 multiplets of particles (see § 10.1); the vector currents also form an octet (see § 15.2).

An octet is described by a mixed spinor of second rank Φ_i^j $(i, j = 1, 2, 3)$ with trace equal to zero: $\Phi_i^i = 0$, so that Φ_i^j has 8 independent components. Under the SU_3 transformation B, an octet component Φ_i^j is replaced by the linear combination

$$\Phi'^{j'}_{i'} = B_{i'}^i \Phi_i{}^j B_j^{+j'} . \tag{49}$$

This shows that the Φ_i^j can be considered the elements of an octet matrix Φ (Tr $\Phi = 0$), for which the SU_3 transformation is

$$\Phi' = B\Phi B^+ . \tag{50}$$

A matrix Φ with zero trace may be expanded in a complete set of linearly independent matrices λ_a:

$$\Phi = \sum_a \lambda_a \Phi_a \qquad (a = 1 \dots 8), \tag{51}$$

where the coefficients Φ_a are components of an SU_3 vector. Formulae (50) and (51) are the analog of the formulae $\mathrm{p} = \sigma_\mu p^\mu$ and $\mathrm{p}' = A\mathrm{p}A^+$ for the matrices p of the SU_2 vector p and their transformation rule under the group SU_2. The SU_3 vector Φ_a is in one-to-one correspondence with the matrix Φ, since, according to (4),

$$\Phi_a = \tfrac{1}{2}\,\mathrm{Tr}\,(\lambda_a \Phi). \tag{52}$$

To find the 8 by 8 matrix $d_{ab}(B)$ of the representation $D(1, 1)$, one may insert the matrix (50) into eqn. (52) for Φ_a':

$$\Phi_a' = \sum_b d_{ab}(B)\Phi_b, \qquad d_{ab} = \tfrac{1}{2}\,\mathrm{Tr}\,(\lambda_a B \lambda_b B^+). \tag{53}$$

From (53) it is clear that d_{ab} is a self-adjoint matrix. $D(1, 1)$ is the adjoint, or regular, representation of the group SU_3.

The eight generators F_a of SU_3 transform according to the rule (53). Since the matrix B is unitary, the trace of the product of any number of octet matrices will be invariant [see (50)]. Consequently, the quantities

$$z_n = \mathrm{Tr}\,\hat{F}^n, \quad \hat{F} = \sum_a F_a \lambda_a \qquad (n = 2, 3, \dots) \tag{54}$$

are invariants of the groups SU_3. Since the second-rank group SU_3 has two independent invariants (e.g. z_2, z_3), then, in principle, any SU_3 multiplet may be denoted by the values of z_2 and z_3. The quantities z_n with $n > 3$ will then be functions of z_2 and z_3. By virtue of (4), (7), and (10),

$$z_2 = 2\sum_a F_a^2, \quad z_3 = 2\sum_{a,\,b,\,c} d_{abc} F_a F_b F_c\,. \tag{55}$$

From this it also follows that, in contrast to SU_2, one can construct not one but two invariant scalar products (55) in SU_3 from the components of Φ_a.

Let us introduce the commutation relations between the generators F_a and the components of the vector Φ_a. Using the infinitesimal form $B = 1 + (i/2)\lambda_a\omega_a$ for the matrices B in (53) and the definition (7) for the structure constants, we find that

$$F_a = -if_{abc}\Phi_b \frac{\partial}{\partial \Phi_c}, \tag{56}$$

so that

$$[F_a, \Phi_b] = if_{abc}\Phi_c\,. \tag{57}$$

Since the commutation relations do not depend on the explicit form of F_a, (57) holds as well when the generators F_a are matrices or operators in Hilbert space. Relations (57) characterize the properties of Φ_a as components of an SU_3 vector. The analog of (57) in the case of SU_2 is the well-known relation $[p_k, M_j] = i\varepsilon_{kji}p_i$, characterizing the vector properties of p.

Using eqs. (57) and the second Jacobi identity (12), one may easily check that the quantities

$$D_a = \tfrac{2}{3} d_{abc} F_b F_c \qquad (a, b, c = 1 \ldots 8) \tag{58}$$

are components of an SU_3 vector. Although the vector D_a is constructed from the generators F_a, eqn. (55) implies that D_a and F_a are linearly independent.

Let us count up the number of independent components $n(p, q)$ for an irreducible SU_3 spinor $\Phi^{i_1 \ldots i_q}_{j_1 \ldots j_p}$ (the representation $D(p, q)$), i.e. the number of terms in a multiplet described by this function. We begin with the symmetric SU_3 spinor $\Phi_{\{j_1 \ldots j_p\}}$, or the representation $D(p, 0)$. $n(p, 0)$ is equal to the number of ways of distributing p indices among the three values 1, 2, 3. If one writes the indices in the order 11 . . ., 22 . . ., 33 . . ., then $n(p, 0)$ is just the number of ways of inserting two commas between p symbols, i.e.,

$$n(p, 0) = \frac{(p+1)(p+2)}{2}. \tag{59}$$

Analogously, in the case $\Phi^{\{i_1 \ldots i_q\}}$ the number of components is equal to

$$n(0, q) = \frac{(q+1)(q+2)}{2}.$$

In the case of the irreducible representation $D(p, q)$ the number of components will be less than in $\Phi^{\{i_1 \ldots i_q\}}_{\{j_1 \ldots j_p\}}$ as a consequence of the supplementary conditions $\Phi^{\{i_1 \ldots i_q\}}_{\{i_1 \ldots i_p\}} = 0$ guaranteeing the absence of representations of lower rank $D(p-1, 0)\, D(0, q)$. According to (59), the number of components in the case of a representation $D(p-1, 0)\, D(0, q-1)$ is equal to $n(p-1, 0)\, n(0, q-1)$. From this, we find the multiplicity of $D(p, q)$:

$$n(p, q) = n(p, 0)\, n(0, q) - n(p-1, 0)\, n(0, q-1) = \tfrac{1}{2}(p+1)(q+1)(p+q+2). \tag{60}$$

The multiplicity $n(p, q)$ of the simplest representations, according to (60), is equal to

$$\left.\begin{array}{ll} n(0, 0) = 1, & n(2, 0) = 6, \\ n(0, 1) = n(1, 0) = 3, & n(3, 0) = 10, \\ n(1, 1) = 8, & n(0, 3) = 10, \\ n(2, 1) = n(1, 2) = 15, & n(2, 2) = 27. \end{array}\right\} \tag{61}$$

Consequently, the simplest representations may be characterized uniquely by their multiplicity, which is equal to $n = n(p, q)$ for $p \geqslant q$ and $n^* = n(p, q)$ for $p < q$. The components of SU_3 multiplets may thus be characterized by the quantum number n (or n^*) and I, $I_3' = t, Y$.

Among the various SU_3 multiplets we shall be particularly interested in those in which the electric charges Q and hypercharges Y of particles are integral. Since Q is a multiple of $\frac{1}{3}$ for a quark, the condition that Y and Q be integral has the form

$$\frac{p-q}{3} = 0, \pm 1, \pm 2, \pm \ldots = \text{integer}. \tag{62}$$

The simplest multiplets containing particles with integral charges Q and Y are $\underline{1}$, $\underline{8}$, $\underline{10}$, $\underline{10}^*$, and $\underline{27}$.

Reduction of products of representations

Rules (44) and (45) are also the basis for reducing products of representations. We shall limit ourselves to several examples, since the general method is sufficiently clear.

The product of two quarks ξ_k and η_i gives the representations $D(2, 0)$ and $D(0, 1)$:

$$\xi_k\eta_i = \tfrac{1}{2}(\xi_k\eta_i+\xi_i\eta_k)+\tfrac{1}{2}\varepsilon_{kil}\xi'^l, \tag{63}$$

where the quark with upper index ξ'^l is

$$\xi'^l = \tfrac{1}{2}\varepsilon^{lki}(\xi_k\eta_i-\xi_i\eta_k), \tag{64}$$

or

$$\underline{3}\times\underline{3} = \underline{6}+\underline{3}^*. \tag{65}$$

The product of the representations $\underline{3}$ and $\underline{3}^*$ decomposes, according to (48), into an octet $\underline{8}$ and a singlet $\underline{1}$:

$$\underline{3}\times\underline{3}^* = \underline{8}+\underline{1}. \tag{66}$$

Let us note an important formula for the product of two octets φ_i^j and Ψ_k^l. The highest representation one can construct out of

$$G_{jk}^{jl} = \varphi_i^j\Psi_k^l,$$

is $\underline{27}$, or $\Phi_{ik}^{jl}(\underline{27})$. It is obtained, in correspondence with the rules (44) and (45), by symmetrization in upper and lower indices separately: $G_{ik}^{jl} \to G_{\{ik\}}^{\{jl\}}$ and subtraction of the trace:

$$\Phi_{ik}^{jl}(\underline{27}) = G_{\{ik\}}^{\{jl\}}-\tfrac{1}{5}\big[\delta_i^jG_{\{rk\}}^{\{rl\}}+\delta_k^jG_{\{ri\}}^{\{rl\}}+\delta_i^lG_{\{rk\}}^{\{rj\}}+\delta_k^lG_{\{ri\}}^{\{rj\}}\big]+\tfrac{1}{20}(\delta_k^j\delta_i^l+\delta_i^j\delta_k^l)G_{\{rm\}}^{\{rm\}}. \tag{67}$$

Antisymmetrizing in upper indices $G_{ik}^{jl}\to G_{ik}^{[jl]}$ and multiplying by ε_{njl} we obtain a tensor with three lower indices $\Psi_{nik} = \varepsilon_{njl}G_{ik}^{[jl]}$. Symmetrizing Ψ_{nik} in these indices, we obtain a decimet:

$$\Phi_{nik}(\underline{10}) = \Psi_{\{nik\}}. \tag{68}$$

Analogously one can obtain the conjugate decimet $\Phi^{nik}(\underline{10}^*)$.

Moreover, from G_{jk}^{il} one can construct two octet functions:

$$\left.\begin{aligned}\Phi_k^i(\underline{8}_S) &= G_{jk}^{ij}+G_{kj}^{ji}-\tfrac{2}{3}\delta_k^iG_{jl}^{lj},\\ \Phi_i^j(\underline{8}_A) &= G_{ki}^{jk}-G_{ik}^{kj},\end{aligned}\right\} \tag{69}$$

describing the symmetric $\underline{8}_S$ and antisymmetric $\underline{8}_A$ octets. Summing on both pairs of indices in G_{ik}^{jl}, we obtain an SU_3 scalar $\underline{1}$, or

$$\Phi(\underline{1}) = G_{ik}^{ki}.$$

Thus the product of two octets may be written in the following manner:

$$\underline{8}\times\underline{8} = \underline{27}+\underline{10}+\underline{10}^*+\underline{8}_S+\underline{8}_A+\underline{1}. \tag{70}$$

To distinguish the components of the octets $\underline{8}_S$ and $\underline{8}_A$ from one another, one must introduce an additional quantum number ν whose meaning depends on the specific problem.

The product of three quarks, or the tensor with three lower indices Ψ_{ijk}, may be decomposed into irreducible parts in an analogous manner with the result

$$\underline{3}\times\underline{3}\times\underline{3} = \underline{10}+2(\underline{8})+\underline{1}. \tag{71}$$

The reduction of a direct product in SU_3 may also be performed in the canonical way using Clebsch–Gordan coefficients for SU_3. Denoting the state of an SU_3 multiplet $\underline{n}$ by $|\underline{n}, \alpha\rangle$, where $\alpha = (I, t, Y)$, we find from the general quantum mechanical formula ($t = I_3'$)

$$\left.\begin{aligned} |n_1, \alpha_1\rangle\,|n_2, \alpha_2\rangle &= \sum_{n,\alpha,\gamma} \begin{pmatrix} n_1 & n_2 & n_\gamma \\ \alpha_1 & \alpha_2 & \alpha \end{pmatrix} |n_\gamma, \alpha\rangle, \\ t = t_1+t_2, \quad Y = Y_1+Y_2, &\quad I = |I_1-I_2| \ldots I_1+I_2. \end{aligned}\right\} \tag{72}$$

The Clebsch–Gordan coefficients for the group SU_3 [117]

$$\begin{pmatrix} n_1 & n_2 & n_\gamma \\ \alpha_1 & \alpha_2 & \alpha \end{pmatrix} = \langle I_1 I_2 t_1 t_2 \,|\, I_1 I_2 I t\rangle \left(\begin{array}{cc|c} n_1 & n_2 & n_\gamma \\ I_1 Y_1 & I_2 Y_2 & IY \end{array}\right) \tag{73}$$

are equal to the product of the usual Clebsch–Gordan coefficients for isospin by an isoscalar factor. The coefficients (73) form a real orthogonal matrix. The isoscalar factors are shown in Table A.5 (pp. 365-367).

Isopin content of SU_3 multiplets

The reduction formulae of the type (66) and (70) also may be used to find the isospin content of SU_3 multiplets. Let us denote an isospin multiplet by (I, Y). The product of two isospin multiplets (I_1, Y_1) and (I_2, Y_2) contains multiplets with hypercharge (Y_1+Y_2) and isospins I_1-I_2, I_1-I_2+1, ..., I_1+I_2, defined by the rule for vector addition of angular momenta:

$$(I_1, Y_1)(I_2, Y_2) = (|I_1-I_2|, Y_1+Y_2)+(|I_1-I_2|+1, Y_1+Y_2)+ \ldots (I_1+I_2, Y_1+Y_2). \tag{74}$$

Knowing the isospin content of $\underline{3}$ and $\underline{3}^*$ [formulae (28) and (29)], we find

$$\underline{3}\times\underline{3}^* = \underline{1}+\underline{8} = (\tfrac{1}{2}, 1)+2(0, 0)+(1, 0)+(\tfrac{1}{2}, -1),$$

from which

$$\underline{8} = (\tfrac{1}{2}, 1)+(\tfrac{1}{2}, -1)+(0, 0)+(1, 0), \tag{75}$$

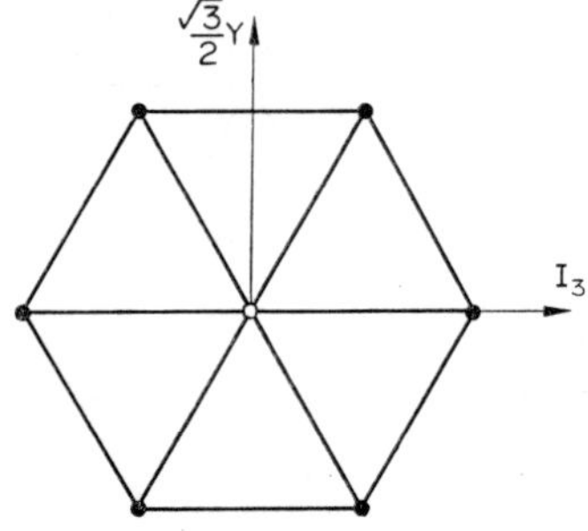

FIG. 7. Weight diagram for the octet representation of SU_3. The point at the center is a doubled state.

i.e. the octet contains an isotriplet with $Y = 0$, an isospinor with $Y = 1$, an isospinor with $Y = -1$, and an isosinglet with $Y = 0$. The weight diagram for the octet is shown in Fig. 7.

Let us calculate the (I, Y) content of the representations contained in $\underline{3}\times\underline{3}\times\underline{3}$. According to (28) and (74),

$$\underline{3}\times\underline{3}\times\underline{3} = (\tfrac{3}{2}, 1)+2(\tfrac{1}{2}, 1)+3(1, 0)+3(0, 0)+3(\tfrac{1}{2}, -1)+(0, -2).$$

But according to (71)

$$\underline{3}\times\underline{3}\times\underline{3} = \underline{10}+2(\underline{8})+\underline{1}.$$

Consequently, taking account of (75), we find the (I, Y) content of the decimet:

$$\underline{10} = (\tfrac{3}{2}, 1)+(1, 0)+(\tfrac{1}{2}, -1)+(0, 2), \tag{76}$$

i.e. the decimet $\underline{10}$ includes an isoquartet with $Y = 1$, an isotriplet with $Y = 0$, an isospinor with $Y = -1$, and an isosinglet with $Y = -2$ (Fig. 8).

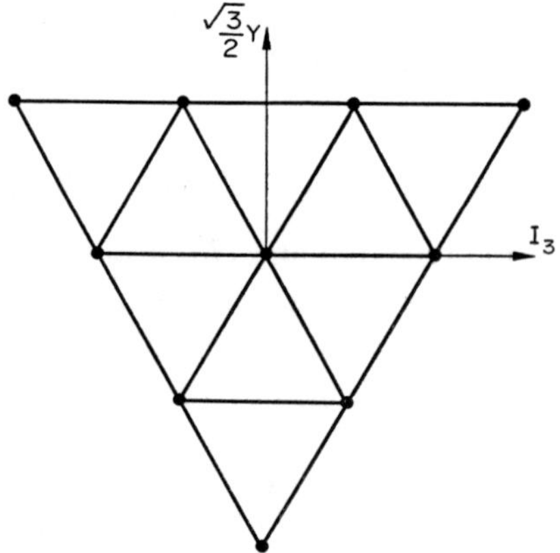

FIG. 8. Weight diagram for the decimet.

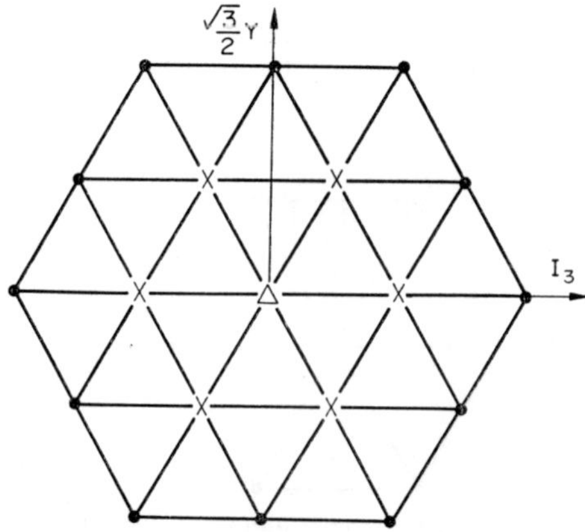

FIG. 9. Weight diagram for the $\underline{27}$-plet. Crosses denote doubled states, the triangle a tripled state.

Analogously, using (71), (75), and (76), it is not hard to find the (I, Y) content of the $\underline{27}$-plet:

$$\underline{27} = (1, 2)+(\tfrac{3}{2}, 1)+(\tfrac{1}{2}, 1)+(0, 0)+(1, 0)+(2, 0)+(\tfrac{1}{2}, -1)+(\tfrac{3}{2}, -1)+(1, -2). \quad (77)$$

The weight diagram of the $\underline{27}$-plet is shown in Fig. 9.

In contrast to the decimet, some states are doubled in the $\underline{27}$-plet; they are shown in Fig. 9 by crosses. The state with $Y = I_3 = 0$ in the $\underline{27}$-plet is contained three times (in Fig. 9 it is denoted by a triangle).

CHAPTER 10

SU_3 SYMMETRY AND THE CLASSIFICATION OF PARTICLES AND RESONANCES

THE primary evidence for SU_3 symmetry of the strong interactions is the correspondence between observed particle multiplets and irreducible representations of the group SU_3. However, not all of the simplest possible multiplets correspond to physical particles. Aside from the singlets $\underline{1}$, the particles and resonances belong only to octets $\underline{8}$ (mesons and baryons) and decimets $\underline{10}$, $\underline{10}^*$ (baryons) leaving, for example, the triplets, sextets, $\underline{15}$-plets, and $\underline{27}$-plets empty [see (9.61)]. In view of the essential role of the octet $\underline{8}$, one sometimes says that Nature follows the Eightfold Way.

SU_3 symmetry is only approximate. If one judges the degree of its breaking by mass splitting in the baryon multiplets $\underline{8}$ and $\underline{10}$, then the medium strong interaction (which breaks SU_3 symmetry but conserves isospin and hypercharge) is approximately an order of magnitude weaker than the SU_3 symmetric interaction and an order stronger than electromagnetism. In the case of mesonic multiplets, the relative mass splitting is larger. The success of the SU_3 classification of single-particle states hence depends not only on the identification of particles with components of a unitary multiplet, but also on being able to explain the mass splitting. The final confirmation of the assignment of particles to unitary multiplets also includes experimental agreement with other consequences of unitary symmetry such as relations between decay constants and electromagnetic properties.

§ 10.1. Unitary representations and multiplets

Let us assume that SU_3 is an exact internal symmetry group, if only for single-particle states. Then resonant particles must belong to multiplets corresponding to irreducible representations of SU_3. As in the case of isospin symmetry, all particles of a multiplet must have the same space–time properties (mass, spin, parity), so that the full symmetry (of one-particle states) may be described as $\overline{\mathcal{P}}^\dagger \times SU_3$, where $\overline{\mathcal{P}}^\dagger$ includes parity as well. The transformations of the Poincaré group $\overline{\mathcal{P}}^\dagger$ do not affect internal symmetry variables, so that the single-particle state may be written in the form of a product of a space–time part $|\boldsymbol{p}, \lambda; m, J\rangle$ and an SU_3 part $|\underline{n}; I, t, Y\rangle$, where $\underline{n}$ is the multiplicity. For brevity we shall write only the SU_3 part of the state vector.

The connection between SU_3 multiplets and isospin multiplets was made in the previous chapter by assuming that the isospin and hypercharge may be expressed in terms of SU_3 generators via eqns. (9.25):

$$I_k = F_k \qquad (k = 1, 2, 3), \qquad Y = \frac{2}{\sqrt{3}} F_8 . \tag{1}$$

Let us now verify that eqns. (1) really allows one to classify particles with respect to SU_3 multiplets. With this identification of SU_3 generators, the fact that the known particles have integral electric charge and hypercharge singles out (as we saw in § 9.3) several of the simplest SU_3 multiplets; specifically the singlet $\underline{1}$, the octet $\underline{8}$, the decimets $\underline{10}$ and $\underline{10}^*$, and the multiplet $\underline{27}$.

One must distinguish two types of SU_3 states: the "tensor" states $\left|\Phi^{i_1 \dots i_q}_{k_1 \dots k_p}\right\rangle \equiv \Phi^{i_1 \dots i_q}_{k_1 \dots k_p}$ and the basis states $|\underline{n}; I, t, Y\rangle$. The states of the first type are characterized by their transformation properties, which are defined by the number and type of indices (the representation $D(p, q)$); they are used for constructing fields (after passing to creation and annihilation operators) and SU_3 invariants. The basis states differ from states of the first type by a phase factor. If one writes a basis state in tensor form: $|\underline{n}; I, t, Y\rangle = \left\{\left|\tilde{\Phi}^{ij \dots}_{kl \dots}\right\rangle\right\}$, then the lower indices in $\tilde{\Phi}$ transform according to the same representation as in Φ, while the upper indices of $\tilde{\Phi}$ transform according to a unitarily equivalent representation $\tilde{\Phi}^{ij} = W^i{}_{i'} W^j{}_{j'} \Phi^{i'j'}$.

The unitary operator W was defined in § 9.2. The basis states are convenient since their isospin multiplets all have lower indices. Consequently, one may use Clebsch–Gordan coefficients (see § 9.3) to reduce products of representations. It is for the basis states that the tables of crossing matrices are constructed (see § 11.3).

Let us find the quantum numbers $t = I_3'$ and Y of the components $\left|\Phi^{i_1 \dots i_q}_{k_1 \dots k_p}\right\rangle$. The values of the additive quantum numbers of $\left|\Phi^{i_1 \dots i_q}_{k_1 \dots k_p}\right\rangle$ will be the same as for the product of quark and antiquark states $|\xi^{i_1}\rangle|\xi^{i_2}\rangle \dots |\xi_{k_1}\rangle|\xi_{k_2}\rangle \dots$ The values of t and Y thus may be found by reading off the upper and lower quark indices. Let us denote by $p(k)$ the number of lower indices with value k, and by $q(i)$ the number of upper (i.e. antiquark) indices with value i. Then, using (9.27) we find that the component $\left|\Phi^{i_1 \dots i_q}_{k_1 \dots k_p}\right\rangle$ has the following Y, t, and Q:

$$Y = -p(3)+q(3)+\tfrac{1}{3}(p-q), \tag{2}$$
$$t = \tfrac{1}{2}[p(1)-q(1)-p(2)+q(2)], \tag{3}$$
$$Q = p(1)-q(1)-\tfrac{1}{3}(p-q), \tag{4}$$

where $p = p(1)+p(2)+p(3)$ and $q = q(1)+q(2)+q(3)$.

The octet

Let us consider the octet tensor state $\left|\Phi_i^j\right\rangle \equiv \Phi_i^j$. By subdividing the values of the index j into isospinor ($j = \alpha = 1, 2$) and isosinglet ($j = 3$) and using (2)–(4), it is easy to construct the isomultiplets: Φ_3^3 is an isosinglet, Φ_α^3 is an isospinor with lower index, Φ_3^α is an isospinor with upper index; the components $\Phi_\alpha^{\alpha'}$ form an isotriplet if $\Phi_\alpha^\alpha = 0$. Since $\Phi_j^j = 0$, the component of the triplet with $t = 0$ is proportional to $\Phi_1^1 - \Phi_2^2$. The components of the octet Φ_i^j form the matrix elements of a 3 by 3 matrix Φ (see § 9.3), whose lower index labels the row and the upper index the column.

The pseudo-scalar mesons (spin-parity 0^-), i.e. the "stable particles" (see Table 8.1 (p. 176) or Table A.1 (p. 359) may be combined into four multiplets:

an isotriplet $\pi_{\alpha\beta} \sim \pi_i^{I=1}(I = 1, Y = 0)$: the states [see (8.22)]

$$|\pi_{11}\rangle = -|\pi^+\rangle, \quad |\pi_{21}\rangle = |\pi_{12}\rangle = \frac{1}{\sqrt{2}}|\pi^0\rangle, \quad |\pi_{22}\rangle = |\pi^-\rangle;$$

an isodoublet $K_\alpha(I = \frac{1}{2},\ Y = 1)$: the states $|K_1\rangle = |K^+\rangle, |K_2\rangle = |K^0\rangle$ (these are not to be confused with $|K_1^0\rangle, |K_2^0\rangle$ of Chapters 6 and 15);

an isodoublet of antiparticles $K_\alpha(I = \frac{1}{2}, Y = -1)$: the states $|\overline{K}_1\rangle = -|\overline{K}_0\rangle, |K_2\rangle = |K^-\rangle$;

an isosinglet $(I = Y = 0)$: the state $|\eta\rangle$.

These isomultiplets form the octet $\underline{8}$. Thus, neglecting the meson mass differences, we may form the octet P out of π, K, $\overline{\text{K}}$, and η.

Let us expand $|P_i^j\rangle \equiv P_i^j$ in terms of isomultiplets:

$$P_i^j = \{P_\alpha^\beta - \tfrac{1}{2}(P_1^1 + P_2^2)\delta_\alpha^\beta,\ P_\alpha^3,\ P_3^\beta,\ P_3^3 - \tfrac{1}{2}(P_1^1 + P_2^2)\}, \quad (\alpha, \beta = 1, 2)$$

and identify them respectively with $\pi_\alpha^\beta = (C^{-1})^{\beta\beta'}\pi_{\alpha\beta'}$, K_α, $\overline{K}^\beta = (C^{-1})^{\beta\beta'}\,\overline{K}_{\beta'}$, and $-2\eta/\sqrt{6}$. We then obtain the octet matrix of the 0^- mesons:

$$P = \begin{pmatrix} \frac{\pi^0}{\sqrt{2}} + \frac{\eta}{\sqrt{6}} & \pi^+ & K^+ \\ \pi^- & -\frac{\pi^0}{\sqrt{2}} + \frac{\eta}{\sqrt{6}} & K^0 \\ K^- & \overline{K}^0 & -\frac{2\eta}{\sqrt{6}} \end{pmatrix}, \quad \mathrm{Tr}\, P = 0, \tag{5}$$

where the particle symbols π, K, ..., denote the states $|\pi\rangle$, $|K\rangle$. ... The factors in (5) are such that all states appear with identical weight in the completeness condition for octet states

$$\tfrac{1}{2}\,\mathrm{Tr}\,(PP^+) = 1.$$

The choice of phases between different isomultiplets in the octet (5) is the same for any octet. We may thus write an analogous matrix B for the baryons ($J^P = \frac{1}{2}^+$). The baryons-octet consists of the isospinors $N_\alpha = (p, n)$ and $\Xi^\alpha = (\Xi^-, \Xi^0)$, the isotriplet $\Sigma^\pm$, Σ^0 and the isosinglet Λ. Consequently,

$$B = \begin{pmatrix} \frac{\Sigma^0}{\sqrt{2}} + \frac{\Lambda}{\sqrt{6}} & \Sigma^+ & p \\ \Sigma^- & -\frac{\Sigma^0}{\sqrt{2}} + \frac{\Lambda}{\sqrt{6}} & n \\ \Xi^- & \Xi^0 & -\frac{2\Lambda}{\sqrt{6}} \end{pmatrix}. \tag{6}$$

Thus relations (1) certainly hold for the stable particles—the 0^- mesons and the $\frac{1}{2}^+$ baryons. Consequently we have every right to apply them to the classification of the remaining particles in SU_3 multiplets.

If, in addition to the octet, there exists an SU_3 singlet with mass close to the octet, all these particles can be considered together (see § 10.2). This is, in fact, the case for the vector mesons ($J^P = 1^-$), where not eight but nine mesons with mass close to one another are known. Seven of them—the isotriplet $\varrho^\pm$, ϱ^0, the isodoublet K^{*+}, K^{*0} and their antiparticles $\overline{K^{*+}} = K^{*-}$ and $\overline{K^{*0}}$—are members of the vector octet V. As an isosinglet with $I = Y = 0$, one may choose either the ω- or φ-meson. The other of these particles then must be considered as an SU_3 singlet. Let us denote for the moment the isosinglet member of the octet V by ω_8, and the SU_3 singlet by $\omega^{(0)}$, deferring judgement on the properties of φ and ω until we discuss the question of mass splitting. Then the vector meson octet may be written together with the singlet in the form

$$V = \begin{pmatrix} \frac{1}{\sqrt{3}}\omega^{(0)} \frac{1}{\sqrt{2}}\varrho^0 + \frac{1}{\sqrt{6}}\omega_8 & \varrho^+ & K^{*+} \\ \varrho^- & \frac{1}{\sqrt{3}}\omega^{(0)} - \frac{1}{\sqrt{2}}\varrho^0 + \frac{1}{\sqrt{6}}\omega_8 & K^{*0} \\ K^{*-} & \overline{K^{*0}} & \frac{1}{\sqrt{3}}\omega^{(0)} - \frac{2}{\sqrt{6}}\omega_8 \end{pmatrix}. \tag{7}$$

Under charge conjugation C, the meson octets transform into themselves, since they contain both particles (e.g., $\pi^+, \pi^0, K^+, K^0, \eta$), and antiparticles (e.g., $\pi^-, \pi^0, \overline{K^0}, K^-, \eta$).

Keeping in mind the definition (8.27) of charge conjugation for isospin multiplets containing a neutral particle, we may specify the phases of the kaon states under charge conjugation in such a way that the whole SU_3 multiplet has the same phase. Then

$$CPC^{-1} = P^T, \quad CVC^{-1} = -V^T, \tag{8}$$

i.e. charge conjugation is equivalent to transposition of meson octet matrices. The neutral members of the octets (8) have charge parity $\eta_C(\pi) = +1$, $\eta_C(\omega) = -1$. We shall thus speak of the octets P, V as having charge parity $\eta_C(P) = +1$, $\eta_C(V) = -1$.

For antibaryons one must choose $\eta_C = 1$ to correspond to $C\,|\,\text{particle}\rangle = \eta_C\,|\,\text{antiparticle}\rangle$.

The octet matrix for antibaryons is thus

$$\bar{B} = \begin{pmatrix} \frac{\bar{\Sigma}^0}{\sqrt{2}} + \frac{\bar{\Lambda}}{\sqrt{6}} & \bar{\Sigma}^- & \bar{\Xi}^- \\ \bar{\Sigma}^+ & -\frac{\bar{\Sigma}^0}{\sqrt{2}} + \frac{\bar{\Lambda}}{\sqrt{6}} & \bar{\Xi}^0 \\ \bar{p} & \bar{n} & -\frac{2\bar{\Lambda}}{\sqrt{6}} \end{pmatrix} = \overline{B^T} = CBC^{-1}. \tag{9}$$

Basis states for the octet

These states are denoted by $|\underline{8}; I, t, Y\rangle$; the set of quantum numbers I, t, Y denotes the specific particle. The basis states are constructed out of the isospin states $|I, t\rangle$ without additional phase factors. The choice of the factor η_A in the relation

$$|A\rangle = \eta_A |\underline{8}; I, t, Y\rangle, \tag{10}$$

between the basis state and the particle state A is defined with respect to isospin multiplets or formulae (8.31) and (8.32). Consequently, for the basis states of the P-octet, we write all isomultiplets consecutively—the self-adjoint pion isotriplet, the kaon isodoublet, the antikaon isodoublet, and the singlet η:

$$|\underline{8}; I, t, Y\rangle = \{-|\pi^+\rangle, |\pi^0\rangle, |\pi^-\rangle, K^+\rangle, |K^0\rangle, -|\overline{K}^0\rangle, |\overline{K}^-\rangle, |\eta^0\rangle\}. \tag{11}$$

Here (see § 8.1) the isomultiplets in (11) are described by isospinors with lower indices $\pi_{\alpha\beta}$ (or π_t), K_α, and $\overline{K}_\alpha$.

The set of states (11) forms a matrix P with elements

$$\tilde{P}_i^j = W^j{}_k P_i^k, \tag{12}$$

where W is the matrix (9.30) with which one lowers the isospin index (but not the SU_3 index). The components of (12) transform according to the representation $\tilde{D}(1, 1)$, unitarily equivalent to $D(1, 1)$ (but distinct from it).

To obtain the isospin multiplets of antiparticles, described by isospinors with lower indices, we must use formulae (8.31) and (8.32). We thus obtain the octet (11) again. The phase factors (10) for particles and antiparticles are connected by

$$\eta_A \eta_{\bar{A}} = (-1)^{Q_A}. \tag{13}$$

If one denotes by $|-A\rangle$ the state obtained from $|A\rangle$ by reversing the signs of t and Y: $|-A\rangle = \eta_A |8; I_A, -t_A, Y_A\rangle$, then the basis SU_3 state for an antiparticle is

$$|\bar{A}\rangle = (-1)^{Q_A} |-A\rangle. \tag{14}$$

The states (14) form a basis for the conjugate octet representation.

Rules (13) or (14) also hold for baryon octets. They are essential for multiplying representations and constructing crossing-symmetric amplitudes. One may verify that they hold in this form for any self-adjoint multiplet, such as $\underline{27}$.

The decimet

In the baryon decimet $\underline{10}$ there are four isospin multiplets of particles with spin-parity $J^P = \frac{3}{2}^+$: the quartet of Δ-resonances, the triplet of Y_1^*'s, the Ξ^* doublet, and the singlet Ω^-.

The decimet is described by the symmetric SU_3 spinor $D_{\{ijk\}}$. Let us decompose D into isospin components by singling out the values $i, j, k = 3$. Then, using (2)–(4) and the rules for constructing isospin states, we find that ($\alpha, \beta, \delta = 1, 2$):

D_{333} describes an isosinglet ($I = 0$, $Y = -2$) or Ω^-;
$D_{\{\alpha 33\}}$ describes an isodoublet ($I = \frac{1}{2}$, $Y = -1$) or Ξ^*;
$D_{\{\alpha\beta 3\}}$ describes an isotriplet ($I = 1$, $Y = 0$) or Y_1^*;
$D_{\{\alpha\beta\delta\}}$ describes a quartet ($I = \frac{3}{2}$, $Y = +1$) or Δ.

Choosing the phases of all states positive, one may set

$$\left.\begin{array}{llll} D_{111} = \Delta^{++}, & D_{112} = \dfrac{1}{\sqrt{3}}\Delta^{+}, & D_{122} = \dfrac{1}{\sqrt{3}}\Delta^{0}, & D_{222} = \Delta^{-}, \\ D_{113} = \dfrac{1}{\sqrt{3}} Y_1^{*+}, & D_{123} = \dfrac{1}{\sqrt{6}} Y_1^{*0}, & D_{223} = \dfrac{1}{\sqrt{3}} Y_1^{*-}, & \\ D_{133} = \dfrac{1}{\sqrt{3}} \Xi^{*0}, & D_{233} = \dfrac{1}{\sqrt{3}} \Xi^{*-}, & D_{333} = \Omega^{-}. & \end{array}\right\} \quad (15)$$

The coefficients in (15) are chosen so that in the normalized sum

$$D_{\{ijk\}}\bar{D}^{\{ijk\}} = \Omega^-\bar{\Omega}^- + \Xi^{*0}\bar{\Xi}^{*0} + \dots \quad (16)$$

each state enters with unit weight. The baryon decimet is shown in Fig. 10.

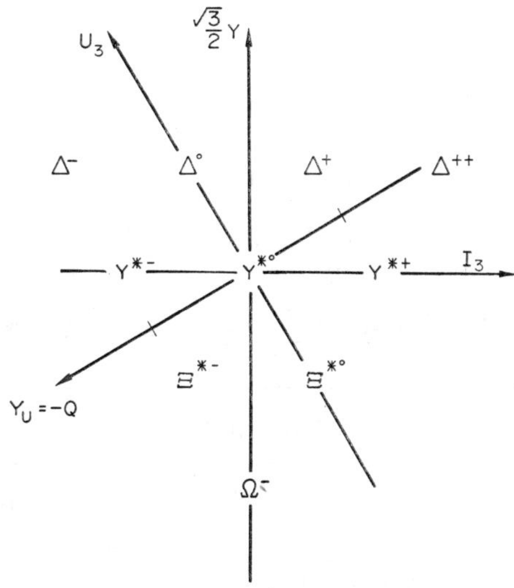

FIG. 10. Weight diagram for the $J^P = \frac{3}{2}^+$ baryon decimet. The U-spin axis, U_3, and the "U-hypercharge" perpendicular to it, Y_U, are also shown.

In the case of the decimet $\underline{10}$ the tensor states $|D_{\{ijk\}}\rangle \equiv D_{\{ijk\}}$ have only lower indices and hence coincide with the respective basis states $|\underline{10}; I, t, Y\rangle$. For the decimet $\underline{10}^*$ (containing antiparticles), the tensor states $D^{\{ijk\}}$ differ from the basis states $|\underline{10}^*; I, t, Y\rangle$. The basis states for antiparticles are also defined using rules (13) or (14):

$$|\bar{A}\rangle = |\underline{10}^*; I_{\bar{A}}, t_{\bar{A}}, Y_{\bar{A}}\rangle = (-1)^{Q_A}|\underline{10}; I_A - t_A, -Y_A\rangle. \quad (17)$$

These states transform according to the representation $\tilde{D}(0, 3)$ which is unitarily equivalent to the representation $D(0, 3)$. It is for the basis (17) that the tables of crossing matrices and Clebsch–Gordan coefficients are constructed.

U- and V-spin multiplets

The group SU_3 is symmetric with respect to its three SU_2 subgroups—the isospin, group the U-spin group, and the V-spin group (see § 9.2).

Particles belonging to an SU_3 multiplet may be classified both with respect to isospin multiplets and with respect to U- and V-spin multiplets. Along with the (I, Y) content of an SU_3 multiplet one may speak of its (U, Y_U) and (V, Y_V) contents (here Y_U and Y_V are the analogs of hypercharge for U- and V-spin). In § 9.2 it was shown that the (U, Y_U) and (V, Y_V) contents of the quark triplets $\underline{3}$ and $\underline{3}^*$ are identical to their (I, Y) content. Consequently, the (U, Y_U) and (V, Y_V) contents of any SU_3 multiplet coincide with its (I, Y) content.

The U- and V-spin multiplets may be conveniently defined with respect to the weight diagram of SU_3 multiplets by introducing the axes (U_3, Y_U) and (V_3, Y_V), i.e. the axes $(-\frac{1}{2}I_3+\frac{3}{4}Y, -Q)$ and $(\frac{1}{2}I_3+\frac{3}{4}Y, I_3-\frac{1}{2}Y)$. The assignment of particles in the baryon and meson octets to U- and V-spin multiplets is given in Tables 10.1 and 10.2. Applying the

TABLE 10.1

U	Y_U	Baryons	Mesons
0	0	$-\frac{\sqrt{3}}{2}\Sigma^0-\frac{1}{2}\Lambda$	$-\frac{\sqrt{3}}{2}\pi^0-\frac{1}{2}\eta$
$\frac{1}{2}$	1	Σ^-, Ξ^-	π^-, K^-
$\frac{1}{2}$	−1	p, Σ^+	K^+, π^+
1	0	n, $-\frac{1}{2}\Sigma^0+\frac{\sqrt{3}}{2}\Lambda$, Ξ^0	K^0, $-\frac{1}{2}\pi^0+\frac{\sqrt{3}}{2}\eta$, $\overline{K^0}$

TABLE 10.2

V	Y_V	Baryons	Mesons
0	0	$\frac{\sqrt{3}}{2}\Sigma^0-\frac{1}{2}\Lambda$	$\frac{\sqrt{3}}{2}\pi^0-\frac{1}{2}\eta$
$\frac{1}{2}$	1	Σ^+, Ξ^0	π^+, $\overline{K^0}$
$\frac{1}{2}$	−1	n, Σ^-	K^0, π^-
1	0	p, $\frac{1}{2}\Sigma^0+\frac{\sqrt{3}}{2}\Lambda$, Ξ^-	K^+, $\frac{1}{2}\pi^0+\frac{\sqrt{3}}{2}\eta$, K^-

axes (U_3, Y_U) to the weight diagram of the decimet (Fig. 10) we can write the U-spin multiplets at once:

quartet $(U = \frac{3}{2}, Y_U = 1)$: $\Omega^-, \Xi^{*-}, Y^{*-}, \Delta^-$,
triplet $(U = \frac{1}{2}, Y_U = 0)$: $\Xi^{*0}, Y^{*0}, \Delta^0$,
doublet $(U = \frac{1}{2}, Y_U = -1)$: Y^{*+}, Δ^+,
singlet $(U = 0, Y_U = -2)$: Δ^{++}.

The classifications with respect to isospin, U and V-spin are equivalent only in the limit of exact SU_3 symmetry. The U- and V-spin transformations interchange particles with different isospin and hypercharge, i.e. the U-and V-spin subgroups contain those characteristic additional transformations generated by F_4, F_5, F_6 and F_7 which distinguish SU_3 from the group $(SU_2)_I \times U_Y(1)$. Isospin classification hence plays a special role in discussing the approximate nature of SU_3, as isospin is also conserved at a lower symmetry level.

Summary of SU_3 classification of particles

Besides the SU_3 multiplets mentioned above—the octets P, V, B, and $\bar{B}$ and the decimets D, $\bar{D}$—there exist a whole series of other full or partially full baryon and meson multiplets.

TABLE 10.3

J^{PC}	$I = 1$	$I = \frac{1}{2}$	$I = 0$	$I = 0$
0^{-+}	π	K	η	η'
1^{--}	ϱ	K*(892)	ω	ϕ
0^{++}	δ (970)	K_N (1100)	ε (700) } ε' (1240) }	S* (993)
1^{++}	A_1 (1100)	K_A (1240)	D (1285)	?
2^{++}	A_2 (1310)	K_N (1420)	f (1270)	f′ (1514)
1^{+-}	B (1235)	?	?	?

TABLE 10.4

J^P	N	Λ	Σ	Ξ
$\frac{1}{2}^+$	940	1115	1190	1320
$\frac{3}{2}^-$	1520	1690	1670	1820?
$\frac{5}{2}^-$	1670	1830	1765	1940?
$\frac{5}{2}^+$	1688	1815	1915	2030?

TABLE 10.5

J^P	Δ	Σ	Ξ	Ω
$\frac{3}{2}^+$	1236	1385	1530	1672
$\frac{7}{2}^+$	1950	2030	?	?

Nov 15

A summary of the data on unitary multiplets is shown in Tables 10.3–10.5. The symbols N, Λ, Σ, Ξ, Δ, and Ω denote the isospin baryon states (without dependence on spin), while the symbols π, K, η denote the isospin states of mesons with any spin.

A comparison with the full table of particles (see Tables A.1–A.3, pp. 359–364) allows one to draw the following conclusions about SU_3 classification.

1. The overwhelming majority of hadrons are well classified in the scheme of unitary multiplets.
2. It is sufficient to use only the representations $\underline{1}$ and $\underline{8}$ for mesons and $\underline{1}$, $\underline{8}$, and $\underline{10}$ for baryons.
3. Mesons evidently have only definite combinations of parity and charge-parity J^{PC}:

$$J^{PC} = 0^{++}, 0^{-+}, 1^{--}, 1^{++}, \ldots$$

The fact that only a small number of the possible unitary multiplets are used in classification indicates the limited nature of the SU_3 scheme.

Despite persistent searches for particles with fractional charge, especially quarks, such particles have not been observed up to now. The experimental observation of quarks would provide great motivation for a theory of elementary particles based on quarks as the "most" fundamental objects.

Resonances with integral charges Q and Y and with quantum numbers B, I, or Y different from those encountered in observed octets and decimets are called exotic. Particles with integral charges Q and Y may belong to higher representations of SU_3, e.g., to a $\underline{27}$-plet. The characteristic feature of a $\underline{27}$-plet is the presence of isomultiplets with $I = 2$ (which have so far not been observed) and with $I = \frac{3}{2}$ (which have so far not been observed among mesons). Resonances with "anomalous" values of J^{PC} (i.e. different from those indicated above in paragraph 3), also are exotic. Exotic resonances have not yet been observed conclusively. As we shall see in § 10.4, such resonances are absent in the quark model.

§ 10.2. Symmetry breaking and mass splitting

The unification of several isomultiplets with different hypercharge into one multiplet, performed in the previous section, assumes that the strongest part of the interaction is SU_3 invariant. This interaction does not distinguish, for example, the pion from the kaon or the nucleon from the Σ-hyperon. If the whole strong interaction, and not only its strongest part, were really SU_3 symmetric, then the masses of all particles in an SU_3 multiplet (such as pions, kaons, and the η-meson) would have to be equal.

Experimentally, the masses of particles in all SU_3 multiplets are different. Since the interaction is also responsible for the internal structure of particles, the mass difference in multiplets is evidence for a medium–strong interaction which breaks SU_3 but preserves isospin and hypercharge. This interaction splits apart the isospin multiplets within SU_3 multiplets. From the value of the mass splitting one may conclude that for the baryons this interaction is an order of magnitude weaker than the SU_3-symmetric one. However, in the meson octet P, the mass splitting is large: $m_\pi \approx 140$ MeV, $m_K \approx 495$ MeV, and $m_\eta \approx 550$ MeV. In order that the classification of particles into SU_3 multiplets be convincing, it is necessary to explain this splitting using the same mechanism of SU_3 breaking for baryons and mesons.

Let $|\underline{n}, \alpha\rangle$ be an SU_3 state of dimension $\underline{n}$, where $\alpha \equiv I, t, Y$ distinguishes the separate basis states (i.e. the particles) of the multiplet from one another. In the case of exact SU_3, the masses of all particles in an SU_3 multiplet will be the same, so that the mass operator for the multiplet may be represented as a unit $\underline{n}$ by $\underline{n}$ matrix $m_0 \times 1^{(n)}$. If the masses of the isomultiplets in an SU_3 multiplet become different from one another (but the mass splitting inside an isomultiplet is equal to zero), then the mass matrix $m^{(n)}$ will contain the term $m'^{(n)}$ depending on isospin and hypercharge:

$$m^{(n)} = m_0 \cdot 1^{(n)} + m'^{(n)}(I, Y). \tag{18}$$

The matrix $m^{(n)}$ is diagonal since the mass of a physical particle by definition cannot have nondiagonal elements. [For unstable particles the mass does not have a definite value, and a particle is characterized by a mass spectrum (see § 2.3). However, the mass operator still commutes with I and Y.] The matrix $m'^{(n)}$ characterizes the breaking of SU_3 symmetry and determines the mass spectrum inside the multiplet $\underline{n}$.

Let us assume that the term $m'^{(n)}$ breaking SU_3 symmetry has definite and unique transformation properties with respect to SU_3 in all representations $\underline{n}$. The matrix $m'^{(n)}$ may contain the parts $m'^{(n)}(\underline{8})$, $m'^{(n)}(\underline{27})$, $m'^{(n)}(\underline{64})$, ..., transforming as components of the self-adjoint representations $\underline{8}$, $\underline{27}$, $\underline{64}$, ...

Let us assume, moreover, that $m'^{(n)}$ is a component of the simplest possible representation—the octet. Since $m'^{(n)}$ must commute with isospin and hypercharge, it must be the eighth component of the octet: $m'^{(n)} = m_8^{(n)}(\underline{8})$.

Now, in SU_3, two independent vectors (of the octet) may be constructed from generators F_a: specifically, F_a and $D_a = \frac{2}{3} d_{abc} F_b F_c$. Consequently,

$$m^{(n)} = m_0 \cdot 1^{(n)} + c_1^{(n)} F_8^{(n)} + c_2^{(n)} D_8^{(n)}, \tag{19}$$

where the coefficients $c_1^{(n)}$ and $c_2^{(n)}$ depend on the representation $\underline{n}$, while $F_8^{(n)}$ and $D_8^{(n)}$ are the components of the vectors F_a and D_a in this representation.

$D_8^{(n)}$ may be computed using (9.10) and (9.58). Bearing in mind the invariant properties of the operator (9.55), we find

$$D_8^{(n)} = a^{(n)} \cdot 1^{(n)} + \frac{1}{\sqrt{3}} (I^2 - \tfrac{1}{4} Y^2). \tag{20}$$

Insertion of this expression into (19) yields the Gell-Mann–Okubo formula for the values of masses in a multiplet:[113, 119]

$$m = m_0 + aY + b[I(I+1) - \tfrac{1}{4} Y^2], \tag{21}$$

where the real constants are different for different multiplets. Formula (21) gives the mass splitting for baryons.

In the case of the baryon octet B there are four different masses: m_N, m_Ξ, m_Σ, and m_Λ. Eliminating the three parameters m_0, a, and b from (21), we obtain the relation

$$\tfrac{1}{4}(m_\Sigma + 3m_\Lambda) = \tfrac{1}{2}(m_N + m_\Xi). \tag{22}$$

Table A.1 (p. 359) shows that (22) holds with an accuracy of 0.5 percent.

In the baryon decimet the quadratic terms in (21) cancel. Consequently, the mass in a decimet depends linearly on Y:

$$m^{(10)} = m_0 + a'Y. \tag{23}$$

From Table 10.5 (p. 211) it is clear that the masses in the decimet are indeed spaced by the same interval $a' \approx 146$ MeV.

The meson multiplets contain both particles and antiparticles, which differ, in particular, with respect to the sign of the hypercharge. But the masses of particles and antiparticles are equal by virtue of CPT invariance (see § 6.1). Hence the mass must be an even function of Y in the meson multiplets, and $a = 0$ in (21).

One usually writes (21) for the squares of masses m^2 in the case of mesons:

$$m^2 = m_0^2 + b[I(I+1) - \tfrac{1}{4}Y^2]. \tag{24}$$

This way of writing mass formulae implicitly assumes that they may be derived in perturbation theory from a Lagrangian (in which the baryon mass appears linearly, while the meson mass appears quadratically). Of course, the large value of the mass splitting excludes the application of perturbation theory. From the point of view of the Poincaré group it is natural to write both formulae for the invariant m^2. Equation (21) for m^2 is satisfied in the fermion case almost as well as the formula for m. Equations (21) and (24) express a hypothesis about the nature of SU_3 breaking.

For the pseudo-scalar octet P, the Gell-Mann–Okubo formula (24) gives

$$4m_K^2 = 3m_\eta^2 + m_\pi^2, \tag{25}$$

which holds to about 6 per cent. The error in this case exceeds the limits of uncertainty in the mass related to the neglect of electromagnetic effects.

In the case of the vector meson octet V, using relation (24), one may find the mass of the isosinglet component ω_8:

$$m^2(\omega_8) = \tfrac{1}{3}(4m_{K^*}^2 - m_\varrho^2) \approx (930 \text{ MeV})^2 \tag{26}$$

This value agrees neither with the mass of the ω-meson ($m_\omega \approx 784$ MeV) nor with the mass of the φ-meson ($m_\varphi \approx 1020$ MeV), each of which has the same isospin and hypercharge $I = Y = 0$ and spin $J = 1$. But from the point of view of the medium-strong interaction, which conserves only isospin and hypercharge, the quantum numbers of these particles are the same. Hence, in the presence of SU_3 breaking, the meson states ω and φ may be mixtures of the unitary singlet $\omega^{(0)}$ and the eighth component of the octet ω_8. Let us introduce the mixing angle θ and set[120]

$$\left.\begin{aligned} \varphi &= \omega_8 \cos\theta + \omega^{(0)} \sin\theta, \\ \omega &= -\omega_8 \sin\theta + \omega^{(0)} \cos\theta. \end{aligned}\right\} \tag{27}$$

As a consequence of unitary symmetry breaking, transitions between ω_8 and $\omega^{(0)}$ are possible, so that the matrix of the squares of the masses (in the space of states ω_8 and $\omega^{(0)}$) may have nondiagonal elements

$$M^2 = \begin{pmatrix} m_8^2 & m_{08}^2 \\ m_{08}^2 & m_0^2 \end{pmatrix}. \tag{28}$$

The mass of the ω_8 component m_8^2 may be found using the Gell-Mann–Okubo formula (26). The eigenstates of the matrix (28) are the physical states ω and φ, while its eigenvalues are equal to the observed masses m_ω^2 and m_φ^2.

Diagonalizing (28) and inserting the experimental values m_ω^2 and m_φ^2 into (26) we determine the parameter m_0 and the mixing angle

$$\theta(\omega, \varphi) = 39.6^\circ. \tag{29}$$

Relations (27) now become a formula relating the squares of masses of particles in the vector nonet:

$$m_\varphi^2 \cos^2\theta + m_\omega^2 \sin^2\theta = \tfrac{1}{3}[4m_{\mathrm{K}^*}^2 - m_\varrho^2]. \tag{30}$$

The mixing of a unitary singlet and the eighth component of an octet must be considered whenever both these particles exist.

For the nonet of mesons with spin-parity $J^{\mathrm{P}} = 2^+$ the mixing angle between f and f′ is equal to

$$\theta(\mathrm{f}, \mathrm{f}') = 29.9^\circ. \tag{31}$$

n the case of the pseudo-scalar mesons, one must add the unitary singlet η' with mass $m_{\eta'} = 958$ MeV to the octet P. In view of the large mass difference, the angle of $(\eta\eta')$-mixing is not large: $\theta(\eta, \eta') = 10.4^\circ$. If one assumes that the pseudo-scalar nonet includes the E-meson with mass $m_E = 1422$ MeV instead of the η', the mixing angle will be $\theta(\eta, \mathrm{E}) = 6.2^\circ$.

§ 10.3. Relations between transition amplitudes

An SU_3-invariant theory describes the scattering, decays, and production of SU_3 multiplets. As in the case of isospin symmetry, by examining separate components of multiplets one may find relations between amplitudes for processes with different particles in these multiplets.

Let a, b, c, and d be SU_3 multiplets participating in the reaction

$$a+b \to c+d; \tag{32}$$

we shall consider the physical multiplets $\underline{8}$, $\underline{10}$, or $\underline{10}^*$. Let us assume for definiteness that the initial states a and b are octets. Then the initial state $a+b$ contains the following irreducible representations:

$$\underline{8}\times\underline{8} = \underline{1}+\underline{8}_A+\underline{8}_S+\underline{10}+\underline{10}^*+\underline{27}. \tag{33}$$

In an SU_3-invariant theory, transitions are possible only between states of the same irreducible representation. The transition to the state $c+d$ thus is possible only when at least one of the irreducible SU_3 representations in $c+d$ coincides with one of those in (33). The reduction of products of octets and decimets gives formulae (33) and

$$\left.\begin{aligned} \underline{8}\times\underline{10} &= \underline{8}+\underline{10}+\underline{27}+\underline{35}, \\ \underline{10}^*\times\underline{10} &= \underline{1}+\underline{8}+\underline{27}+\underline{64}, \\ \underline{10}\times\underline{10} &= \underline{10}^*+\underline{27}+\underline{28}+\underline{35}. \end{aligned}\right\} \tag{34}$$

From (33) and (34) it follows that the final state $c+d$ may belong to any of the classes (33) and (34).

The number of independent amplitudes in the process (32) is determined by the number and the multiplicity of the irreducible representations shared by the initial and final states. According to (33) and (34) there are only two SU_3 amplitudes in the case $\underline{8}+\underline{8} \to \underline{10}+\underline{10}$: $T(\underline{10}^*)$ and $T(\underline{27})$; in the cases $\underline{8}+\underline{8} \to \underline{10}+\underline{10}^*$ and $\underline{8}+\underline{8} \to \underline{8}+\underline{10}$ there are four amplitudes. The scattering $\underline{8}+\underline{8} \to \underline{8}+\underline{8}$ will be described by eight SU_3 amplitudes, corresponding to the four transitions between the representations $\underline{1}$, $\underline{10}$, $\underline{10}^*$, $\underline{27}$, and the four octet transitions $\underline{8}_S \to \underline{8}_S$, $\underline{8}_A \to \underline{8}_A$, $\underline{8}_S \to \underline{8}_A$, and $\underline{8}_A \to \underline{8}_S$ (with amplitudes $T(\underline{8}_{SS})$, $T(\underline{8}_{AA})$, $T(\underline{8}_{SA})$, $T(\underline{8}_{AS})$). The large number of independent amplitudes greatly complicates relations among cross-sections.

The expansion of the transition amplitude in terms of invariant SU_3 amplitudes may be found by using the Clebsch–Gordan coefficients (9.72) for the group SU_3. These coefficients are constructed and tabulated for the product of basis states $|\underline{n}, \alpha\rangle$, $\alpha \equiv I, t, Y$. The particle states must first be expressed in terms of the basis states $|A\rangle = \eta_A |\underline{n}_A, \alpha_A\rangle$, defining η_A for particles via the rules (9.31) and (9.32) [see (11) and (15)]. In the case of antiparticles, according to (14) and (17), one will have $|\bar{A}\rangle = (-1)^{Q_A} |\underline{n}_A^*, -\alpha_A\rangle$. We then obtain

$$\langle \underline{n}_c, \alpha_c; \underline{n}_d, \alpha_d | T | \underline{n}_a, \alpha_a; \underline{n}_b, \alpha_b \rangle = \sum_{n,\gamma} \begin{pmatrix} n_c & n_d & n_\gamma \\ \alpha_c & \alpha_d & \alpha \end{pmatrix} T(n, \gamma', \gamma') \begin{pmatrix} n_a & n_b & n_{\gamma'} \\ \alpha_a & \alpha_b & \alpha \end{pmatrix}. \tag{35}$$

The SU_3 amplitude $T(n, \gamma, \gamma')$ depends on the representation $\underline{n}$ and the parameters γ, γ' distinguishing multiple representations $\underline{n}$ from one another in the initial and final states. In practice γ is the quantum number which distinguishes the octets $\underline{8}_A$ and $\underline{8}_S$.

The general formula (35) contains the Clebsch–Gordan coefficients of SU_3. In a number of cases, however, one may dispense with them if one uses U- or V-spin conservation along with isospin conservation in an SU_3-invariant theory. Since all generators of SU_3 may be expressed in terms of generators of the I-, U-, and V-spin groups, invariance with respect to I- and U-spin or V- and I-spin is equivalent to SU_3 invariance. Moreover, reduction with respect to any of these three subgroups requires only the usual SU_2 Clebsch–Gordan coefficients, leading to a great simplification in calculation.[121] For example, from the fact that the quantities (π^-, K^-), (K^+, π^+), (p, Σ^+), (Σ^-, Ξ^-) form U-spin doublets, one obtains at once the equality of amplitudes for the processes

$$\pi^- + \Sigma^+ \to K^+ + \Xi^-, \quad K^- + p \to \pi^+ + \Sigma^-. \tag{36}$$

The results obtained by studying isospin symmetry may be applied directly in the case of U- and V-spin. In the case of isospin one may easily derive the relation

$$T(K^+ + p \to K^+ + p) = T(K^0 + p \to K^0 + p) + T(K^0 + p \to K^+ + n). \tag{37}$$

for the scattering amplitude of particles in the isodoublets $N = (p, n)$ and $K = (K^+, K^0)$. Using (37) one may write at once a relation between the amplitudes for the scattering of particles described by the U-spinors (π^-, K^-) and $(p, -\Sigma^+)$:

$$T(\pi^- + p \to \pi^- + p) = T(K^- + p \to K^- + p) - T(K^- + p \to \pi^- + \Sigma^+). \tag{38}$$

In contrast to isospin, U- and V-spins are not preserved by the medium–strong interactions transforming as the eighth component of an SU_3 vector.

Since SU_3 symmetry is only approximate, one must pay attention to the effects of symmetry breaking when trying to compare equations of the type (36)–(38) with experiment. In such calculations one usually considers only the mass splitting, since it usually has an important effect on the phase space. As a consequence of the large mass splitting, a reaction possible for some particles in the multiplet may be energetically forbidden for other particles of the same multiplet. This leads to the question of the energies and momentum transfers at which one may expect the relations of SU_3 symmetry to hold approximately. Comparison with experiment shows that such relations hold poorly for scattering amplitudes even when mass splittings are taken into account.

Effective Lagrangian and decays

Relations between rapid decays may be found using Clebsch–Gordan coefficients for SU_3 via a formula of the type (35). However, it is simpler and more graphic to deal with an effective Lagrangian $\mathcal{L}_{\text{eff}}$. In the $\mathcal{L}_{\text{eff}}$ approach, trilinear products of fields (interactions of the Yukawa type), describing an elementary vertex, characterize both real and virtual decays, i.e. the interactions of fields.

In the case of unitary symmetry, $\mathcal{L}_{\text{eff}}$ must be a relativistic and unitary invariant product of three unitary multiplets. By virtue of the properties of the strong interactions, $\mathcal{L}_{\text{eff}}$ must also be invariant with respect to C-, P- and T-transformations separately.

Let us consider the vertex $\bar{B}BP$, where B and P are the nucleon and pseudo-scalar octet fields. The octet matrices of the fields B and P [formulae (5), (6), and (9)] transform via (9.49), and, consequently, $\text{Tr}\,(\bar{B}BP)$ is invariant under SU_3 transformations. But in SU_3 there are two independent invariant scalar products, $\text{Tr}\,([\bar{B}, B]P)$ and $\text{Tr}\,(\{\bar{B}, B\}P)$, or the antisymmetric and symmetric sets of structure constants f_{abc} and d_{abc}. $\mathcal{L}_{\text{eff}}$, hence, for $\bar{B}BP$ interactions, will contain two constants g_F and g_D:

$$\mathcal{L}_{\text{eff}}(\bar{B}BP) = -i\tfrac{1}{4}g_F\,\text{Tr}\,(\bar{B}\gamma_5[B,P]) \\ +i\tfrac{1}{4}g_D\,\text{Tr}\,(\bar{B}\gamma_5\{B,P\}) = g_F f_{abc}\bar{B}_a\gamma_5 B_b P_c + g_D i d_{abc}\bar{B}_a\gamma_5 B_b P_c. \tag{39}$$

(We recall that $i\bar{B}\gamma_5 B$ is a Hermitian operator.)

One may construct the interactions of the remaining octets analogously:

1. Interactions of the vector and pseudo-scalar octets V and P:

$$\mathcal{L}_{\text{eff}}(VPP) = g_F(VPP)\,\text{Tr}\,(V_\mu[P, \partial^\mu P]), \tag{40}$$

$$\mathcal{L}_{\text{eff}}(VVP) = g_D(VVP)\,\text{Tr}\,(\varepsilon_{\mu\nu\varrho\sigma}P\{\partial^\mu V^\nu, \partial^\varrho V^\sigma\}). \tag{41}$$

The constant $g_D(VPP)$ vanishes by virtue of charge conjugation invariance.

2. The interaction of the vector and baryon octets V and B:

$$\mathcal{L}_{\text{eff}} = g_F(\bar{B}BV)\,\text{Tr}\,(\bar{B}\gamma_\mu[B, V^\mu]) + g_D(\bar{B}BV)\,\text{Tr}\,(\bar{B}\gamma_\mu\{B, V^\mu\}). \tag{42}$$

3. Interactions of the axial and baryon octets A and B:

$$\mathcal{L}_{\text{eff}} = g_F(\bar{B}BA)\,\text{Tr}\,(\bar{B}\gamma_\mu\gamma_{5\mu5}[B, A_\mu]]) + g_D(\bar{B}BA)\,\text{Tr}\,(\bar{B}\gamma_\mu\gamma_5\{B, A_\mu\}). \tag{43}$$

4. Interaction of the tensor and pseudo-scalar octets T and P

$$\mathcal{L}_{\text{eff}} = g_D(TPP)\,\text{Tr}\,(T_{\mu\nu}\,\partial^\mu P\,\partial^\nu P) \tag{44}$$

etc.

If one passes from the unitary multiplets B, P, V, ..., to the isospin multiplets N, Σ, π, K, ..., the above formulae for $\mathcal{L}_{\text{eff}}$ imply relations among coupling constants for different isomultiplets. The four isomultiplets in B and the three isomultiplets in P are connected in the case of isospin symmetry by 12 coupling constants. Under unitary symmetry these 12 constants may be expressed in terms of g_F and g_D:

$$\left.\begin{array}{l} g_{\text{NN}\pi} \\ g_{\Sigma\Xi\text{K}} \end{array}\right\} \sim g_D+g_F \qquad \left.\begin{array}{l} g_{\text{NN}\eta} \\ g_{\Xi\Lambda\text{K}} \end{array}\right\} \sim -g_D+3g_F \qquad \left.\begin{array}{l} g_{\Lambda\Lambda\eta} \\ g_{\Sigma\Sigma\eta} \\ g_{\Sigma\Lambda\eta} \end{array}\right\} \sim g_D,$$
$$\left.\begin{array}{l} g_{\Sigma\text{NK}} \\ g_{\Xi\Xi\pi} \end{array}\right\} \sim -g_D+g_F \qquad \left.\begin{array}{l} g_{\text{N}\Lambda\text{K}} \\ g_{\Xi\Xi\eta} \end{array}\right\} \sim g_D+3g_F \qquad g_{\Sigma\Sigma\pi} \sim g_F, \tag{45}$$

where the specific coefficients depend on charge states in general. Consequently, for purely F-type coupling, the interaction vanishes for $\Sigma \to \Sigma\eta$, $\Sigma \to \Lambda\pi$, and $\Lambda \to \Lambda\eta$, while for purely D-type coupling the transitions $\Sigma \to \Sigma+\pi$ are forbidden.

Interactions between the decimet and octets are most conveniently written in tensor form. For example, for the decimet D_{ijk} the baryon octet B and the octet P, $\mathcal{L}_{\text{eff}}$ has the form

$$\mathcal{L}_{\text{eff}}(BDP) = g_{BDP}\bar{D}_\mu^{ijk}\gamma_5 B_i{}^{i'}\partial^\mu P_j{}^{j'}\varepsilon_{ki'j'} + \text{H. c.} \tag{46}$$

Here the baryon decimet is described by the field $D_{\alpha\mu}$ with one spinor and one vector index (spin $\frac{3}{2}^+$); (46) takes account of the positive parity of the decimet with respect to the octet B. The BDP interaction depends on only one parameter g_{BDP}. In the case of the decimet D the decay $D \to B+P$ is energetically possible. Thus relations between the coupling constants implied by (46) are also relations between decay constants. Expressing (46) in terms of specific components of isomultiplets, we find

$$g(\Delta^{++} \to p\pi^+) : g(Y^{*+} \to \Sigma^+\pi^0) : g(Y^{*+} \to \Lambda\pi^+) : g(\Xi^{*0} \to \Xi^-\pi^+)$$
$$= 1 : \frac{1}{\sqrt{2}} : \frac{1}{\sqrt{6}} : \frac{3}{\sqrt{3}}. \tag{47}$$

The decay of vector mesons V into two pseudo-scalar mesons P, according to (41), also depends on one decay constant. Passing from unitary octets to individual particles via formulae (9), (12), and (28), we obtain the relation

$$g(\varrho^+ \to \pi^+\pi^0) : g(\text{K}^{*+} \to \text{K}^+\pi^0) : g(\varphi \to K^+\text{K}^-) = 1 : \frac{1}{2} : \frac{\sqrt{3}}{2}\cos\theta. \tag{48}$$

Relations between decay constants refer to the idealized case of exact unitary symmetry. In reality, as we have seen, the mass splitting implies that unitary symmetry is only approximate. When comparing relations obtained with experiment, it is necessary to take symmetry breaking into account. To first order it is sufficient to allow for the mass splitting, i.e. to insert the physical masses into the phase space expressions and the wave functions of particles, using SU_3-symmetric matrix elements. This gives reasonable agreement with experiment. The next step in taking account of SU_3 breaking consists of introducing corrections to the matrix elements (i.e. to the effective Lagrangians). New parameters thus appear in the theory, connected with breaking terms. In view of the large number of new parameters such calculations often turn out not to be very predictive.

Details of the applications of SU_3 symmetry may be found in several reviews.[122-124]

§ 10.4. The quark model

The number of observed resonances is growing rapidly: around 30 meson isospin multiplets and nearly 50 baryon isomultiplets have been confirmed, with a large number of additional candidates. In the scheme of unitary multiplets, the stable particles (the octet of pseudo-scalar mesons P and the baryon octet B, including the nucleon) do not form a separate class, since the resonances are also classified with respect to octet representations of SU_3. The level scheme of resonances is reminiscent in its complexity of the level scheme of a nucleus or of any other complex system—with the octets P and B and the decimet D as ground states. The question arises: Do there exist "more elementary" particles whose bound states form both stable particles and resonances?

In the quark model[125] it is assumed that these subparticles are quarks q with spin $\frac{1}{2}$. In this model the quarks may be considered physical particles, and the fact that they have not been observed up till now ascribed to their large mass, or they may be regarded merely as convenient mathematical entities. A quark–antiquark pair $q\bar{q}$, or any number of them $(q\bar{q})^n$, have integral spins and charges Q, Y. To construct fermions with integral charges Q and Y, one needs at least three quarks: qqq. The product of three quarks, $\underline{3}\times\underline{3}\times\underline{3} = \underline{1}+\underline{8}+\underline{8}+\underline{10}$, contains both the octet and the decimet: the three-quark combination hence suffices to describe all known baryons.

In the usual quark model, mesons appear as bound states of only one quark–antiquark pair $q\bar{q}$, while the baryons are considered as bound states of the three-quark system qqq. States of the type $(q\bar{q})^n$, $n > 1$, in the case of mesons and of the type $qqq(q\bar{q})^n$, $n \geqslant 1$, in the case of baryons, are assumed not to exist. In other words, in the quark model the mesons fall only into singlets and octets, while the baryons may be in singlets, octets, and decimets; the quark model distinguishes these multiplets from all others. Consequently, in the quark model there do not exist states with $I = 2$ nor baryons with $Y \geqslant 2$ and mesons with $|Y| > 1$. Such states are among the exotic ones mentioned in § 10.1.

It is possible that these exotic resonances exist but have very large masses and thus are not observed in experiment. But even if such resonances are found at higher masses, the singlets, octets, and decimets will still remain the most prominent multiplets of SU_3.

Let us assume that quarks behave nonrelativistically in hadrons. The nonrelativistic quark model of hadrons provides a classification scheme for hadronic states. In this model, the spin of a meson or baryon J is written as the sum of a total quark spin $\boldsymbol{S}$ and an orbital angular momentum $\boldsymbol{L}(\boldsymbol{J} = \boldsymbol{L}+\boldsymbol{S})$, and one neglects $\boldsymbol{LS}$ coupling and dependence of the mass on $\boldsymbol{S}^2$. In this approximation L and S may be thought of as additional quantum numbers for classifying states.

The allowed mesonic states of the quark model are the octets and singlets J^{PC} which may be formed by a quark–antiquark pair with given orbital L and total spin S angular momenta. The mesonic octets form two sequences with $S = 0$ and $S = 1$. In the first, $J = L$, and in the second $J = L-1$, L, $L+1$. The spatial η_P and charge η_C parities of the octet J^{PC} are determined by the total spin and orbital angular momenta of the quark–antiquark system; here, as usual, η_C is equal to the charge parity of the neutral members of the octet.

The nonrelativistic wave function Ψ of the quark–antiquark system with given L and S consists of three factors, referring to the coordinate φ, the spin χ, and the charge ζ parts of the wave function:

$$\Psi = \varphi^L(x_{12})\,\chi^S(\sigma_1, \sigma_2)\,\zeta^{I,\,t,\,Y}(\alpha, \beta).$$

Under interchange of the particle coordinates ($x_1 \leftrightarrow x_2$), the functions φ^L change sign for odd L: $\varphi^L(-x_{12}) = (-1)^L\varphi^L(x_{12})$. The application of spatial reflection P to Ψ thus gives

$$\mathrm{P}\Psi(1, 2) = \eta_P(q)\,\eta_P(\bar{q})\,(-1)^L\,\Psi(1, 2) = (-1)^{L+1}\,\Psi(1, 2),$$

since the product of the parities of fermion and antifermion $\eta_P(q)$ is always negative (see § 6.2).

The spin wave functions $\chi^S(\sigma_1, \sigma_2)$ of a system of two particles with spin $\frac{1}{2}$ is antisymmetric for $S = 0$ and symmetric for $S = 1$, or $\chi^S(\sigma_2, \sigma_1) = (-1)^{S+1}\chi^S(\sigma_1, \sigma_2)$. Since the total wave function of two fermions $\Psi(1, 2)$ always changes sign under interchange of all variables, $\Psi(1, 2) = -\Psi(2, 1)$, the interchange of all variables alone gives

$$\zeta(\beta, \alpha) = (-)^{L+S}\,\zeta(\alpha, \beta).$$

But for a quark–antiquark system the charge conjugation operator C is

$$\mathrm{C}\zeta^{I,\,t,\,Y}(\alpha, \beta) = \eta_C\zeta^{I,\,-t,\,-Y}(-\alpha, -\beta),$$

which is equivalent to interchanging the charge variables for $\alpha = -\beta$, $t = Y = 0$, i.e. for the neutral members of the octet. Hence $\eta_C = (-1)^{L+S}$.

Consequently, in the nonrelativistic quark model, the quantum numbers S, L, J, P, and C of the meson octets are connected by the relations $\eta_P = (-1)^{L+1}$, $\eta_C = (-1)^{L+S}$, $J = |L-S|$, $|L-S|+1, \ldots, L+S$. These relations hold for the following octets:

$$\begin{aligned}
S = 0: L = J \quad & J^{PC} = 0^{-+}, 1^{+-}, 2^{-+}, 3^{+-}, \ldots;\\
S = 1: L = 0, \quad & J^{PC} = 1^{--};\\
L = 1, \quad & J^{PC} = 0^{++}, 1^{++}, 2^{++};\\
L = 2, \quad & J^{PC} = 1^{--}, 2^{--}, 3^{--};\\
L = 3, \quad & J^{PC} = 2^{++}, 3^{++}, 4^{++}; \text{etc.}
\end{aligned}$$

These meson octets do not include $J^{PC} = 0^{--}; 0^{+-}, 1^{-+}, 2^{+-}, \ldots$, which, thus, are forbidden from the point of view of the quark model of mesons as $q\bar{q}$ states. These resonances have not, in fact, been observed, and are also among the exotics.

The ground state of the meson octet levels corresponds to $S = 0$, $L = 0$, i.e. to the octet of stable mesons 0^{-+}. Since the energy in the nonrelativistic theory does not depend on spin, the quark model predicts not only new meson octets but also the degeneracy of their mass for fixed L.

The wave functions of three-quark baryon states may also be constructed using the rules of the nonrelativistic theory. The total wave function is a product of the coordinate function $\varphi(x_1, x_2, x_3)$, the spin wave function $\chi(\sigma_1, \sigma_2, \sigma_3)$, and the SU_3 function $\zeta(\alpha_1, \alpha_2, \alpha_3)$. In the case of the octet B with spin $\frac{1}{2}$ and the decimet D with spin $\frac{3}{2}$, which correspond to the ground states of the octet and decimet series, the product $\chi(\sigma_1, \sigma_2, \sigma_3)\,\zeta(\alpha_1, \alpha_2, \alpha_3)$ is symmetric with respect to the interchange of the variables $\sigma_1\alpha_1, \sigma_2\alpha_2, \sigma_3\alpha_3$. Consequently, if quarks satisfy Fermi–Dirac statistics, the ground state coordinate function $\varphi(x_1, x_2, x_3)$ must be antisymmetric, whereas one might expect it to be symmetric. This fact is an important difficulty of the quark model. It may be avoided by assigning additional degrees of freedom to the quarks.[(126)]

The set of baryon states is most simply described in terms of partial-wave amplitudes for $0^- - \frac{1}{2}^+$scattering. In this description one uses the usual spectroscopic designation S_{2J}, P_{2J}, ..., where J is the spin of the baryon. Then the following states are possible: S_1, P_1, P_3, D_3, D_5, F_5, F_7 .... The parities of states with neighboring values of L are different; for example, D_5 describes $J^P = \frac{5}{2}^-$, while F_5 describes $J^P = \frac{5}{2}^+$.

In the nonrelativistic quark model a description in terms of the internal orbital angular momentum and total quark spin is also convenient. Here again one may use s, p, d, ..., to describe internal $L = 0, 1, 2, \ldots$, with a superscript $2S_q$ to denote quark spin and a subscript J to denote total spin. For example, the nucleon would be written ${}^2s_{1/2}$ in this notation.

Besides the orbital and spin quantum numbers L, S, and J, hadronic levels also may be classified with respect to radial quantum numbers; there are many orbital wave functions with the same L and internal quantum numbers. Experimentally, aside from one or two candidates, only one such series of L-excitations has been observed for the mesons. It is possible that other series have higher mass; they may also be very unstable. In the case of the baryons there exist several candidates for radial excitations.

§ 10.5. SU_6 multiplets

The internal degrees of freedom of particles include not only isospin or unitary spin but also the usual (mechanical) spin. At first glance, it would seem very attractive to broaden SU_3 so that it contained spin. Then one might hope that the new supermultiplet could combine particles of different spins, giving rise to formulae describing the splitting of masses with respect to spins as well as with respect to isospins.

Such a program involves almost insurmountable difficulties in a relativistic theory. According to very general theorems,[(127)] any such extended symmetry group must consist merely of a direct product of the internal symmetry group and the Poincaré group. Multiplets containing particles of different spins hence may be obtained only by giving up the relativistic nature of the theory.

Let us assume that the motion of quarks in a hadron in the center of mass system is nonrelativistic. Let us further assume that the nature of the symmetry of the strongest part of the interaction which determines the properties of the single-particle states and the composition of multiplets is conserved in a strictly nonrelativistic statistical limit. We are interested in an extended internal symmetry group combining both SU_3 and the statistical space–time symmetry group—the spin group SU_2. The minimal semi-simple group containing the direct product $SU_3 \times SU_2$ is SU_6. Under the above assumptions, the group SU_6 may describe a symmetry of the interaction. If the interaction really possesses SU_6 symmetry, the observed multiplets of particles will correspond to irreducible representations of SU_6. The group SU_6 was proposed by Gürsey and Radicati[128] and Sakita.[129] In nuclear physics an analogous extension of the spin–isospin group $(SU_2)_J \times (SU_2)_I$ to SU_4 was made some time ago by Wigner.[130]

Let us consider a quark with spin $\frac{1}{2}$. Its six components $q_A (A = 1 \ldots 6)$ may be denoted by two indices: $A = (\alpha, j)$, one of which, $\alpha = 1, 2$, refers to the spin state, while the other, $j = 1, 2, 3$ refers to the SU_3 component.

The unitary transformations of the quark functions q_A are described by 6 by 6 matrices U:

$$q_A = U_A{}^B q_B, \quad UU^+ = 1. \tag{49}$$

The transformations U connect both spin and unitary components of a quark.

The matrices U form the fundamental representation $\underline{6}$ of the group SU_6. The infinitesimal form of U may be written in the form

$$U = 1 + i\tfrac{1}{2} N_N \omega_N, \quad N_N^+ = N_N, \quad \mathrm{Tr}\, \Lambda_N = 0, \tag{50}$$

where the Hermitian matrices Λ_N, together with $\Lambda_0 = \sqrt{\frac{3}{2}} \cdot \mathbf{1}$, are the basis 6 by 6 matrices. As a consequence of the restriction $\mathrm{Tr}\, \Lambda_N = 0$ for $N \neq 0$, the number of independent matrices Λ_N is equal to 35; $N = 1 \ldots 35$. The matrices Λ_N play in SU_6 the same role as the matrices λ_a in SU_3 and the Pauli matrices τ_k in SU_2.

The matrices Λ_N may be expressed as a direct product of the matrices σ_k, $\sigma_0 = 1$, acting on the spin variables of the quark α, and the matrices λ_a, $\lambda_0 = \sqrt{\frac{3}{2}} \cdot \mathbf{1}$ $(a = 1 \ldots 8)$ acting on the SU_3 index of the quark j. Let us replace N by a double index: $N = (k, a)$. Then

$$\left.\begin{aligned} \Lambda_{0a} &= \sigma_0 \times \lambda_a, \\ \Lambda_{k0} &= \sigma_k \times \lambda_0, \\ \Lambda_{ka} &= \sigma_k \times \lambda_a. \end{aligned}\right\} \tag{51}$$

For example,

$$\Lambda_{k3} = \begin{pmatrix} \sigma_k & \mathbf{0} & \mathbf{0} \\ \mathbf{0} & -\sigma_k & \mathbf{0} \\ \mathbf{0} & \mathbf{0} & \mathbf{0} \end{pmatrix}, \tag{52}$$

where $\mathbf{0}$ denotes the null 2 by 2 matrix. The matrices Λ_N are orthonormal:

$$\mathrm{Tr}\,(\Lambda_N \Lambda_{N'}) = 4\delta_{NN'}. \tag{53}$$

The commutation relations for Λ_N may be obtained using expression (51):

$$[\Lambda_N, \Lambda_M] = 2iF_{NML} \Lambda_L \qquad (N, M, L = 1 \ldots 35); \tag{54}$$

the structure constants F_{NML} will be defined in terms of f_{abc} and ε_{ijk}.

From the explicit form (51) of the matrices Λ_N it follows that there are five diagonal matrices among them: Λ_{03}, Λ_{30}, Λ_{08}, Λ_{38}, Λ_{33}. The matrices Λ_{03} and Λ_{08} are related to the electric charge Q and the quark hypercharge. Y:

$$Q = \frac{1}{2}\Lambda_{03} + \frac{1}{\sqrt{3}}\Lambda_{08}, \quad Y = \frac{1}{\sqrt{3}}\Lambda_{08}, \tag{55}$$

while Λ_{30} is related to the third component of the quark spin: $J_3 = \frac{1}{2}\Lambda_{30}$. Similarly one can show that Λ_{33} characterizes the quark magnetic moment. The eigenvalues of these diagonal matrices uniquely distinguish the various components q_A of the quark.

From (51) and (55) it also follows that the matrices $\frac{1}{2}\Lambda_{0a}$ generate SU_3 transformations on the quark, while the matrices $\frac{1}{2}\Lambda_{k0}$ generate spin rotations. It is sometimes convenient to consider the strange quark spin J_k^S:

$$J_k^S = \frac{1}{\sqrt{3}}\left(\frac{1}{\sqrt{2}}\Lambda_{k0} - \Lambda_{k8}\right). \tag{56}$$

along with the spin $\frac{1}{2}\Lambda_k$.

As in the case of SU_3, the matrices U and U^* in SU_6 are not unitarily equivalent, so that quarks q^A with an upper SU_6 index transform according to the independent conjugate representation $\underline{6}^*$:

$$q'^A = (U^*)^A{}_B\, q^B = q^B(U^+)_B{}^A. \tag{57}$$

Using q^A, one may describe antiquarks.

The irreducible representations of the group SU_6 may be constructed using the tensor method used earlier for SU_2 and SU_3. However, in SU_6 the isotropic tensors are δ_A^B and two antisymmetric tensors of sixth rank ε^{ABCDEF} and ε_{ABCDEF}. As in the case of unitary symmetry, the representations of the group SU_6 with integral charge Q and hypercharge Y are described by functions transforming as the product of $3n$ quarks (n is an integer) and any number of quark–antiquark pairs. The simplest representations will correspond to three quarks qqq and the pair $q\bar{q}$. The product of a quark q_A and an antiquark $\bar{q}^B$

$$q_A\bar{q}^B = \tfrac{1}{6}\delta_A^B q_C\bar{q}^C + (q_A\bar{q}^B - \tfrac{1}{6}\delta_A^B q_C\bar{q}^C) \tag{58}$$

may be decomposed into an SU_6 singlet $q_C\bar{q}^C$ and a $\underline{35}$-plet:

$$\underline{6}\times\underline{6}^* = \underline{1} + \underline{35}. \tag{59}$$

Thus the wave function of the $\underline{35}$-plet is a mixed SU_6 tensor of second rank $\Phi_A{}^B$ with zero trace: $\Phi_A^A = 0$. Φ_A^B transforms according to the regular representation of SU_6: according to (9.49)

$$\Phi' = \mathrm{U}\Phi\mathrm{U}^{-1}. \tag{60}$$

Let $(2J+1, \underline{n})$ denote the SU_3 multiplet $\underline{n}$ containing particles with spin J. The quark $\underline{6}$ and the antiquart $\underline{6}^*$ have spin–unitary spin content:

$$\underline{6} = (\underline{2}, \underline{3}), \quad \underline{6}^* = (\underline{2}, \underline{3}^*). \tag{61}$$

The product $\underline{6}\times\underline{6}^*$ then contains the following particles:

$$\underline{6}\times\underline{6}^* = (\underline{2}, \underline{3})\times(\underline{2}, \underline{3}^*) = (\underline{1}+\underline{3}, \underline{1}+\underline{8}) = (\underline{1}, \underline{1}) + [(\underline{1}, \underline{8}) + (\underline{3}, \underline{1}) + (\underline{3}, \underline{8})] = \underline{1} + \underline{35}, \tag{62}$$

from which it follows that the $\underline{35}$-plet contains a meson octet with spin 0, and a singlet and an octet of mesons with spin 1. All particles of the $\underline{35}$-plet must have negative spatial parity, as the relative parity of a quark and antiquark is negative (since the quarks are of spin $\frac{1}{2}$). An explicit expression for the $\underline{35}$-plet $\Phi_A{}^B$ in terms of the unitary octets $P_i{}^j$ and $P^{\beta j}_{\alpha i}$ and the singlet $V^{0\beta}_{\alpha}$ may be obtained by passing to the spin–unitary spin index $A = (\alpha, i)$:

$$\Phi^{\beta j}_{\alpha i} = \frac{1}{\sqrt{2}}\left[\delta^{\beta}_{\alpha}P^{j}_{i} + \boldsymbol{\sigma}^{\beta}_{\alpha}\cdot\left(V^{j}_{i} + \frac{1}{\sqrt{3}}V^{0}\delta^{j}_{i}\right)\right]. \tag{63}$$

Thus the multiplet $\underline{35}$ contains the mesons π, K, η, ϱ, K*, ω, and φ, where the vector mesons enter naturally as a nonet and not an octet.

From the product of three quarks q_A, q'_B, q''_C one may form the following irreducible representations: (a) a symmetric SU_6 spinor $\Phi_{\{ABC\}}$ with 56 components; (b) an antisymmetric SU_6 spinor $\Phi_{[ABC]}$ with 20 components; (c) a spinor $\Phi_{\{A[BC]\}}$ of mixed symmetry with 70 components.

In the reduction of the product of three quarks the $\underline{70}$-plet occurs twice:

$$\underline{6}\times\underline{6}\times\underline{6} = \underline{20}+\underline{56}+\underline{2}\times\underline{70},$$

The three-quark combinations qqq may have spin $\frac{3}{2}$ or $\frac{1}{2}$. Thus if one assigns baryon number $B_q = \frac{1}{3}$ to the quarks, such combinations can describe the lowest-lying baryons. To combine the lowest baryonic SU_3 multiplets into one SU_6 multiplet, one must choose an SU_6 representation which contains both an octet and a decimet. Now the baryon decimet with spin $\frac{3}{2}$ is contained only in one three-quark combination, corresponding to the representation $\underline{56}$:

$$\underline{56} = (\underline{2}\times\underline{8})+(\underline{4}\times\underline{10}). \tag{64}$$

Aside from the decimet, $\underline{56}$ contains also the baryon octet. Consequently, the representation $\underline{56}$ may be chosen for the baryons. Let us write the $\underline{56}$-plet in terms of the decimet and octet states:

$$\Phi_{\{ABC\}} \equiv \Phi_{\{\alpha i,\, \beta j,\, \gamma k\}} = \Psi_{\{\alpha\beta\gamma\}}D_{\{ijk\}} + \frac{1}{3\sqrt{2}}\left\{\varepsilon_{ijl}B^{l}_{k}C_{\alpha\beta}u_{\gamma} + \varepsilon_{jkl}B^{l}_{i}C_{\beta\gamma}u_{\alpha} + \varepsilon_{kil}B^{l}_{j}C_{\gamma\alpha}u_{\beta}\right\}, \tag{65}$$

Here $\Psi_{\{\alpha\beta\gamma\}}$ and u_α are the spin wave functions for particles with spins $\frac{3}{2}$ and $\frac{1}{2}$, $D_{\{ijk\}}$ is the decimet (Δ, Y^*, Ξ^*, Ω^-) and $B_i{}^j$ is the octet (N, Σ, Ξ, Λ).

Thus hadrons with different spins and SU_3 properties may be combined into meson and baryon SU_6 multiplets. In SU_6 multiplets, all particles have the same parity. The meson octet $J^P = 0^-$ (eight components) and the nonet of vector mesons $J^P = 1^-$ (27 components) form a $\underline{35}$-plet. The baryon octet $J^P = \frac{1}{2}^+$ (16 components) and the decimet $J^P = \frac{3}{2}^+$ (40 components) form the $\underline{56}$-plet; the conjugate multiplet $\underline{56}^*$ consists of the antibaryons. Orbitally excited multiplets (see § 10.4) also exist: for example, there is evidence for multiplets of $\underline{35}$, $L = 1$ mesons ($J^P = 0^+, 1^+, 2^+$); $\underline{35}$, $L = 2$ mesons ($J^P = 1^-, 2^-, 3^-$); $\underline{70}$, $L = 1$ baryons ($J^P = \frac{1}{2}^-, \frac{3}{2}^-, \frac{5}{2}^-$); $\underline{56}$, $L = 2$ baryons ($J^P = \frac{1}{2}^+, \frac{3}{2}^+, \frac{5}{2}^+, \frac{7}{2}^+$); and possibly some others. While none of these multiplets is completely filled, the missing states are predominantly hyperons or $I = Y = 0$ mesons, both of which are more difficult to detect than nonstrange baryons or $I = 1$, $Y = 0$ mesons. Candidates for all of these latter two types have, in fact, been found for the multiplets just mentioned.

Some recent developments†

The quarks just used to construct hadrons and those by which current algebra is realized (§ 15.2) cannot be the same.(131) A transformation between the two types of quark has recently been constructed by Melosh,(132) with many interesting consequences.(132) Among these we may note: (a) a new discussion of the magnetic moment ratio $\mu_p/\mu_n = -\frac{3}{2}$, one of the early results of SU_6,(133) (b) a relaxation of the (wrong) result $|G_A/G_V| = \frac{5}{3}$ usually obtained in the quark model (see §§ 15.3, 15.4: experimentally, $|G_A/G_V| = 1.25$), and (c) a new understanding of the decays and photoproduction of resonances. The reader is referred to the literature(132) for details.

† [*Translator's note:* This brief guide to recent references replaces a subsection in the original text on magnetic moments.]

PART IV

ELEMENTS OF DYNAMICAL THEORY

AS WE showed in Part II, relativistic kinematics and reflection invariance allow one to write the scattering amplitude in terms of several invariant amplitudes which characterize the dynamics of the process. In quantum field theory these amplitudes could be calculated in principle if we knew the explicit form of the interaction between fields, could restrict ourselves to a finite number of fields, and could avoid using perturbation theory. However, we still do not know the equations of motion in the presence of interactions or how to solve them for particles. Moreover, any theory of interactions based on a fixed set of interacting particles is probably incorrect. The interactions of particles are closely linked with the question of which particles can exist. Because of the large number of interactions, the spectrum of particles and the description of their interactions are really two different aspects of the same problem.

There are two approaches to the problem of dynamics. In one of them—the group theoretic approach—the aim is to find the dynamical group which would give both the level scheme of the system and the degeneracy of each level. In Part III we considered internal symmetry groups which, in the exact symmetry limit, describe the degeneracy of levels. This role is played, to some extent, by the nonrelativistic quark model and the corresponding $U(6)\times U(6)\times O(3)$ symmetry.

Another way to deal with dynamics is by dispersion methods. This approach relies on general principles of quantum field theory and reasonable assumptions. In Part IV we shall consider the fundamentals of the dispersion approach and related phenomenological models. The last chapter deals with the algebra of currents, which is a local generalization of the group-theoretic considerations of Part III. The applications of current algebra make great use of concepts of the dispersion approach.

CHAPTER 11

THE S-MATRIX, CURRENTS, AND CROSSING SYMMETRY

IN THIS chapter we shall study interpolating fields, currents, and properties of crossing-symmetric amplitudes.

By introducing interpolating fields one may formulate the principle of microcausality in terms of a locality condition for these fields. In a Lagrangian formalism, the interpolating fields $\Phi(x)$ should satisfy equations of motion for interacting fields. We shall examine their local values, which have the same Lorentz transformation and internal symmetry properties as the asymptotic fields. The matrix elements of the interpolating fields are directly related to observable quantities. They may be connected with elements of the S-matrix using a reduction formula.

The local quantities having direct dynamical meaning are currents. In the Lagrangian formalism, currents are constructed explicitly from fields. We shall examine currents (similarly to fields) as local quantities characterized by their transformation properties (which are those of the corresponding interpolating fields).

Interpolating fields and currents are a convenient basis in which to derive crossing symmetry. As a consequence of this general property of the amplitude, the same function of momenta describes the amplitude for different (crossing-conjugate) processes. This property restricts the range of independent dynamical quantities in a theory. The invariant amplitudes F_l (see § 7.5), as functions of the momentum variables s, t, u, will now refer to all crossing-conjugate processes. For readers in need of specifics, we also indicate in this chapter how to construct crossing matrices for the groups SU_2 and SU_3.

§ 11.1. Interpolating fields, currents, and the reduction formula

The scattering matrix is a unitary operator connecting the asymptotic in- and out-states: $|\,\text{in}\rangle = S\,|\,\text{out}\rangle$, which describe the initial and final configurations of particles under scattering. The S-matrix contains all the information about the interactions of particles.

Asymptotic states may be constructed using creation operators applied to the vacuum (see § 4.5). For example, in the case of the scattering $1+2 \to 3+4$ of particles with spin J^r and masses m_r ($r = 1, 2, 3, 4$), we may write the S-matrix element as

$$\langle \boldsymbol{p}_3, \sigma_3; \boldsymbol{p}_4, \sigma_4\,|\,S\,|\,\boldsymbol{p}_1, \sigma_1; \boldsymbol{p}_2, \sigma_2\rangle = \langle 0\,|\,c_{3\,\text{out}}(\boldsymbol{p}_3, \sigma_3)\,c_{4\,\text{out}}(\boldsymbol{p}_4, \sigma_4)\,c^{+}_{1\,\text{in}}(\boldsymbol{p}_1, \sigma_1)\,c^{+}_{2\,\text{in}}(\boldsymbol{p}_2, \sigma_2)\,|\,0\rangle, \quad (1)$$

where $c^+_{r\,\substack{\text{in}\\\text{out}}}(\boldsymbol{p}, \sigma)$ and $c_{r\,\substack{\text{in}\\\text{out}}}(\boldsymbol{p}, \sigma)$ are the creation and annihilation operators for the rth particle: $p_r^2 = m_r^2$.

Let us pass from creation and annihilation operators to local in- and out-fields $\Phi_{\substack{\text{in}\\\text{out}}}(x)$. These fields satisfy the equations of motion for free fields. They are connected with one another by the same relation as the in- and out- creation and annihilation operators:

$$\Phi_{\text{in}}(x) = S\Phi_{\text{out}}(x)S^{-1}$$

Expressions for free fields in terms of creation and annihilation operators were obtained in Chapter 5 for arbitrary spin and for various higher spin wave functions. The formulae of Chapter 5 also allow one to find the creation and annihilation operators.

We shall limit the discussion to the fields $\Phi_{\substack{\text{in}\\\text{out}}}(x)$ and $\psi_{\substack{\text{in}\\\text{out}}}(x)$ for particles with spins $J = 0$ and $J = \frac{1}{2}$. The creation and annihilation operators for neutral spinless particles are related to fields by

$$\left.\begin{aligned} c^+_{\substack{\text{in}\\\text{out}}}(k) &= \frac{i}{(2\pi)^{3/2}} \int d^3x \Phi_{\substack{\text{in}\\\text{out}}}(x)\, \overleftrightarrow{\partial}_0 e^{ikx}, \\ c_{\substack{\text{in}\\\text{out}}}(k) &= \frac{i}{(2\pi)^{3/2}} \int d^3x \Phi_{\substack{\text{in}\\\text{out}}}(x)\, \overleftrightarrow{\partial}_0 e^{-ikx}, \end{aligned}\right\} \tag{2}$$

where $f_1 \overleftrightarrow{\partial} f_2 = f_1(\partial f_2) - (\partial f_1) f_2$. For a particle with spin $\frac{1}{2}$, the creation operators a^+, b^+ and annihilation operators a, b are

$$\left.\begin{aligned} a^+_{\substack{\text{in}\\\text{out}}}(\boldsymbol{p}, \sigma) &= \frac{1}{(2\pi)^{3/2}} \int d^3x \bar{\psi}_{\substack{\text{in}\\\text{out}}}(x)\, \gamma_4 u(\boldsymbol{p}, \sigma) e^{-ipx}, \\ b^+_{\substack{\text{in}\\\text{out}}}(\boldsymbol{p}, \sigma) &= \frac{1}{(2\pi)^{3/2}} \int d^3x \bar{v}(\boldsymbol{p}, \sigma)\, \gamma_4 \psi_{\substack{\text{in}\\\text{out}}}(x) e^{-ipx}, \\ a_{\substack{\text{in}\\\text{out}}}(\boldsymbol{p}, \sigma) &= \frac{1}{(2\pi)^{3/2}} \int d^3x \bar{u}(\boldsymbol{p}, \sigma)\, \gamma_4 \psi_{\substack{\text{in}\\\text{out}}}(x) e^{ipx}, \\ b_{\substack{\text{in}\\\text{out}}}(\boldsymbol{p}, \sigma) &= \frac{1}{(2\pi)^{3/2}} \int d^3x \bar{\psi}_{\substack{\text{in}\\\text{out}}}(x)\, \gamma_4 v(\boldsymbol{p}, \sigma) e^{ipx}, \end{aligned}\right\} \tag{3}$$

where u and v are the Dirac wave functions (see § 5.2). Equations (2) may be verified by inserting the explicit expressions for $\Phi_{\substack{\text{in}\\\text{out}}}(x)$ (e.g., from (4.93) with $a = b = c$). When checking (3) using (5.24) and (5.28), one must use the orthogonality condition $\bar{v}(\boldsymbol{p}, \sigma)\gamma_4 \times u(-\boldsymbol{p}, \sigma) = 0$.

Let us introduce the interpolating local fields $\Phi(x)$ and $\psi(x)$. The locality of an interpolating field means, first of all, that under transformations of the Poincaré group it transforms in the same way as the corresponding in- and out-fields; for example:

$$\left.\begin{aligned} e^{iPa}\psi(x)e^{-iPa} &= \psi(x+a), \\ \mathrm{U}(0, \Lambda)\, \psi(x)\, \mathrm{U}^{-1}(0, \Lambda) &= S(\Lambda^{-1})\, \psi(\Lambda(\Lambda)x), \end{aligned}\right\} \tag{4}$$

and, secondly, that the causality condition holds in the form

$$\{\psi(x), \psi^+(y)\} = 0, \quad [\Phi(x), \Phi(y)] = 0 \tag{5}$$

for $(x-y)^2 < 0$. In (4), $S(A)$ is the Lorentz transformation matrix (5.22) for the Dirac field.

It is assumed that the interpolating fields $\Phi(x)$ and $\psi(x)$ have the same internal symmetry properties as the asymptotic fields $\Phi_{\substack{\text{in}\\ \text{out}}}(x)$ and $\psi_{\substack{\text{in}\\ \text{out}}}(x)$. The interpolating fields are taken to describe the interaction in the scattering process, even when the Lagrangian is not written down explicitly.

Let us consider a scattering process as a function of time and identify the asymptotic in-state with the state of the system for $x_0 \to -\infty$, and the out-state with the state for $x_0 \to +\infty$. The fields $\Phi_{\substack{\text{in}\\ \text{out}}}(x)$ are then limits of the field $\Phi(x)$ for $x_0 \to -\infty(+\infty)$. These limits are taken in the weak convergence sense, i.e. in the sense of asymptotic equality of matrix elements of $\Phi(x)$ and $\Phi_{\substack{\text{in}\\ \text{out}}}(x)$ on the mass shell $k^2 = m^2$:

$$\left.\begin{aligned} \lim_{x_0 \to +\infty} \int d^3x \langle A \,|\, \Phi(x) - \Phi_{\text{out}}(x) \,|\, B\rangle \overleftrightarrow{\partial}_0 e^{\pm ikx} &= 0, \\ \lim_{x_0 \to -\infty} \int d^3x \langle A \,|\, \Phi(x) - \Phi_{\text{in}}(x) \,|\, B\rangle \overleftrightarrow{\partial}_0 e^{\pm ikx} &= 0. \end{aligned}\right\} \tag{6}$$

Here $|A\rangle$ and $|B\rangle$ are normalized states (i.e. wave packets). The conditions (6) apply to all matrix elements of Φ_{in} and Φ_{out}. For example, when the upper sign is chosen, the first of eqns. (6) is equivalent to the condition

$$\lim_{x_0 \to +\infty} \langle A \,|\, c^+(k, x_0) - c^+_{\text{out}}(k) \,|\, B\rangle = 0,$$

where the operator $c^+(k, x_0)$ is obtained from the interpolating field $\Phi(x)$ by the same rule (2) used to obtain the creation operator $c^+_{\text{out}}(k)$ from $\Phi_{\text{out}}(x)$. One may interpret the remaining cases in (6) analogously.

The condition corresponding to (6) for the spinor fields $\psi(x), \bar{\psi}(x)$ and $\psi_{\substack{\text{in}\\ \text{out}}}(x)$, $\bar{\psi}_{\substack{\text{in}\\ \text{out}}}(x)$ must, according to (3), be written in the form

$$\left.\begin{aligned} \lim_{x_0 \to \pm\infty} \int d^3x \langle A' \,|\, \bar{\psi}(x) - \bar{\psi}_{\substack{\text{out}\\ \text{in}}}(x) \,|\, B'\rangle \gamma_4 u(\boldsymbol{p}, \sigma) e^{-ipx} &= 0, \\ \lim_{x_0 \to \pm\infty} \int d_3x \langle A' \,|\, \bar{\psi}(x) - \bar{\psi}_{\substack{\text{out}\\ \text{in}}}(x) \,|\, B'\rangle \gamma_4 v(\boldsymbol{p}, \sigma) e^{ipx} &= 0, \\ \lim_{x_0 \to \pm\infty} \int d^3x \bar{u}(\boldsymbol{p}, \sigma) \gamma_4 \langle A' \,|\, \psi(x) - \psi_{\substack{\text{out}\\ \text{in}}}(x) \,|\, B'\rangle e^{ipx} &= 0, \\ \lim_{x_0 \to \pm\infty} \int d^3x \bar{v}(\boldsymbol{p}, \sigma) \gamma_4 \langle A' \,|\, \psi(x) - \psi_{\substack{\text{out}\\ \text{in}}}(x) \,|\, B'\rangle e^{-ipx} &= 0. \end{aligned}\right\} \tag{7}$$

Definitions (6) and (7) have meaning only for stable particles, since unstable particles do not have asymptotic in- and out-fields (see § 2.3). However, if one neglects the interaction giving rise to the decay, these formulae also may be used for unstable particles. The stability of a single-particle state may be expressed in the form of the condition

$$|\boldsymbol{p}, \sigma; \text{in}\rangle = |\boldsymbol{p}, \sigma; \text{out}\rangle = |\boldsymbol{p}, \sigma\rangle. \tag{8}$$

We shall normalize interpolating fields in the same way as asymptotic fields—via the matrix element of the interpolating field between the vacuum $|0\rangle$ and a single-particle state. In view of the stability of the single-particle state, this matrix element is the same as for the free field. Hence

$$\left.\begin{aligned}\langle 0|\Phi(0)|k\rangle &= \langle 0|\Phi_{\substack{\text{in}\\\text{out}}}(0)|k\rangle = \frac{1}{(2\pi)^{3/2}},\\ \langle 0|\psi(0)|\boldsymbol{p},\sigma\rangle &= \langle 0|\psi_{\substack{\text{in}\\\text{out}}}(0)|\boldsymbol{p},\sigma\rangle = \frac{1}{(2\pi)^{3/2}}u(\boldsymbol{p},\sigma).\end{aligned}\right\} \tag{9}$$

Interpolating fields may not satisfy the same equations as free fields. We thus introduce scalar and spinor currents:

$$\left.\begin{aligned}J(x) &= -(\partial^2+m^2)\Phi(x) \equiv K(x)\Phi(x),\\ \eta(x) &= (i\gamma^\mu\partial_\mu-\varkappa)\psi(x) \equiv \vec{\mathcal{D}}(x)\psi(x),\\ \bar{\eta}(x) &= -(i\,\partial_\mu\bar{\psi}(x)\gamma^\mu+\varkappa\bar{\psi}(x)) \equiv \bar{\psi}(x)\overleftarrow{\mathcal{D}}(x).\end{aligned}\right\} \tag{10}$$

The matrix element of the current between the vacuum and a single-particle state (corresponding to the interpolating field) vanishes:

$$\langle 0|J(x)|k\rangle = 0, \quad \langle 0|\eta(x)|\boldsymbol{p},\sigma\rangle = 0, \tag{11}$$

since single-particle states describe free particles.

As a consequence of the locality of the fields, the currents are also local. They have a local transformation law analogous to (4) and commute or anticommute for space-like intervals:

$$[J(x), J(y)] = \{\eta(x), \bar{\eta}(y)\} = 0 \quad \text{when} \quad (x-y)^2 < 0. \tag{12}$$

We shall not consider equal-time commutators of currents at present.

Let us now derive an expression connecting the scattering matrix with the matrix elements of currents. In the scattering process $1+2 \to 3+4$ let particle 2 have spin 0. Then the matrix elements of $S-1$, according to (1), (2), and (6), is equal to

$$\begin{aligned}&\langle 3,4,\text{out}|c^+_{\text{in}}(\boldsymbol{p}_2)|1\rangle - \langle 3,4,\text{out}|c^+_{\text{out}}(\boldsymbol{p}_2)|1\rangle\\ &\quad = -\frac{i}{(2\pi)^{3/2}}\int d^3x\langle 3,4,\text{out}|\Phi(x)-\Phi_{\text{out}}(x)|1\rangle\overleftrightarrow{\partial}_0 e^{-ip_2x}\\ &\quad = -\frac{i}{(2\pi)^{3/2}}\int d^4x\langle 3,4,\text{out}|J(x)|1\rangle e^{-ip_2x}.\end{aligned} \tag{13}$$

In passing to the final formula in (13) we use the mass shell condition $p^2 = m^2$, the condition (8), the definition (10), and the fact that integrals over space-like surfaces must vanish for normalized states (i.e. for wave packets).

In expression (13) one may separate out a δ-function:

$$\langle 3,4|S-1|1,2\rangle = -\frac{i(2\pi)^4}{(2\pi)^{3/2}}\delta^4(p_1+p_2-p_3-p_4)\langle 3,4,\text{out}|J(0)|1\rangle,$$

so that the T-matrix element is equal to

$$\langle 3,4|T|1,2\rangle = -\frac{1}{(2\pi)^{3/2}}\langle 3,4,\text{out}|J(0)|1\rangle. \tag{14}$$

Analogously, one may obtain a formula for the case that particle 1 has spin $\frac{1}{2}$:

$$\langle 3, 4 | T | 1, 2 \rangle = -\frac{i}{(2\pi)^{3/2}} \langle 3, 4, \text{out} | \bar{\eta}(0) | 2 \rangle. \tag{15}$$

Let us return to (13) and repeat the operation of replacing the creation or annhilation operator by an interpolating field and then by a current. Let particles 2 and 4 be identical and have spin $J = 0$. We then find using (2) and (6) that

$$\langle 3, 4, \text{out} | J(0) | 1 \rangle = \langle 3 | c_{\text{out}}(p_4) J(0) | 1 \rangle = -\frac{i}{(2\pi)^{3/2}} \int d^4y e^{ip_4y} K(y) \langle 3 | T(\Phi(y) J(0)) | 1 \rangle, \tag{16}$$

since, by virtue of (11), the term with $c_{\text{in}}(p_4) | 1 \rangle$ does not contribute to (16), even when particles 1 and 4 are identical.

The symbol $T(A(x_0) B(y_0))$ introduced in formula (16) denotes the time-ordered product (or T-product) of fields:

$$T(A(x_0) B(y_0)) = \begin{cases} A(x_0) B(y_0) & \text{for} \quad x_0 > y_0, \\ \pm B(y_0) A(x_0) & \text{for} \quad y_0 > x_0, \end{cases} \tag{17}$$

and the sign $\pm$ distinguishes boson fields (+) from fermion fields (−). The matrix element of the operator $S-1$ may now be written in a form symmetric with respect to $\Phi(x)$ and $\Phi(y)$:

$$\langle 3, 4 | S-1 | 1, 2 \rangle = -\frac{1}{(2\pi)^{3/2}} \int d^4x \, d^4y e^{-ip_2x+ip_4y} \vec{K}(x) \langle 3 | T(\Phi(x)\Phi(y)) | 1 \rangle \overleftarrow{K}(y). \tag{18}$$

Formulae (13), (15), and (18) are examples of the Lehmann–Symanzik–Zimmermann reduction formula.[134] If one applies these formulae consecutively to all creation and annihilation operators in the matrix element $\langle 3, 4, \text{out} | 1, 2, \text{in} \rangle$, the quantity $\langle 3, 4 | S-1 | 1, 2 \rangle$ may be expressed in terms of the vacuum expectation value of T-products of interpolating fields.

If all particles are identical and spinless, but are in different states, one finds without trouble that

$$\begin{aligned} \langle 3, 4 | S | 1, 2 \rangle = & \int d^4x_1 \ldots d^4x_4 f^*_{p_3}(x_3) f^*_{p_4}(x_4) \vec{K}(x_3) \vec{K}(x_4) \\ & \times \langle 0 | T(\Phi(x_1)\Phi(x_2)\Phi(x_3)\Phi(x_4)) | 0 \rangle \overleftarrow{K}(x_1) \overleftarrow{K}(x_2) f_{p_1}(x_1) f_{p_2}(x_2), \end{aligned} \tag{19}$$

where $\vec{K}$ acts to the right and $\overleftarrow{K}$ acts to the left. The wave function $f_p(x)$ of a particle with momentum p is

$$f_p(x) = \frac{1}{(2\pi)^{3/2}} e^{-ipx}.$$

Formula (19) has the form that would follow if each creation operator $c^+_{\text{in}}(p)$ and annihilation operator $c_{\text{out}}(p)$ were replaced by

$$\left.\begin{aligned} c^+_{\text{in}}(p) & \sim -i \int d^4x \Phi(x) \overleftarrow{K}(x) f_p(x), \\ c_{\text{out}}(p) & \sim -i \int d^4x f^*_p(x) \vec{K}(x) \Phi(x) \end{aligned}\right\} \tag{20}$$

with differentiation performed after calculating the expectation value of the T-product of the operators $\Phi(x)$.

For scattering of fermions one must fix the order of fermion operators from the beginning, since it affects the sign of the expression. We shall write the fermion operators in the same order as they appear in the matrix element: $\langle 3, 4, \text{out}| = \langle 0| a_{\text{out}}(3)\, a_{\text{out}}(4)$ and $|1, 2, \text{in}\rangle = a_{\text{in}}^{+}(1)\, a_{\text{in}}^{+}(2)|0\rangle$.

When all particles are identical and have spin $\frac{1}{2}$ the scattering matrix is

$$\begin{aligned}\langle 3, 4|S|1, 2\rangle = \int d^4x_1 \ldots d^4x_4 \bar{u}(\boldsymbol{p}_3, \sigma_3)\, \bar{u}(\boldsymbol{p}_4, \sigma_4)\, \vec{\mathcal{D}}(x_3)\, \vec{\mathcal{D}}(x_4) \\ \times \langle 0|T(\psi(x_3)\, \psi(x_4)\, \bar{\psi}(x_1)\, \bar{\psi}(x_2))|0\rangle\, \overleftarrow{\mathcal{D}}(x_1)\, \overleftarrow{\mathcal{D}}(x_2)\, u(\boldsymbol{p}_1, \sigma_1)\, u(\boldsymbol{p}_2, \sigma_2) \\ \times \frac{1}{(2\pi)^6} \exp\{-i(p_1x_1 + p_2x_2 - p_3x_3 - p_4x_4)\}.\end{aligned} \tag{21}$$

Equation (21) has the same form as if each annihilation operator a_{out} were replaced by

$$a_{\text{out}}(\boldsymbol{p}, \sigma) \sim -\frac{i}{(2\pi)^{3/2}} \int d^4x \bar{u}(\boldsymbol{p}, \sigma)\, e^{ipx} \vec{\mathcal{D}}(x)\, \psi(x), \tag{22}$$

and each creation operator a_{in}^{+} by

$$a_{\text{in}}^{+}(\boldsymbol{p}, \sigma) \sim -\frac{i}{(2\pi)^{3/2}} \int d^4x \bar{\psi}(x)\, \overleftarrow{\mathcal{D}}(x)\, u(\boldsymbol{p}, \sigma) e^{-ipx}. \tag{23}$$

In the case of antiparticles, the rule changes in form. Repeating the derivation, but substituting b, b^{+} for the original operators a, a^{+}, we find that for antiparticles each annihilation operator must be represented in (22) by the expression

$$b_{\text{out}}(\boldsymbol{p}, \sigma) \sim \frac{i}{(2\pi)^3} \int d^4x \bar{\psi}(x)\, \overleftarrow{\mathcal{D}}(x)\, v(\boldsymbol{p}, \sigma) e^{ipx}, \tag{24}$$

and each creation operator by the expression

$$b_{\text{in}}^{+}(\boldsymbol{p}, \sigma) \sim \frac{i}{(2\pi)^{3/2}} \int d^4x \bar{v}(\boldsymbol{p}, \sigma)\, e^{-ipx} \vec{\mathcal{D}}(x)\, \psi(x). \tag{25}$$

It is assumed here that the operators $\mathcal{D}(x)$ are applied after calculating the T-product of the operators ψ and $\bar{\psi}$.

Analogously it is not hard to show that when all particles are identical and have spin 1 one must replace the creation and annihilation operators according to the rule

$$\left.\begin{aligned} c_{\text{in}}^{+}(\boldsymbol{p}, \lambda) &\sim \frac{i}{(2\pi)^{3/2}} \int d^4x e_{\mu}(\boldsymbol{p}, \lambda)\, e^{ipx} \vec{K}(x)\, B^{\mu}(x), \\ c_{\text{out}}(\boldsymbol{p}, \lambda) &\sim -\frac{i}{(2\pi)^{3/2}} \int d^4x B_{\mu}^{+}(x)\, \overleftarrow{K}(x)\, e^{\mu}(\boldsymbol{p}, \lambda) e^{-ipx}, \end{aligned}\right\} \tag{26}$$

where $B^{\mu}(x)$ is an interpolating vector field, and $e_{\mu}(\boldsymbol{p}, \lambda)$ is a wave function for a particle of spin 1 (see § 5.4).

As in the case of particles with spin 0 and $\frac{1}{2}$, one must first calculate the T-product of the operators B_μ, B_ν^+, after which one performs the differentiation $K(x)$ and finds the vacuum expectation value. The reduction formulae (19) and (21) for the scattering process $1+2 \to 3+4$ are easily generalized to the case of any number of particles of different types (if one assumes that identical particles are in different states). In the general case, one must use the substitution rules (20), (22)–(26), performing the differentiation operations $K(x)$ and $\mathcal{D}(x)$ after calculating the T-product of all interpolating field operators. One then calculates the vacuum expectation value of the T-product.

Since the wave functions $f_p(x)$, $u(\boldsymbol{p}, \sigma)$, and $e_\mu(\boldsymbol{p}, \lambda)$ are known, all dynamical information is contained in the vacuum expectation values of the products of interpolating field operators. The differentiation operators $K(x)$ and $\mathcal{D}(x)$ lead us to currents and equal-time commutators.

We note that in the two-particle scattering case $1+2 \to 3+4$, the T-product in (18) may be reexpressed in terms of the retarded commutator $\vartheta(x-y)\,[\Phi(x), \Phi(y)]$, where $\vartheta(x)$ is the step function: $\vartheta(x) = 1$ for $x_0 > 0$; $\vartheta(x) = 0$ for $x_0 < 0$. Then

$$\langle 3, 4 | S-1 | 1, 2\rangle = -\frac{1}{(2\pi)^3}\int d^4x\, d^4y e^{-ip_2x+ip_4y}$$

$$\times \vec{K}(x) \langle 3 | \vartheta(x-y)\,[\Phi(x), \Phi(y)] | 1\rangle \overleftarrow{K}(y). \tag{27}$$

We shall leave it to the reader to verify that the difference between expressions (27) and (18) really vanishes if the momenta p_2 and p_4 are in the physical region.

§ 11.2. Crossing symmetry

Let us consider, along with the reaction

$$a+b \to c+d \qquad (s\text{-channel}) \tag{28}$$

two other crossed reactions obtained from (28) by replacing a particle in the initial (final) state by an antiparticle in the final (initial) state:

$$a+\bar{c} \to \bar{b}+d \qquad (t\text{-channel}), \tag{29}$$

$$a+\bar{d} \to c+\bar{b} \qquad (u\text{-channel}). \tag{30}$$

In our study of the kinematics of the S-matrix (§ 7.2) it was mentioned that in the s-, t-, and u-channels the square of the energy in the c.m.s. was given respectively by the invariant variables

$$s = (p_a+p_b)^2, \quad t = (p_a-p_c)^2, \quad u = (p_a-p_d)^2. \tag{31}$$

When passing from the s-channel reaction (28) to the crossed t-channel reaction (29), the momenta of particles p_b and p_c with $p_{b0}, p_{c0} > 0$ are replaced by the antiparticle momenta $-p_c$ and $-p_b$ with $p_{c0}, p_{b0} < 0$. The crossing transformation from (28) to (29) must be distinguished from the total reflection θ, under which all particles are replaced by antiparticles and the order of the process reversed: $\bar{c}+\bar{d} \to \bar{a}+\bar{b}$.

Let all particles a, b, c, and d be charged bosons with spin 0 (e.g., $\pi^+ + K^- \rightarrow \pi^+ + K^-$. Let us denote the π- and K-fields by $\pi(x)$ and $\Phi(x)$. Consider the function

$$\mathcal{M}(p_a \ldots p_d) = \int d^4x_1 \ldots d^4x_4 e^{i(p_a x_1 + p_b x_2 - p_c x_3 - p_d x_4)} \left(-\frac{i}{(2\pi)^6}\right)$$
$$\times \{\vec{K}(x_1)\,\vec{K}(x_2)\langle 0 | T(\pi^+(x_1)\Phi(x_2)\pi(x_3)\Phi^+(x_4)) | 0\rangle \overleftarrow{K}(x_3)\,\overleftarrow{K}(x_4)\}$$
$$\equiv (2\pi)^4\,\delta(p_a + p_b - p_c - p_d)\,F(s, t, u), \tag{32}$$

assuming that it is defined for both signs of the time component p_{i0} of each of the four-vectors p_a, p_b, p_c, p_d. Then, according to (20), for momenta in the physical s-channel region ($p_i^2 = m_i^2$; $p_i^0 > 0$) this function is proportional to the amplitude for the process (28), or

$$\pi^+ + K^- \rightarrow \pi^+ + K^-.$$

For momenta in the physical t-channel region ($p_i^2 = m_i^2$; $p_{a0}, p_{d0} > 0$; $p_{b0}, p_{c0} < 0$), the function (32) is proportional to the amplitude for the process (29), or

$$\pi^+ + \pi^- \rightarrow K^+ + K^-,$$

and for momenta in the physical u-channel region ($p_i^2 = m_i^2$; $p_{a0}, p_{c0} > 0$, $p_{b0}, p_{d0} < 0$), the function (32) is proportional to the amplitude for the process (30), or

$$\pi^+ + K^+ \rightarrow \pi^+ + K^+.$$

Moreover, changing the sign of all time components for each of the three cases, we obtain three more physical regions referring to the three processes connected with those listed above by total reflection θ:

$$\pi^- + K^+ \rightarrow \pi^- + K^+,$$
$$K^- + K^+ \rightarrow \pi^- + \pi^+,$$
$$\pi^- + K^- \rightarrow \pi^- + K^-.$$

Of course, the physical regions in the (s, t, u)-plane for reactions related by θ are identical.

Thus the expression (32) describes six different processes depending on the range of the variables p_{i0}. When C-invariance holds (i.e. when one neglects the weak interactions), expression (32) will then describe 12 processes.

Consequently, (32), obtained using a local theory for $\mathcal{M}(p_a \ldots p_d)$, is consistent with its being an analytic function of momenta in a connected region of complex momentum space that includes all physical regions of crossed processes. This analyticity property has been proven in perturbation theory.[135] The function $\mathcal{M}(p_a \ldots p_d)$ is an analytic amplitude which will be called simply "the amplitude" in what follows.

For scalar particles the analytic amplitude (32) depends on momenta through the invariant variables s, t, u connected by the relation $s+t+u = 2m_\pi^2 + 2m_K^2$, and is the invariant amplitude in the sense of § 7.5:

$$\mathcal{M}(p_a \ldots p_d) = F(s, t, u).$$

Let us denote the invariant amplitudes in the s-, t-, and u- channels by $F^s(s, t, u)$, $F^t(t, u, s)$, and $F^u(u, s, t)$, always writing the square of the energy in the c.m.s. as the first variable and the momentum transfer as the second. These amplitudes are values of the same function —the analytic amplitude $F(s, t, u)$—in the physical regions of the respective channels:

$$\begin{aligned} F^s(s, t, u) &= F(s, t, u) && \text{in the } s\text{-channel,} \\ F^t(t, s, u) &= F(s, t, u) && \text{in the } t\text{-channel,} \\ F^u(u, t, s) &= F(s, t, u) && \text{in the } u\text{-channel.} \end{aligned}$$

On the other hand, if we know the amplitude in the s-channel $F^s(s, t, u)$, by performing an analytic continuation with respect to p_{b0} and p_{c0} into the region $p_{b0}, p_{c0} < 0$, we obtain the amplitude for the process in the t-channel, and analytically continuing into the region p_{b0}, $p_{d0} < 0$, we obtain the amplitude in the u-channel.

If particles 1, 2, 3, and 4 are baryons with spin $\frac{1}{2}$, we start with (21), which is written in the s-channel as

$$\langle 3, 4 | S-1 | 1, 2\rangle = \bar{u}^\alpha(\boldsymbol{p}_3, \sigma_3)\, \bar{u}^\beta(\boldsymbol{p}_4, \sigma_4)\, \mathcal{M}^{\gamma\delta}_{\alpha\beta}(p_1 \ldots p_4)\, u_\gamma(\boldsymbol{p}_1, \sigma_1)\, u_\delta(\boldsymbol{p}_2, \sigma_2),$$

where we have separated out the spinor amplitude matrix [see (7.70)]

$$\begin{aligned} \mathcal{M}^{\gamma\delta}_{\alpha\beta}(p_1 \ldots p_4) = \frac{1}{(2\pi)^6} \int d^4x_1 \ldots d^4(x_4)\, \vec{\mathcal{D}}^{\alpha'}_{\alpha}(x_3)\, \vec{\mathcal{D}}^{\beta'}_{\beta}(x_4) \\ \times \langle 0 | T(\psi_{\alpha'}(x_3)\, \psi_{\beta'}(x_4)\, \bar{\psi}^{\gamma'}(x_1)\, \bar{\psi}^{\delta'}(x_2)) | 0\rangle \\ \times \vec{\mathcal{D}}^{\gamma}_{\gamma'}(x_1)\, \vec{\mathcal{D}}^{\delta}_{\delta'}(x_2) \exp\{-i(p_1x_1 + p_2x_2 - p_3x_3 - p_4x_4)\}. \end{aligned} \tag{33}$$

One may easily verify that if the reaction in the s-channel is proton–proton scattering:

$$\mathrm{p}+\mathrm{p} \to \mathrm{p}+\mathrm{p},$$

then the matrix $\mathcal{M}^{\gamma\delta}_{\alpha\beta}(p_1 \ldots p_4)$ for $p_{20}, p_{30} < 0$, contracted with the Dirac spinors $\bar{v}^\alpha(-\boldsymbol{p}_3, \sigma_3)$, $\bar{u}^\beta(\boldsymbol{p}_4, \sigma_4)$, $v_\delta(-\boldsymbol{p}_2, \sigma_2)$, and $u_\gamma(\boldsymbol{p}_1, \sigma_1)$, will describe the crossing-conjugate proton–antiproton reaction in the t-channel:†

$$\mathrm{p}+\bar{\mathrm{p}} \to \bar{\mathrm{p}}+\mathrm{p}.$$

Here the momenta of the physical particles will be $p_1, -p_2, -p_3, p_4$.

The u- and t-channel reactions coincide in this case, and the amplitude in the u-channel may be obtained from the same spinor amplitude (33) contracted with the Dirac spinors $\bar{v}^\beta(-\boldsymbol{p}_4, \sigma_4)$, $\bar{u}^\alpha(\boldsymbol{p}_3, \sigma_3)$, $v_\delta(-\boldsymbol{p}_2, \sigma_2)$, and $u_\gamma(\boldsymbol{p}_1, \sigma_1)$. Thus all of these crossed (and also θ- and C-conjugate) reactions are characterized by one spinor analytic amplitude (33).

If particles 1, 2, 3, and 4 have arbitrary spins, then relying on the formulae of Chapter 5 for localized fields and wave functions with arbitrary spin, we may easily construct an analogous expression of the type (33). If one chooses the wave functions $u_r(\boldsymbol{p}_r, \sigma_r)$ $(r = 1, 2, 3, 4)$, to describe particles, it is obvious that such an expression will define the spinor amplitude $\mathcal{M}^{r_1r_2}_{r_3r_4}(p_1 \ldots p_4)$ introduced in § 7.5. Concrete formulae of the type (32) connecting a spinor amplitude with a vacuum expectation value of interpolating fields were really necessary only to illustrate the analyticity of the amplitude. Thus we shall consider the spinor amplitude $\mathcal{M}^{r_1r_2}_{r_3r_4}$ as analytic without appealing to concrete formulae.

† In reactions with baryons the channels are usually designated so that t-channel is bosonic, i.e. contains a baryon and an antibaryon in the initial and final states.

The expansion of the spinor amplitude $\mathcal{M}^{r_1r_2}_{r_3r_4}(p_1 \ldots p_4)$ in terms of invariant amplitudes $F(s, t, u)$ was considered in general in § 7.5 [see (7.77)]. For an analytic amplitude, one may use the kinematic covariants $X_l(p_1 \ldots p_4)$ associated with any of the crossed channels, whichever is most convenient for expanding the amplitude in a given channel.

The momentum variables for a given channel usually include the total momentum of the states and both relevant momenta (in the initial and final states of the channel), allowing one both to write the invariant energy easily and to take account very simply of particle identity. Thus "natural" sets of momenta are:

$$\left.\begin{array}{ll}
s\text{-channel:} & \\
P = p_1+p_2, & s = P^2, \\
p = \tfrac{1}{2}(p_1-p_2), & p' = \tfrac{1}{2}(p_3-p_4); \\
t\text{-channel:} & \\
\Delta = p_3-p_1 = p_2-p_4, & t = \Delta^2, \\
K = \tfrac{1}{2}(p_1+p_3), & Q = \tfrac{1}{2}(p_2+p_4), \\
KQ \equiv \nu = \tfrac{1}{4}(s-u), & K^2+Q^2 = \tfrac{1}{2}(s+u); \\
u\text{-channel:} & \\
\bar{\Delta} = p_1-p_4 = p_3-p_2, & u = \bar{\Delta}^2, \\
\bar{K} = \tfrac{1}{2}(p_1+p_4), & \bar{Q} = \tfrac{1}{2}(p_2+p_3).
\end{array}\right\} \quad (34)$$

The corresponding $X_l(P, p, p')$ $X_l(\Delta, K, Q)$, ... are called s-channel covariants, t-channel covariants, etc. If they have matrix indices [as in (33) or (7.80)], it is convenient to group them separately for the initial and final states of the channel. In particular, the covariants (7.80) are a natural set for the t-channel; the indices of each of the two independent sets of γ-matrices [see (7.80′)] refer to the wave functions of the same (initial or final) state.

The spinor amplitude may be expanded in terms of covariants in any channel:

$$\begin{aligned}\mathcal{M}(p_1 \ldots p_4) &= \sum_l F_l^{(s)}(s, t, u)\, X_l(P, p, p') \\ &= \sum_l F_l^{(t)}(t, u, s)\, X_l(\Delta, K, Q) = \sum_l F_l^{(u)}(u, s, t)\, X_l(\bar{\Delta}, \bar{K}, \bar{Q}).\end{aligned} \quad (35)$$

The expansion in terms of t-channel invariants is convenient in studying elastic scattering (in the s-channel) and reflections. The sets of invariant amplitudes $F^{(s)}$, $F^{(t)}$, and $F^{(u)}$ are connected by linear relations whose coefficients form crossing matrices. To calculate a crossing matrix $\beta(s, t)$ one must find the expansion of s-channel covariants in terms of t-channel ones:

$$X_l(P, p, p') = \sum_{l'} \beta_{ll'}(s, t)\, X_{l'}(\Delta, K, Q). \quad (36)$$

The expansions (35) and (36) must be such that the invariant amplitudes do not contain kinematic singularities (see § 7.5). This holds in particular if the elements of the crossing matrix are constants. The calculation of spinor crossing matrices may be found in the literature.[108]

Crossing symmetry and identical particles

If any of the crossed reactions coincide or differ by a θ-reflection, the amplitudes in these channels must be equal (up to a sign in the case of crossed fermions). This entails an additional crossing symmetry of the amplitude. For example, the s-channel reaction $\pi^+ + \pi^- \to \pi^+ + \pi^-$ coincides with the t-channel reaction. Consequently, the analytic (invariant) amplitude will be symmetric in s and t:

$$F(s, t, u) = F(t, s, u). \tag{37}$$

For particles with spin, it is convenient to use the expansion (35) in which the covariants have simple behavior under crossing (e.g. are multiplied by ± 1).

Crossing symmetry is a consequence of the identity of particles in the initial or final states of at least one of the crossed processes. For instance, the reactions $1+2 \to \bar{1}+4$ (s-channel) and $1+\bar{4} \to \bar{1}+\bar{2}$ (u-channel) differ only by a θ-reflection; in the t-channel $(1+1 \to \bar{2}+4)$, the initial state contains identical particles. In the process $1+2 \to 3+4$, crossing symmetry is possible only if at least one of the pairs 1 and 2, 1 and $\bar{3}$, 1 and $\bar{4}$, $\bar{2}$ and 4, 3 and 4, or $\bar{2}$ and 3 consists of identical particles. In the example considered above, the u-channel contains four identical particles: $\pi^+ + \pi^+ \to \pi^+ + \pi^+$.

Let the initial state in the s-channel have two identical particles: $1+1 \to 3+4$; interchanging them can only change the sign of the amplitude if 1 is a fermion. Under this interchange, $p_1 \leftrightarrow p_2$ and $t \leftrightarrow u$. Consequently, if one chooses s-channel covariants to be even or odd under this interchange:

$$X^{r_1 r_2}_{l r_3 r_4}(P, p, p') = \pm \xi_l X^{r_2 r_1}_{l r_3 r_4}(P, -p, p') \tag{38}$$

($-$ for fermions), the invariant amplitudes will possess crossing symmetry:

$$F_l^{(s)}(s, t, u) = \xi_l F_l^{(s)}(s, u, t). \tag{39}$$

If the theory is invariant with respect to charge conjugation, crossing symmetry is also possible when in the s-channel particles 2 and 4 are identical and particles 1 and 3 are neutral. For example, if the s-channel is $\pi^0 + \mathrm{p} \to \omega^0 + \mathrm{p}$, the u-channel is $\pi^0 + \bar{\mathrm{p}} \to \omega^0 + \bar{\mathrm{p}}$ which is related to the s-channel process by charge conjugation.

Helicity amplitudes have more complex analytic properties[136–138] and behavior under crossing[139–141] than spinor amplitudes. The source of these complications may be seen from (33) for the spinor amplitude in baryon–baryon scattering. To obtain the helicity amplitude in the s-channel one must contract the spinor amplitude (33) with Dirac "helicity" wave functions $u(\boldsymbol{p}, \lambda)$ and $\bar{u}(\boldsymbol{p}, \lambda)$, while the t-channel helicity amplitude is obtained by contraction of the same analytic amplitude (33) with other helicity bispinors: $\bar{v}^{\beta}(-\boldsymbol{p}_4, \lambda_4)$, $u^{\alpha}(\boldsymbol{p}_3, \lambda_3)$, etc. [Helicity wave functions are given in (5.29), (5.31), and (5.38), in which the matrix α must be written in the helicity basis, according to (4.61) and (4.63); see also §§ 4.5 and 7.4.]

Multiplication by wave functions introduces additional kinematic factors; one thus loses the local transformation law for $\mathcal{M}$. When passing from one channel to another the sets of helicity amplitudes transform among themselves linearly (via "helicity" crossing matrices). The analytic form of helicity amplitudes is the same as that of invariant amplitudes only up to kinematic singularities.[136–141] (These are additional conditions at thresholds and pseudo-thresholds and for forward scattering; see § 7.5.)

§ 11.3. Crossing matrices for SU_2 and SU_3

Crossing symmetry relates amplitudes in different channels. An amplitude in one channel, when analytically continued into the physical region of another channel, becomes an amplitude for a process in this channel. If the interaction has isospin or unitary symmetry, one must relate amplitudes with given spinor or unitary multiplicity in different channels. This relation is given by crossing matrices.

Let us consider the isospin-conserving scattering process

$$a+b \to c+d \quad (s\text{-channel})$$

The relativistically invariant s-channel amplitude

$$\langle c, d|T|a, b\rangle = \sum_I T_s^{(I)}(s, t, u)\, \lambda_I(c, d\,|\,a, b) \tag{40}$$

may be expanded in terms of scattering amplitudes $T_s^{(I)}$ with definite isospin I; the quantities λ_I may be expressed in terms of Clebsch–Gordan coefficients and the phases ε_a (see § 8.3):

$$\lambda_I(a, b\,|\,c, d) = \varepsilon_a\varepsilon_b\varepsilon_c\varepsilon_d c(a, b; I)\, c(c, d; I), \tag{41}$$

where the Clebsch–Gordan coefficient is

$$c(a, b; I) = \langle I_a I_b I_{3a} I_{3b}\,|\,I_a I_b I, I_{3a}+I_{3b}\rangle. \tag{42}$$

A similar expansion may be written for the amplitude in the t-channel:

$$\langle \bar{b}, d|T|a, \bar{c}\rangle = \sum_I T_t^{(I)}(t, s, u)\, \lambda_I(\bar{b}, d\,|\,a, \bar{c}), \tag{43}$$

and in the u-channel:

$$\langle \bar{b}, c|T|a, \bar{d}\rangle = \sum_I T_u^{(I)}(u, t, s)\, \lambda_I(c, \bar{b}\,|\,a, \bar{d}). \tag{44}$$

As a consequence of crossing symmetry, all three amplitudes (40), (43), and (44) are values of the same function of momenta in different physical regions, and each of these amplitudes when continued into another physical region becomes the amplitude of the crossed process. Thus, performing the analytic continuation, we must set (40) = (43) = (44), or

$$\sum_I T_s^{(I)}(s, t, u)\, \lambda_I(c, d\,|\,a, b) = \sum_I T_u^{(I)}(u, t, s)\, \lambda_I(c, \bar{b}\,|\,a, \bar{d}) = \sum_I T_t^{(I)}(t, s, u)\, \lambda_I(\bar{b}, d\,|\,a, \bar{c}). \tag{45}$$

The isospin crossing matrixes X_{st} and X_{su} relate the isospin amplitudes $T_s^{(I)}$ and $T_t^{(I')}$, $T_s^{(I)}$, and $T_u^{(I')}$:

$$T_s^{(I)} = \sum_{I'} (X_{st})_{II'}\, T_t^{(I')}, \tag{46}$$

$$T_s^{(I)} = \sum_{I'} (X_{su})_{II'}\, T_u^{(I')}. \tag{47}$$

To calculate crossing matrices one must choose the phases of single-particle states with respect to those basis isospin states $|I, I_3\rangle$ (transforming as $\Phi_{I_3}^I$) for which one may use Clebsch–Gordan coefficients (see § 8.3). When calculating X_{st} and X_{su} one must first pass from the single-particle states $|a\rangle$ and $|b\rangle$ to the basis isospin states $|I, I_3\rangle$. Replacing $|b\rangle$ by $\langle\bar{b}|$ introduces the phase

$$\varepsilon_b \varepsilon_{\bar{b}} = \eta_b,$$

which was defined above in (8.31) and (8.32).

Using the orthogonality properties of Clebsch–Gordan coefficients, we find, from (45),

$$(X_{st})_{II'} = \sum_{abcd} \eta_b \eta_c c(a, b; I)\, c(a, d; I)\, c(a, -c; I')\, c(-b, d; I'). \tag{48}$$

X_{su} and X_{tu} are defined analogously. If the order of multiplication of states is reversed, so that instead of $|a, b\rangle$ one considers $|b, a\rangle$, the crossing matrices acquire a phase factor [from the symmetry property of the covariants (3.8)]:

$$c(a, b; I) = (-1)^{I-I_a-I_b}\, c(b, a; I). \tag{49}$$

Let us consider the following crossed processes:

$$\left.\begin{aligned} &\Xi + \mathrm{K} \to \Sigma + \pi && (s\text{-channel}),\\ &\Xi + \bar{\Sigma} \to \bar{\mathrm{K}} + \pi && (t\text{-channel}),\\ &\Xi + \pi \to \Sigma + \bar{\mathrm{K}} && (u\text{-channel}). \end{aligned}\right\} \tag{50}$$

Recalling that the particle isospins are $I_\Xi = \frac{1}{2}$, $I_\mathrm{K} = \frac{1}{2}$, $I_\Sigma = 1$, $I_\pi = 1$, we may write the isotopic structure of these processes as:

$$\begin{aligned} &(s)\quad \tfrac{1}{2} + \tfrac{1}{2}' \to 1 + 1', && T_s^{(0)}, T_s^{(1)};\\ &(t)\quad \tfrac{1}{2} + 1 \to \overline{\tfrac{1}{2}}' + 1', && T_t^{(1/2)}, T_t^{(3/2)};\\ &(u)\quad \tfrac{1}{2} + 1' \to 1 + \overline{\tfrac{1}{2}}', && T_u^{(1/2)}, T_u^{(3/2)}. \end{aligned} \tag{51}$$

We have also written the isospin amplitudes in each channel in (51). Inserting the values of the Clebsch–Gordan coefficients, we find, from (45),

$$\left.\begin{aligned} T_s^{(0)} &= -\sqrt{\tfrac{2}{3}}\, T_t^{(1/2)} - 2\sqrt{\tfrac{2}{3}}\, T_t^{(3/2)},\\ T_s^{(1)} &= \tfrac{2}{3}(T_t^{(3/2)} - T_t^{(1/2)}),\\ X_{st} &= \begin{pmatrix} -\sqrt{\frac{2}{3}} & -2\sqrt{\frac{2}{3}} \\ -\frac{2}{3} & \frac{2}{3} \end{pmatrix}. \end{aligned}\right\} \tag{52}$$

The phase in the antiparticle multiplets $\bar{\Sigma}$ and $\bar{\mathrm{K}}$ has been taken into account; see (8.31), (8.32), and § 10.1. It is often convenient to calculate using specific members of isospin multiplets. The isospin crossing matrices are given in Table A.6.[142]

In the case of SU_3, the calculation of the crossing matrices follows the same scheme as in the case of the isospin group SU_2. The symbol I in (40)–(47) must be replaced by the index $\underline{n}$ denoting the multiplets $\underline{1}$, $\underline{8}_A$, $\underline{8}_S$, $\underline{10}$, $\underline{10}^*$, of the group SU_3, while instead of the Clebsch–Gordan coefficients $c(a, b; I)$ one must insert the coefficients of the group SU_3 considered in § 10.3:

$$c(a, b; \underline{n}) = \begin{pmatrix} \mu_a & \mu_b & \mu_{n\gamma} \\ \nu_a & \nu_b & \nu_n \end{pmatrix} = \xi(a \times b, \underline{n})\, c(b, a; \underline{n}). \tag{53}$$

Replacing the particle b by the antiparticle $\bar{c}$ entails replacing μ_b by μ_c^* and ν_b by $-\nu_c$ in (53). Since the SU_3 coefficients $c(a, b; \underline{n})$ have different symmetry from that of SU_2 coefficients, in a symmetry relation of the type (49) a different phase ξ_I enters[142, 143]. The factors ξ_I depend upon the representations a, b, and the multiplet $\underline{n}$ in the product $a \times b$ (Table 11.1).

TABLE 11.1

	$\underline{1}$	$\underline{8}$	$\underline{8}$	$\underline{10}$	$\underline{10}^*$	$\underline{27}$	$\underline{28}$	$\underline{35}$	$\underline{35}^*$	$\underline{64}$
$\underline{8} \times \underline{8}$	1	1	−1	−1	−1	1				
$\underline{8} \times \underline{10}$		1		−1		−1		1		
$\underline{8} \times \underline{10}^*$		1			−1	−1			−1	
$\underline{10} \times \underline{10}$					−1	1	1	−1		
$\underline{10} \times \underline{10}^*$	−1	1				1				1

If particles $a, b, \ldots,$ have spin, then the amplitude in (40) is one of the invariant amplitudes in the expansion of the spinor amplitude (35). The full crossing matrix in this case is the direct product of the isospin or SU_3 crossing matrix and the crossing matrix related to the usual spin [see § 11.2 and Table A.7 (p. 371)].

§ 11.4. Properties of vertex parts

A current $j_\mu(x)$ in this section will be understood as a Hermitian vector or axial-vector local operator having definite transformation properties under P and C and belonging to a unitary octet. This definition of a current differs from that in § 11.1. First of all, in this section we shall consider only currents with a vector index. Secondly, we shall endow them with internal symmetry properties and shall not discuss their connection with interpolating fields.

The vector nature of a current means that under a Lorentz transformation $g = (0, \Lambda)$

$$U(g)\, j_\mu(0)\, U^{-1}(g) = \Lambda_\mu{}^\nu(\Lambda)\, j_\nu(0).$$

The spatial reflection operator P distinguishes the vector current v_μ from the axial-vector current a_μ:

$$\left.\begin{aligned} P\boldsymbol{v}(x^0, \boldsymbol{x})P^{-1} &= -\boldsymbol{v}(x^0 - \boldsymbol{x}), \qquad & Pv^0(x^0, \boldsymbol{x})P^{-1} &= v^0(x^0, \boldsymbol{x}), \\ P\boldsymbol{a}(x^0, \boldsymbol{x})P^{-1} &= \boldsymbol{a}(x^0, \boldsymbol{x}), & Pa^0(x^0, \boldsymbol{x})P^{-1} &= -a^0(x^0, \boldsymbol{x}). \end{aligned}\right\} \tag{54}$$

The octet nature of the current is expressed by

$$[F_a, j_{\mu b}(x)] = if_{abc}j_{\mu c}(x) \qquad (a, b, c = 1 \ldots 8), \tag{55}$$

where F_a are the generators of SU_3.

The transformation properties of currents do not depend upon their explicit form. When calculating these properties, for simplicity we thus may take currents constructed of Dirac fields (see Table 6.1, p. 144). To account for SU_3 properties, it is convenient to introduce the

Dirac quark fields $\psi_{\alpha i}(x)$, $i = 1, 2, 3$. Then

$$v_{\mu a} = \tfrac{1}{2}\bar{\psi}\gamma_\mu\lambda_a\psi, \quad a_{\mu a} = \tfrac{1}{2}\bar{\psi}\gamma_\mu\gamma_5\lambda_a\psi. \tag{56}$$

Under SU_3 transformations $u = \exp(iF_a\alpha_a)$, the fields ψ transform according to $u\psi u^{-1} = \exp(i\lambda_a\alpha_a/2)\psi$ (see § 9.2), leading to (55).

Currents fall into two classes with respect to charge conjugation C. "First class currents" behave like the quantities (56):

$$\mathrm{C}v_{\mu a}\mathrm{C}^{-1} = -\sigma_a v_{\mu a}^+, \quad \mathrm{C}a_{\mu a}\mathrm{C}^{-1} = \sigma_a a_{\mu a}^+, \tag{57}$$

where

$$\sigma_a = -1\,(a = 2, 5, 7), \qquad \sigma_a = 1\,(a = 1, 3, 4, 6, 8).$$

"Second class currents" $k_{\mu\alpha}$ and (the axial current) $r_{\mu\alpha}$ have the opposite sign under C:

$$\mathrm{C}k_{\mu a}\mathrm{C}^{-1} = +\sigma_a k_{\mu a}^+, \quad \mathrm{C}r_{\mu a}\mathrm{C}^{-1} = -\sigma_a r_{\mu a}^+. \tag{58}$$

Relations (57) and (58) also hold for non-Hermitian currents [as do the previous equations, (54) and (55)]. Thus second-class currents behave like the quantities

$$k_{\mu a}(x) = \tfrac{1}{2}i\bar{\psi}(x)\,\lambda_a\,\partial_\mu\psi(x), \quad r_{\mu a}(x) = \tfrac{1}{2}i\bar{\psi}(x)\,\gamma_5\lambda_a\psi(x). \tag{59}$$

Under total reflections θ currents of both classes transform in the same way:

$$\theta j_\mu(x)\theta^{-1} = -j_\mu^+(-x). \tag{60}$$

A vertex part will be understood as a matrix element of the current j_μ between single-particle states. In the case of helicity states

$$\langle 2\,|\,j_\mu\,|\,1\rangle \equiv \langle \boldsymbol{p}_2, \lambda_2; m_2, J_2\,|\,j_\mu(0)\,|\,\boldsymbol{p}_1, \lambda_1; m_1, J_1\rangle, \tag{61}$$

where m_1, m_2 and J_1, J_2 are the masses and spins of particles 1 and 2.

Vertex parts appear when one studies the transition amplitude in perturbation theory. Such transitions are induced by the weak or electromagnetic interactions (see also Chapter 15). For identical particles 1 and 2 the vertex part (61) characterizes electron–hadron scattering or leptonic decays of baryons inside a given multiplet, depending on the type of the current. When $J_1 \neq J_2$, the vertex part (61) appears in the amplitude for electroproduction or photoproduction of resonances or leptonic decays of hadrons.

Transformation properties

The rules for transformation of the vertex part by an element of the Poincaré group $g = (a, A)$ may easily be found using two formulae of Chapter 4:

under translations $x \to x+a$

$$\langle 2\,|\,j_\mu(x+a)\,|\,1\rangle = e^{i(p_2-p_1)a}\langle 2\,|\,j_\mu(x)\,|\,1\rangle;$$

under the Lorentz transformation $g = (0, A)$

$$\begin{aligned}\langle \boldsymbol{p}_2, \lambda_2\,|\,j_\mu(0)\,|\,\boldsymbol{p}_1, \lambda_1\rangle &= \langle \boldsymbol{p}_2, \lambda_2\,|\,\mathrm{U}^{-1}(g)\,\mathrm{U}(g)\,j_\mu\mathrm{U}^{-1}(g)\,\mathrm{U}(g)\,|\,\boldsymbol{p}_1, \lambda_1\rangle \\ &= \sum_{\lambda_1', \lambda_2'} \mathcal{D}^{J_2*}_{\lambda_2'\lambda_2}(\tilde{A}_2)\,\langle \boldsymbol{p}_2', \lambda_2'\,|\,\Lambda_\mu^{-1\nu} j_\nu(0)\,|\,\boldsymbol{p}_1', \lambda_1'\rangle\,\mathcal{D}^{J_1}_{\lambda_1'\lambda_1}(\tilde{A}_1),\end{aligned} \tag{62}$$

where $\tilde{A}_i = \alpha^{-1}(\boldsymbol{p}_i')\,A\alpha(\boldsymbol{p}_i)$ is the Wigner rotation (4.45), and $p_\mu' = \Lambda_\mu{}^\nu p_\nu$.

The reflection properties of vertex parts may be found using the formulae for the transformation of the state vectors and currents (§6.2).

Under spatial reflection P

$$\langle \boldsymbol{p}_2, \lambda_2 | j_\mu(0) | \boldsymbol{p}_1, \lambda_1 \rangle = \eta_P(1)\, \eta_P^*(2) \langle -\boldsymbol{p}_2, -\lambda_2 | j_{\mu P}(0) | -\boldsymbol{p}_1, -\lambda_1 \rangle, \tag{63}$$

where $j_{\mu P} = (j_0, -\boldsymbol{j})$ for the vector current and $j_{\mu P} = (-j_0, \boldsymbol{j})$ for the axial current, while $\eta_P(i)$ is the intrinsic parity of particle i.

Under time-reversal T

$$\langle \boldsymbol{p}_2, \lambda_2 | j_\mu(x^0, \boldsymbol{x}) | \boldsymbol{p}_1, \lambda_1 \rangle = \eta_T^*(1)\, \eta_T(2) e^{2i\varphi_1\lambda_1 - 2i\varphi_2\lambda_2} \langle \boldsymbol{p}_1, \lambda_1 | j_{\mu T}(-x^0, \boldsymbol{x}) | \boldsymbol{p}_2, \lambda_2 \rangle, \tag{64}$$

where $j_{\mu T}(-x^0, \boldsymbol{x}) = (\mathrm{T} j_\mu(x^0, \boldsymbol{x}) \mathrm{T}^{-1})^+ = \left(\tilde{j}_0^+, -\tilde{\boldsymbol{j}}^+\right)$ both for the vector and for the axial vector currents. For first-class currents, $\tilde{j}_{\mu a} = \sigma_a j_{\mu a}$, while for second-class currents, $\tilde{j}_{\mu a} = -\sigma_a j_{\mu a}$. In formula (64) φ_i is the azimuthal angle of the vector $\boldsymbol{p}_i$.

Crossing-conjugate vertices

Along with the vertex (61) we shall also consider the "crossing-conjugate" vertex

$$\langle 2, \bar{1} | j_\mu(0) | 0 \rangle = \langle \boldsymbol{p}_2, \lambda_2\, 2; \boldsymbol{p}_{\bar{1}}, \lambda_{\bar{1}}, \bar{1} | j_\mu(0) | 0 \rangle, \tag{65}$$

in which particle 1 on the right in the matrix element is replaced by an antiparticle on the left. The state $\langle 2, \bar{1}|$ is to be understood as the state $\langle 2, \bar{1};\, \text{out}|$. Analogously one may introduce the quantity

$$\langle 0 | j_\mu(0) | \bar{2}, 1, \text{in} \rangle = \langle 0 | j_\mu(0) | \boldsymbol{p}_{\bar{2}}, \lambda_{\bar{2}}, 2; \boldsymbol{p}_1, \lambda_1, 1; \text{in} \rangle. \tag{66}$$

For vertices of the type $\langle 2 | j_\mu | 1 \rangle$ the physical region of the invariant variable $t = (p_1 - p_2)^2$ is

$$t \leqslant (m_1 - m_2)^2; \qquad p_1^0, p_2^0 > 0. \tag{67}$$

For "annihilation" vertices $\langle 2, \bar{1} | \boldsymbol{j} | 0 \rangle$ the physical region of t is defined by the inequality

$$t = (p_{\bar{1}} + p_2)^2 = (-p_1 + p_2)^2 \geqslant (m_1 + m_2)^2; \qquad p_1^0 < 0, \quad p_2^0 > 0. \tag{68}$$

In the case of the vertex (66), we have, in analogy with (68),

$$t = (p_{\bar{1}} + p_2)^2 = (-p_1 + p_2)^2 \geqslant (m_1 + m_2)^2; \qquad p_1^0 > 0, \quad p_2^0 < 0. \tag{69}$$

The crossing-conjugate vertices (65) and (66) may be obtained from the vertex (61) by analytically continuing with respect to momenta. To verify this, let us consider the vertex function (61) for spinless particles $J_1 = J_2 = 0$:

$$\Gamma_\mu(p_1, p_2; 1, 2) = -\frac{1}{(2\pi)^3} \int d^4x\, d^4y\, e^{i(p_2 y - p_1 x)} \vec{K}_{2y} \langle 0 | T(\varphi_2(y)\, j_\mu(0)\, \varphi_1(x)) | 0 \rangle \overleftarrow{K}_{1x}. \tag{70}$$

Here, $\varphi_1(x)$ and $\varphi_2(y)$ are the interpolating fields related asymptotically to particles 1 and 2. The right-hand side of this formula in the region (67) determines the vertex part (61) by virtue of the reduction formula (26). It is easy to verify using this same reduction formula that in the region (68) we obtain the vertex part (65) from (70), while in the region (69) we obtain (66). Thus all the three crossing-conjugate matrix elements (61), (65), and (66) may be described by the same function of momenta $\Gamma_\mu(p_1, p_2; 1, 2)$, which may be called the analytic vertex part. We shall henceforth speak of Γ_μ as a vertex function.

As in the case of crossing-conjugate amplitudes (see § 11.3), for particles with spin we shall first pass to spinor states using wave functions, and then introduce the spinor vertex functions Γ_μ:

$$\langle \boldsymbol{p}_2, \lambda_2 | j_\mu(0) | p_1, \lambda_1 \rangle = \bar{u}^{r_2}(\bar{\boldsymbol{p}}_2, \lambda_2) \Gamma_{\mu r_2}{}^{r_1}(p_1, p_2; 1, 2) u_{r_1}(\boldsymbol{p}_1, \lambda_1). \tag{71}$$

The functions $\Gamma_\mu(p_1, p_2; J_1, J_2)$ describe all three crossing-conjugate vertex parts (61), (65), and (66) in their respective physical regions. As in the case of the spinor amplitudes (see § 7.5), under Lorentz transformations $g = (0, A)$, the functions Γ_μ transform via momentum-independent matrices. The explicit form of these matrices is determined by the choice of the wave functions $u(\boldsymbol{p}_1, \lambda_1)$, $u(\boldsymbol{p}_2, \lambda_2)$.

Invariant form-factors

The expansion of the spinor vertex part $\Gamma_\mu(p_1, p_2; J_1, J_2)$ in terms of the invariant form-factors $F_l(t)$ depending only on $t = (p_1, -p_2)^2$ may be performed using the same rules as for the expansion of the spinor amplitude in terms of invariant amplitudes (see § 7.5). If we construct a set of independent covariants $X_\mu(p_1, p_2)$ with the same Lorentz transformation properties as $\Gamma_\mu(p_1, p_2; J_1, J_2)$, then

$$\Gamma^\mu(p_1, p_2; J_1, J_2) = \sum_l X_l^\mu(p_1, p_2) F_l(t). \tag{72}$$

Both momenta p_1, p_2 are on the mass shell: $p_i^2 = m_i^2$.

The covariants X^μ are defined in terms of relativistic kinematics, i.e. by the choice of the wave functions for particles 1 and 2. The form factors contain information on the dynamics of the process. Calculation of form factors requires a dynamical theory (see § 12.5). Experimental results are analyzed in the language of form factors.

The expansion (72) of the spinor vertex function in terms of invariant form factors must satisfy conditions of the same type as the expansion of the spinor amplitude. The choice of independent covariants X_μ must guarantee the covariance of the expansion, the absence of kinematic singularities, and the presence of reflection invariance when required. Special care must be taken in excluding the supplementary conditions which are imposed on wave functions with higher spins ($J \geqslant 1$). The supplementary conditions are absent for the choice of $(2J+1)$-component wave functions. However, such a parametrization[144, 145] of the vertex function is not so convenient, since in this language it is difficult to deal with reflection invariance.[146] Such difficulties are not present when using Rarita–Schwinger or Bargmann–Wigner wave functions.[147, 148] The possibility of such a choice of wave functions, for which the form factors do not have kinematic singularities, has been proven[103] for $(2J+1)$-component wave functions and extended to the case of other wave functions using a theorem on the analytic equivalence[111] of wave functions describing the same particle.

Since finding an invariant set of covariants is elementary in the simplest cases and is a particular case of the method examined earlier for introducing invariant amplitudes, we shall refer the reader to §§ 7.5 and 11.3. In general, the procedure for expanding the vertex function is very awkward, and we recommend the original articles mentioned above to the reader.

When discussing the expansion of the vertex function in terms of covariants we have not considered internal symmetry properties of currents. The question of the isospin or SU_3 expansion of the vertex function was in fact considered in §§ 8.3 and 10.2 in the discussion of rapid decays and phenomenological Lagrangians. For example, the SU_3 properties of the matrix element of the axial current $j_{\mu a}$

$$\langle \boldsymbol{p}_2, \lambda_2;\ \underline{10}, \alpha_2 | j_{\mu a} | \boldsymbol{p}_1, \lambda_1;\ \underline{8}, \alpha_1 \rangle$$

(here $\alpha = I, Y, t$ denotes the state of the particle in the multiplet $\underline{n}$) will be the same as the amplitude for the decay of a decimet into two octets: $\underline{10} \to \underline{8} + \underline{8}$.

CHAPTER 12

ANALYTIC PROPERTIES OF THE SCATTERING AMPLITUDE

IN THIS chapter we shall introduce the basic concepts related to the analytic properties of the scattering amplitude. For simplicity, we shall consider the case of spinless particles.

A self-consistent description of the analytic properties of the amplitude should be based on postulating analyticity properties or in deriving them using perturbation theory but not in relying on interpolating fields whose use permits the proof of analytic properties only in idealized cases. However, we shall begin with formulae involving interpolating fields (which are not themselves necessary after analytic properties have been established). This allows one to illustrate the role of the locality of fields, or microcausality, in analytic properties of the amplitude and the role of the unitary condition in determining singularities of the amplitude. We shall not consider Landau-type singularities.

The analytic properties of the amplitude and its asymptotic behavior may be expressed using dispersion relations, which are an important step toward the understanding of dynamics in elementary particle theory. We shall not perform explicit calculations based on the dispersion approach which are developed at length in many books.[149–153] The dispersion approach is indispensable for the development of new approximation methods and for the introduction of approximate concepts. In this respect the dispersion approach replaces equations of motion in the Lagrangian treatment of quantum field theory. As a prelude to Regge-pole theory (Chapter 13) and the dual approach (Chapter 14), we shall dwell in some detail on the Gribov–Froissart formula and continuation of the partial-wave amplitude into the angular momentum plane. In this context we also discuss the pole approximation and the connection between poles of the amplitude on the unphysical sheet with resonances. Analytic properties of form factors are examined briefly.

§ 12.1. Unitarity and the absorptive part

The complex s-, t-, and u-planes. The physical sheet

Let us consider the simplest model for the scattering of scalar particles $1+2 \to 3+4$ of identical mass m, assuming that these particles are the lightest in the theory (e.g. $\pi+\pi \to \pi+\pi$).

Let $t = t' < 0$ be fixed so that $u = 4m^2 - t' - s$. In the complex plane (Fig. 11) the physical region of the s-channel corresponds to a segment along the positive real axis from the threshold value $s_0 = 4m^2 - t'$ to ∞. If there exist stable single-particle states with quantum numbers of the system $1+2$ and energy $\sqrt{s_k}$, they will correspond to the points $m^2 < s_k < 4m^2$ along the real axis. The physical region of the u-channel is $u_0 \leqslant u < \infty$, where $u_0(t')$ is the threshold of the two particle process in the u-channel. Let us denote by u_l the points corresponding to the single-particle states with quantum numbers of the u-channel, i.e. the system

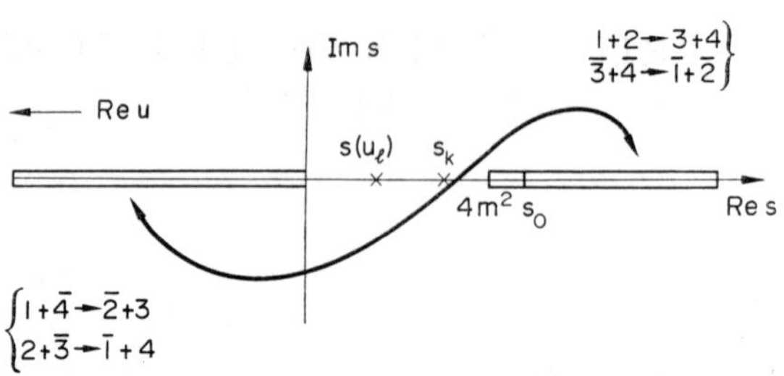

FIG. 11. Singularities of the scattering amplitude in the complex s-plane.

$1+\bar{4}$. In the complex s-plane the physical region of the u-channel is a segment along the negative real axis $-\infty < s \leqslant s(u_0)$, where $s(u_0) = 4m^2 - t' - u_0$. The points $s(u_l)$ refer to particles in the u-channel. For forward scattering, $t' = 0$, and the first threshold point of the s-channel is $s_0(0) = s_{\min} = 4m^2$.

Analogously, one may find the segments of the real axis of the complex t-plane corresponding (for fixed $s < 0$) to the physical regions of the t- and u-channels and the points of the axis corresponding to particles with quantum numbers of the t- and u-channel.

Before turning to the analytic properties of the amplitude $F(s, t') \equiv F(s, t', u(s))$, let us recall the properties of the amplitude $f(E)$ in potential scattering. The nonrelativistic amplitude $f(E)$ is analytic in the whole complex w-plane with a cut along the real axis beginning from the scattering threshold $E_0 = 0$ to ∞, with the exception of simple poles (on the real axis) corresponding to bound (i.e. single-particle) states. The physical scattering region corresponds to the upper edge of the cut, where $E \to E+i\varepsilon$, $\varepsilon \to 0$; the scattering amplitude is the boundary value $f(E+i\varepsilon)$, $\varepsilon \to 0$, of the analytic amplitude $f(w)$ on the real axis. In contrast to the relativistic case, the nonrelativistic amplitude in potential scattering has only one physical region $E > 0$. Its continuation into the lower half w-plane is performed using the symmetry principle:

$$f(w^*) = f^*(w), \tag{1}$$

or, on the lower edge of the cut,

$$f(E-i\varepsilon) = f^*(E+i\varepsilon),$$

so that the discontinuity across the cut

$$\underset{\varepsilon \to 0}{f(E+i\varepsilon) - f(E-i\varepsilon)} = 2i \operatorname{Im} f(E) \tag{2}$$

is equal to the imaginary part of the amplitude and may be found from the unitarity condition for $f(E)$.

In the case of potential scattering one may prove that on the first Riemann sheet (with the exception of the point $E+i\varepsilon$) complex poles are impossible, since they describe created (and growing) waves, whose existence contradicts probability conservation.

The analytic properties of the nonrelativistic amplitude $f(E)$ allow one to draw some conclusions about the properties of the analytic relativistic amplitude $F(s, t')$. For forward scattering the right-hand cut $E > 0$ of the amplitude $f(w)$ in the complex w-plane corresponds to the right hand cut $s_{min} \leqslant s \leqslant \infty$ of the relativistic amplitude $F(s, 0)$, beginning with the two-particle threshold branch point s_{min}. One maintains the definition of the s-channel amplitude $F^s(s, 0)$ as the boundary value of an analytic amplitude $F(s, 0)$ on the upper edge of the cut:

$$F^s(s, 0) = F(s+i\varepsilon, 0), \quad s > s_{min}. \tag{3}$$

Analogously the relation between a pole in the amplitude f and a bound state also continues to hold, having the form

(pole of $F(s, t')$ for $0 < s < 4m^2$) $\leftrightarrow$ (particle with s-channel quantum numbers). (4)

Because of crossing relations (see § 11.2), the relativistic amplitude $F(s, t, u)$ describes (in different physical regions) the amplitude for all crossed reactions, i.e. in all channels. The nature of the limiting procedure (3) and the relation (4) hence must be the same in the s-, t-, and u-channels.

In particular, in the u-plane there is a right-hand cut along the real axis for $u_{min} \leqslant u < \infty$, $\text{Im}\, u = 0$, where

$$F^u(u, s(u), 0) = \underset{\varepsilon \to 0}{F}(s(u), 0, u+i\varepsilon) \tag{5}$$

—the amplitude in the u-channel—will also be the boundary value of the function $F(s, 0, u+i\varepsilon)$ when approaching this cut from above. Consequently, the forward-scattering amplitude $F(s, 0)$ must have not only a right-hand cut in the complex s-plane, corresponding to the physical region of the s-channel, but also a left-hand cut, $-\infty < s \leqslant s(u_{min}) = 0$, associated with the physical region of the u-channel. This cut begins at the branch point $s(u_{min}) = 0$, corresponding to the threshold $u_{min} = 4m^2$ for two-particle reactions in the u-channel (see Fig. 11). According to (5) the physical region of the u-channel must be identified with the lower edge of the left-hand cut. Moreover, if there exist stable particles with masses m_k and m_l and quantum numbers of the s- and u-channels, then there must be poles at the points $s_k = m_k^2$ and $s(u_l) = 4m^2 - m_l^2$. Figure 11 also shows one possible route of analytic continuation of the amplitude from the u-channel into the s-channel, which was discussed in § 11.2. The existence of this continuation is proven in axiomatic field theory.[135]

If we now turn to the amplitude $F(s', t, u(t))$ for fixed $s' = 0$, by using a similar combination of the analytic properties of the nonrelativistic amplitude and the crossing properties of $F(s, t, u)$ one can obtain a reasonable picture of poles and "two-particle" cuts of $F(s', t, u)$ in the complex t-plane. This picture refers only to those cuts and poles which may be associated with physical regions and stable paticles, and avoids the question of whether there are other cuts and poles. In the present model (equal masses, forward scattering of the lightest particles), the position of the cuts in the complex t-plane has the same form as in the s-plane (see Fig. 11). The cuts will change in general if the masses of the particles are different, if the lightest particles are not the ones scattering, or if unstable particles exist.

Unitarity and discontinuities of the amplitude

In the nonrelativistic theory, the analytic continuation (1) of the amplitude into the lower half plane and the calculation of the discontinuity (2) of the amplitude across the cut are performed using the unitarity condition. We shall thus use explicit expressions for the scattering amplitude of scalar particles to illustrate the relation between unitarity and discontinuities of the amplitude across cuts in the relativistic theory.

In § 11.1 we obtained (11.18) and (11.27) for the scattering amplitude of scalar particles. These two expressions coincide for momenta in the physical region but differ in the non-physical region. Equation (11.18) is meaningless outside the physical region, since the integral diverges there. Formula (11.27) contains the retarded commutator of fields instead of the T-product. As a result of microcausality, this commutator $[\Phi(x), \Phi(y)]$ vanishes for space-like intervals: $(x-y)^2 < 0$. This is important in proving the existence of a region of analyticity of the amplitude (11.27) as a function of s. For these reasons we shall work with (11.27):

$$\langle 3, 4|S-1|1, 2\rangle = -\int d^4x\, d^4y f_4^*(x)\, \vec{K}(x) \langle 3|\theta(x-y)[\Phi(x), \Phi(y)]|1\rangle \overleftarrow{K}(y) f_2(y), \tag{6}$$

assuming for simplicity that particles 2 and 4 are identical.

Let us write the unitarity condition:

$$(S^+ - 1)+(S-1) = -(S-1)(S^+ -1),$$

and calculate the matrix element according to (11.27):

$$\begin{aligned}\langle 3, 4|S^+ -1|1, 2\rangle &= \langle 1, 2|S-1|3, 4\rangle^* \\ &= -\int d^4x\, d^4y f_4^*(x)\, \vec{K}(x) \langle 3|\theta(y-x)[\Phi(x), \Phi(y)]|1\rangle \overleftarrow{K}(y) f_2(y).\end{aligned} \tag{7}$$

Then the right-hand side of the unitarity condition will contain only the usual commutator of the currents $J(x) = K(x)\Phi(x)$ and will not contain the singular functions $\theta(x)$:

$$\langle 3, 4|(S-1)(S^+ -1)|1, 2\rangle = \int d^4x\, d^4y f_4^*(x) \langle 3|[J_4(x), J_2(y)]|1\rangle f_2(y). \tag{8}$$

This formula simplifies if one introduces the relative coordinates ξ:

$$x = X+\tfrac{1}{2}\xi, \quad y = X-\tfrac{1}{2}\xi, \tag{9}$$

and performs a shift by X in the matrix element (8):

$$J(x) = e^{iPX} J(x-X) e^{-iPX}. \tag{10}$$

Integration with respect to X in (8) gives a δ-function which, when omitted, leads to the amplitude

$$\langle 3, 4|T-T^+|1, 2\rangle = i\int d^4\xi f_4^*(\tfrac{1}{2}\xi) \langle 3|[J_4(\tfrac{1}{2}\xi), J_2(-\tfrac{1}{2}\xi)]|1\rangle f_2(-\tfrac{1}{2}\xi). \tag{11}$$

Expanding the right-hand side of (11) in terms of a complete set of intermediate states,

$$\begin{aligned}\langle 3, 4|T-T^+|1, 2\rangle &= i2\pi \sum_n \delta(p_1+p_2-p_n) \langle 3|J_4(0)|n\rangle \langle n|J_2(0)|1\rangle \\ &\quad -i2\pi \sum_{n'} \delta(p_1-p_4-p_{n'}) \langle 3|J_2(0)|n'\rangle \langle n'|J_4(0)|1\rangle \equiv 2i(A_s - A_u).\end{aligned} \tag{12}$$

The quantum numbers of the states $|n\rangle$ are the same as in the system 1+2, i.e. in the s-channel. If the isospin of the states $|1, 2\rangle$ or $|3, 4\rangle$ is well defined, all states $|n\rangle$ have the same values of isospin I_3, hypercharge, parity, and G-parity under the strong interactions. The quantum numbers of the states $|n'\rangle$ coincide with the quantum numbers in the u-channel, i.e. the system $1+\bar{4}$. Thus the first term on the right-hand side, $2iA_s$, in (12) is defined by the spectrum of particles and states in the s-channel, and the second term $2iA_u$ by the spectrum of particles and states in the u-channel.

Since the energy spectrum is positive A_s contributes to the right-hand side of (12) starting at the two-particle threshold $s_0(t')$, i.e. in the physical region of the right-hand cut $s_0 \leqslant s < \infty$ in the complex s-plane. Moreover, for $m^2 < s < s_{\min}$, one may have δ-function contributions to A_s from stable particles in the s-channel. In the u-channel region, i.e. for $s(u_0) \geqslant s > -\infty$, A_s is equal to zero. $A_s(s, t')$ is called the absorptive part of the amplitude in the s-channel.

The absorptive part A_s in (12) may easily be transformed into the canonical form (1.68) (for the s-channel):

$$A_s = \frac{(2\pi)^4}{2} \sum_n \langle 3, 4 | T^+ | n\rangle \langle n | T | 1, 2\rangle \, \delta^4(p_n - p_1 - p_2), \tag{13}$$

if one bears in mind the relation (11.14) between the amplitude for the process $1+2 \to n$ and the matrix element of the current:

$$\langle n, \text{out} | J_2(0) | 1\rangle = -(2\pi)^{3/2} \langle n | T | 1, 2\rangle, \tag{14}$$

and identifies $|n\rangle$ with $|n, \text{out}\rangle$.

The terms A_u in the sum (12) are nonzero at the points $u_l = m_l^2 = (p_1 - p_4)^2$, corresponding to particles with mass m_l in the u-channel, and above the two-particle $(1+\bar{4})$—threshold u_0, i.e. A_u is nonzero along the segment $s(u_0) \geqslant s > -\infty$ of the complex s-plane, but A_u vanishes in the physical region of the s-channel. A_u is the absorptive part of the amplitude in the u-channel; it determines the right-hand side of the unitarity condition in the u-channel:

$$A_u = \frac{(2\pi)^4}{2} \sum_{n'} \langle 3, \bar{2} | T^+ | n'\rangle \langle n' | T | 1, \bar{4}\rangle \, \delta^4(p_{n'} - p_1 - p_4). \tag{15}$$

In relation (12), written for the s-channel, we could set $A_u = 0$. However, if on the left-hand side of (12) we passed from the s-channel amplitudes $\langle 3, 4 | T | 1, 2\rangle$ and $\langle 1, 2 | T | 3, 4\rangle^*$ to the single analytic amplitude $F(s, t, u)$, then (12) would be an expression for the discontinuities of the analytic amplitude across the segments of the real s-axis corresponding to the physical regions of the s- and u-channels.

Let us trace the basic steps of such an approach. For the moment, let us introduce the notation

$$\langle 3, 4 | T | 1, 2\rangle = F^+(s, t'), \quad \langle 3, 4 | T^+ | 1, 2\rangle = F^-(s, t') \tag{16}$$

for s-channel amplitudes which are given respectively by the integrals (6) and (7) after isolating the δ-functions expressing the conservation of energy-momentum using (9) and (10). The integrand in the expression for F^+ contains the retarded commutator of local fields

$$F^+(s, t') = \frac{i}{(2\pi)^6} \int d^4\xi \exp\left[\frac{i}{2}(p_2+p_4)\xi\right]$$
$$\times \vec{K}\left(\frac{1}{2}\xi\right) \langle 3 | \theta(\xi) \left[\Phi\left(\frac{1}{2}\xi\right), \Phi\left(-\frac{1}{2}\xi\right)\right] | 1 \rangle \overleftarrow{K}\left(-\frac{1}{2}\xi\right), \tag{17}$$

i.e. the integration in (17) is performed only inside the backward light cone. Consequently,(149) the integral (17) may be continued from the real axis $s > s_0$ to the upper half-plane Im $s > 0$. More precisely, the value of F^+ along the real axis $s > s_0$ is the boundary value of the function $F^+(s+i\varepsilon, t')$, $\varepsilon \to 0$, when approached from above. This characterizes the relation between micro-causality [the retarded commutator in (17)] and analyticity.

The expression for $F^-(s, t')$ arising from (7) differs from (17) by the replacement of $\theta(\xi)$ by $\theta(-\xi)$, i.e. it contains the advanced commutator. The integral representation for $F^-(s, t')$ for Re $s > s_0$ thus will be defined in the lower half plane Im $s < 0$, while the amplitude F^- in (16) is the boundary value of the function $F^-(s-i\varepsilon, t')$, $\varepsilon \to 0$, with the real axis approached from below. (The proof in ref. 149 is limited to small values of $t > 0$). Thus neither $F^+(s, t')$ nor $F^-(s, t')$ may be identified with the analytic amplitude $F(s, t')$, which must be defined in the whole complex s-plane. In the upper half s-plane above the physical s-channel region the functions $F(s, t')$ and $F^+(s, t')$ coincide, and it is necessary to find an analytic continuation of $F^+(s, t')$ into the lower half-plane.

Let us assume that the analytic continuation of $F^+(s, t')$ along the path encircling the threshold point $s_{\min}$ (Fig. 12) is $F^-(s, t')$, so that

$$F(s, t') = \begin{cases} F^+(s, t'), & \text{Im } s \geqslant 0, \\ F^-(s, t'), & \text{Im } s < 0, \end{cases} \qquad \text{Re } s \geqslant s_{\min} - l. \tag{18}$$

The possibility of such a continuation assumes the existence of some finite segment l of the real axis between the left- and right-hand cuts, along which $F(s, t')$ is regular and $F^+(s, t') = F^-(s, t')$. The existence of a finite segment l between the cuts is nontrivial for unequal masses or in the presence of unstable particles; it has been proven in general.(101) The analyticity of $F^-(s, t')$ and $F^+(s, t')$ along the segment l may be proven for some region of values of t' in the simplest cases.(149) In the general case (arbitrary masses and spins) the existence of the amplitude $F(s, t')$ with properties (18) must be postulated. Consequently, the difference

$$\underset{\varepsilon \to +0}{F(s+i\varepsilon, t') - F(s-i\varepsilon, t')} = 2iA_s(s, t'), \qquad (s \geqslant s_0), \tag{19}$$

coinciding by virtue of (16) with the unitarity condition (12), is the discontinuity of the amplitude $F(s, t')$ along the right-hand cut. To pass to the left-hand cut one must perform an analytic continuation into the physical u-channel region using crossing relations [see § 11.2 and Fig. 11, as well as eqns. (6) and (7)]. Then for $s < s(u_0)$ we find that the second term on the right-hand side of (12)

$$\underset{\varepsilon \to +0}{F(s-i\varepsilon, t') - F(s+i\varepsilon, t')} = 2iA_u(s, t'), \qquad (s \leqslant s(u_0)), \tag{20}$$

gives the discontinuity of $F(s, t')$ across the left-hand cut.

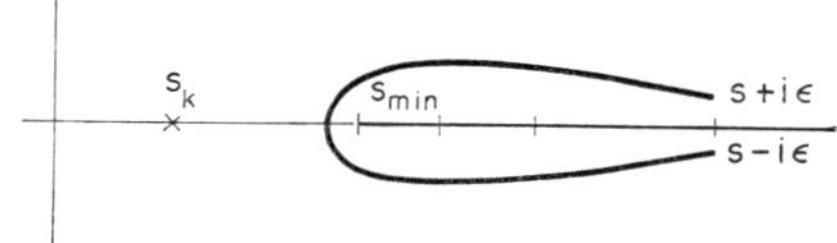

FIG. 12. Analytic continuation from $F^{+}(s, t)$ to $F^{-}(s, t)$ via a path around the threshold branch point in the complex s-plane.

The discontinuities A_s and A_u [see (13) and (15)] will be real if, in addition to particles 2 and 4, particles 1 and 3 are also identical, i.e. for elastic scattering. Then (18) will define an analytic continuation on the basis of the Schwarz reflection principle

$$F(s^*, t') = F^*(s, t'), \tag{21}$$

i.e. $F(s, t)$ is a real-analytic function.

In general the discontinuities (13) and (15) are real only when the amplitude is T-invariant. For scalar particles,

$$\begin{aligned}\langle \boldsymbol{p}_3, \boldsymbol{p}_4, \text{in} | S^+ | \boldsymbol{p}_1, \boldsymbol{p}_2, \text{in}\rangle &= \langle \boldsymbol{p}_3, \boldsymbol{p}_4, \text{in} | \mathrm{T}^{-1}\mathrm{T}S^+\mathrm{T}^{-1}\mathrm{T} | \boldsymbol{p}_1, \boldsymbol{p}_2, \text{in}\rangle \\ &= \eta_{1\mathrm{T}}\eta_{2\mathrm{T}}\eta^*_{3\mathrm{T}}\eta^*_{4\mathrm{T}}\langle -\boldsymbol{p}_3, -\boldsymbol{p}_4, \text{out} | S | -\boldsymbol{p}_1, -\boldsymbol{p}_2, \text{out}\rangle^* \\ &= \eta_{1\mathrm{T}}\eta_{2\mathrm{T}}\eta^*_{3\mathrm{T}}\eta^*_{4\mathrm{T}}\langle -\boldsymbol{p}_3, -\boldsymbol{p}_4, \text{in} | S | -\boldsymbol{p}_1, -\boldsymbol{p}_2, \text{in}\rangle^*,\end{aligned}$$

where $\mathrm{T}|\boldsymbol{p}_r, \text{in}\rangle = \eta_{r\mathrm{T}}| -\boldsymbol{p}_r, \text{out}\rangle$. Since the T-transformation is antiunitary, one may set all $\eta_{r\mathrm{T}}$ to unity, multiplying the single-particle state vectors by a phase factor. Passing to the c.m.s. and performing a rotation by 180° in the plane of the vectors $\boldsymbol{p}_3 - \boldsymbol{p}_4$ and $\boldsymbol{p}_1 - \boldsymbol{p}_2$,

$$\langle \boldsymbol{p}_3, \boldsymbol{p}_4 | S^+ - 1 | \boldsymbol{p}_1, \boldsymbol{p}_2\rangle = \langle \boldsymbol{p}_3, \boldsymbol{p}_4 | S - 1 | \boldsymbol{p}_1, \boldsymbol{p}_2\rangle^*.$$

From this it follows that eqns. (13) and (15) for A_s and A_u are real. If the theory is PT-invariant, then this result (21) may be obtained using the PT transformation.

The number of states $|n\rangle$ in the sum (13) for the absorptive part A_s depends on the energy in the s-channel, i.e. on s. As the energy is increased, more and more new processes $1+2 \to n$ become possible. As each threshold s_n is passed, the absorptive part of A_s acquires new terms so that s_n correspond to new branch points. In the present model, with equal-mass particles, the threshold branch points occur at $s_n = 4m^2, 9m^2, 16m^2, \ldots$, not counting thresholds for production of particles of different mass. From each such branch point at $s = s_n$ we may pass a cut along the positive real s-axis to infinity. Analogously, from each branch point associated with a u-channel threshold, one must pass a cut along the negative real s-axis to $-\infty$. On the segment of the real s-axis between the threshold values s_0 and $s(u_0)$ for two-particle states, both A_s and A_u, according to (18), will vanish, aside from points giving the contribution of single-particle states.

From the form of the δ-function in (13) it is clear that the single-particle state $|n_a\rangle$ contributes in the s-channel if there exists a particle with mass $m_a^2 < s_0$, spin 0, and quantum numbers of the s-channel. The contribution of this particle to A_s is equal to

$$A_s^{(1)} = \frac{\pi}{(2\pi)^6}\,\delta(s - m_a^2) g_{a34} g^*_{a12}. \tag{22}$$

Similarly, a particle with mass $m_b^2 < 4m^2$ and quantum numbers of the u-channel yields a contribution to A_u equal to

$$A_u^{(1)} = \frac{\pi}{(2\pi)^6}\,\delta(u-m_b^2)g_{b1\bar{4}}g^*_{b\bar{2}3}\,.$$

The constant g_{abc} describes the effective coupling of the spinless particles a, b, c:

$$g_{abc} = -(2\pi)^3\,\langle b\,|\,J_c(0)\,|\,a\rangle, \tag{23}$$

where the momenta are related by $p_a = p_b+p_c$. The factor $(2\pi)^3$ in (23) is introduced to preserve the relation with perturbation theory based on the Lagrangian formalism. Let the interaction Lagrangian for scalar neutral fields φ_i $(i = a, b, c)$ of mass m have the form $\mathcal{L}_I = g_{abc}\varphi_a\varphi_b\varphi_c$, and the current be defined by (11.10) or $J_c(x) = K(x)\,\varphi_c(x) = -(\partial_\mu\partial^\mu - m^2)\times \varphi_c(x)$. Then (23) holds to lowest order in $\mathcal{L}_I$.

§ 12.2. Maximal analyticity

In the simplest model for forward scattering of equal-mass scalar particles, we discussed the assumed connection between analyticity and causality and the role of unitarity in defining the discontinuities of the analytic amplitude. In this section we turn to the analytic properties of the amplitude in the general case.

In the preceding section we obtained the singularities of the relativistic amplitude from the analytic properties of the nonrelativistic amplitude and crossing symmetry relations. These singularities are fully characterized by the spectrum of states of the physical system. The existence of such singularities (poles and cuts along the real axis) has been proven for the simplest cases.[149]

Let us assume that for arbitrary masses and spins the analytic properties of the amplitude are also fully defined by the spectrum of physical states. More precisely, we shall consider the principle of maximal analyticity to hold:[150–152] An amplitude is an analytic function, possessing only those singularities necessary for the unitarity condition (in all channels and for all sets of amplitudes).

This principle is somewhat indeterminate if the existence of particles is a result of overall "self-consistency", and appears only at the end of calculations based on the unitarity condition. If the number and type of elementary particles are fixed, the principle of maximal analyticity is in practice equivalent to the assumption that an analytic amplitude has only singularities characteristic of all Feynman integrals. The analysis of Feynman integrals in fact has allowed one to systematize singularities of the amplitude with some confidence. [151, 152]

Let us first mention some additional features of the amplitude $F(s, t, u)$ for the scattering $1+2 \to 3+4$ of spinless particles that arise when one passes from the simplest model of § 12.1 to the real case. We shall now consider the scattering of particles of different masses at an arbitrary angle. Particles 1, 2, 3, and 4 are called "external", taking "internal" particles $a, b, c, \ldots$, as those which appear only in the unitarity condition (for all crossed channels).

In the general case, in contrast to the simplest model of § 12.1, the boundary of the physical region does not coincide with the beginning of the cut or with the first two-particle threshold branch point. The threshold branch points are always fixed by the masses of the external and internal particles taking part in the interaction; the position of these points does not depend on the variables s, t, u. The first threshold point in the s-channel, $s = (m+M)^2$, corresponds to the two-particle state with internal quantum numbers of the s-channel (i.e. with quantum numbers of the system of particles 1+2) having the lowest mass. The particles with masses m and M may be either external or internal. The sequence of thresholds $s_n = M_n^2$ consists of the smallest-mass s-channel multi-particle states $|n\rangle$, arranged in increasing order.

The physical region was studied in § 7.2. The boundary of the physical region depends only on the mass of the external particles [see (7.22)]. Moreover, if one of the variables s, t, and u is fixed (e.g. $t = t'$), the lower boundaries of the physical region for the two other variables $s_0(t')$ and $u_0(t')$ will differ from the absolute lower boundaries of the physical region s_{min} and u_{min}. This fact is directly visible on the Mandelstam plane. (See Figs. 4 (p. 152) and 15 (p. 262), where the line $t = t'$ is dotted. Its intersection with the boundaries of the physical regions determines the points s_0 and u_0.) Thus the physical region will coincide with the cut beginning at the first threshold branch point only for forward scattering under the idealized conditions of the model of § 12.1; in the general case the cut contains an unphysical part (the normal unphysical cut) from the first branch point to the boundary of the physical region. An unphysical cut related to the difference between $s_0(t')$ or $u_0(t')$ and the absolute lower boundaries of the physical region may be present in any reaction, and we shall not discuss it further.

In the reaction $N+\overline{N} \rightarrow \pi^+ + \pi^-$ the beginning of the physical s-channel region is $4m_N^2$, but the threshold branch points on the axis $\text{Re}\, s > 0$ will occur at $s_n = (2nm_\pi)^2$, $n = 1, 2, \ldots$ The normal unphysical cut will consist of the segment $4m_\pi^2 \leq s \leq 4m_N^2$. We have assumed that the character of this reaction is defined only by the external particles N, π. If we also allow for interactions with K-mesons, one must accommodate the K-meson branch points $s_{n'} = (2n'm_K)^2$, $n' = 1, 2, \ldots$, and the combined branch points $4(m_K+m_\pi)^2$ along the axis $\text{Re}\, s > 0$. In the t- and u-channels of this reaction the lowest branch points occur at $s = (m_\pi + m_N)^2$ and coincide with the absolute lower boundary of the physical region.

In the case of the scattering of nucleons $N+N \rightarrow N+N$ there is also an extensive unphysical cut in the t-channel due to the fact that the masses of the external particles are large in comparison with the masses of the internal particles. The internal particles which are important for the interactions of nucleons are pions, K-mesons etc. The lightest of these are pions, whose mass determines the beginning of the cut $t = (2m_\pi)^2$.

The study of Feynman integrals and the unitarity condition shows that in addition to the special points of the two types considered above (poles and threshold branch points), the amplitude $F(s, t, u)$ may have singularities of a third type as well—Landau singularities.[(154)] A point s_a of this type lies below the first threshold; hence s_a is sometimes called an anomalous threshold or anomalous branch point. [One should not confuse the threshold s_a with the anomalous thresholds introduced in § 7.5, eqn. (7.78).] The position of the anomalous threshold s_a may depend on two invariant variables. An anomalous branch point may arise when the masses of the external particle 1 and the internal particles a, b interacting with it are related by $m_1^2 > (m_a+m_b)^2$; here, as usual, the presence of the three-point interaction $1 \rightarrow a+b$ assumes the conservation of internal quantum numbers. The presence of an anomalous threshold may have important effects on phenomena near threshold. One may expect its influence far from the normal threshold to be unimportant. In any case, in what follows we shall assume that the amplitude $F(s, t, u)$ does not have Landau singularities.

Thus we shall postulate that the amplitude $F(s, t', u(s))$, for given t' in the physical region, is analytic in the whole complex s-plane with left and right-hand cuts, whose positions are determined by the normal thresholds in the s- and u-channels (see Fig. 11). If, moreover, there exist stable particles in the s- and u-channels, the corresponding poles must lie on the real s-axis between the cuts. Since s, t, u are equivalent, a similar structure must be postulated for the complex t- and u-planes.

For the scattering of particles with spin, one must examine spinor amplitudes (§§ 7.5 and 11.2) or helicity amplitudes (§ 7.4). The spinor amplitudes $\mathcal{M}$ may be expanded in terms of invariant amplitudes $F_l(s, t, u)$ free from kinematic singularities. Specifically, the invariant amplitudes $F_l(s, t, u)$ have the same analytic properties as the amplitude $F(s, t, u)$ for scalar particles. For a reasonable choice of covariants X_l, the invariant amplitudes F_l will have simple behavior under reflections (6.2) and crossing transformations [see eqns. (11.35) and (11.36)]. However, the condition for the discontinuity of the amplitude F_l across the cut arising from the unitarity condition is complicated in general and may contain other invariant amplitudes $F_{l'}$. Moreover, the functions $F_l(s, t', u(s))$ will be real-analytic only for a specific set of covariants X_l.[107, 108]

In each of the crossed channels $r = s, t, u$ the invariant amplitude F_l may be represented in the form

$$F_l(s, t, u) = D_{rl}(s, t, u) + iA_{rl}(s, t, u),$$

where the absorptive part A_{rl} determines the discontinuity of F_l across the right-hand cut in the complex r-plane. If the function F_l is real-analytic, A_{rl} will be equal to the imaginary part of the amplitude F_l; in this case the dispersive part D_{rl} of the amplitude F_l in the r-channel coincides with the real part of the amplitude. Consequently, for real-analytic invariant amplitudes F_l the expansion of the absorptive part of the spinor amplitude $\mathcal{A}$ in the s-channel in terms of the covariants X_l will have the form

$$\mathcal{A} = \sum_l A_{sl} X_l = \sum_l \operatorname{Im} F_l \cdot X_l. \tag{24}$$

The covariants X_l in the expansion (24) may be chosen by studying the left-hand side of the unitarity condition (1.68) for the reaction $1+2 \to 3+4$:

$$\begin{aligned} &\tfrac{1}{2}i(\langle 3, 4 | T | 1, 2\rangle - \langle 1, 2 | T | 3, 4\rangle^*) \\ &\quad = \tfrac{1}{2}i\bar{u}^{r_3}\bar{u}^{r_4}(\mathcal{M}^{r_1 r_2}_{r_3 r_4}(p_1 \ldots p_4) - \overline{\mathcal{M}}^{\mathrm{T} r_1 r_2}_{r_3 r_4}(p_1 \ldots p_4))u_{r_1} u_{r_2}. \end{aligned} \tag{25}$$

Here $\mathcal{M}^{\mathrm{T} r_1 r_2}_{r_3 r_4} = \mathcal{M}^{r_3 r_4}_{r_1 r_2}$. The adjoint matrix $\overline{\mathcal{M}}$ may be obtained from the matrix $\mathcal{M}^+$ by multiplying by the matrix $\bar{g}(i)$ (see § 5.1) in each index. In particular, for the Dirac matrix $\mathcal{M}_\alpha{}^\beta$ we will have $\overline{\mathcal{M}} = \gamma_0 \mathcal{M}^+ \gamma_0$.

Let us expand the spinor amplitudes in terms of covariants using (7.77) and the relation

$$\overline{\mathcal{M}}^{\mathrm{T}}(p_1 \ldots p_4) = \sum F_l(s, t, u)\, \overline{X}_l^{\mathrm{T}}(p_1 \ldots p_4).$$

The expansion (24) will be a consequence of (25) if the covariants satisfy the condition

$$X(p_1, p_2, p_3, p_4) = \bar{X}^{\mathrm{T}}(p_3, p_4, p_1, p_2). \tag{26}$$

These covariants are related to each other by the reflection PT. Reflections are easily taken into account if one uses an expansion in terms of t-channel invariants $X_l(K, Q)$ for the s-channel amplitude (see §§ 7.5 and 11.2).

The unitarity condition (1.68), rewritten for the spinor amplitudes (25), leads to the following expression for the absorptive part in the s-channel (24):

$$\sum_l A_{sl}(s, t, u)\, X_l(p_1, p_2, p_3, p_4) = \tfrac{1}{2} \sum_n \bar{\mathcal{M}}(n \mid p_3, p_4)\, \mathcal{M}(p_1, p_2 \mid n)\, \delta^4(p_1+p_2-p_3-p_4), \tag{27}$$

where $\mathcal{M}(p_1, p_2 \mid n)\, u(\boldsymbol{p}_1)\, u(\boldsymbol{p}_2) = \langle n \mid T \mid p_1, p_2 \rangle$ and $\bar{u}(\boldsymbol{p}_3)\, \bar{u}(\boldsymbol{p}_4)\, \bar{\mathcal{M}}(n \mid p_3, p_4) = \langle n \mid T \mid p_3, p_4 \rangle^*$ depend on the intermediate s-channel states $\mid n \rangle$. The sum in (27) is over a complete set of states $\mid n \rangle$, i.e. both over states with different numbers of particles, and over all possible states with a given number of particles. While the unitarity condition holds only in the physical region, eqn. (27) also defines the absorptive part along the real s-axis below the first s-channel threshold. The masses of stable particles $s = m_a^2$ may lie in this region $0 < s < s_{\min}$. The corresponding term in (27) is

$$\bar{\mathcal{M}}(a \mid p_3, p_4)\, \mathcal{M}(p_1, p_2 \mid a)\, \delta(s-m_a^2). \tag{28}$$

The single-particle state (28) has the following properties.

1. *Selection rules*

The quantities $\bar{\mathcal{M}}(p_1, p_2 \mid a)$ and $\bar{\mathcal{M}}(a \mid p_3, p_4)$ in (28) may be obtained from the matrix elements ("vertex parts") of the transitions $1+2 \to a$ and $a \to 3+4$ by continuation in the mass m_a^2 from the physical region $m_a^2 > s_{\min}$. Hence these quantities satisfy all the usual selection rules (aside from energy conservation).

2. *Factorizability*

The coefficient of $\delta(s-m_a^2)$ in (28) consists of two factors—the vertex parts referring to the interactions $a \leftrightarrow 1+2$ and $a \leftrightarrow 3+4$ of the internal particles a with the initial and final particles.

If the S-matrix is isospin- or SU_3-invariant, the invariant amplitudes F_l must be expanded in terms of isospin or SU_3 amplitudes using methods considered in §§ 8.3 and 9.3.

§ 12.3. Dispersion relations

Dispersion relations for the amplitude $F_s(s, t, u)$ in one of the variables s, t, u for fixed values of the other variable may be obtained from the Cauchy formula

$$f(z) = \frac{1}{2\pi i} \int_C \frac{f(z')\, dz'}{z'-z}$$

by choosing a contour C surrounding the cuts and consisting of a circle at infinity (Fig. 13). This contour also includes small circles around the poles. If the amplitude $F(s, t, u)$ vanishes sufficiently rapidly when one of the variables is large,

$$F(s, t_1, u) \sim s^{-a}, \quad (a > 0, s \to \infty), \tag{29}$$

then one may choose this amplitude as the function $f(s)$.

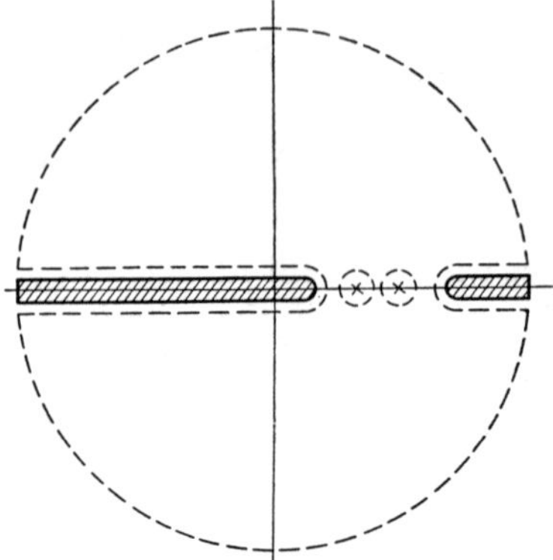

FIG. 13. Contour C (dotted line) defining the amplitude in the complex s-, t-, or u-plane in terms of Cauchy's theorem.

Thus if the singularities of the amplitude are cuts determined by normal thresholds and poles associated with stable particles (Fig. 13), and the amplitude itself vanishes with increasing energy (in the physical region) according to (29), the amplitude for the scattering of scalar particles $F(s, t_1, u)$ satisfies the following dispersion relation† in s for fixed $t = t_1 < 0$ (the pole terms are usually not written):

$$\begin{aligned} F(s, t_1, u) &= \frac{1}{\pi}\int\limits_{s_{\min}}^{\infty} \frac{A_s(s', t_1, u(s'))\, ds'}{s'-s} + \frac{1}{\pi}\int\limits_{u_{\min}}^{\infty} \frac{A_u(s(u'), t_1, u')\, du'}{u'-u} \\ &= \frac{1}{\pi}\int\limits_{s_{\min}}^{\infty} \frac{A(s', t_1, u(s'))}{s'-s}\, ds' + \frac{1}{\pi}\int\limits_{-\infty}^{s(u_{\min})} \frac{A_u(s', t_1, u(s'))}{s'-s}\, ds', \end{aligned} \tag{30}$$

where the discontinuities across the cuts, or the absorptive parts A_s, A_u, were defined in § 12.1. Analogously, if one fixes $s = s_1$ and $F(s_1, t, u) \sim t^{-a}$ for $|t| \to \infty$, one finds dispersion relations in t for fixed s:

$$F(s_1, t, u(t)) = \frac{1}{\pi}\int\limits_{t_{\min}}^{\infty} \frac{A_t(s_1, t', u(t'))}{t'-t}\, dt' + \frac{1}{\pi}\int\limits_{u_{\min}}^{\infty} \frac{A_u(s_1, t(u'), u')}{u'-u}\, du'. \tag{31}$$

† In quantum field theory, dispersion relations were first examined by Gell-Mann *et al.*[155]

For symmetry in s, t, and u we also write the dispersion relation for fixed $u = u_1$:

$$F(s, t, u_1) = \frac{1}{\pi} \int_{s_{min}}^{\infty} \frac{A_s(s', t(s'), u_1)}{s'-s} ds' + \frac{1}{\pi} \int_{t_{min}}^{\infty} \frac{A_t(s(t'), t', u_1)}{t'-t} dt', \tag{32}$$

which may be easily derived if the function $F(s, t, u)$ is regular in the complex s- (or t-) plane with its respective cuts along the real axis.

In these formulae $A_t(s, t, u)$ is the absorptive part in the t-channel or the discontinuity of the amplitude $F(s, t, u)$ across the cut for $t > t_{min}$ in the complex t-plane for fixed s or u:

$$A_t(s_1, t, u(s_1)) = \underset{\varepsilon \to +0}{} F(s_1, t+i\varepsilon, u(s_1)) - F(s_1, t-i\varepsilon, u(s_1)). \tag{33}$$

We note that if the amplitude vanishes as (29) along the real axis, it will also satisfy (29) when $|s| \to \infty$ in any direction in the complex plane [here $t = t_1$, the value for which the dispersion relation (30) is written]. Conditions of the type (29) in t or u must hold when we are interested in dispersion relations in these variables.

From general requirements of quantum field theory, it follows[149] that the amplitude $F(s, t, u)$ is polynomially bounded in s:

$$|F(s, t_1, u)| < s^{N(t_1)}, \qquad (s \to \infty).$$

For $t_1 < 0$ the asymptotic behavior in s is bounded by the Froissart theorem (see § 13.2).

It has been experimentally established that the total cross-section σ_{tot} does not fall at large energies (it may grow logarithmically) which, by virtue of $\sigma_{tot} \sim (1/s) \operatorname{Im} F(s, 0, u)$, means $N(0) \approx 1$.

If (29) does not hold, the dispersion relation may be written with a subtraction, i.e. for the function $F(s)/((s-s_1)(s-s_2)\ldots)$, whose rate of growth is reduced to that demanded by (29) by introducing factors $(s-s_i)^{-1}$. The number of such factors is called the number of subtractions. For one subtraction, in place of (25),

$$F(s, t_1, u) = (s_1, t_1, u(t_1, s)) + \frac{s-s_1}{\pi} \int_{s_{min}}^{\infty} \frac{A_s(s', t_1)\, ds'}{(s'-s_1)(s'-s)} + \frac{s-s_1}{\pi} \int_{-\infty}^{s(u_{min})} \frac{A_u(s', t_1)\, ds'}{(s'-s_1)(s'-s)}, \tag{34}$$

which contains an additional function—the value of the amplitude at the point s_1, depending on t_1. As above, we do not write the pole terms in (34).

The dispersion relation (30) in s for fixed t is correct not only for $t = t_1$, lying in the physical region of the s-channel, but for a broader region as well; the absorptive parts A_s and A_u may be analytically continued in t from $t < 0$ to $t > 0$ and then into the region of the t-channel. Crossing relations indicate the possibility of this continuation. This also applies to dispersion relations in other variables, of course. However, absorptive parts have physical meaning only along the physical part of the cut in the corresponding channel, where they are defined by the unitarity condition [see (13) and (15)] and can equal the imaginary part of the amplitude.

Dispersion relations allow one in principle to define the real part of the amplitude in terms of the imaginary part (in the physical region), if there is no unphysical cut or if the absorptive part along it is known from other considerations. When comparing dispersion relations with experiment it is thus convenient to take processes with no unphysical part of the cut. As we saw in §§ 12.1 and 12.2, one must examine the elastic forward scattering of the lightest possible particles for this to be true. For elastic forward scattering the imaginary part of the amplitude may be expressed using the optical theorem (7.53) in terms of the total cross-section, which may be obtained from experiment. This simplifies tests of dispersion relations considerably.

Dispersion relations for forward $\pi+N \to \pi+N$ scattering have been compared in detail with experiment, and have proven successful in this process. For pion–nucleon scattering they are treated in detail in a number of books.[1, 2, 11, 149, 156]

The dispersion relations (30)–(32) were written without taking account of isospin or SU_3. In an isospin-symmetric theory the transition amplitude $\langle 3, 4|T|1, 2\rangle$ must be expanded in terms of isospin amplitudes $T^I(s, t, u)$ (see §§ 8.3 and 11.3), for each of which one may write dispersion relations. The dynamics of the process $1+2 \to 3+4$ will be determined only by the value of the total isospin, but not by its projection. The threshold points may depend on isospin. Since one may write an expansion in terms of isospin amplitudes for each crossed channel, while the integration region in dispersion relations covers two channels, isospin crossing matrices will occur. Similarly, in an SU_3-invariant theory, the amplitude must be expanded in terms of SU_3 amplitudes (see §§ 9.3 and 11.3), and the dispersion relations postulated for these amplitudes.

If the external particles have spin, one must express the matrix element $\langle 3, 4|T|1, 2\rangle$ in terms of the spinor amplitude via (7.70). This amplitude then must be expanded in terms of invariant amplitudes F_l (see § 7.5), preserving crossing symmetry (see § 11.2) and real analyticity of F_l (see § 11.3). The amplitudes F_l contain information about the dynamics of the process; they have simple analytic properties, and dispersion relations may be written for them.

Pole terms and stable particles

As an example, let us write a pole term in (30). We insert the absorptive part $A_s^{(1)}$, corresponding to an internal stable particle a with mass $s = m_a^2$ below the physical s-channel threshold, into the first integral in (30). For the process $1+2 \to 3+4$, when all particles are spinless, we find from (22) and (30):

$$F^{(1)}(s, t) = \frac{1}{(2\pi)^6}\,\frac{g_{a34}g_{a12}}{m_a^2-s}. \tag{35}$$

This term describes an s-channel pole of F with residue consisting of two factors which refer to the interaction of the internal particle a with the final and initial particles, respectively. For elastic scattering, $g_{a34} = g_{a12}$, so the residue is positive.

If the external and internal particles have isospin, then (see § 12.2) the pole at $s = m_a^2$ will be present only in the amplitude corresponding to the isospin of the internal particle. Similar considerations hold for SU_3.

If an internal particle has spin J_a, then, since the invariant w^2 of the Poincaré group is conserved, the pole at $s = m_a^2$ appears in the partial wave amplitude $a_l(s)$ with $l = J_a$. Assuming that the amplitude $a_l(s)$ may be continued into the unphysical region of real s below the first threshold [see (53) and (54)], we may write the expansion (7.58) for $F(s, z_s)$

in this region as

$$F(s, z_s) = B \sum_{l=0}^{\infty} (2l+1)\, a_l(s)\, P_l(z_s).$$

Here B is a normalizing factor, and $z_s = 1+2t/(s-4m^2)$ (see § 7.2); thus the physical region in z_s will correspond to unphysical values $t > 0$. From this expansion it is clear that if the partial wave amplitude $a_l(s)$ has a pole $1/(m_a^2-s)$, the residue of $F(s, z_s)$ at this pole contains the factor $P_l(z_s(m_a^2))$. The factorizability of the residue at the pole follows directly from expressions (13) and (28) for absorptive parts. This property is very general, since it relies only on being able to define the absorptive parts (13) and (28), i.e. on being able to extend the unitarity condition into the unphysical region of real s below the first threshold.

By crossing symmetry, expression (35) also holds in the t- and u-channels. In the t-channel s is a momentum transfer. For identical masses $m_1 = m_2 = m_3 = m_4 = m$, $s = -(1-z_t)\times(t-4m^2)/2$. Here $z_t = \cos\theta_t$, where θ_t is the t-channel scattering angle (see § 7.2). The pole term (35) thus may be rewritten as

$$F^{(1)}(s, t) = \frac{1}{(2\pi)^6}\,\frac{2g_{a34}g_{a12}}{t-4m^2}\,\frac{1}{z_t(m_a^2)-z_t}, \tag{36}$$

where $z_t(m_a^2) = 1+2m_a^2/(t-4m^2)$ corresponds to the pole at $s = m_a^2$. In the physical t-channel region the denominator of (36) does not vanish since $z_t(m_a^2) > 1$. The amplitude (36) is maximum for forward scattering in the t-channel, where $s = 0$ and $F^{(1)}(0, t) = g_{a34}g_{a12}m_a^{-2}(2\pi)^6$.

In the pole approximation it is assumed that terms (35) describe the amplitude F in the physical region around a pole. The pole approximation corresponds to the Born approximation for channels crossed with respect to the one containing the pole. Thus for a pole in the s-channel $1+2 \to 3+4$, the amplitude (35) is the Born approximation for the t-channel $1+\bar{3} \to \bar{2}+4$, which corresponds to the exchange of particle a. The pole approximations (35) and (36) are shown in Fig. 14.

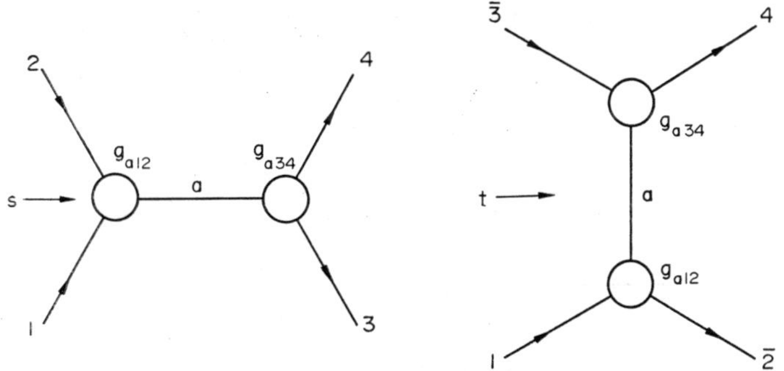

FIG.14. One-pole terms in the s- and t-channels describing two-particle scattering.

The Mandelstam representation

The double-dispersion relation, or the Mandelstam representation, is a hypothesis regarding analytic properties of the amplitude $F(s, t, u)$ as a function of two independent variables, e.g. s and t. It is thus a hypothesis regarding analytic properties of absorptive parts (e.g. regarding $A_s(s, t)$ as a function of t). The Mandelstam representation was first found by studying the unitarity condition to fourth order in perturbation theory.[102] The Mandelstam representation for the scattering amplitude of scalar particles $F(s, t, u)$ is

$$F(s, t, u) = \frac{1}{\pi^2}\int_{s_{\min}}^{\infty} ds' \int_{t_{\min}}^{\infty} dt' \frac{\varrho_{st}(s', t')}{(s'-s)(t'-t)}$$

$$+\frac{1}{\pi^2}\int_{t_{\min}}^{\infty} dt' \int_{u_{\min}}^{\infty} du' \frac{\varrho_{tu}(t', u')}{(t'-t)(u'-u)} + \frac{1}{\pi^2}\int_{u_{\min}}^{\infty} du' \int_{s_{\min}}^{\infty} ds' \frac{\varrho_{us}(u', s)}{(u'-u)(s'-s)}, \tag{37}$$

where the pole terms have not been written and $s'+t'+u' = \Sigma m_i^2$ (m_i are the masses of the external particles). Here $s_{\min}$, $t_{\min}$, and $u_{\min}$ correspond to the first two-particle states with quantum numbers of the respective channels; they are defined by the lightest masses of internal and external particles taking part in the interaction, and, for the model of § 12.1, are equal to $4m^2$. The representation (37) is crossing symmetric.

The functions ϱ_{st}, . . ., in the representation (37) are called spectral functions. They are real. The properties of the spectral functions determine the behavior of the amplitude $F(s, t, u)$. The spectral functions may be found in principle from the unitarity relation—in perturbation theory, at any rate. In (37) it is assumed that the spectral functions vanish rapidly enough that the integrals in (37) are well defined.

One may bound the region of nonzero spectral functions via the limits of integration in (37). Each of the variables in (37) varies along the physical edge of the right-hand cut, i.e. according to the rule $s \to s+i\varepsilon$, etc. Thus, for example, the spectral function may differ from zero only in the unphysical region $s > 4m^2$, $t > 4m^2$. More precisely, the lower boundary of the region in which the spectral function is nonzero may be calculated from the unitarity condition using perturbation theory. For the simplest model of equal-mass particles (§ 12.1) this region is shown in Fig. 15.

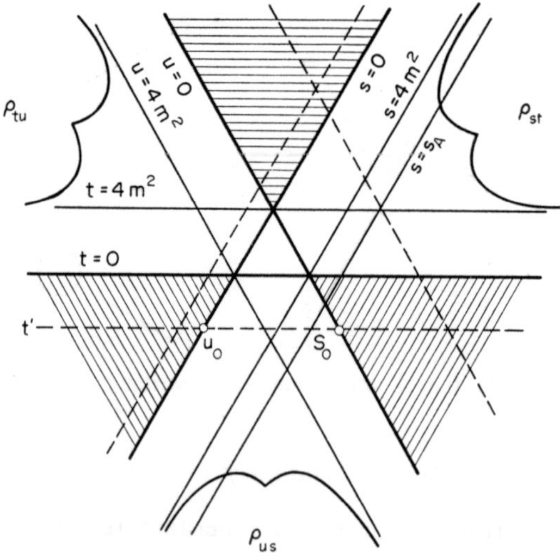

FIG. 15. Regions of nonzero double spectral functions ϱ_{st}, ϱ_{tu}, and ϱ_{us} in the Mandelstam *stu* plane, for equal-mass particles. Physical regions are shaded.

If the Mandelstam representation holds for the amplitude $F(s, t, u)$ then single-dispersion relations hold for the absorptive parts. For example, for the absorptive part A_s in the physical s-channel region,

$$A_s(s, t, u) = \frac{1}{2i}(F(s+i\varepsilon, t)-F(s-i\varepsilon, t)) = \frac{1}{\pi}\int_{t_{\min}}^{\infty} dt' \frac{\varrho_{st}(s, t')}{t'-t} + \frac{1}{\pi}\int_{u_{\min}}^{\infty} du' \frac{\varrho_{us}(u', s)}{u'-u}$$

$$= \frac{1}{\pi}\int_{t_{\min}}^{\infty} dt' \frac{\varrho_{st}(s, t')}{t'-t} + \frac{1}{\pi}\int_{-\infty}^{-s} dt' \frac{\varrho_{us}(4m^2-s-t', s)}{t'-t}. \tag{38}$$

This formula allows one to analytically continue A_s as a function of t or u. The representation (38) defines A_s as an analytic function in the complex t-plane with cuts from the boundary of ϱ_{st} to $t = \infty$ and from $t = -\infty$ to the boundary of the spectral function ϱ_{us}. Figure 15 shows the line $s = s_A > 0$ along which one integrates in (38). In deriving (38) we have the well-known relation $1/(x-i\varepsilon) = P(1/x)+i\pi\delta(x)$. One may find representations for A_t and A_u analogously. Using the Mandelstam representation, one may obtain a dispersion relation for the partial wave amplitude $a_l(s)$ as a function of s (see § 12.4).

If the spectral functions vanish too slowly, the Mandelstam representation [and also the dispersion relation (38)] must be written with subtractions. It is assumed here that the number of subtractions is finite.

§ 12.4. Partial wave amplitudes and fixed-energy dispersion relations. The Gribov–Froissart formula

As above, we shall consider the elastic scattering of scalar particles of equal mass m. Let us expand the invariant amplitude $F(s, t, u) = F(s, z)$ in terms of partial waves $a_l(s)$ in the s-channel†

$$\left.\begin{aligned} F(s, z) &= B\sum_{l=0}^{\infty}(2l+1)\, a_l(s)\, P_l(z) = \langle 3, 4|T|1, 2\rangle, \\ a_l(s) &= \frac{1}{2B}\int_{-1}^{1} dz F(s, z)\, P_l(z), \qquad B = \frac{8}{(2\pi)^5}, \end{aligned}\right\} \tag{39}$$

and study properties of $a_l(s)$ related to the existence of a dispersion relation for $F(s, z)$ at fixed s. Let the mass of the internal particles be $m_i > m$.

Let us first assume that this dispersion relation may be written without subtractions for $s_1 > 0$:

$$F(s_1, t, u) = \frac{1}{\pi}\int_{t_{\min}}^{\infty} \frac{A_t(s_1, t', u(t'))}{t'-t} dt' + \frac{1}{\pi}\int_{u_{\min}}^{\infty} \frac{A_u(s_1, t(u'), u')}{u'-u} du \tag{40}$$

$$(t_{\min} = u_{\min} = 4m^2),$$

† The coefficient B leads to the correct form of the unitarity condition (41).

where A_t and A_u are the respective absorptive parts in the t- and u-channels. We shall transform to the variable

$$z = 1+\frac{2t}{s-4m^2} = -1-\frac{2u}{s-4m^2} \tag{41}$$

(in the physical region $z = \cos\theta_s$). The dispersion relation (40) may be written in the form

$$\left.\begin{aligned} F(s, z) &= \frac{1}{\pi}\int_{z_0}^{\infty}\frac{dz'\, A_t(s, z')}{z'-z}+\frac{1}{\pi}\int_{-z_0}^{-\infty}\frac{dz'\, A_u(s, z')}{z'-z}, \\ z_0 &= \frac{s+4m^2}{s-4m^2} > 1. \end{aligned}\right\} \tag{42}$$

Let us now extract the partial wave $a_l(s)$ in (39) from the amplitude (42), using

$$\frac{1}{2}\int_{-1}^{1}\frac{P_l(z')}{z-z'}\,dz' = Q_l(z), \qquad (l = 0, 1, 2, \ldots), \tag{43}$$

which relates $P_l(z)$ to the Legendre function of the second kind $Q_l(z)$ (for all z except in the segment $-1 < z < 1$). Since $P_l(-z) = (-1)^l P_l(z)$,

$$\left.\begin{aligned} a_l(s) &= \frac{1}{\pi}\int_{z_0}^{\infty}(A_t(s, z')+(-1)^l A_u(s, z'))\, Q_l(z')\, dz', \\ A_u(s, z) &\equiv A_u(s, u(-z)), \quad u(x) = -\tfrac{1}{2}(s-4m^2)(1+x). \end{aligned}\right\} \tag{44}$$

Equation (44) for $a_l(s)$ is the basis for analyzing analytic properties of partial wave amplitudes. The analyticity in z of $F(s, z)$ allows one to replace it by the absorptive parts A_t and A_u, at the same time passing from the functions $P_l(z)$ to $Q_l(z)$ by integrating over values of z which are unphysical in the s-channel.

Starting with (44) and the properties of the Q_l, one may define a partial wave amplitude for complex l. The function $Q_l(z)$ is analytic in l in the right-half plane, vanishing exponentially for large Re l:

$$Q_l(z) \approx \frac{1}{\sqrt{l}}\,e^{-(l+\frac{1}{2})\xi}, \qquad (\xi = \cosh^{-1} z,\ \operatorname{Re} l > 0,\ |l| \to \infty). \tag{45}$$

When passing to infinity along a line parallel to the imaginary axis, the function $Q_l(z)$, according to (45), is oscillatory.

A number of consequences follow from (44).

1. For large real (integral) angular momenta l, the amplitude $a_l(s)$ vanishes exponentially with l. In this case, the largest contribution to $a_l(s)$ comes from the region of smallest z', i.e. from those values of z' which are closest to the boundary of the physical region $z = 1$. This contribution may come from poles with the smallest mass either in the u-channel for $u = m_u^2$ (the absorptive part $A_u(s, z)$), or in the t-channel for $t = m_t^2$ (the absorptive part $A_t(s, z)$). The position of these poles at m_t^2 and m_u^2 is related to poles in z of A_t and A_u by

$z^{(t)} = 1+2m_t^2/(s-4m^2)$ and $z^{(u)} = 1+2m_u^2/(s-4m^2)$. The partial waves with very large l describe peripheral interactions.

2. Passing to complex angular momenta l, one may find the analytic continuation of $a_l(s)$ by starting from (44). However, the analyticity of $Q_l(z)$ in the right-half l-plane, Re $l > 0$, is still insufficient to continue $a_l(s)$ into the region of complex angular momenta l via (44). The right-hand side of (44) depends on l not only through Q_l but also via the factor $(-1)^l = e^{i\pi l}$, which grows without bound when $|l| \to \infty$, Re $l > 0$. To circumvent this difficulty, one must perform separate analytic continuations for the functions a^+ and a^-. We then obtain the Gribov–Froissart formula:[157, 158]

$$a^{\pm}(l, s) = \frac{1}{\pi B} \int_{z_0}^{\infty} dz'[A_t(s, z') \pm A_u(s, z')]\, Q_l(z'). \tag{46}$$

By virtue of the properties of the Q_l these functions are analytic in the right half l-plane. Since they are defined for all complex angular momenta l with Re $l > 0$, the functions $a^{\pm}(l, s)$ have meaning only for integral values of l. For even $l = 0, 2, \ldots, a^+(l, s)$ is the partial wave amplitude $a_l(s)$, and for odd $l = 1, 3, \ldots, a^-(l, s)$ is equal to $a_l(s)$.

Thus in the expansion (38) the functions $a^{\pm}(l, s)$ lead respectively to the symmetric and antisymmetric parts of $F^{\pm}(s, z)$ with respect to z:

$$\begin{aligned} F^{\pm}(s, z) &= B \sum_{l=0}^{\infty} (2l+1)\, P_l(z)\, a^{\pm}(l, z) \\ &= B \sum_{l=0}^{\infty} (2l+1)\, [P_l(z) \pm P_l(-z)]\, a_l(s) = F^+(s, z) \pm F^-(s, -z), \end{aligned} \tag{47}$$

i.e. F^+ contains only even $l = 0, 2, \ldots$, while F^- contains only odd $l = 1, 3, \ldots$.

Up to now it has been assumed that all integrals in (40) and (46) converge, so that subtractions in the dispersion relation (42) are unnecessary. If, however, the absorptive parts A_t and A_u behave as polynomials of degree N (for fixed s), then N subtractions are necessary, and (46) will define $a^{\pm}(l, s)$ as an analytic function of l in the region Re $l > N$. In view of the asymptotic behavior of $Q_l(z)$ for large z,

$$\begin{gathered} Q_l(z) \to z^{-(l+1)}, \quad |z| \to \infty, \quad |\arg z| < \pi, \\ 2l \neq -3, -5, -7, \ldots, \end{gathered} \tag{48}$$

the integrals (46) converge only for Re $l > N$. The subtraction terms in the dispersion relation (42) do not contribute to $a^{\pm}(l, s)$ because of their polynomial nature.

We now discuss the uniqueness of the Gribov–Froissart formula (46) for analytically continuing the partial wave amplitude to complex l when it is given for integral positive l. At first glance it seems obvious that the Gribov–Froissart function $a^{\pm}(l, s)$ may always have terms added to it of the type $f(l, s) \sin \pi l$, vanishing for integral l and giving a different continuation of the partial wave amplitude. However, the uniqueness of the analytic continuation via (46) is guaranteed by Carlson's theorem.

CARLSON'S THEOREM.[159] *Let* $F(l)$ *be analytic for* $\mathrm{Re}\, l > N$ *and* $F(l) < Ce^{\lambda|l|}$ for $|l| \to \infty$, $\mathrm{Re}\, l > 0$, *where* $\lambda < \pi$. *Then, if* $F(l) = 0$ *for* $l = 1, 2, 3, \ldots$, *the function* $F(l)$ *is identically equal to zero.*

According to Carlson's theorem, functions of the type $f(l, s) \sin \pi l$ are inadmissible since (for polynomial behavior of $f(l, s)$) they grow as $e^{\pi |l|}$ along a straight line parallel to the imaginary l-axis. Consequently, if the Gribov–Froissart amplitude (46) satisfies the condition

$$a(l, s) < Ce^{\lambda|l|}, \quad |l| \to \infty, \quad \mathrm{Re}\, l > N, \quad \lambda < \pi, \tag{49}$$

the analytic continuation via (8) will be unique. From (46) it is clear that $a^{\pm}(l, s)$ vanishes most slowly with $|l|$ along the line $\mathrm{Re}\, l = N$, where

$$a^{\pm}(l, s) \sim \int^{\infty} (A_t \pm A_u)\, e^{i\beta\xi - (N - \frac{1}{2})\xi}\, dz, \quad (\beta = \mathrm{Im}\, l \to \infty). \tag{50}$$

Since by assumption the absorptive parts $A_t \pm A_u$ are polynomially behaved (as z^N for $z \to \infty$), the integral in (50) in general cannot grow exponentially as $\beta \to \infty$.

Thus under the assumption of polynomial boundedness of the growth of the absorptive parts A_t and A_u, the Gribov–Froissart formula (46) provides a unique continuation of $a_l(s)$ into the region $\mathrm{Re}\, l \geqslant N$, specifically: except for $a^{\pm}(l, s)$, given by (46), there exist no functions $f^{\pm}(l, s)$ coinciding with the physical partial waves $a_l(s)$ for even (odd) values of $l \geqslant N$ and satisfying the condition (49).

Proof. The difference $f^{\pm}(l, s) - a^{\pm}(l, s)$ vanishes identically by virtue of Carlson's theorem.

Let us turn to the unitarity condition for partial wave amplitudes. Let $s = s_\delta$ be the inelastic threshold so that in the region $4m^2 < s < s_\delta$ only elastic scattering is possible (for pion–pion scattering $s_\delta = 16m_\pi^2$). Then in this region the partial wave amplitude $a_l(s)$ for $l = 0, 1, 2, \ldots$, satisfies the "elastic" unitarity condition (7.68):

$$a_l(s) - a_l^*(s) = 2i \sqrt{\frac{s - 4m^2}{s}}\, a_l(s)\, a_l^*(s). \tag{51}$$

Let us now write the general unitarity condition for the functions $a^{\pm}(l, s)$ starting from (51). Using Carlson's theorem one may verify that the analytic continuation of (51) is

$$a^{\pm}(l, s) - (a^{\pm}(l^*, s))^* - 2i \sqrt{\frac{s - 4m^2}{s}}\, a^{\pm}(l, s)\, (a^{\pm}(l^*, s))^* = 0, \tag{52}$$

since the left-hand side of (52) is an analytic function of l, which vanishes for even (odd) $l > N$ and has asymptotic behavior given by (46). The unitarity condition (52) holds separately for the functions $a^+(l, s)$ and $a^-(l, s)$, since (51) holds for each value $l = 0, 1, 2, \ldots$, separately.

The existence of separate unitarity conditions for $a^+(l, s)$ and $a^-(l, s)$ means that the functions $a^+(l, s)$ and $a^-(l, s)$ may have different (and independent) sets of poles.

Analytic properties of $a_l(s)$ in the s-plane

These properties may easily be found when the Mandelstam representation holds for the amplitude $F(s, t, u)$ in (39) (for equal-mass particles), allowing one to write a dispersion relation of the type (38) for the absorptive parts A_t and A_u. We shall start with (44), expressing the partial wave amplitude $a_l^{\pm}(s)$ in terms of the function Q_l and the absorptive parts $A_t \pm A_u$. This formula will also define $a_l(s)$ outside the physical s-channel region.

The functions $Q_l(1+2t/(s-4m^2))$ in (44) have a cut from $s = -\infty$ to $s = 4m^2 - t$ for integral l. By virtue of the Mandelstam representation or the dispersion relation (38), the functions A_t and A_u have a right-hand cut in the complex s-plane from $4m^2$ to ∞ and a left-hand cut from $-\infty$ to the boundary of the spectral function ϱ_{tu} (see Fig. 15). We assume here that the amplitude $F(s, t, u)$ has no poles in the t- or u-channels. In the absence of such poles, the partial wave amplitude $a_l(s)$ defined via (44) will thus have two cuts: a right-hand cut from $s = 4m^2$ to ∞ and a left-hand cut from $s = -\infty$ to $s = 0$. In the interval between the cuts $a_l(s)$ is real, so that $a_l(s^*) = a_l^*(s)$.

Consequently, if $a_l(s) \to 0$ for $|s| \to \infty$, then one may write an unsubtracted dispersion relation for $a_l(s)$

$$a_l(s) = \frac{1}{\pi}\int_{4m^2}^{\infty} ds' \frac{b_l(s')}{s'-s} + \frac{1}{\pi}\int_{-\infty}^{0} ds' \frac{b_l(s')}{s'-s}, \tag{53}$$

where $b_l(s)$ is the discontinuity of $a_l(s)$ across the cut.

The discontinuity in the physical region across the right-hand cut in principle may be found from the unitarity condition. This cut, however, contains amplitudes of more complicated processes for $s \geqslant 16m^2$. In the elastic scattering region $4m^2 < s < 16m^2$ the discontinuity may be expressed by virtue of (51) in terms of the partial wave amplitudes

$$\underset{\varepsilon \to 0}{b_l(s)} = a_l(s+i\varepsilon) - a_l(s-i\varepsilon) = 2i\sqrt{\frac{s-4m^2}{s}}\, a_l(s+i\varepsilon)\, a_l(s-i\varepsilon). \tag{54}$$

The discontinuity of the function $a_l(s)$ across the left-hand cut is associated with crossed reactions. Equations (53) and (54) suffice for an approximate definition of the partial wave amplitudes for low energies.[160, 161]

Resonances and poles on the unphysical sheet

Let us now turn to the concept of an unstable particle, or a resonance, in S-matrix theory. The notion of a resonance in relativistic S-matrix theory is related, as in the nonrelativistic theory, to the well-known Breit–Wigner formula. In view of the importance of this question in elementary particle theory, we present the basic relations below.

The state of an unstable particle, or resonance, was characterized in § 2.3 by its mass or energy distribution. This distribution may be obtained if the analytic amplitude has a pole s_r in the complex s-plane. This pole may occur only on the second, unphysical sheet, since complex poles on the physical sheet are physically inadmissible.

Let us show that the existence of a pole of the amplitude on the second Riemann sheet is consistent with unitarity. We shall use (54) as the analytic continuation of the unitarity condition to unphysical s. Let us call the partial wave amplitude on the physical sheet a_l^{I}, and the partial wave amplitude on the second sheet a_l^{II}; the functions a_l^{I} and a_l^{II} are the values of the analytic function a_l defined on all sheets. The function a_l^{II} is

$$a_l^{\mathrm{II}}(s) = a_l^{\mathrm{I}}(s^*).$$

We shall now write (54) in a form containing functions on both sheets (for the same complex argument s):

$$a_l^{\mathrm{I}}(s) - a_l^{\mathrm{II}}(s) = 2i\sqrt{\frac{s-4m^2}{s}}\, a_l^{\mathrm{I}}(s)\, a_l^{\mathrm{II}}(s), \tag{55}$$

so that

$$a_l^{\mathrm{II}}(s) = \frac{a_l^{\mathrm{I}}(s)}{1+2i\sqrt{(s-4m^2)/s}\, a_l^{\mathrm{I}}(s)} = \frac{a_l^{\mathrm{I}}(s)}{S_l^{\mathrm{I}}(s)}, \tag{56}$$

where

$$S_l^{\mathrm{I}}(s) = e^{2i\delta_l(s)} \tag{57}$$

is the S-matrix element (for real s) corresponding to angular momentum l. Formula (56) gives the analytic properties of the amplitude a_l^{II} on the second sheet. The function a_l^{II} has the same branch points as a_l^{I}. The poles of a_l^{II} occur at the points at which S_l^{I} has zeroes, i.e. they can occur at complex s. At points where a_l^{I} has poles (for real $s < s_{\mathrm{min}}$), the function a_l^{II} has no poles.

Let a pole of the function $a_l(s)$ occur on the second sheet at the point $s_r^{\mathrm{II}} = (\mu - i\Gamma/2)^2$, where μ and Γ are positive:

$$a_l(s) = \frac{1}{(2\pi)^6}\, \frac{g^2}{s_l^{\mathrm{II}} - s}. \tag{58}$$

This expression is just the Breit–Wigner formula when $\Gamma/\mu \ll 1$:

$$\frac{g^2}{s_r^{\mathrm{II}} - s} \approx -\frac{g^2}{2\mu}\, \frac{1}{\sqrt{s} + \sqrt{s_r^{\mathrm{II}}}} = \frac{\Gamma'}{\sqrt{s} - \mu + i\Gamma/2},$$

if one sets $\Gamma' = g^2/2\mu$ and remembers that $\sqrt{s}$ is the energy in the centre of mass system.

Thus a pole of the partial wave amplitude $a_l(s)$ on the unphysical sheet at $s_r = (\mu - i\Gamma/2)^2$ must be interpreted as a consequence of the existence of an unstable particle with spin $J = l$, (mean) mass μ, and lifetime $1/\Gamma$. These poles are of dynamical origin, and the question of the arbitrariness of their position remains open. A pole (58) in the amplitude shows up experimentally as a resonance in the total cross-section $\sigma_{\mathrm{tot}}(s)$, where the form of the curve $\sigma_{\mathrm{tot}}(s)$ near $s = \mu^2$ will be governed by

$$|a_l(s)|^2 \sim \frac{g^4}{(s-\mu^2+\Gamma^2/4)^2 + \Gamma^2\mu^2}. \tag{59}$$

Distributions of this type have already been considered in § 2.3.

When $\Gamma \to 0$ the pole s_r moves on to the real s-axis, and in this case must be associated with a stable particle; for self-consistency of such a treatment the cut to the left of the point $s_r + \varepsilon$, $\varepsilon > 0$ must be neglected. The residues at the poles of stable particles always factorize;

we assume that this property also holds for poles of the type (58) corresponding to unstable particles.

The elastic scattering of equal-mass particles that we have just considered is the simplest example in which one can illustrate the connection between poles on the unphysical sheet and resonances. For any process $1+2 \rightarrow 3+4$ we shall postulate that the pole term (in the invariant amplitude), corresponding to a resonant particle R, has the form

$$F(s, t) = \frac{1}{(2\pi)^6} \frac{g_{R12} g_{R34}}{s_R^{\mathrm{II}} - s}, \qquad (s \sim \mu); \tag{60}$$

here the ratio Γ/μ need not be small. As in the case of stable particles, the values of g_{R12} and g_{R34} characterize the interaction in the processes $R \leftrightarrow 1+2$ and $R \leftrightarrow 3+4$, i.e. the respective vertex parts.

Let us list the properties of unstable particles (see § 2.3).

1. Unstable particles may have definite values of spin J and internal symmetry and reflection quantum numbers. The set of quantum numbers denoting a particle is defined by the symmetry group of the interaction in which the particle is produced. For the strongly interacting resonant particles, this set includes the baryon number B, the isospin I, the hypercharge Y, the SU_3 multiplicity $\underline{n}$, the parity η_P, and the G-parity.

2. An unstable particle is associated with a pole (60) of the amplitude on the second sheet in the s-channel with quantum numbers J, B, I, $\underline{n}$, η_P, G. Here the position of the pole (i.e. the average mass μ and with Γ) does not depend on the reaction in which the resonance is observed.

3. The residue of the amplitude at a pole associated with an unstable particle factorizes according to the same rule as in the case of a pole corresponding to a stable particle.

As we stressed in § 2.3, the concept of an unstable particle is strictly approximate in nature. The properties of unstable particles mentioned above, and eqn. (60), may be proven to first order in perturbation theory, which, however, is unconvincing in view of the strength of the interaction. The basis of (60) is the assertion that an unstable particle has all properties of a stable one except for the reality of the point s_r at which the pole of the amplitude occurs.

Resonant behavior of an amplitude near $s = \mu$ entails not only the existence of a sharp maximum of the cross-section (59), but also a characteristic variation of the phase shift $\delta(s)$ when passing through the resonance.

The simultaneous variation of δ and $|a_l|$ in the resonance region may be depicted on polar Argand diagrams.

In the elastic region, when the energy is too low for inelastic processes, the partial wave amplitude a_l, in accord with (51), may be written in the form

$$a_l(s) \equiv \sqrt{\frac{s}{s-4m^2}}\, f_l(s) = \sqrt{\frac{s}{s-4m^2}}\, \frac{e^{i2\delta_l(s)}-1}{2i}. \tag{61}$$

If the value of s is above the inelastic threshold (e.g. $s > 16m_\pi^2$ for pion–pion scattering), then the opening of inelastic channels in the unitarity condition may be accounted for phenomenologically by introducing the factor η_l in (61):

$$f_l(s) = (\eta_l e^{i2\delta_l(s)} - 1)/2i. \tag{62}$$

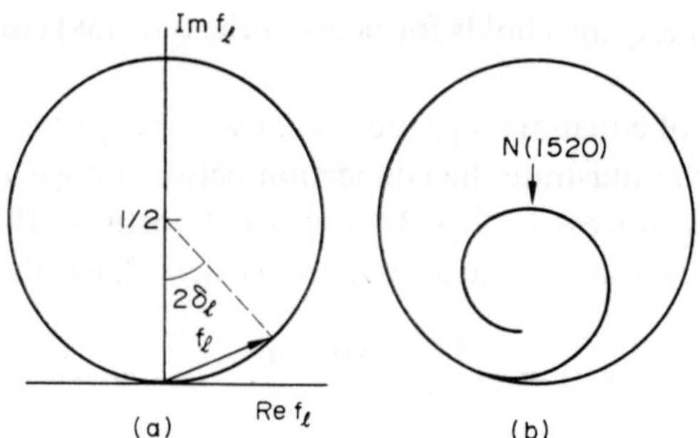

FIG. 16. Argand circles describing partial-wave scattering amplitudes. (a) Interpretation of phase shift δ_l. (b) Argand circle for a typical resonance, $N(1520, 3/2^-)$.

In the plane with axes Re f_l and Im f_l, the amplitude (62) is depicted by a point which describes a trajectory when s varies. For a resonance this trajectory (on the Argand diagram), according to (61) and (62), is a circle with radius depending on η_l (Fig. 16a), and with maximum $(0, f_{max})$ for $\delta_l = \pi/2$ at the resonance. Here the derivative $d\delta_l/ds$ is positive and proportional to the lifetime $1/\Gamma$, so that the more stable the particle the more rapidly δ_l rises as a function of s.

In actual cases the radius of the trajectory on the Argand diagram will depend on energy, since, as energy increases, new inelastic channels open (Fig. 16b). The form of the trajectory will be more complicated in the case of resonances in the reaction $1+2 \to 3+4$ with different particles 1, 2, 3, 4.

§ 12.5. Analytic properties of form factors. The pion form factor

In this section we shall follow the same train of logic as in the previous sections of Chapter 12 dealing with properties of the amplitude. Instead of invariant amplitudes we shall now consider the invariant form factors $F_i(t)$ [see (11.72)], which are functions of one variable. The form factors $F_i(t)$ contain all the information about the dynamics; the study of analytic properties of $F_i(t)$ greatly increases one's understanding of interactions.

Let us consider the electromagnetic vertex function of the pion π^+ as the simplest example. This function occurs in the elastic electron–pion scattering amplitude $e^- + \pi^+ \to e^- + \pi^+$ in the matrix element of the electromagnetic current:

$$\Gamma_\mu(p_1, p_2; 1, 2) = \langle \boldsymbol{p}_2 | j_\mu(0) | \boldsymbol{p}_1 \rangle = \frac{1}{(2\pi)^3} F_1^s(t)(p_1+p_2)_\mu + \frac{1}{(2\pi)^3} F_2^s(t)(p_1-p_2)_\mu, \quad (63)$$

where F_1^s and F_2^s are the invariant form factors depending on $t = (p_1-p_2)^2$; here $t \leq 0$. The current j_μ is Hermitian. In the process $e^- + \pi^+ \to e^- + \pi^+$, the quantity t is the square of the momentum transfer. Because the electromagnetic current obeys the continuity equation $\partial_\mu j^\mu = 0$, the form factor F_2^s is equal to zero:

$$\langle \boldsymbol{p}_2 | \partial_\mu j^\mu(0) | \boldsymbol{p}_1 \rangle = -i \langle \boldsymbol{p}_2 | [P_\mu, j^\mu(0)] | \boldsymbol{p}_1 \rangle = -i(p_2-p_1)_\mu \langle \boldsymbol{p}_2 | j^\mu(0) | \boldsymbol{p}_1 \rangle.$$

The vertex function Γ_μ also is involved (for a different region of t) in the amplitude for the annihilation process $e^- + e^+ \to \pi^- + \pi^+$, which is the crossed reaction of the elastic $e^-\pi^+$

scattering process. The matrix element of the current in this amplitude (by virtue of the continuity equation) is equal to

$$\Gamma_\mu(p_1, p_2; 1, 2) = \langle \boldsymbol{p}_1, \boldsymbol{p}_2; \text{out} \,|\, j_\mu(0) \,|\, 0\rangle = \frac{1}{(2\pi)^3}(-p_{\bar{1}}+p_2)_\mu F_{\bar{1}}^a(t),$$

$$t = (p_2+p_{\bar{1}})^2, \quad p_1 = -p_{\bar{1}}. \tag{64}$$

In § 11.4 we saw that the crossing-conjugate matrix elements (63) and (64) are values of the same function Γ_μ in different physical regions, $F_{\bar{1}}^a(t) = F_{\bar{1}}^s(t) = F(t)$.

We now turn to the analytic properties of $F(t)$ in the complex t-plane. Let us write the matrix element (64) in terms of interpolating fields $\varphi(x)$ using the reduction formula

$$\langle \boldsymbol{p}_{\bar{1}}, \boldsymbol{p}_2; \text{out} \,|\, j_\mu(0) \,|\, 0\rangle = \frac{1}{(2\pi)^3}(p_2-p_{\bar{1}})_\mu F(t)$$

$$\equiv -\frac{i}{(2\pi)^{3/2}} \int d^4x e^{ip_2x} \vec{K}_x \langle \boldsymbol{p}_{\bar{1}} \,|\, \theta(x^0)\, [\varphi(x), j_\mu(0)] \,|\, 0\rangle. \tag{65}$$

The matrix element occurring in the amplitude for the process $\pi^- + \pi^+ \to e^- + e^+$ may be written in analogous form:

$$\langle 0 \,|\, j_\mu(0) \,|\, \boldsymbol{p}_{\bar{1}}, \boldsymbol{p}_2; \text{in}\rangle = \frac{1}{(2\pi)^3}(p_2-p_{\bar{1}})_\mu F'(t)$$

$$\equiv -\frac{i}{(2\pi)^{3/2}} \int d^4x e^{-ip_2x} \vec{K}_x \langle 0 \,|\, \theta(-x^0)\, [j_\mu(0), \varphi^+(x)] \,|\, \boldsymbol{p}_{\bar{1}}\rangle, \tag{66}$$

where $F'(t)$ is an invariant form factor. Particles 1 and 2 are the same in formulae (65) and (66).

Let us form the difference between (65) and the complex conjugate expression (66), bearing in mind the Hermiticity of the current j_μ:

$$\langle \boldsymbol{p}_{\bar{1}}, \boldsymbol{p}_2; \text{out} \,|\, j_\mu(0) \,|\, 0\rangle - \langle 0 \,|\, j_\mu(0) \,|\, \boldsymbol{p}_1, \boldsymbol{p}_{\bar{2}}; \text{in}\rangle^*$$

$$= -\frac{i}{(2\pi)^{3/2}} \int d^4x e^{ip_2x} \vec{K}_x \langle \boldsymbol{p}_1 \,|\, [\varphi(x), j_\mu(0)] \,|\, 0\rangle. \tag{67}$$

The right-hand side of (65) is an analytic function of $t = (p_{\bar{1}}+p_2)^2$ in the upper-half plane. This fact may be proven rigorously[103, 162] from the locality of fields and currents and the positivity of the energy spectrum. Thus the value of the form factor $F(t)$ in the physical region (along the positive real axis) is the boundary value of a function analytic in the upper-half plane. From (65) and (66) it is clear that if the function $F(t)$ is analytic in the upper-half plane, the function $F'^*(t)$ will be analytic in the lower-half plane.

Formula (67) has meaning for real $t > 0$ lying in the physical region of the process $\pi^+\pi^- \to e^-e^+$. It tells us the difference of the boundary values of the functions $F(t)$ and $F'^*(t)$, defined on opposite sides of the real axis. By analytic continuation of the right-hand side of (67) with respect to masses of the vertex particles (i.e. with respect to the pion masses), we can pass along the real axis to negative t. But for sufficiently large $t > 0$, the right-hand side of (67) will vanish. This means that $F(t)$ and $F'^*(t)$ define a single analytic function with a cut from some $t_{\min}$ to ∞, while (67) defines the discontinuity of this function. Thus the discontinuity of $F(t)$ is

$$\frac{1}{(2\pi)^3}(p_2-p_{\bar{1}})_\mu (F(t+i\varepsilon)-F(t-i\varepsilon))$$

$$= -\frac{i}{(2\pi)^{3/2}}\sum_n \int d^4x e^{ip_2x}\vec{K}_x\langle p_{\bar{1}}\,|\,\varphi(x)\,|\,n\rangle\langle n\,|\,j_\mu(0)\,|\,0\rangle$$

$$= \sum_n \langle \boldsymbol{p}_{\bar{1}}, \boldsymbol{p}_2;\ \text{out}\,|\,n, \text{in}\rangle\langle n, \text{in}\,|\,j_\mu(0)\,|\,0\rangle$$

$$= i(2\pi)^4\sum_n \int \langle \boldsymbol{p}_{\bar{1}}, \boldsymbol{p}_2\,|\,T\,|\,n\rangle\, dR_n(p_{\bar{1}}+p_2)\langle n, \text{in}\,|\,j_\mu(0)\,|\,0\rangle. \tag{68}$$

Here we have introduced a sum over a complete set of intermediate states $|\,n, \text{in}\rangle$, and have allowed for the fact that the second term in the commutator (67) leads to the matrix element $\delta^4(p_2-p_n)\,\langle p_n\,|\,J(0)\,|\,0\rangle$, where $J(x) = K_x\varphi(x)$ is the pion current. This matrix element vanishes, since $\langle p_n|$ is a one-pion state. According to (68) the first term $|\,n\rangle$ must be a two-particle term (since otherwise $\langle \boldsymbol{p}_1, \boldsymbol{p}_2\,|\,T\,|\,n\rangle$ would vanish). Relation (68) may be called the unitarity condition for the form factor. It defines the branch points of the function $F(t)$. The first branch point corresponds to the two-pion threshold, so that $t_{\min} = 4m^2$. The next threshold point is at $t = 16m^2$ (by G-parity conservation), etc.

Let us now use PT invariance (see § 11.4):

$$\langle 0\,|\,j_\mu(0)\,|\,\boldsymbol{p}_{\bar{1}}, \boldsymbol{p}_2;\ \text{in}\rangle = \langle \boldsymbol{p}_{\bar{1}}, \boldsymbol{p}_2;\ \text{out}\,|\,j_\mu(0)\,|\,0\rangle. \tag{69}$$

From this and from (67) it follows that the discontinuity of the form factor $F(t)$ in (68) is purely imaginary, and the left-hand side of the unitarity condition may be written $2i(p_2-p_1)_\mu$ $\text{Im}\,F(t)/(2\pi)^3$. Since the discontinuity of $F(t)$ is purely imaginary, we will have

$$F(t^*) = F^+(t), \tag{70}$$

so that the form factor $F(t)$ is a real analytic function of t. On the real t axis for $t < 4m^2$ (outside the cut) $F(t)$ is real.

Thus the pion form factor $F(t)$ is an analytic function in the whole complex t-plane with a cut along the real axis from the threshold branch point $t = 4m^2$ to ∞. The discontinuity of $F(t)$ along the cut may be determined from the unitarity condition (68).

For the asymptotic behavior of the form factor $F(t)$ as $|t| \to \infty$ there are no rigorous theoretical estimates. From scattering experiments it is known that electromagnetic form factors vanish rapidly as $t \to -\infty$. Hence one may assume that the value of $F(\infty)$ is finite. Then, for the form factor $F(t)$, one may write a dispersion relation[163, 164] with one subtraction

$$F(t) = F(0)+t\int_{4m^2}^{\infty}\frac{\text{Im}\,F(t')}{t'(t'-t)}\,. \tag{71}$$

The integrand contains only the contribution of the annihilation process $e^+ + e^- \to \pi^+ + \pi^-$ (since $t \geqslant 4m^2$). $\text{Im}\,F(t)$ may in principle be found from the unitarity condition. Relation (71) defines the form factor for all complex t, and, in particular, along the negative real t-axis, where F is equal to the form factor F_1^s in (63). In this region, the form factor F determines elastic $e^-\pi^+$ scattering.

The value of $F(0)$ may be found from

$$\langle \boldsymbol{p}_2, \pi^+ \,|\, Q \,|\, \boldsymbol{p}_1, \pi^+\rangle = 2p_0\delta(\boldsymbol{p}_1-\boldsymbol{p}_2),$$

where Q is the electric charge (in units of $|e|$). Writing Q in the form of a spatial integral over $j_0(x)$, we find

$$\langle \boldsymbol{p}_2, \pi^+ \,|\, \int j_0(x)\, d^3x \,|\, \boldsymbol{p}_1, \pi^+\rangle = (2\pi)^3\, \delta(\boldsymbol{p}_1-\boldsymbol{p}_2) \langle \boldsymbol{p}_1 \,|\, j(0, x_0) \boldsymbol{p}_1\rangle,$$

which together with (63) gives $F(0) = 1$.

Up to now we have considered specific terms of the pion triplet. The isospin structure of the vertex function for pions with isospin indices i_1 and i_2 ($i = 1, 2, 3$) may be obtainep from (63) or (64) using the rules of § 8.3. Under isospin rotations the electromagnetic current j_μ transforms as the charge Q, i.e. as the sum of an isosinglet $j_\mu^{I=0}$ and the third component of an isotriplet $j_{\mu 3}^{I=1}$ (see § 8.1 and the Gell-Mann–Nishijima formula $Q = I_3+\frac{1}{2}Y$). We find from (64)

$$\langle p_{\bar{1}}, i_1; p_2, i_2 \,|\, j_\mu(0) \,|\, 0\rangle = \frac{1}{(2\pi)^3}\, \varepsilon_{3i_1i_2}(-p_{\bar{1}}+p_2)_\mu F(t). \tag{72}$$

The term $\varepsilon_{3i_1i_2}$ is easily understood from the fact that both sides of (72) must be symmetric with respect to interchange of the variables denoting the pions. According to (72), the electromagnetic vertex function of pions is the third component of an isovector.

Generalizations. Electromagnetic form factors of the nucleon

The case of the electromagnetic vertex function of the pion is in many respects the simplest. Since pions are spinless and since the electromagnetic current is conserved, this vertex function is described by only one form factor $F(t)$. The existence of isospin does not increase the number of form factors (as a consequence of current conservation). Since the pion mass is the lowest of the masses of external and internal particles that one would consider in a theory of strong interactions, the cut of the function $F(t)$ coincides in its entirety with the physical region $4m_\pi^2 < t < \infty$ of the annihilation reaction $e^+e^- \to \pi\pi$. Since there are no stable particles with mass $0 < m^2 < 4m_\pi^2$ and internal quantum numbers of the electromagnetic current, the function $F(t)$ has no poles along this segment.

For particles with spin, the vertex function of the vector or axial-vector current depends, first of all, on several form factors (see § 11.4). Following the same steps as for spinless particles, we may obtain an expression of the type (69) for the discontinuities of form factors. For a suitable choice of the covariants X^μ in the expression (11.72) for the vertex function, the form factors $F_l(t)$ may be made real-analytic: $F_l(t^*) = F_l^*(t)$ (see also § 12.3). However, in the case of particles with spin, one must allow for the appearance of kinematic singularities in $F_l(t)$. After removal of the kinematic singularities, one may postulate analytic properties for the form factors $F_l(t)$. The function $F_l(t)$ is assumed analytic in the whole complex t plane with a cut from $t_{\min} > 0$ to $t = \infty$, where $t_{\min}$ is the first threshold for a multi-particle annihilation reaction depending on the quantum numbers of the form factor $F_i(t)$ in the t-channel. The threshold branch point $t_{\min}$ may not coincide with the beginning of the physical region if the mass m of the external particles is larger than the mass of the internal particles contained in the states $|n\rangle$ in the unitarity condition (69). Then the cut of the function $F_i(t)$ will consist of the (normal) unphysical part $t_l < t < t_0$, in which the unitarity condition is inapplicable, and the physical region $t_0 < t < \infty$.

Aside from the normal threshold branch points, vertex functions may also have anomalous threshold branch points for $t_a < t_{\min}$. These points arise for the following mass relation between the external particle a and the internal particles b and c:

$$m_a^2 > m_b^2 + m_c^2,$$

if particles a, b, and c may interact directly with one another (i.e., for example, if one can write a Lagrangian for the abc interaction).

Moreover, if there are no zero-mass particles in the theory, so that the beginning of the cut is at $t_{\min} > 0$, there may be poles in the interval $0 < t < t_{\min}$ corresponding to stable particles with the same internal quantum numbers as those of the current j_μ. For example, the matrix element $\langle\pi|a_\mu(0)|0\rangle$ of the axial current vertex function between the vacuum and the pion state has a pion pole at $t = m_\pi^2$.

Let us consider as an example the electromagnetic form factors of the nucleon. The matrix element of the current between one-nucleon states contains four invariant form factors—two isoscalar form-factors F_1^S and F_2^S and two isovector form-factors F_1^V and F_2^V in the combinations $F_i = \frac{1}{2}(F_i^S + \tau_3 F_i^V)$, $i = 1, 2$:

$$\begin{aligned}\langle \boldsymbol{p}_2, \sigma_2; |\mathrm{N}|j_\mu|\boldsymbol{p}_1, \sigma_1; \mathrm{N}'\rangle &\equiv \bar{u}(\boldsymbol{p}_2, \sigma_2)\,\Gamma_\mu u(\boldsymbol{p}_1, \sigma_1)\\ &= \frac{1}{(2\pi)^3}\bar{u}(\boldsymbol{p}_2, \sigma_2)\{\gamma_\mu F_1(t) - i\sigma_{\mu\nu}q^\nu F_2(t)\}u(\boldsymbol{p}_1, \sigma_1),\end{aligned} \tag{73}$$

where $q = p_1 - p_2$ and $t = q^2$. A possible term $q_\mu F_3$ is omitted since it contradicts current conservation: $\partial_\mu j^\mu = 0$ or $q^\mu \Gamma_\mu = 0$. The wave functions $u(\boldsymbol{p}, \sigma)$ are the eight-component Dirac functions describing the isospin doublet with isospin components N, N′ = p, n; here the masses of proton and neutron are considered equal. Using the Dirac equation (or the explicit expression for u and $\bar{u}$), one may easily check that the term $\bar{u}(p_1 + p_2)^\mu u$ is not independent but leads to a linear combination of terms of the type $\bar{u}\gamma^\mu u$ and $i\bar{u}\sigma^{\mu\nu}q_\nu u$. By virtue of hermiticity of the current the form factors F_1 and F_2 are real.

The electric charge of the nucleon (in units of e) is equal to $Q_\mathrm{p} = 1$ for the proton and $Q_\mathrm{n} = 0$ for the neutron. We shall use this fact in normalizing the form factors F_1. From

$$\langle \boldsymbol{p}_2, \sigma_2; \mathrm{N}|\int j_0(x)\,d^3x|\boldsymbol{p}_1, \sigma_1; \mathrm{N}'\rangle = Q_\mathrm{N}\delta_{\mathrm{NN}'}2p_0\delta(\boldsymbol{p}_1 - \boldsymbol{p}_2)$$

and the normalization of the wave functions $\bar{u}\gamma_4 u = 2p_0$ it follows that the charges Q_p and Q_n may be expressed in terms of the form factors $F_1^S(0)$ and $F_1^V(0)$, so that

$$F_1^S(0) = 1, \quad F_1^V(0) = 1; \tag{74}$$

$F_1(t)$ is called the electric, and $F_2(t)$ the magnetic form factor. The form factor $F_2(t)$ at $t = 0$ may be expressed in terms of the anomalous magnetic moments of the proton and neutron μ_p and μ_n:

$$e(F_2^S(0)+F_2^V(0)) = \mu_p, \quad e(F_2^S(0)-F_2^V(0)) = \mu_n. \tag{75}$$

Indeed, the matrix element of the magnetic moment operator $\boldsymbol{\mu} = (e/2)\int (\boldsymbol{x}\times\boldsymbol{j})\, d^3x$ is

$$\begin{aligned}\langle \boldsymbol{p}_2, \sigma_2; \mathrm{N}\,|\,\mu^3\,|\,\boldsymbol{p}_1, \sigma_1; \mathrm{N}\rangle &\\ = \frac{c}{2}(2\pi)^3\,\delta(\boldsymbol{p}_1-\boldsymbol{p}_2)\,\varepsilon_{3jk}\left(-\frac{1}{i}\frac{\partial}{\partial q_j}\right)&\langle \boldsymbol{p}_2, \sigma_2; \mathrm{N}\,|\,j^k(0)\,|\,\boldsymbol{p}_1, \sigma_1; \mathrm{N}\rangle\\ = -\frac{e}{2}\bar{u}(\boldsymbol{p}_2, \sigma_2)\,\varepsilon_{3jk}\sigma^{kj}u(\boldsymbol{p}_1, \sigma_1)\,F_2(0)\,\delta(\boldsymbol{p}_2-\boldsymbol{p}_1).&\end{aligned}$$

where one sets $q = 0$ after performing the derivative. Let us now set $\boldsymbol{p}_1 = 0$, $\sigma_1 = \sigma_2 = \frac{1}{2}$. Since

$$-\tfrac{1}{2}\varepsilon_{3jk}\sigma^{kj} = \begin{pmatrix}\sigma_3 & 0\\ 0 & \sigma_3\end{pmatrix}$$

(by (5.32) $\sigma^{kj} = \frac{i}{2}[\gamma^k, \gamma^j]$), then in this case

$$\langle \boldsymbol{p}_2, \sigma_2; \mathrm{N}\,|\,\mu^3\,|\,\boldsymbol{p}_1, \sigma_1; \mathrm{N}\rangle = 2m_{\mathrm{N}}eF_2(0)\,\delta(\boldsymbol{p}_1-\boldsymbol{p}_2).$$

On the other hand, for $\boldsymbol{p}_1 = 0$, $\sigma_1 = \sigma_2 = \frac{1}{2}$, the state $|\,\boldsymbol{p}_1, \sigma_1; \mathrm{N}\rangle$ will be an eigenstate of the operator μ^3, so that

$$\langle \boldsymbol{p}_2, \sigma_2; \mathrm{N}\,|\,\mu^3\,|\,\boldsymbol{p}_1 = 0, \sigma_1; \mathrm{N}\rangle = \mu_{\mathrm{N}}2m_{\mathrm{N}}\delta(\boldsymbol{p}_1-\boldsymbol{p}_2)$$

we thus obtain $eF_2(0) = \mu_{\mathrm{N}}$, i.e. eqn. (75).

The expansion (73) expresses the electromagnetic vertex function of the nucleon Γ_μ in terms of the electric and magnetic form factors $F_1(t)$ and $F_2(t)$ in a manner free from kinematic singularities. We may consequently postulate that these form factors are analytic in the complex t-plane with a cut along the positive real t-axis, where the beginning of the cut $t_{\min}$ is determined by the masses of the particles interacting strongly with the nucleon. The lightest of these are pions. The value of $t_{\min}$ is the smallest m^2 of a state with quantum numbers of the current j_μ. But like the charge Q, the electromagnetic current is the sum of an isoscalar and an isovector. The isoscalar part of the current j_μ^S has isospin $I = 0$ and negative charge parity (the electromagnetic current is a first-class current): $\mathrm{C}j_\mu^S\mathrm{C}^{-1} = -j_\mu^S$. Consequently, the G-parity of this current is also negative: $\mathrm{G}j_\mu^S\mathrm{G}^{-1} = -j_\mu^S$. The lowest state with $I = 0$ and $\eta_{\mathrm{G}} = -1$ is the three-pion state, so that $t_{\min}^S = 9m_\pi^2$. For the isovector current j_μ^V one will have $I = 1$ and $\mathrm{G}j_\mu^V\mathrm{G}^{-1} = j_\mu^V$. The lowest state in this case contains two pions with isospin $I = 1$, G-parity $\eta_{\mathrm{G}} = +1$, and lowest mass $t_{\min}^V = 4m_\pi^2$.

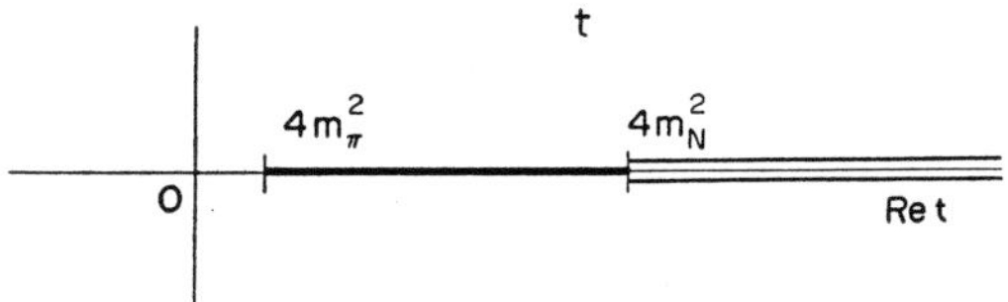

FIG. 17. Singularities of the isovector electromagnetic form factor of the nucleon in the complex t-plane

The physical region of the annihilation vertex (coinciding with the physical region of the reaction $e^+ + e^- \to N + \overline{N}$) begins at $t = 4m_N^2$. Consequently, the isoscalar form factor will have an unphysical cut from $t^S_{min} = 9m_\pi^2$ to $t = 4m_N^2$, while the isovector form factor will have an unphysical cut from $t^V_{min} = 4m_\pi^2$ to $t = 4m_N^2$ (Fig. 17). The region $t < 0$ will correspond to the vertex part $\langle \boldsymbol{p}_2, \sigma_2; N \,|\, j(0) \,|\, \boldsymbol{p}_1, \sigma_1; N \rangle$, which contributes to the elastic e^-N-scattering amplitude.

CHAPTER 13

ASYMPTOTIC BEHAVIOR OF THE SCATTERING AMPLITUDE AT HIGH ENERGIES. REGGE POLES

THE asymptotic behavior of the amplitude at high energies is one of its most important features. Experiments in the high-energy region exhibit simple and striking regularities. While rigorous bounds based on analytic properties are rather close to experiment, they cannot describe the asymptotic régime of the amplitude. The study of the asymptotic behavior of the amplitude as a function of complex momentum and angular momentum variables and of its crossing-symmetry properties has led to the phenomenological concept of Regge poles or trajectories. An introduction to Regge-pole theory will be presented in this chapter.

As in Chapters 11 and 12, we shall discuss only spinless equal-mass particles, indicating how to generalize these calculations. Special attention is paid to introducing the concept of a Regge trajectory and to discussing the properties of trajectories. Although the theory of Regge asymptotics is still incomplete, the concept of a trajectory is of fundamental importance in elementary particle physics.

§ 13.1. Scattering at high energies (experiment)

High-energy scattering ($s \gg m_N^2$, or $s \gg 1$ GeV^2) has a number of simple properties. The high-energy region is thus convenient for checking the basis of any theory of strong interactions.

Experimentally the best-studied processes are two-particle reactions of the type $a+b \to c+d$ and total cross-sections $\sigma_{tot}(a, b)$.

Two-particle processes may be subdivided into: (a) elastic scattering $a+b \to a+b$; (b) diffraction dissociation $a+b \to c+d$, in which the internal quantum numbers (B, Y, Q, I, etc.) of particles c and a, d and b coincide (while spins, in particular, may be different); (c) charge-exchange reactions $a+b \to c+d$, where particle c (or d) has internal quantum numbers different from the quantum numbers of particles a and b. Of course, all charges of the systems $a+b$ and $c+d$ are identical.

Examples of diffraction dissociation are the "excitation" of the resonance K* with spin-parity $J^P = 1^+$ and the resonance N* with $J^P = \frac{3}{2}^-$:

$$K^{\pm}+N \to K^{*\pm}(1^+)+N, \quad N+N \to N^*(\tfrac{3}{2}^-)+N$$

Examples of charge-exchange reactions are the processes

$$\pi^- + p \to \pi^0 + n, \quad K^- + p \to \Lambda^0 + \omega.$$

The charges exchanged in charge-exchange reactions are those of the system $a+\bar{c}$ in the t-channel. For diffraction dissociation, the system $a+\bar{c}$ is neutral with respect to I, Y, etc. From the standpoint of internal symmetries, diffraction dissociation is equivalent to elastic scattering.

Aside from diffraction dissociation and charge-exchange scattering, inelastic processes also include the multi-particle reactions

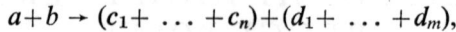

$$a+b \to (c_1 + \ldots + c_n) + (d_1 + \ldots + d_m),$$

in which there are more than two particles in the final state.

The analysis of experimental data yields the following characteristic features of high-energy scattering.

Total cross-sections $\sigma_{tot}(a, b)$

Figure 18 shows some experimental curves[165, 166] of τ_{tot}, including those obtained at the Serpukhov accelerator,[167] at the Fermi National Accelerator Laboratory (FNAL),[168] and at the CERN Intersecting Storage Rings (ISR).[169] The total K$^+$p cross-sections rise slowly above $s \simeq 40$ GeV, while the pp total cross-section rises by about 4 mb within the ISR energy range, and others (except $\bar{p}$p) rise at FNAL energies.

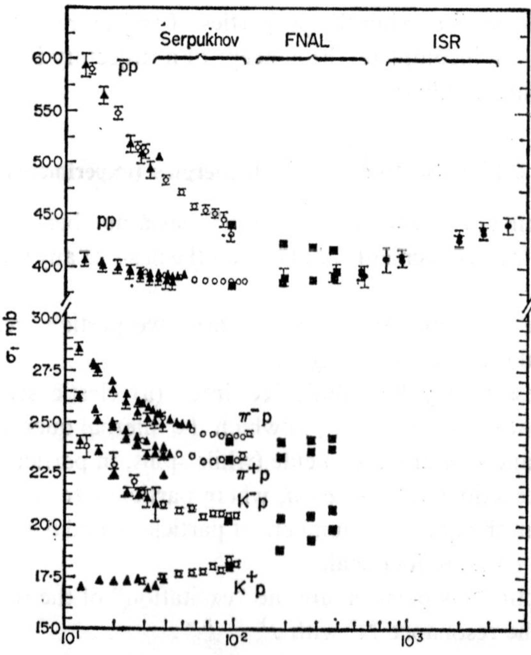

FIG. 18. Total cross-section σ_{tot} as a function of squared c.m. energy s.[170]

In what follows, for reasons of simplicity, we shall sometimes assume that at high energies the total cross-sections are constant:

$$\sigma_{\text{tot}}(a, b) = \text{const.} \tag{1}$$

Under the condition (1), the Pomeranchuk theorem, to be discussed in § 13.2, states that the total cross-sections (on the same target) for a particle and an antiparticle approach the same value as $s \to \infty$: $\sigma_{\text{tot}}(a, b) = \sigma_{\text{tot}}(\bar{a}, b)$.

Elastic scattering

With increasing s, the scattered particles are concentrated more and more in the region of small scattering angles ($\theta \approx 0$), or in the region of finite momentum transfer t. At high energies, the elastic scattering proceeds mainly at $\theta \approx 0$ (the diffraction peak), and with increasing momentum transfer $|t|$ the differential cross-section falls exponentially (Fig. 19):

$$\frac{d\sigma}{dt} \approx \left(\frac{d\sigma}{dt}\right)_{t=0} e^{at}, \quad s \to \infty, \quad t < 0, \tag{2}$$

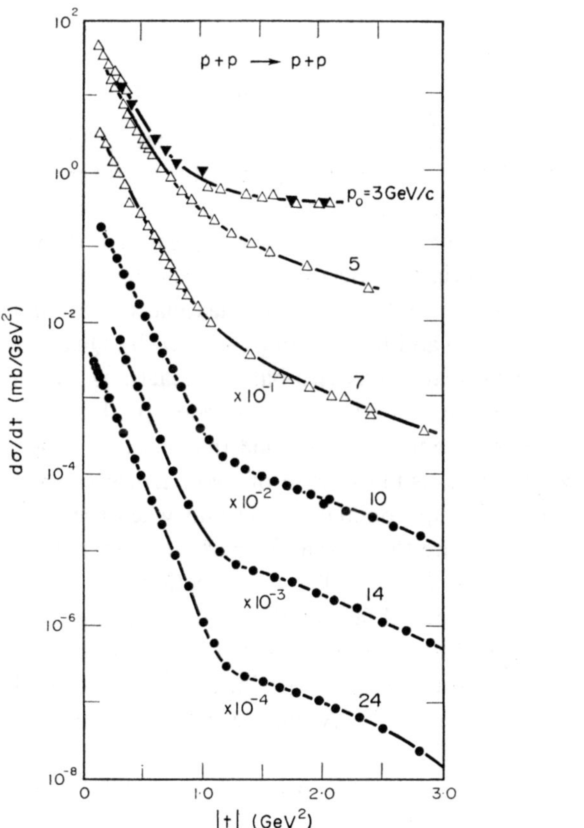

FIG. 19. Differential cross-sections for *pp* scattering as a function of *t*. (From J. V. Allaby *et al., Nucl. Phys.* **B52**, 316 (1972).)

where a lies between 7 and 13 GeV^{-2} for all elastic processes. The elastic scattering cross-section $\sigma_{el}(a+b \to a+b)$ falls with increasing energy but levels off and begins to rise again above $s \simeq 1000$ GeV^2.[169]

Diffraction dissociation

The behavior of the differential cross-section in this case is close to that observed in elastic scattering. The differential cross-section has a diffraction peak, while the total cross-section for the process, $\sigma(a+b \to c+d)$, falls relatively slowly with increasing s.

Charge-exchange scattering

Processes of this type are characterized by a rapid fall of the cross-section $\sigma(a+b \to c+d)$ (with increasing s) in comparison with the elastic scattering cross-section $\sigma(a+b \to a+b)$: near $t = 0$, as s gets large,

$$\frac{d\sigma(a+b \to c+d)/dt}{d\sigma(a+b \to a+b)/dt} \to 0. \tag{3}$$

Multi-particle processes

An important case is one in which the internal quantum numbers of the groups $(c_1+ \ldots +c_n)$ and $(d_1+ \ldots +d_n)$ are the same as for particles a and b respectively. As in the two-particle case, these reactions are called diffraction dissociation. At high energies such reactions are an important part of inelastic processes. For example, the process $\pi+\text{p} \to 2\pi+\Delta$ proceeds mainly via "dissociation" of the proton into $\pi+\Delta$, i.e. $\pi+\text{p} \to \pi+(\pi+\Delta)$. The process $\pi^-+\text{p} \to 2\pi^-+\pi^++\text{p}$ has two maxima, corresponding to the "dissociations" of the pion $\pi^- \to (2\pi^-\pi^+)$ and the proton $\text{p} \to (\pi^-\pi^+\text{p})$. Quasi-two-particle processes have the same characteristic features as two-particle reactions. As in the case of two-particle processes, the particles are emitted (at high energies) predominantly forward and backward in the centre of mass system. Let us assume that $c_1 \ldots c_n$ emerge forward, and particles $d_1 \ldots d_u$ backward. Then, as in the case of two-particle processes, the asymptotic cross-section for the process is approximately constant when the internal quantum numbers of the group $(c_1+ \ldots +c_n)$ are the same as for particle a, or, in the absence of quantum number exchange, $a \leftrightarrow (c_1+ \ldots +c_n)$. The cross-sections for processes in which the system $\bar{a}+(c_1+ \ldots +c_n)$ is not neutral fall rapidly with increasing energy.

One sometimes measures only the number of particles n of a given sort c, i.e. one studies the process $a+b \to nc+$(anything). Experiments of this type determine the multiplicity n of particles c as a function of s. At high energies the multiplicity grows slowly with s:

$$n \approx \text{const} \ln s \quad \text{or} \quad n \approx \text{const}\, s^{1/4}.$$

Experiments in which all particles are observed are called exclusive, while those in which only one particle of a given sort is observed are called inclusive.

§ 13.2. Bounds on the amplitude at high energies

The asymptotic behavior of the amplitude at high energies is of interest from two points of view. First, one may expect that at high energies, as in any other limiting case, the behavior of the amplitude will be governed by simple laws, allowing one to make a clean comparison between theory and experiment. It is assumed, of course, that the theory can predict the asymptotic behavior. Secondly, the asymptotic behavior is needed in calculations; the dispersion relations of § 12.3 were written with a number of subtractions determined by the asymptotic behavior in s, t, or u. We shall present below the simplest derivation[(171)] of the Froissart[(172)] bound for the elastic scattering of spinless particles of identical mass m.

A restriction on the growth of an amplitude with energy at high energies follows from the unitarity condition for an amplitude analytic in a region of $z = \cos\theta$. This bound is related to the existence of a unitary limit on partial wave amplitudes (see the end of § 12.4; also Fig. 16). By virtue of the unitarity condition, the asymptotic régime of the analytic amplitude may vary only within some (rather broad) bounds. The Froissart bound is an upper limit on the rate of growth of the amplitude. Using the optical theorem, a bound of this type may be expressed in terms of the total cross-section.

Let us assume that the amplitude $F(s, t, u) \equiv F(s, z)$ is analytic in $z \equiv z_s = \cos\theta_s$ in an ellipse C with semi-major axis $z_0 = 1+2t_M/(s-4m^2)$ and with foci $z = \pm 1$. The quantity $t_M > 0$ does not depend on s and coincides with the lowest singularity in the t-channel. In the absence of poles, one will have $t_M = t_{\min}$: in this model, $t_M = 4m^2$. This region of analyticity in z (the Lehmann–Martin ellipse) is larger than the physical region $-1 \leqslant z \leqslant 1$. Analyticity in this region may be proven[(173)] from axiomatic field theory assuming that the minimum mass in the theory is nonzero.

Let us further assume that, as in the case of dispersion relations (§ 12.3), the amplitude $F(s, z)$ grows with s no faster than a polynomial:

$$|F(s, z)| \leqslant a s^{N(z)} \qquad (s \to \infty) \tag{4}$$

for all z on the edge of the Lehmann–Martin ellipse. Let us verify that these assumptions, together with the unitarity condition, bound the possible growth of $F(s, z)$ as $s \to \infty$.

Since $F(s, z)$ is analytic in z, we may use the Cauchy formula

$$F(s, z) = \frac{1}{2\pi i} \oint\limits_C \frac{F(s, z')\, dz'}{z'-z},$$

integrating around the ellipse C.

Let us pass from $F(s, z)$ to the partial wave amplitudes $a_l(s)$:

$$a_l(s) = \frac{1}{2\pi i}\,\frac{1}{2B} \int\limits_{-1}^{1} dz \oint\limits_C \frac{P_l(z)\, F(s, z')\, dz'}{z'-z} = \frac{1}{2\pi i B} \oint\limits_C F(s, z')\, Q_l(z')\, dz', \tag{5}$$

where $B = 8/(2\pi)^5$ and $Q_l(z')$ is the Legendre function of the second kind. From (5) one finds at once the bound

$$|a_l(s)| \leq \frac{1}{2\pi B} |F(s, z)|_{\max} |Q_l(z)|_{\max} 4\xi, \tag{6}$$

where ξ is the sum of the semi-major and semi-minor axes of the ellipse: $\xi = z_0 + \sqrt{z_0^2 - 1}$. The asymptotic behavior as $s \to \infty$ will be

$$\xi \sim 1 + 2\sqrt{\frac{t_M}{s}}.$$

(The sign $\sim$ denotes asymptotic behavior.) Inserting the value of $|Q_l(z)|_{\max}$ on the ellipse into (6)

$$|Q_l(z)|_{\max} \leq \sqrt{\frac{\pi}{l}} \frac{\xi^{-l}}{\sqrt{\xi^2 - 1}}$$

and the preliminary bound (4) for $|F(s, z)|_{\max}$, we find

$$|a_l(s)| \leq \frac{2a}{\pi B} \sqrt{\frac{\pi}{l}} s^N \frac{\xi^{-l+1}}{\sqrt{\xi^2 - 1}} \qquad (s \to \infty),$$

or

$$|a_l(s)| \leq R(s) \frac{\xi^{-l}}{\sqrt{l}} \sim s^{N'} \frac{\xi^{-l}}{\sqrt{l}} \left[N' = N + \frac{1}{4}\right]. \tag{7}$$

From (7) it is clear that at high s the partial wave amplitude $a_l(s)$ falls rapidly with l. Let us now turn to the unitarity condition for $a_l(s)$. According to this condition [see (7.68) and (§ 12.4)], the value of $|a_l(s)|$ cannot exceed the unitary limit:

$$|a_l(s)| \leq \left(\frac{s}{s - 4m^2}\right)^{1/2} \sim 1. \tag{8}$$

We now separate the sum over partial wave amplitudes in $F(s, z)$ into two parts. The first part $F_1(s, z)$ contains the partial waves from $l = 0$ up to some L, and the second part $F_2(s, z)$ the remainder. Let us choose the angular momentum L so that the bounds (7) and (8) coincide. Comparing (7) and (8) in the asymptotic region:

$$L^{1/2} \xi^L \sim s^{N'}. \tag{9}$$

But as $s \to \infty$ one has $\ln \xi \sim 2(t_M/s)^{1/2}$. Consequently, for $s \to \infty$ one may choose

$$L = \frac{N'}{2\sqrt{t_M}} s^{1/2} \ln \frac{s}{s_1}, \tag{10}$$

where s_1 fixes the scale of s. The first part $F_1(s, z)$ is bounded by inserting the unitary limit of the partial wave amplitude (8). For forward scattering ($z = 1$)

$$|F_1(s, 1)| = B\left|\sum_{l=0}^{L-1}(2l+1)\,a_l(s)\right| \leqslant B\left(\frac{s}{s-4m^2}\right)^{1/2}\sum_{l=0}^{L-1}(2l+1) = B\left(\frac{s}{s-4m^2}\right)^{1/2}L^2. \quad (11)$$

The second part $F_2(s, z)$ is bounded using (7), obtained from the assumption of analyticity of $F(s, z)$ in z. For forward scattering ($z = 1$)

$$|F_2(s, 1)| = B\left|\sum_{l=L}^{\infty}(2l+1)\,a_l(s)\right| \leqslant BR(s)\xi^{-L}\sum_{l=L}^{\infty} l^{-1/2}(2l+1)\xi^{-(l-L)}. \quad (12)$$

Let us bound the sum in this expression

$$\sum_{l=L}^{\infty} l^{-1/2}(2l+1)\xi^{-(l-L)} \leqslant \sum_{l=L}^{\infty}(2l+1)\xi^{-(l-L)}$$
$$= \sum_{n=0}^{\infty}(2L+1+2n)\xi^{-n} = (2L+1)\frac{\xi}{\xi-1} - 2\xi\frac{d[\xi/(\xi-1)]}{d\xi} = \frac{(2L+1)\xi}{\xi-1} - \frac{2\xi}{(\xi-1)^2}.$$

Asymptotically, this sum is

$$\sim \frac{s^{1/2}}{2t_{\mathrm{M}}^{1/2}}(2L+1) + \frac{s}{t_{\mathrm{M}}} \sim \frac{N'}{2t_{\mathrm{M}}}\, s\ln s.$$

Consequently, as $s \to \infty$, the bound on the second part is

$$|F_2(s, 1)| \leqslant Bs^N\xi^{-L}\frac{N'}{2t_{\mathrm{M}}}\, s\ln s \sim \text{const}\; s^{N'+1}\ln s\left(1+2\left(\frac{t_{\mathrm{M}}}{s}\right)^{1/2}\right)^{-(N'/2)(s/t_{\mathrm{M}})^{1/2}\ln s}$$
$$= \text{const}\; s\ln s. \quad (13)$$

The function $F_2(s, 1)$ grows more slowly than $F_1(s, 1)$, and the asymptotic behavior of the amplitude is characterized only by the part $F_1(s, 1)$. We thus find from (10) and (11) the Froissart bound

$$|F(s, 1)| \leqslant \text{const}\, L^2 = \text{const}\; s\ln^2 s. \quad (14)$$

This bound was first obtained using the Mandelstam representation. From (14), using (7.49) and (7.53), one may derive a bound for the total cross-section:

$$\sigma_{\text{tot}} \sim \frac{16\pi}{Bs}\,\text{Im}\, F(s, 1) \leqslant \frac{16\pi}{s}L^2 \sim \text{const}\ln^2 s, \quad (15)$$

and for the elastic scattering cross-section:

$$\left(\frac{d\sigma_{\text{el}}}{dt}\right)_{t=0} \sim \left|\frac{1}{s}F(s, 1)\right|^2 \leqslant \text{const}\ln^4 s. \quad (16)$$

To obtain a bound on the growth of the amplitude for $t \neq 0$, one must take account of the inequality

$$|P_l(\cos\theta)| \leqslant (2/\pi l\sin\theta)^{1/2} \qquad (0 < \theta < \pi).$$

One then obtains a bound on the asymptotic behavior of the scattering amplitude at fixed angle:

$$|F(s, z)| \leqslant \text{const}\,(\theta) s^{3/4} \ln^{3/2} s \qquad (0 < \theta < \pi). \tag{17}$$

The above bounds rely on the choice of angular momentum L, which in turn depends on the assumed region of analyticity. By extending the region of analyticity in z one may obtain [174, 175] a stronger bound on the growth of $F(s, z)$ when $\theta \neq 0$. For forward scattering, $\theta = 0$, the Lehmann–Martin ellipse gives the maximum possible region of analyticity in z, and a further strengthening of the bounds is impossible.

The Froissart bound does not allow one to decide whether the cross-section grows without bound at high energy or approaches a constant limit. If one assumes that the amplitude grows polynomially (without terms of the type $\ln s$) then, according to the Froissart bound, the degree of growth of s^N is bounded by $N \leqslant 1$ and the total cross-section will approach a constant asymptotically. This means that the number of subtractions in the dispersion relation in s cannot exceed two, so that we can set $N = 2$ in (4) or,

$$L = \left(\frac{s}{t_M}\right)^{1/2} \ln s = \frac{s^{1/2}}{2m} \ln s,$$

and then, according to (15),

$$\sigma_{tot} \leqslant \frac{4\pi}{m^2} \ln^2 s. \tag{18}$$

The general considerations presented above refer to the asymptotic region $s \to \infty$ and cannot fix the scale s_1 connected with the beginning of the asymptotic region. In the simplest case of $\pi\pi$ scattering, one may actually find the number s_1.[176]

The diffraction peak

For large energies the angular distribution of elastic scattering has a sharp maximum (a diffraction peak) for small angles $\theta \approx 0$ or $t \approx 0$. The width of the diffraction peak may be characterized by the quantity

$$\Delta = \frac{\sigma_{el}}{(d\sigma_{el}/dt)_{t=0}}. \tag{19}$$

The width of the diffraction peak is exactly Δ when $d\sigma_{el}/dt$ vanishes exponentially with t; i.e. Δ is the interval of the square of the momentum transfer over which the differential cross-section falls by e.

A bound on $d\sigma_{el}/dt$ at $t = 0$ was obtained in (16). Let us modify the derivation to obtain a bound on Δ. We start with the formula

$$\left(\frac{d\sigma_{el}}{dt}\right)_{t=0} \sim B\left|\frac{1}{s} F(s, 1)\right|^2 = B\left|\frac{1}{s}\sum_{l=0}^{\infty}(2l+1)\, a_l(s)\right|^2.$$

As before, in deriving an estimate for $F(s, 1)$ we may replace the infinite sum over l by a sum from 0 to $L \sim \text{const}\, s \ln s$, since the remaining part grows more slowly with increasing s. Moreover, we shall use the Bunyakovskiĭ–Schwartz inequality, which gives

$$\left(\frac{d\sigma_{el}}{dt}\right)_{t=0} \leqslant B^2 \frac{1}{s} \sum_{l=0}^{L} (2l+1) \frac{1}{s} \sum_{l'}^{L} (2l'+1) \, |a_{l'}(s)|^2 \leqslant \text{const} \frac{1}{s} L^2 \sigma_{el}. \tag{20}$$

Hence one finds a bound on the width of the diffraction peak:[177]

$$\Delta \geqslant \text{const} \ln^{-2} s. \tag{21}$$

The lower bound on the diffraction peak thus decreases with increasing s (the diffraction peak may shrink with increasing energy).

Let us return to eqn. (20) and use it to obtain a relation between the asymptotic behavior of the total cross-section and the elastic scattering cross-section. Replacing F by Im F, we obtain, using (10) and (15), the inequality

$$\sigma_{el} \geqslant \text{const} \frac{\sigma_{tot}^2}{\ln^2 s} \qquad (s \to \infty). \tag{22}$$

According to (22), the asymptotic behavior of σ_{el} cannot differ sharply from the asymptotic behavior of σ_{tot}. If $\sigma_{tot}(s) \approx$ const, then σ_{el} cannot vanish more rapidly than $\ln^{-2} s$.

The Pomeranchuk theorem

Let us examine $\sigma_{tot}(a, b)$ and $\sigma_{tot}(\bar{a}, b)$—the total cross-sections for scattering of the particle a and the antiparticle $\bar{a}$ on the same target b. Let us assume that the asymptotic behavior as $s \to \infty$ of these cross-sections is finite and nonzero. Then they must be equal:

$$\sigma_{tot}(a, b) = \sigma_{tot}(\bar{a}, b) = \text{const}.$$

In contrast to the above asymptotic bounds on the amplitude, the Pomeranchuk theorem cannot be derived from general assumptions alone (locality of fields or micro-causality, positivity of the spectrum, unitarity). An additional assumption is the demand that the real part of the amplitude for elastic forward scattering vanish more rapidly than its imaginary part:

$$\frac{\text{Re } F(s)}{\text{Im } F(s)} \frac{1}{\ln s} \to 0 \qquad (s \to \pm \infty). \tag{23}$$

The function $F(s)$ is the elastic scattering amplitude for $b+a \to b+a$ (in the s-channel) on the upper edge of the right-hand cut $s \geqslant s_{min}$ in the complex s-plane; on the lower edge of the left-hand cut $s \leqslant s(u_{min})$ the function $F(s)$ is the forward elastic scattering amplitude for the reaction with antiparticles $b+\bar{a} \to b+\bar{a}$ (the u-channel).

If assumption (23) holds, the amplitude $F(s)$ will be imaginary at high energies. But, by virtue of the assumed constancy of the total cross-sections and the optical theorem, the imaginary part of the amplitude is proportional to s at both ends $s \to \pm \infty$ of the real axis:

$$\text{Im } F(s) = \lambda \sigma_{tot}(a, b) s \qquad (s \to +\infty),$$
$$\text{Im } F(s) = -\lambda \sigma_{tot}(\bar{a}, b) s \qquad (s \to -\infty),$$

where the real factor λ is the same for both limits. Consequently, $F(s) = i\lambda\sigma_{tot}(a, b)s$ on the upper edge of the cut on the right for $s \to \infty$ and $F(s) = -i\lambda\sigma_{tot}(\bar{a}, b)s$ on the lower edge of the left-hand cut for $s \to -\infty$. The Pomeranchuk theorem then may be proven by analytically continuing $F(s)$ around a circle of infinite radius. The existence of this continuation may be seen by writing a once-subtracted dispersion relation for $F(s)$.

Generalizations of the Pomeranchuk theorem for total cross-sections which do not approach constants are discussed in ref. 178.

§ 13.3. The Regge-pole hypothesis and the asymptotic form of the amplitude

Let us consider the asymptotic form of the elastic scattering amplitude $F(s, t, u)$ for scalar equal-mass particles at high energies and small angles, i.e. for $s \to \infty$ and fixed $t \leqslant 0$. By virtue of crossing symmetry and analyticity of $F(s, t, u)$, its asymptotic behavior in the s-channel at $s \to \infty$ and $t \leqslant 0$ is connected with its asymptotic behavior in the t-channel for finite $t > 0$ and large (unphysical) values of $z_t = \cos\theta_t \to \infty$ (which corresponds to $s \to \infty$).

At first glance one does not seem to gain much by transforming from the asymptotic behavior with respect to s in the s-channel to the asymptotic behavior with respect to z_t in the t-channel. The importance of the latter was first shown by Regge.[179]

Regge found a way of uniquely continuing the partial waves $a_l(E)$ into the region of complex l in the nonrelativistic case. He then showed, for this case, that the asymptotic behavior of the scattering amplitude in $z = \cos\theta$ was fully determined by the poles $l = \alpha(E)$ of the function $a(l, E)$ in the complex angular momentum plane. The function $a(l, E)$ is the analytic continuation of $a_l(E)$ in the complex l-plane; its poles in l came to be known as Regge poles.

In the nonrelativistic case, $a(l, E)$ is an analytic function of two variables l and E (and not just an analytic function of l for fixed E). While initially defined only for positive energies $E > 0$ (in the scattering region), this function hence may be analytically continued into the region $E < 0$.

The position of a Regge pole $l = \alpha(E)$ depends on the energy E; as energy varies the Regge pole describes a trajectory $\alpha(E)$. As a consequence of the analytic properties of $a(l, E)$, the Regge-pole trajectory $\alpha(E)$ is an analytic function of E, which is determined both for $E > 0$ and for $E < 0$. The segment of the trajectory with $E < 0$ is real: $\text{Im}\,\alpha(E) = 0$, while in the scattering region $E > 0$ the poles are complex: $\text{Im}\,\alpha(E) \neq 0$. The trajectory points with integral values $\text{Re}\,\alpha(E) = l_0$ thus describe bound states with angular momentum l_0 for $E < 0$ and resonances with angular momentum l_0 for $E > 0$. Consequently, according to Regge, the asymptotic behavior of the nonrelativistic amplitude in z is determined by bound states and resonances.

If one assumes that Regge analysis basically retains its form in the relativistic theory ("the Regge-pole hypothesis"),[180, 181] then, according to crossing symmetry, one expects the asymptotic behavior of the amplitude $F(s, t, u)$ in the s-channel to be determined by the resonances and bound states in the t-channel. This simple relation is borne out by experiment.

We shall not consider the nonrelativistic case and begin at once with the relativistic theory. We shall rely on the analytic properties of the total amplitude established in Chapter 12 (in the complex s- and t-planes) and of the partial wave amplitudes (in the complex l- and s-planes), on the crossing symmetry properties of the amplitudes, on the unitarity condition, and, of course, on experimental data at high energies.

We first note the steps in finding the asymptotic behavior of the amplitude $F(s, t)$ for $t \to \infty$ and $s < 0$. Let us consider the amplitude F as a function of s and z_s in the s-channel:

$$F(s, z) = B \sum_{l=0}^{\infty} (2l+1)\, a_l(s)\, P_l(z) \qquad (z \equiv z_s),$$

and ask for the asymptotic behavior of $F(s, z)$ at high (unphysical) values $z \to \infty$ and fixed $s > 4m^2$. This series in l converges only for values of z lying inside the ellipse with foci $z = \pm 1$, not containing the singularities of $F(s, z)$ (the Lehmann–Martin ellipse, see § 13.2). For this reason, to find the asymptotic behavior of $F(s, z)$ for $z \to \infty$, $s > 4m^2$ one first needs an analytic continuation of $F(s, z)$ into the region of large z. The function $F(s, t)$ thus obtained will give the asymptotic behavior in the t-channel, $t \to \infty$, for unphysical $s > 4m^2$, so that the final step consists of the analytic continuation of this function from the region $s > 4m^2$ into the region $s < 0$.

Let us list the assumptions regarding the analytic properties of $F(s, t)$ necessary to obtain its asymptotic behavior in the t-channel.

1. It is assumed that the amplitude satisfies dispersion relations in t with $N'(s)$ subtractions, if s is fixed and lies in the region

$$-a < \operatorname{Re} s < 4m^2+a, \quad 0 < \operatorname{Im} s < \varepsilon, \quad a > 0, \quad \varepsilon \to +0, \tag{24}$$

which contains an interval between the upper edges of both cuts in the s-plane. The number of subtractions N' is determined by the power N (see §§ 12.3, 12.4, and (4)), so that $N' = N$ for integral N' and $N-1 < N' < N$ for nonintegral N'. We shall assume $N' = N$ for simplicity.

2. It is assumed that the amplitude and its discontinuities in the t- and u-channels $A_t(t, s)$ and $A_u(u, s)$ are analytic in s in the region (24). This condition allows us to continue the function $a^{\pm}(l, s)$ from the region $\operatorname{Re} s > 4m^2$ into the region $\operatorname{Re} s < 0$ by a route passing over the real axis and avoiding the branch points of the amplitude.

Let us now find the asymptotic behavior of the amplitude $F(s, z)$ for $z \to \infty$ when s lies on the upper edge of the s-channel cut (i.e., $\operatorname{Re} s > 4m^2$, $0 < \operatorname{Im} s < \varepsilon$). We shall use the analytic properties of the Gribov–Froissart functions $a^{\pm}(l, s)$ (see § 12.4) in the complex angular momentum variable l. The functions $a^{\pm}(l, s)$ are analytic in l in a half-plane lying to the right of the line $\operatorname{Re} l = N(s)$. In this region, the functions $a^{\pm}(l, s)$ behave for $|l| \to \infty$ as

$$\left.\begin{aligned} a^{\pm}(l, s) &\sim e^{-l\xi}, \\ \xi = z_0+\sqrt{z_0^2-1}, &\quad z_0 = 1+2t_M/(s-4m^2), \end{aligned}\right\} \tag{25}$$

where $t_M = 4m^2$ is the first singularity in the t- or u-channel, and is independent of s. For integral values of $l \geqslant N(s)$, the functions $a^{\pm}(l, s)$ are uniquely determined in terms of the physical partial wave amplitudes $a_l(s)$:

$$a^+(l, s) = a_l(s) \quad \text{for even} \quad l \geqslant N,$$
$$a^-(l, s) = a_l(s) \quad \text{for odd} \quad l \geqslant N.$$

From the properties of the functions $a^{\pm}(l, s)$ it is clear that instead of the asymptotic behavior of $F(s, z)$ one finds the asymptotic behavior of the functions

$$F^{\pm}(s, z) = \tfrac{1}{2}B\sum_{l=0}^{N}(2l+1)\,a_l(s)\,[P_l(z)\pm P_l(-z)]+\tfrac{1}{2}B\sum_{l=N+1}^{\infty}(2l+1)\,a^{\pm}(l, s)\,[P_l(z)\pm P_l(-z)]. \tag{26}$$

If z is in the physical region, $-1 \leqslant z \leqslant 1$, then, by virtue of the properties of $a(l, s)$, eqn. (26) may be written as a contour integral:

$$F^{\pm}(s, z) = \tfrac{1}{2}B\sum_{l=0}^{N}(2l+1)\,a_l(s)\,[P_l(z)\pm P_l(-z)] + \frac{i}{2}B\int_C \frac{(2l+1)\,a^{\pm}(l, s)\,[P_l(z)\pm P_l(-z)]}{\sin \pi l}\,dl. \tag{27}$$

The contour C (Fig. 20) encloses only poles of the integrand along the real l-axis arising from $\sin \pi l$. Evaluation of the integral in (27) via its residues gives (26). The sum from $l = 0$ to $l = N$ in (26) and (27) is displayed explicitly to reflect the fact that the analytic properties of $a^{\pm}(l, s)$ in l are known only for $\mathrm{Re}\, l > N$.

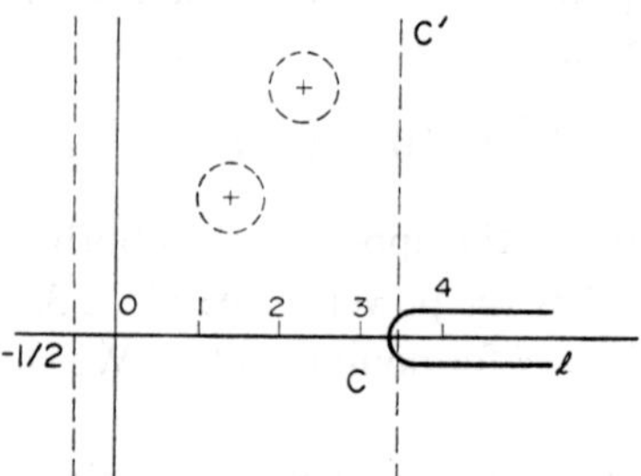

FIG. 20. Contours of integration in complex l-plane defining scattering amplitudes $F^{\pm}(s, z)$ in terms of partial wave amplitudes $a^{\pm}(l, s)$.

Let us now replace the contour C by the contour C' consisting of the line $\mathrm{Re}\, l > N$ (see Fig. 20). The integral in (27) does not change since $a^{\pm}(l, s)$ and $P_l(z)$ do not have singularities inside C'', while the integral along the large semicircle R_C is equal to zero by virtue of the property (25) of the Gribov–Froissart functions and the asymptotic bound on the behavior of Legendre polynomials in l:

$$\left|\frac{(2l+1)\,P_l(-\cos\theta)}{\cos \pi(l+\frac{1}{2})}\right| \leqslant \frac{c\lambda}{|\sin\theta|^{1/2}} \exp -(|\mathrm{Re}\,\theta\ \mathrm{Im}\,\lambda|+|\mathrm{Im}\,\theta\ \mathrm{Re}\,\lambda|) \tag{28}$$

where $\lambda = l+\frac{1}{2}$. Consequently, for $-1 \leqslant z \leqslant 1$ the sum

$$F_N^{\pm}(s, z) = \tfrac{1}{2}B\sum_{l=N+1}^{\infty}(2l+1)\,a^{\pm}(l, s)\,[P_l(z)\pm P_l(-z)] \tag{29}$$

is identically equal to the integral

$$F_N^{\pm}(s, z) = \frac{i}{2}B\int_{C'} \frac{(2l+1)\,a^{\pm}(l, s)\,[P_l(z)\pm P_l(-z)]}{\sin \pi l}\,dl \tag{30}$$

along the line C' parallel to the imaginary axis (see Fig. 20). In contrast to the integral in (27), the integral (30) also converges for $|z| > 1$, so that it can serve as an analytic continuation of $F_N^{\pm}(s, z)$ to large z. However, $F_N^{\pm}(s, z)$ still is not the amplitude $F^{\pm}(s, z)$. To obtain an integral representation of the type (30) for the amplitude, one needs to know analytic properties of the functions $a^{\pm}(l, s)$ in the region of the complex l-plane to the left of the line $\text{Re}\, l > N(s)$.

According to the Regge-pole hypothesis, the singularities of $a^{\pm}(l, s)$ in the complex l-plane are simple poles (Regge poles), whose position depends on s. Let the poles of the function $a^{\pm}(l, s)$ for $\text{Re}\, l < N(s)$ be at the points $l = \alpha^{\pm}(s)$ with residues $\beta^{\pm}(s)$, so that each pole contributes an amount

$$\frac{\beta^{\pm}(s)}{l-\alpha^{\pm}(s)} \tag{31}$$

(in Fig. 20 these poles are denoted by crosses).

To pick out the contribution of the Regge poles in (27), let us move the contour C' to the left until it coincides with the line $\text{Re}\, l = -\frac{1}{2}$. Then, aside from the integral along this line, one picks up the contributions of the Regge poles $\alpha^{\pm}(s)$, and also the contributions from the poles of the integrand for $l = 0, 1, 2, \ldots, N(s)$. Expression (27) then takes the form

$$\begin{aligned} F^{\pm}(s, z) = {} & \frac{i}{2} B \int_{-\frac{1}{2}+i\infty}^{-\frac{1}{2}+i\infty} dl \frac{(2l+1)\, a^{\pm}(l, s)\, [P_l(z) \pm P_l(-z)]}{\sin \pi l} \\ & -\pi B \sum_n \left(2\alpha_n^{\pm}(s)+1\right) [P_{\alpha_n^{\pm}}(-z) \pm P_{\alpha_n^{\pm}}(z)]\, \frac{\beta_n^{\pm}(s)}{\sin \pi\alpha_n^{\pm}(s)} \\ & +\tfrac{1}{2} B \sum_{l=0}^{N(s)} (2l+1)\, [a_l(s) - a^{\pm}(l, s)]\, [P_l(z) \pm P_l(-z)]. \end{aligned} \tag{32}$$

Here the functions $a^{\pm}(l, s)$ are the result of analytic continuation of $a^{\pm}(l, s)$ from the region $\text{Re}\, l > N(s)$. These functions may not coincide with $a_l(s)$ for even (odd) $l < N(s)$, since Carlson's theorem (see § 12.4) guarantees the uniqueness of the continuation only if $a^{\pm}(l, s)$ is sufficiently bounded for $|l| \to \infty$, i.e. for $\text{Re}\, l > N(s)$. Expression (32) is applicable so far only to the region $s > 4m^2$. The second term in (32) is the contribution from the poles of the functions $a^{\pm}(l, s)$. The contour C' cannot be moved to the left of $\text{Re}\, l = -\frac{1}{2}$ without further special considerations, since the function $Q_l(z)$ in $a(l, s)$ grows exponentially as $|l| \to \infty$ to the left of $\text{Re}\, l = -\frac{1}{2}$ [see (12.45)].

Let us find the asymptotic behavior of expression (32) for $|z| \to \infty$. Since $z = 1+2t/(s-4m^2)$, the limit of large z for $s > 4m^2$ is equivalent to the limit $t \to \infty$. For $\text{Re}\, l > -\frac{1}{2}$ we have, asymptotically,

$$P_l(z) \approx \frac{\Gamma(2l+1)}{2^l[\Gamma(l+1)]^2} z^l + O(z^{l-2}) \qquad (|z| \to \infty). \tag{33}$$

In calculating the asymptotic behavior of $P_l(-z)$, it is important that $s > 4m^2$ in (32), since as $|z| \to \infty$

$$P_l(-z) = P_l(z) \exp(i\pi l \,\text{sgn}\, [\text{Im}\, z]) + O(z^{-l-1}).$$

The physical region of t in the t-channel corresponds to the upper edge of the cut in the t-plane, where $t = \mathrm{Re}\, t + i\varepsilon$, $\varepsilon > 0$. Then, for $s > 4m^2$, we will have

$$\mathrm{Im}\, z = \mathrm{Im}\, (1+2t/(s-4m^2)) > 0,$$

and, consequently,

$$P_l(-z) = e^{-il\pi} P_l(z) + O(z^{-l-1}). \tag{34}$$

From these asymptotic formulae for $P_l(z)$, it is clear that in the absence of a third term the asymptotic expression (32) in z for $s > 4m^2$ is determined by the pole term with the largest value of Re $l > 0$, or the right-most pole $\alpha^{\pm}(s)$ in the l-plane. The contribution of a single pole $\alpha^{\pm}(s)$ to $F(s, z)$ for $|z| \gg 1$ is equal to

$$\pi B(2\alpha^{\pm}(s)+1)\, \beta'^{\pm}(s) z^{\alpha^{\pm}(s)}\, \eta^{\pm}(s), \tag{35}$$

where $\eta^{\pm}$ is the signature factor:

$$\eta^{\pm}\, (1 \pm e^{-i\pi\alpha}) \frac{1}{-\sin \pi\alpha^{\pm}}, \tag{36}$$

i.e. if one introduces the signature $\sigma = \pm$, then

$$\eta^{\sigma} = \begin{cases} i - \cot \dfrac{\pi\alpha}{2}, & \sigma = +, \\[2ex] i + \tan \dfrac{\pi\alpha}{2}, & \sigma = -. \end{cases}$$

In (35) we have introduced the notation

$$\beta' = \beta \frac{\Gamma(2\alpha+1)}{2^{\alpha}[\Gamma(\alpha+1)]^2}.$$

The integral in (32) vanishes for $|z| \gg 1$ as $z^{\varepsilon-\frac{1}{2}}$, $\varepsilon \to +0$, and hence gives no contribution to the asymptotic behavior. In particular, if there are no poles $\alpha(s)$ and the third term in (32) is equal to zero, then the amplitude vanishes asymptotically as $|z| \to \infty$.

Thus the asymptotic regime of the amplitude (32) as $t \to \infty$ will be described by a sum of terms arising from the Regge poles:

$$F(s, t) \sim \sum_n \pi B(2\alpha_n^{\pm}(s)+1)\, \tilde{\beta}_n^{\pm}(s)\, t^{\alpha_n^{\pm}(s)} \eta^{\pm}(s). \tag{37}$$

if one temporarily neglects the third term in (32). Here

$$\tilde{\beta}^{\pm}(s) = \left(\frac{2}{s-4m^2}\right)^{\alpha^{\pm}(s)} \beta'^{\pm}(s).$$

Let us now discuss the third term in (32) and the analytic continuation of the expressions (32) and (37) into the region $s < 0$. (The region $s < 0$, $t > 4m^2$ characterizes the t-channel.)

The functions $F^{\pm}(s, t)$ are analytic in s in the region (24), according to assumption 1 on page 287. Moreover, by virtue of assumption 2, the function $a(l, s)$ may be analytically continued in the region (24) to negative s. One may verify this by writing (12.46) in the form

$$a^{\pm}(l, s) = \frac{2B}{\pi(s-4m^2)} \int_{t_M}^{\infty} dt[A_t(s, t) \pm A_u(u(t), s)]\, Q_l\left(1+\frac{2t}{s-4m^2}\right)$$

and noting that all branch points may be avoided by passing into the upper half plane ($\operatorname{Im} s = \varepsilon > 0$).

Consequently, on the right-hand side of (32), the two first terms and each individual term in the third piece will be analytic in s in the region (24). This means that the number of nonzero terms in the third member of (32) cannot depend on s and must be equal to N_{min}— the smallest value of $N(s)$ in the analyticity region (24). But for $s < 0$ we can use the Froissart bound (for the t-channel) and set $N_{min} = 1$. Then the third term in (32) will contain the partial waves $\varphi_0(s)$ (in the case of F^+) and $\varphi_1(s)$ (in the case of F^-) which are independent of $a^{\pm}(l, s)$ and reflect the fact that, along with Regge poles, there may exist "elementary poles" with $l = 0, 1$.

In the case of inelastic scattering (in the t-channel), according to experiment, $N_{min} < 1$. Moreover, for elastic scattering (in the t-channel), $N(s)$ also seems to be less than 1 when $s < 0$. Then one may take $N_{min} < 1$ and set $\varphi_1(s) = 0$. We shall assume that $\varphi_0(s)$ is also equal to zero. With these assumptions, the asymptotic behavior of $F^{\pm}(s, t)$ for $t \to \infty$ is fully determined (within the framework of the Regge pole hypothesis) by (37). In view of the analytic properties of $F(s, t)$ and $a(s, l)$, this expression, while initially derived for $t > 0$, $s > 4m^2$, also holds in the physical region of the t-channel ($t > 4m^2$, $s < 0$).

Similarly, one may obtain the asymptotic behavior of the amplitude in other crossed channels. In the s-channel, as $s \to \infty$, the contribution of one pole $\alpha^{\pm}(t)$ to the amplitude is equal [in analogy with (37)] to

$$F(s, t) \sim \pi B(2\alpha^{\pm}(t)+1)\, \tilde{\beta}^{\pm}(t)\, s^{\alpha^{\pm}(t)}\, \eta^{\pm}(t), \tag{38}$$

where $\tilde{\beta}^{\pm}(t)$ is the residue of the function

$$\tilde{a}^{\pm}(l, t) = a^{\pm}(l, t)\left(\frac{2}{t-4m^2}\right)^{l} \frac{\Gamma(2l+1)}{[\Gamma(l+1)]^2}$$

at the Regge pole $l = \alpha^{\pm}(t)$.

The position of the pole $\alpha(t)$ in the complex angular momentum plane varies with t. As in the nonrelativistic case, this curve $\alpha(t)$ is called the Regge trajectory. For eqn. (37) the Regge trajectory is $\alpha(s)$.

Physical interpretation of the Regge trajectory

The poles $l = \alpha(s)$ are sometimes called "moving", in contrast to a fixed pole $l = b$ whose position does not depend on s. The existence of fixed poles is inconsistent with the elastic unitarity condition (12.52):

$$a(l, s) - a^*(l^*, s) = 2i\left(\frac{s}{s-4m^2}\right)^{1/2} a(l, s)\, a^*(l^*, s), \tag{39}$$

which holds in the region from $s = 4m^2$ to the first inelastic threshold s_{inel}. If we insert the expression for the amplitude $a(l, s)$ near a fixed pole

$$a(l, s) \approx \frac{g(s)}{l-b}, \tag{40}$$

into (39), it is easy to see that consistency of (39) and (40) at $l \approx b$ demands $g(s) = 0$, i.e. the residue of a fixed pole must vanish for $4m^2 < s < s_{\text{inel}}$. But, like the amplitude $F(s, t)$, the residue $g(s)$ is analytic in s. Thus, if $g(s)$ vanishes in a finite interval of s, then it must vanish in the whole region of analyticity in s. These considerations are true only in the absence of moving Regge cuts [which has been postulated earlier in connection with (31)].

The trajectories $\alpha^+(s)$ and $\alpha^-(s)$ are distinct from one another. Since the unitarity condition (39) is written separately for the functions a^+ and a^- (see § 12.4), the poles of these functions will also be independent. The trajectories α^+ and α^- are called trajectories of positive and negative signature respectively.

The physical interpretation of Regge trajectories $\alpha(s)$ is determined, on the one hand, by their role in the description of the asymptotic régime of the amplitude (37) and, on the other, by the possibility of relating the values of $\alpha(s)$ at individual points to quantities characterizing bound states and resonances. For $s < 0$ the Regge trajectory $\alpha(s)$, according to (37), describes the asymptotic behavior in the t-channel; for $0 < s < 4m^2$ the points s_J, at which the trajectory passes through integral nonnegative values $\alpha(s_J) = J$, may correspond to bound states with spin J and mass $m_J^2 = s_J$. For $s > 4m^2$ at points where $\operatorname{Re}\alpha^{\pm}(s) = J$, the trajectory of the pole passes close to the physical region (if $\operatorname{Im}\alpha^{\pm}(s)$ is not large), generating resonances there (see the discussion on poles and resonances in § 12.4). The imaginary part of $\alpha(s)$ determines the width of the resonance. Here, in correspondence with the meaning of the functions $a^{\pm}(l, s)$, even-signature trajectories $\alpha^{\pm}(s)$ may contain resonances only with even spins $J = 0, 2, 4, \ldots$, while odd-signature trajectories $\alpha^-(s)$ correspond to resonances with odd spins $J = 1, 3, 5. \ldots$

To see that resonances may be associated with points of the trajectory $\operatorname{Re}\alpha(s_J) = J$, let us find the partial wave amplitude $a(l, s)$ in the s-channel when the full amplitude $F(s, z)$ is given by one Regge pole in (32):

$$F(s, z) = \pi B \frac{(2\alpha^{\pm}(s)+1)\,\beta^{\pm}(s)\,[P_{\alpha^{\pm}}(-z) \pm P_{\alpha^{\pm}}(z)]}{\sin \pi\alpha^{\pm}(s)}. \tag{41}$$

Using the formula

$$\frac{1}{2}\int_{-1}^{1} P_J(z)\,P_\alpha(-z)\,dz = \frac{1}{\pi}\,\frac{\sin \pi\alpha}{(\alpha-J)(\alpha+J+1)}$$

we find, at $\operatorname{Re}\alpha(s_J) = J$,

$$a_J(s) = \frac{(2\alpha^{\pm}(s)+1)\,\beta^{\pm}(s)\,(1 \pm (-1)^J)}{(\alpha^{\pm}-J)\,(\alpha^{\pm}+J-1)}. \tag{42}$$

If the signature σ of the function α^σ is positive (negative), there are no poles for odd (even) integral α.

Let us now expand Re α near the resonance J:

$$\operatorname{Re}\alpha = J+(s-s_J)\alpha'(s_J), \quad \alpha'(s) = \frac{d \operatorname{Re}\alpha(s)}{ds}. \tag{43}$$

We have assumed that near a resonance $\operatorname{Im}\alpha(s) \approx \operatorname{Im}\alpha(s_J)$ and $\beta(s) \approx \beta(s_J)$. Then for the partial wave $a_J(s)$ near the pole with $\operatorname{Re}\alpha(s_J) = J$, we obtain the Breit–Wigner formula:

$$a_J(s) = -\frac{(2\alpha^{\pm}(s_J)+1)\beta^{\pm}(s_J)(1\pm(-1)^J)}{(s-s_J+i\Gamma)\alpha'(s_J)(\alpha+J+1)} \approx -\frac{\beta^{\pm}(s_J)}{(s-s_J+i\Gamma)\alpha'(s_J)} \tag{44}$$

with width†

$$\Gamma = \frac{\operatorname{Im}\alpha(s_J)}{\alpha'(s_J)}.$$

The self-consistency of this interpretation of a Regge trajectory $\alpha(s)$ demands that the width Γ be a positive quantity which vanishes below the first threshold $s_{\min} = 4m^2$. Consequently, in such a picture, $\alpha(s)$ must be real for $s < s_{\min} = 4m^2$. Since the masses of the resonances grow with spin, i.e. experimentally, $\alpha'(s) > 0$, one must have $\operatorname{Im}\alpha(s) > 0$ for $s > s_{\min}$.

The reality of $\alpha(s)$ in the region $0 < s \leqslant 4m^2$ may be proven (within the framework of assumptions 1 and 2, made at the beginning of this section, and the Regge-pole hypothesis) starting from the analytic properties of the quantities $(s-4m^2)^{-l}a^{\pm}(l, s)$ in the complex s-plane. On the other hand, reality of $\alpha(s)$ above the threshold $s_{\min} = 4m^2$ would contradict the elastic unitarity condition (39).

The unitarity condition (39) and the analytic properties of the functions $(s-4m^2)^{-l}a^{\pm}(l, s)$ allow one to find[182, 183] the threshold behavior of the Regge trajectories $\alpha(s)$ ($s_{\min} = 4m^2$):

$$\alpha(s) = \alpha(4m^2)+c(4m^2-s)^{\alpha(4m^2)+\frac{1}{2}}+ \dots, \tag{45}$$

where $\alpha(4m^2)$ and the constant c are real and $s \geqslant 4m^2$. From (45) it is clear that in the complex l-plane the Regge trajectory rises into the upper half plane, beginning at the point $\alpha(4m^2)$ on the real axis.

From formula (45) it also follows that the function $\alpha(s)$ has a cut in the complex s-plane along the positive axis beginning at the first threshold: $4m^2 \leqslant s < \infty$. It is usually assumed, on the basis of correspondence with the nonrelativistic theory, that this is the only cut of the function $\alpha(s)$ and, consequently, $\alpha(s)$ does not have a left-hand cut inherent in the partial wave amplitude $a(l, s)$.

Thus one may combine two important features of hadron physics in Regge trajectories $\alpha(s)$. On one hand, for $s < 0$, when s has the meaning of an invariant momentum transfer in the t-channel, the trajectory $\alpha(s)$ determines the asymptotic form of the amplitude as $t \to \infty$.

On the other hand, for $s > 0$, (where $\sqrt{s}$ is the invariant energy), the trajectory $\alpha(s)$ determines the bound states and resonances of the s-channel and can be constructed using experimental data on masses and spins of resonances. Here, the lower the energy of the resonance the closer is this part of the trajectory $\alpha(s)$ to that ($s < 0$) characterizing high-energy scattering in the t-channel.

† This definition differs from that used earlier.

§ 13.4. Simplest consequences of the Regge-pole hypothesis. The diffraction peak and the total cross-section

Let us consider the asymptotic behavior of the elastic scattering amplitude $1+2 \to 1+2$ of spinless equal-mass particles at high energies $\sqrt{s}$ when the behavior of the amplitude in the s-channel is fully determined by one trajectory $\alpha(t)$ with the largest value of Re $\alpha(t)$.

We first write the basic formulae for the asymptotic behavior in the s-channel in the one-pole approximation. As we shall see below, these formulae hold not only for elastic scattering but also for any two-body process $1+2 \to 3+4$. According to (38), the amplitude in the one-pole approximation has the form

$$F^{\pm}(s,t) \approx \text{const}\,(2\alpha^{\pm}(t)+1)\,\tilde{\beta}^{\pm}(t)\,s^{\alpha^{\pm}(t)}\left[i \mp \left(\tan\frac{\pi\alpha^{\pm}(t)}{2}\right)^{\mp 1}\right], \tag{46}$$

for $s \to \infty$, where the upper signs refer to positive signature and the lower to negative. The overall constant and the residue $\beta^{\pm}(t)$ in (46) are real. In the scattering region, the trajectory $\alpha^{\pm}(t)$ is real (see § 13.3).

Using (46), let us find the asymptotic behavior in the s-channel of the total cross-section $\sigma_{tot}(1, 2)$, the differential elastic scattering cross-section $d\sigma_{el}/dt$ and the elastic scattering cross-section σ_{el}. These quantities are directly observed in experiment (see § 13.1), and rigorous bounds exist for them (see § 13.2). By the optical theorem and (46), the contribution to the total cross-section is

$$\begin{aligned}\sigma_{tot} &= \frac{16\pi}{B}\,\frac{1}{[(s-4m^2)s]^{1/2}}\,\text{Im}\,F(s,0)\\ &\approx \text{const}\,(2\alpha^{\pm}(0)+1)s^{\alpha^{\pm}(0)-1}\,\text{Im}\left\{\tilde{\beta}^{\pm}(0)\left[i\mp\left(\tan\frac{\pi\alpha(0)}{2}\right)^{\mp 1}\right]\right\}.\end{aligned} \tag{47}$$

The differential scattering cross-section $d\sigma_{el}/dt$ (in the s-channel) in the one-pole approximation is asymptotically equal to

$$\frac{d\sigma_{el}}{dt} \approx \frac{1}{s^2}\,|F(s,t)|^2 = |f(t)|^2\,s^{2(\alpha(t)-1)}. \tag{48}$$

Experimentally, the asymptotic cross-section $d\sigma_{el}/dt$ has a maximum (the diffraction peak) as a function of t for $t \approx 0$, exponentially falling as the momentum transfer $|t|$ increases (see § 13.1). The region of small t is thus the most important, and we may expand $\alpha(t)$ around $t = 0$:

$$\alpha(t) = \alpha(0)+t\alpha'(0)+\ \ldots$$

If the dependence of $f(t)$ on t in (48) is neglected, assuming that $f(t)$ varies slowly in comparison with exponential behavior, one has

$$\frac{d\sigma}{dt} \approx \left(\frac{d\sigma}{dt}\right)_{t=0} \exp\,[2\alpha'(0)\,t\ln s] \qquad (t<0). \tag{49}$$

Consequently, the amplitude with one Regge trajectory indeed describes an exponentially falling cross-section in t if the trajectory slope is positive: $\alpha'(0) > 0$. For elastic scattering, the positivity of $\alpha'(0)$ also follows from theory.[184] The slope in t of the differential cross-section at small t depends on s, and with increasing s the diffraction peak shrinks.

The position of the trajectory $\alpha_n^{\pm}(t)$ cannot yet be obtained from theory. We must therefore turn to the experimental facts (§ 13.1).

The cross-section σ_{tot} will not grow with s if $\alpha(0) \leqslant 1$. The value $\alpha(0) = 1$ corresponds to the Froissart bound (for polynomial growth of the amplitude). If the interaction is sufficiently strong that $\alpha(0)$ attains the bound $\alpha(0) = 1$, one may write the expression

$$\sigma_{tot} \sim \lim_{t \to 0} \tilde{\beta}(t) \operatorname{Im} \left[i \mp \left(\frac{2}{\pi \alpha'(0) t} \right)^{\mp 1} \right],$$

for σ_{tot}, where we have expanded $\alpha(t)$ for small t: $\alpha(t) = \alpha(0) + \alpha'(0)\, t$ ($\alpha'(0)$ is real).

For positive signature in (46) and (47), the limit as $t \to 0$ exists when $\tilde{\beta}(0) \neq 0$:

$$\sigma_{tot} \approx \tilde{\beta}(0) = \text{const}; \tag{50}$$

Thus the forward scattering amplitude in the asymptotic region will be purely imaginary:

$$F(s, 0) \approx i\tilde{\beta}(0)s, \quad \operatorname{Re} F(s, 0) \approx 0.$$

If the signature of the pole with $\alpha^{-}(0) = 1$ is negative, the amplitude at $t \to 0$ will behave as $\tilde{\beta}(t)/t$, i.e. first of all, it will be real (for real $\tilde{\beta}(t)$), and, secondly, it will be finite only when $\tilde{\beta}(t) = at$ near $t = 0$. This negative signature pole with $\alpha^{-}(0) = 1$ must be rejected. It violates the assumption (23) leading to the Pomeranchuk theorem, and contradicts the semi-empirical notion of a nearly imaginary amplitude $F(s, 0)$ at high energies.†

The trajectory leading to a constant or nearly constant asymptotic total cross-section (47) is called the Pomeranchuk trajectory α_P. It has positive signature and $\alpha_P(0) \approx 1$. From (50) it is also clear that $\tilde{\beta}_P(0) > 0$. (This fact may be proven.[184])

The total elastic cross-section σ_{el} in the asymptotic region may be found by integrating (49) for $\alpha^{+} = \alpha_P$:

$$\sigma_{el} = \left(\frac{d\sigma}{dt} \right)_{t=0} \frac{1}{2\alpha_P'(0) \ln s} \sim \frac{1}{\ln s}. \tag{51}$$

Consequently, when dominated by the Pomeranchuk trajectory, the elastic cross-section vanishes logarithmically with energy at the same time that the total cross-section remains constant. The width of the diffraction peak for elastic scattering $\Delta = \sigma_{el}/(d\sigma_{el}/dt)_{t=0}$ [see (19)] will vanish logarithmically with increasing s: $\Delta \sim (\ln s)^{-1}$. By definition, all remaining trajectories $\alpha_n^{\pm}(t)$ lead to a vanishing total cross-section (47). These trajectories thus must lie lower in the (α, t)-plane than the Pomeranchuk trajectory $\alpha_n^{\pm}(t) < \alpha_P(t)$ for at least a finite range of $t < 0$.

Diffraction dissociation and charge-exhange reactions

If one traces the derivation of the one-Regge-pole term (46) and the assumptions made there, it is easy to see that the restriction to elastic scattering is not essential. The masses of the scattered particles were set equal only in order to avoid unnecessary complications. In the same way, internal particles (in the sense of the unitarity condition) were assumed heavier than the external ones to avoid normal unphysical cuts from s_{min} to the beginning

† Moreover, using (19) and (21), one would find such a pole gave $\sigma_{el} > \sigma_{tot}$, in contradiction with unitarity.

of the physical region. Nothing prevents us from extending the method of § 13.3 to the case of the invariant amplitude for any process $1+2 \rightarrow 3+4$ (in the case of spinless particles) and performing the "Reggeization" of the amplitude, i.e. writing its asymptotic behavior in terms of Regge trajectories in the crossed channel. Here one must pay attention to the quantum numbers of the trajectories α.

As was shown in § 13.3, the trajectory $\alpha(s)$ may pass through resonances in the s-channel for $s > s_{\min}$. The quantum numbers of these resonances, i.e. the quantum numbers of internal symmetries in the s-channel, are also the quantum numbers of the trajectory $\alpha(s)$. However, a trajectory may not be related to resonances in general (if, for example, it varies with s too slowly to attain integral values $\operatorname{Re}\alpha^{\pm} = l$ of the right signature for observable s). In this case the trajectory $\alpha(s)$ still has the quantum numbers of the s-channel. This expresses the invariance of the amplitude with respect to internal symmetry transformations when the Regge amplitude is represented in the pole form (40) (near the pole):

$$a(l, s) \approx \frac{\beta(s)}{l-\alpha(s)} = \frac{g_{\alpha 12}(s)\, g_{\alpha 34}(s)}{l-\alpha(s)}. \tag{52}$$

In assigning quantum numbers to trajectories it is thus assumed that the residue factors into $g_{\alpha 12}$ and $g_{\alpha 34}$, characterizing the interaction of the trajectory $\alpha(s)$ with particles 1, 2, and 3, 4, via the same rule as for a particle a (see § 12.3). Factorizability of residues will be demonstrated below.

Thus the Pomeranchuk pole $\alpha_P(t)$, which determines the asymptotic behavior of the elastic scattering amplitude in the s-channel $1+2 \rightarrow 1+2$, must have the quantum numbers of the t-channel, i.e. the quantum numbers of the particle–antiparticle system $1+\bar{1}$. These are the quantum numbers of the vacuum, and hence $\alpha_P(t)$ is also called the vacuum trajectory.

The question then arises: Is the vacuum trajectory α_P universal, i.e. is it encountered (*i*) for elastic scattering of other particles (e.g. $1+3 \rightarrow 1+3$), and, (ii) for inelastic processes in the s-channel with the same (vacuum) quantum numbers in the t-channel (for diffraction dissociation)? The intuitive answer to this question, of course, is yes. One may assume that Regge trajectories are a consequence of the strong interactions, so that their existence cannot depend on the specific process induced by these interactions. As we shall see below, from the unitarity condition, it follows that Regge trajectories are, indeed, universal.

From the universality of Regge trajectories it follows that the Pomeranchuk trajectory will describe the asymptotic behavior of the amplitudes both for elastic scattering and for diffraction dissociation. Of course, the residues β in diffraction dissociation may differ from those in elastic scattering. Consequently, the asymptotic behavior of the process $\pi+\mathrm{N} \rightarrow A_1+\mathrm{N}$ will be the same [formulae (48), (49) with $\alpha^+ = \alpha_P = 1$] as for elastic pion–nucleon scattering $\pi+\mathrm{N} \rightarrow \pi+\mathrm{N}$.

The asymptotic behavior of processes with charge exchange cannot be characterized by the trajectory α_P, and the cross-sections for these processes will vanish more rapidly than the cross-sections for elastic processes. This conclusion is completely in accord with the experimental situation (see § 13.1).

The factorizability of Regge-pole residues allows one to draw graphs (Fig. 21a) for an amplitude dominated by one Regge pole (52) of the same type as for the usual pole terms in the amplitude $F^{(1)} \approx g_{a12}g_{a34}/(s-m_a^2)$. The resonance a is replaced here by a "Reggeon." Instead of a pole in the amplitude for a fixed mass $s = m_a^2$ and $J = J_a$, determined by the properties of the particle a, in the case of a Regge pole (52) we obtain a series of resonances whose masses are related to their spin by the Regge trajectory function $\alpha(s)$. In the crossed channel, the Reggeon diagram of Fig. 21a determines the asymptotic behavior in the t-channel; it is related to the exchange of the Reggeon $\alpha(s)$. In the case of the Pomeranchuk trajectory, the Reggeon α_P is called the Pomeron.

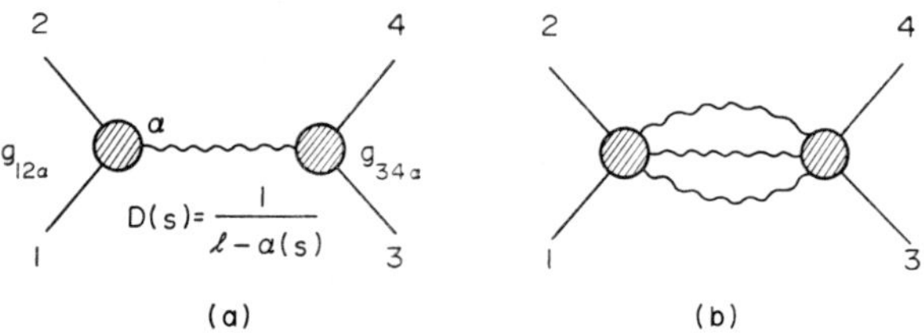

FIG. 21. (a) Exchange of a single s-channel Reggeon $\alpha(s)$ governing asymptotic behavior in the t-channel. (b) Multi-Reggeon exchange, leading to branch points in l.

As in the case of a particle, when introducing the idea of one-Reggeon exchange we should also allow the possibility of multi-Reggeon exchanges (Fig. 21b). The exchange of several Reggeons corresponds to a branch point of the amplitude[185, 186] and gives rise to nonleading terms in the asymptotic behavior of the amplitude [see below, eqns. (58)–(60)].

Let us consider pion–nucleon scattering, $\pi_1+N_1 \to \pi_2+N_2$, (s-channel) as an example. In the t-channel, $\pi_1+\bar{\pi}_2 \to \bar{N}_1+N_2$, one may have amplitudes with internal quantum numbers $I = 0$, $I_3 = 0$, and $I = 1$, $I_3 = \pm 1, 0$, $Y = 0$ (we are assuming isospin symmetry for the strong interactions). These amplitudes correspond to trajectories $\alpha(t)$ with the same quantum numbers, including the vacuum trajectory. All these trajectories determine the s- and u-channel asymptotic behavior. In turn, the s-channel trajectories $\alpha(s)$, which will have the quantum numbers $B = 1$, $I = \frac{1}{2}$, $I_3 = \pm\frac{1}{2}$, $Y = 0$ and $B = 1$, $I = \frac{3}{2}$, $I_3 = \pm\frac{3}{2}, \pm\frac{1}{2}$, $Y = 0$, are important for the asymptotic behavior in the u- and t-channels. In the u-channel $\pi_1+\bar{N}_2 \to \bar{N}_1+\pi_2$ we have antibaryon resonances with $B = -1$ for $I = \frac{1}{2}, \frac{3}{2}$. The trajectories $\alpha(u)$ with these quantum numbers may contribute to the asymptotic behavior in the s- and t-channels. In general, the leading trajectories yield the dominant contribution. By definition, the Pomeranchuk trajectory dominates if it is allowed by quantum numbers.

Factorizability of residues and universality of trajectories

In order to display these properties[187] let us consider reactions in the t-channel involving two types of spinless particles which we may call pions and kaons (we shall neglect complications related to isospin):

$$\pi+\bar{\pi} \to \pi+\bar{\pi} \qquad \text{amplitude } f(l, t),$$

$$K+\bar{K} \to \pi+\bar{\pi} \qquad \text{amplitude } g(l, t),$$

$$K+\bar{K} \to K+\bar{K} \qquad \text{amplitude } h(l, t).$$

The transition $2\pi \to 3\pi$ is considered forbidden, as for real pions.

The unitarity relations in the t-channel in the elastic scattering region $4m^2 < t < 16m^2$, $m \equiv m_\pi$, continued to complex angular momenta l, are [see (12.52)]:

$$\left.\begin{aligned} f(l, t)-f^*(l^*, t) &= 2i\gamma(t) f(l, t) f(l^*, t), \\ g(l, t)-g^*(l^*, t) &= 2i\gamma(t) g(l, t) f^*(l^*, t), \\ h(l, t)-h^*(l^*, t) &= 2i\gamma(t) g(l, t) g^*(l^*, t), \end{aligned}\right\} \tag{53}$$

where $\gamma(t) = ((t-4m^2)/t)^{1/2}$. The second and third of eqns. (53) relate to the unphysical region (in t) of the reactions $K+\bar{K} \to \pi+\bar{\pi}$ and $K+\bar{K} \to K+\bar{K}$, since here $t < 4m_K^2$. (For $t > 4m_K^2$ the right-hand side of the third eqn. (53) contains the term $\gamma_K(t)\, h(l,t)\, h^*(l^*, t)$.)

The unitarity condition (53) may be solved for f, g, h (see also § 12.4):

$$\left.\begin{aligned} f(l, t) &= \frac{f^*(l^*, t)}{1-2i\gamma(t) f^*(l^*, t)}, \\ g(l, t) &= \frac{g^*(l^*, t)}{1-2i\gamma(t) f^*(l^*, t)}, \\ h(l, t) &= h^*(l^*, t)+\frac{2i\gamma(t)\,(g^*(l^*, t))^2}{1-2i\gamma(t) f^*(l^*, t)}. \end{aligned}\right\} \tag{54}$$

Hence all amplitudes f, g, and h have the same pole at $1-2i\gamma(t) f^*(l^*\ t) = 0$. Near the pole $l = \alpha(t)$ the amplitudes have the form

$$\left.\begin{aligned} f(l, t) &= \frac{1}{2i}\, \frac{\beta(t)}{\gamma(t)\,(l-\alpha(t))}, \\ g(l, t) &= \frac{\beta(t)\, g^*(l^*, t)}{l-\alpha(t)}, \\ h(l, t) &= \frac{2i\gamma(t)\, \beta(t)\, (g^*(l^*, t))^2}{l-\alpha(t)}, \end{aligned}\right\} \tag{55}$$

which demonstrates universality of the trajectory $\alpha(t)$.

From (55) it is clear that the residues of the amplitude $r_{\pi\pi} = \beta/2i\gamma$, $r_{K\pi} = \beta g^*(l^*)$, $r_{KK} = 2i\gamma\beta[g^*(l^*)]^2$ may be written in factorized form

$$r_{AB}(t) = g_{AA\alpha}(t)\, g_{BB\alpha}(t), \tag{56}$$

where $g_{AA\alpha}$ characterizes the "interaction" of the trajectory α with the particles $A = \pi$, K. This relation, obtained for $4m^2 < t < 16m^2$, may be analytically continued to any t.

If one trajectory $\alpha(t)$ is sufficient to describe the asymptotic régime, eqn. (56) implies limiting relations between total cross-sections:[187]

$$\sigma_{AC}\sigma_{BD} = \sigma_{AD}\sigma_{BC}, \tag{57}$$

where $\sigma_{AB} = \sigma_{\text{tot}}(A, B)$.

In particular, for the scattering of a particle on an antiparticle $A+\bar{A} \to A+\bar{A}$, relation (57) gives $\sigma_{A\bar{A}} = \sigma_{AA}$ if one bears in mind that $\sigma_{\bar{A}\bar{A}} = \sigma_{AA}$. This is in accord with the

Pomeranchuk theorem (§ 13.2). The inclusion of spin does not change this result. However, for currently attainable energies one must take account of several trajectories in the amplitude, and the simple limiting relations (57) are not obtained.

Unphysical states

If the "intercept" of the trajectories $\alpha(0)$ is positive, then as a result of the positivity of the slope $\alpha'(0)$ a trajectory $\alpha(t)$ may vanish for some $t < 0$, i.e. in the scattering region. Now for a trajectory with positive signature, the point $\alpha^+(t_0) = 0$ must correspond to a particle δ with spin 0 and with mass $m_\delta^2 = t_0$. In the present case $t_0 < 0$, so the value $\alpha^+ = 0$ refers to an unphysical state. In order not to allow such a state in a theory, we must assume that it does not interact with physical particles, i.e. $g_{\delta 12}(t_0) = g_{\delta 34}(t_0) = 0$ at the point $t = t_0 < 0$. But then all one-pole Regge-term amplitudes vanish at this point. Using similar reasoning about the inadmissibility of unphysical states, we conclude that a positive signature residue must vanish at the points $t_n < 0$, where $\alpha^+(t_n) = -2n$, $n = 1, 2, \ldots$, while a negative signature residue must vanish at the points $t_n < 0$, where $\alpha^-(t_n) = -(2n+1)$, $n = 0, 1, 2, \ldots$

The Regge-pole hypothesis and cuts

Let us consider briefly a further consequence of the Regge-pole hypothesis and the unitarity condition according to which the functions $a^\pm(l, s)$ must have branch points as well as simple poles in the complex l-plane. This contradicts the assumption of the meromorphic nature of $a^\pm(l, s)$ in the l-plane ("the Regge-pole hypothesis"). The poles of the function $a^\pm(l, s)$ in the l-plane, nonetheless, may provide the main contribution to the asymptotic part of the amplitude, in which case the Regge-pole hypothesis can serve as a good first approximation.

The need for branch points in the complex l-plane follows from the unitarity condition, with multi-particle intermediate states included. For these states the two-particle term on the right-hand side of (39) or (53) must be supplemented with terms containing integration over the angular momenta and masses of the complexes forming the multi-particle states. The analytic continuation of this multi-particle unitarity condition into the complex l-plane leads to amplitudes containing two or more Regge poles. In analogy with branch points in the s-plane corresponding to the threshold for production of physical particles in the s-channel, the threshold $\alpha_c^{(n)}(t)$ for "production" of n Regge poles corresponds to a (logarithmic) branch point $\alpha_c^{(n)}(t)$ in the l-plane. The location of a branch point corresponding to n identical Regge poles depends in a simple way on the number n:[185, 186]

$$\alpha_c^{(n)}(t) = n\alpha\left(\frac{t}{n^2}\right) - n + 1. \tag{58}$$

These branch points may also be illustrated using Reggeon graphs (Fig. 21b). The branch point corresponds to an additional term in the amplitude referring to the simultaneous exchange of n Reggeons.

Let us consider the asymptotic behavior of the amplitude due to a cut of the function $a^\pm(l, s)$ in the l-plane with right-most branch point $\alpha_c(s) > 0$, $\operatorname{Im} \alpha_c = 0$. To isolate the necessary term in the amplitude, we shall assume that the cut with a branch point at $\alpha_c(s)$ is the only singularity of $a^{\pm}(l, s)$, and return to the steps leading from (27) to (32). Then, instead of a sum over Regge poles in (32), we obtain an integral along the boundaries of the cut:

$$F_c^{\pm}(s, t) = \frac{B}{2} \int^{\alpha_c^{\pm}(s)} \frac{1}{2}(2l+1)\, \Delta a^{\pm}(l, s) t^l \frac{1 \pm e^{-i\pi l}}{\sin \pi l}\, dl, \tag{59}$$

where $\Delta a^{\pm}$ is the discontinuity of the function $a^{\pm}(l, s)$ across the cut:

$$\Delta a^{\pm}(l, s) = \tfrac{1}{2}[a^{\pm}(l_+, s) - a^{\pm}(l_-, s)]$$

(l_+ and l_- are the values of the variable l on the upper and lower edges of the cut).

Comparing $F_c^{\pm}$ with expression (37) for the one-pole term of the amplitude, we see that the contribution $F_c^{\pm}$ corresponds to a continuous distribution of Regge poles. The asymptotic contribution as $t \to \infty$ from $F_c^{\pm}$ will contain an additional ln s in the denominator:

$$F_c^{\pm}(s, t) \approx \text{const} \ln^{-1} s\, s^{\alpha_c^{\pm}(t)}. \tag{60}$$

The function $F_c^{\pm}$ has definite signature, but the signature factor cannot be brought out from under the integral in $F_c^{\pm}$. The phase of the asymptotic expression F_c thus depends on the form of the cut and the discontinuity $\Delta a^{\pm}$. From eqn. (59) for the cut's contribution, it is clear that the "residue" (i.e. the "const" in (60)) cannot factorize here.

Thus the existence of cuts in $a^{\pm}(l, s)$ entails a term $s^{\alpha_c^{\pm}(t)}/\ln s$, in the asymptotic behavior of the amplitude, where the power is determined by the position $\alpha_c^{\pm}$ of the right-most branch point of the cut. When the contribution (60) of the cut vanishes more rapidly in comparison with the pole $\alpha^{\pm}$ (i.e. when $\alpha^{\pm}(t) > \alpha_c^{\pm}(t)$), one may neglect the cuts as long as the poles themselves exist, the energy is high enough, and scattering in the diffraction peak is being considered.

Inclusion of spin. Fermion Regge poles

Up to now we have discussed only spinless equal-mass particles, implicitly assuming that the Regge pole model may be extended to particles with spin. The spins of external particles may introduce complications of two types. First of all, for particles with spin the total amplitude may be decomposed into several independent amplitudes (helicity or invariant amplitudes), which may have kinematic singularities. Each independent amplitude for the scattering of states with definite parity must be Reggeized separately, since the strong interactions conserve parity. Kinematic singularities must be eliminated before Reggeization. Secondly, processes involving fermion Regge poles lead to additional complications. Fermion Regge poles corresponding to states with opposite parity are complex conjugate to one another in the physical scattering region.[188] We shall illustrate these features via the example of fermion Regge-pole exchange in pion–nucleon scattering.

In the process $N_1+\pi_1 \to N_2+\pi_2$, the fermion Regge poles have the quantum numbers of the u-channel $N_1+\bar{\pi}_2 \to N_2+\bar{\pi}_1$. Let the nucleon momenta be p_1, p_2 and the pion momenta be k_1, k_2. The expansion of the spinor amplitude $\mathcal{M}_\alpha^{\ \beta}$ in terms of u-channel covariants has the form

$$\mathcal{M} = a + b\not{q}, \tag{61}$$

where $q = (p_1+p_2-k_1-k_2)/2$, while a and b are invariant amplitudes. In this notation, $u = q^2$.

The fermion Regge poles in the u-channel may describe resonances both with positive and with negative parity (with respect to the neutron). Hence one must relate the invariant amplitudes a and b to the amplitudes f_+ and f_- with definite parity (in the u-channel c.m.s.):

$$\mathscr{M} = f_+(s, t, u)[\not q+\sqrt{u}]+f_-[\not q-\sqrt{u}]. \tag{62}$$

In the u-channel c.m.s. the new covariants $\not q \pm \sqrt{u}$ become $\gamma^0 \pm 1$. However, the new expansion of $\mathscr{M}$ introduces a $\sqrt{u}$ kinematic singularity. The amplitudes a and b do not have this singularity: in the complex u-plane, the dynamical cuts begin at threshold branch points (see § 12.2). In order that the functions $\sqrt{u}(f_+-f_-)$ and f_++f_- be regular and nonzero at $u = 0$ subject to the condition $f_+ \neq f_-$ that there be no parity degeneracy, we must set

$$f_+ = f(\sqrt{u}, s), \quad f_- = f(-\sqrt{u}, s). \tag{63}$$

Thus amplitudes with opposite parity are related to one another. At $u = 0$ they coincide: $f_+(0, s) = f_-(0, s)$. In the u-channel, where $u > (m_N+m_\pi)^2$, the amplitudes $f_\pm$ depend on the total energy $W = \sqrt{u}$; here

$$f_+(W, s) = f_-(-W, s), \quad W > (m_N+m_\pi). \tag{64}$$

Comparing with the spinless case, we see that here there are two independent amplitudes f_+ and f_-. After passing to the Gribov–Froissart amplitudes, continuing them to complex l, and introducing the Regge-pole hypothesis, we finally obtain four amplitudes f_+^σ, f_-^σ, differing from one another with respect to parity and signature σ. Each type of amplitude corresponds to its own trajectory:

$$j_\pm^\sigma(u) = \alpha_\pm^\sigma(u)+\tfrac{1}{2}.$$

Let us compare the properties of the trajectories and the residues for opposite parities. We write the asymptotic expressions for $f_\pm^\sigma$ in the s-channel in the case of one fermion pole $j_\pm^\sigma(u)$:

$$\left.\begin{aligned} f_+^\sigma(s, u) &= \beta_+(u)\, s^{\alpha_+^\sigma(u)}\eta_+^\sigma(\alpha_+^\sigma), \\ f_-^\sigma(s, u) &= \beta_-(u)\, s^{\alpha_-^\sigma(u)}\eta_-^\sigma(\alpha_-^\sigma). \end{aligned}\right\} \tag{65}$$

By virtue of (63) we have, for each signature σ:

$$\alpha_+ = \alpha(\sqrt{u}), \quad \alpha_- = \alpha(-\sqrt{u}),$$
$$\beta_+ = \beta(\sqrt{u}), \quad \beta_- = \beta(-\sqrt{u}).$$

Consequently, for $u = 0$, the trajectories α_+ and α_- and the residues β_+ and β_- will coincide. Let us now show that, for $u < 0$, opposite-parity trajectories and residues will be complex conjugates of one another. To do this we consider the absorptive parts a_1 and b_1 of the amplitudes a and b in the physical s-channel region:

$$a_1 = (\beta_+ s^{\alpha_+} - \beta_- s^{\alpha_-})\sqrt{u}, \quad b_1 = (\beta_+ s^{\alpha_+} + \beta_- s^{\alpha_-}),$$

where a_1 and b_1 are real. For $u < 0$ it then follows that

$$\alpha_+(u) = \alpha_-(u), \quad \beta_+(u) = \beta_-(u). \tag{66}$$

In the region $u > 0$, the trajectories $\alpha_\pm(W)$, as functions of the energy W in the u-channel satisfy the MacDowell symmetry conditions[188a]

$$\alpha_+(W) = \alpha_-(-W). \tag{67}$$

Consequently, linear fermion Regge trajectories $\alpha(u) = \alpha(0)+\alpha' u$ will be parity-doubled.

The parity of the πN system in a state with orbital angular momentum l is $\eta_P(\pi N) = \eta_P(N)(-1)^{l+1}$, where $\eta_P(N)$ is the parity of the nucleon (which is taken as positive). For given total angular momentum J, the πN system may have two values of l: $l_+ = J+\frac{1}{2}$ or $l_- = J-\frac{1}{2}$ i.e. the partial wave amplitudes $a^J_{l_+}$ and $a^J_{l_-}$ have opposite parity. From this it follows that both for bosons and for fermions it is convenient to introduce the concept of normality of a trajectory:

$$\left.\begin{aligned} N &= \eta_P^{(J)}(-1)^J && \text{(bosons)},\\ N &= \eta_P^{(J)}(-1)^{J-\frac{1}{2}} && \text{(fermions)}, \end{aligned}\right\} \tag{68}$$

where $\eta_P^{(J)}$ and J are the parity and spin of the resonance on the trajectory. The value of N is the same both along a boson trajectory and along a parity-doubled fermion trajectory.

§ 13.5. Properties of Regge trajectories

As was shown in the previous section, Regge trajectories $\alpha(t)$ describe the asymptotic behavior of the amplitude in the s-channel at high energies, $s \to \infty$, and finite $t < 0$. On the other hand, for $t > t_{\min}$ in the t-channel region the real part $\operatorname{Re}\alpha(t_J) = J$ is equal to the spin of resonances of mass $m_J^2 = t_J$, while the imaginary part $\operatorname{Im}\alpha(t_J)$ characterizes the resonance width Γ. Thus trajectories lead to a relation between resonances in one channel and asymptotic behavior in another (crossing conjugate) channel. The set of trajectories $\alpha(s)$, $\alpha(t)$, and $\alpha(u)$ in principle contains information both on resonances in all channels (including those at low energies $\approx$ 1–2 GeV), and on the high-energy behavior of a given amplitude for finite momentum transfer. Trajectories α, like particles, are assumed universal in the sense that they are the same for all reactions with the same quantum numbers. Hence, if one knows the trajectories α in this picture, one can interrelate resonances and asymptotic behavior in general, i.e. for all reactions. The basic features of the Regge trajectory description of resonances and of the asymptotic behavior of scattering amplitudes are supported by experiment. The notion of Regge trajectories is thus of fundamental importance in hadron physics.

Let us summarize the properties of Regge trajectories noted above in §§ 13.3 and 13.4, adding experimental information on the known trajectories.

Each trajectory α is labeled by the set of those quantum numbers characterizing particles (charge, isospin, parity, etc.), with the exception of spin and mass. Moreover, a trajectory α is characterized by signature σ. The spins of neighboring resonances on a meson or baryon trajectory differ by 2. Meson trajectories with positive signature $\sigma = +1$ contain only even spins, and trajectories with odd signature $\sigma = -1$ contain only odd spins.

The set of trajectory quantum numbers depends on the symmetry of the strong interactions. When SU_3 holds, this set includes the SU_3 multiplicity n, the isospin I, the additive quantum numbers B, $t = I_3$, Y, and the parity η_P and G-parity η_G. All resonances lying on the same trajectory will have the same values of $B, n, I_3 = t, Y, \eta_G, \eta_P$; i.e. when SU_3 holds the trajectories form SU_3 multiplets. The set of resonances on a trajectory is sometimes called a Regge family of particles.

A trajectory is denoted by its lowest-spin particle. For example, the ϱ-trajectory, or α_ϱ, is the trajectory with the quantum numbers of the ϱ-meson ($I^G = 1^+, \eta_P = -1, \sigma = -1$), containing mesons of odd spins beginning with the ϱ.

If trajectories are indeed universal, they are characterized by quantum numbers without respect to the reactions whose asymptotic behavior they determine.

The same trajectories determine the asymptotic behavior of different reactions. This property follows from the unitarity condition (see § 13.4).

Elastic scattering is always associated with trajectories possessing vacuum values of the charges $Q = I_3 = Y = B = 0$: for elastic scattering $1+2 \to 1+2$ the initial and final state in the t-channel reaction $1+\bar{1} \to 2+\bar{2}$ always contain a particle–antiparticle pair. Experimentally, the existence of two vacuum trajectories has been established—the Pomeranchuk trajectory α_P and the second vacuum trajectory $\alpha_{P'}$. The Pomeranchuk pole $l = \alpha_P$ is by definition the right-most pole in the complex angular momentum plane. The trajectory $\alpha_P(t)$ is the one that guarantees the constancy of the total cross-section at high s. The trajectories α_P and $\alpha_{P'}$ are also important for the asymptotic behavior of diffraction dissociation processes. From experiment,

$$\alpha_{P'}(0) < \alpha_P(0).$$

The quantum numbers of the trajectories $\alpha(t)$ describing the asymptotic behavior in the s-channel ($1+2 \to 3+4$) are determined by the quantum numbers of the invariant amplitudes in the t-channel process ($1+\bar{3} \to \bar{2}+4$). The charges B, I_3, Y, and Q of these trajectories are those of the states ($1+\bar{3}$) or ($\bar{2}+4$). Since the amplitude generally depends on several isospin or SU_3 amplitudes, the asymptotic behavior of such a process may depend on several trajectories. In particular, the asymptotic behavior of πN scattering near the forward direction will be related to the trajectories α_P, $\alpha_{P'}$, and α_ϱ; the asymptotic behavior of the cross-section for $\pi^\pm + \mathrm{N} \to \varrho^\pm + \mathrm{N}$ may depend on the contribution of the trajectories $\alpha_\omega, \alpha_{A_2}, \alpha_\pi, \alpha_{A_1}$, etc.

The residues β at Regge poles $l = \alpha$ factorize. Factorizability of residues is a consequence of the unitarity condition. For a reaction in the s-channel, $1+2 \to 3+4$, the residue at the pole $l = \alpha(t)$,

$$\beta(t) = g_{13\alpha}(t)\, g_{24\alpha}(t)$$

decomposes into factors $g_{ab\alpha}(t)$ characterizing the interaction of particles a, b with a Reggeon α. Since a pole term of an amplitude is invariant with respect to all internal symmetry transformations of the strong interactions, $\beta(t)$ is also invariant with respect to these transformations. Isospin and SU_3 invariants of this type may be constructed using methods studied in Chapters 8 and 10.

To some extent the form of the trajectory $\alpha(t)$ is determined by general considerations.

From unitarity it follows that Im $\alpha(t) \neq 0$ for $t > 4m^2$ (above the elastic threshold), since the poles cannot occur on the real t-axis.

The interpretation of poles as bound states demands that the imaginary part of the trajectory vanish: Im $\alpha = 0$ in the region $0 < t < 4m^2$ (below the elastic threshold).

In the narrow-width approximation for resonances, in which resonances are considered as stable particles, one may set Im $\alpha = 0$. This approximation will be used below.

At the point $t = 0$ two inequalities hold:

$$\alpha(0) \leq 1, \quad \alpha'(0) > 0.$$

The first of these follows from the Froissart bound and the assumption of polynomial growth of the amplitude, while the second is based on the experimental vanishing of the cross-section with increasing momentum transfer t at constant s.

Let us pass the (real) trajectories $\alpha_X(t)$ through the known resonances ($t > 0$), including scattering data for the $t < 0$ region. Then we obtain a series of mesonic and baryonic trajectories. The most important mesonic trajectories are α_ϱ, α_ω, α_{A_2} and $\alpha_{P'}$ (Fig. 22).

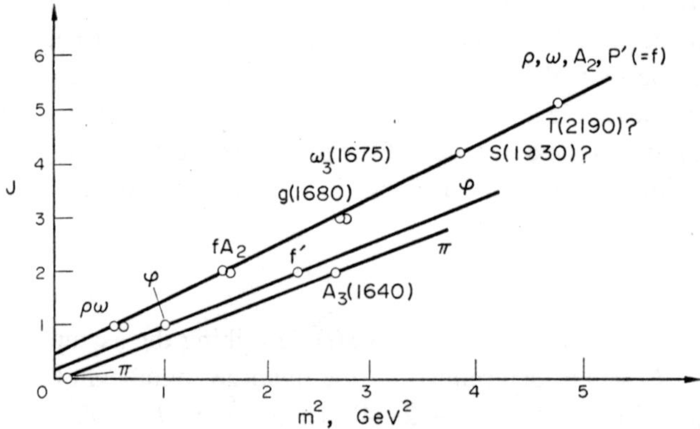

FIG. 22. Mesonic Regge trajectories showing spin J as a function of squared mass m^2.

The trajectory α_ϱ passes through the particles $\varrho(J^P = 1^-, m = 765$ MeV and $g(J^P = 3^-$, $m = 1680$ MeV). As a consequence of isospin symmetry, the trajectory actually consists of three trajectories with $I = 1, I_3 = \pm 1, 0$.

The isoscalar trajectory α_ω contains, besides the particle ω ($J^{PG} = 1^{--}$, $m = 784$ MeV), the particle ω_3 ($J^{PG} = 3^{--}$, $m = 1675$ MeV).

The isovector trajectory A_2 passes only through one established meson A_2 ($J^{PG} = 1^{+-}$, $m = 1310$ MeV). On the second vacuum trajectory (Fig. 22), there is only one meson f ($J^{PG} = 2^{++}$, $m = 1270$ MeV). The trajectories $\alpha_\varrho, \alpha_\omega, \alpha_{A_2}$ and $\alpha_{P'}$ nearly coincide with one another. The explanation of this fact ("exchange degeneracy") will be given below.

Figure 22 also shows the trajectories α_π and α_φ. Besides the pion π (140) with $J^{PG} = 0^{--}$, the trajectory also contains the meson A_3 (1640) with $J^{PG} = 2^{--}$. The trajectory φ, aside from the meson φ (1019) with $J^{PG} = 1^{--}$ and $I^G = 0^-$, at present contains no other particles. The meson f′ (1514) with $J^{PG} = 2^{++}$ and $I^G = 0^+$ begins a new trajectory, which seems to coincide with the φ-trajectory.

Of the baryonic trajectories, the Δ_δ, Σ_δ, N_α, Λ_α, Σ_α, and Ξ_α are well established.

The trajectory Δ_δ begins with the isoquartet $\Delta\,(J^P = \frac{3}{2}+, m = 1236$ MeV) in the decimet; it contains also the resonances $\Delta\,(J^P = \frac{7}{2}+,\ m = 1950$ MeV) and $\Delta\,(J^P = \frac{11}{2}+, m = 2420$ MeV), and possibly also $\Delta\,(m \approx 2850$ MeV) and $\Delta\,(m \approx 3230$ MeV), if the spins of these resonances are found to be $\frac{15}{2}+$ and $\frac{19}{2}+$ (Fig. 23).

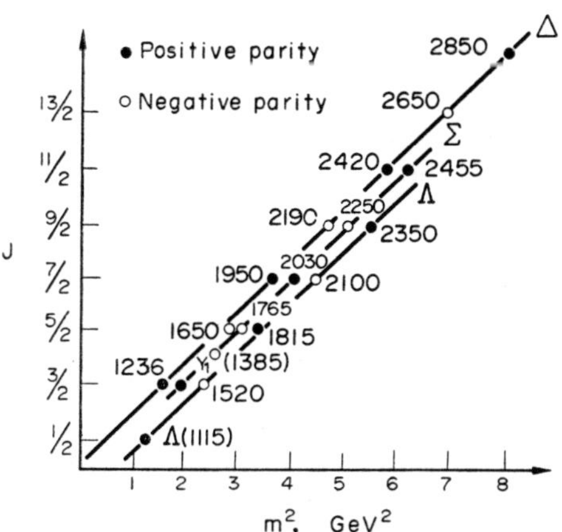

FIG. 23. Baryonic Regge trajectories.

Another trajectory related to the decimet is the isotriplet trajectory Σ_δ which contains the resonances $\Sigma\,(J^P = \frac{3}{2}+, m = 1385$ MeV) and $\Sigma\,(J^P = \frac{7}{2}+, m = 2030$ MeV) (Fig. 23).

The trajectories N_α, Λ_α, Σ_α, and Ξ_α begin with the particles of the baryon octet. The nucleon trajectory contains the resonances $N\,(J^P = \frac{5}{2}+, m = 1688$ MeV) and $N\,(J^P = \frac{9}{2}+$, $m = 2220$ MeV); the Λ_α trajectory also contains two other particles, while the Σ_α and Ξ_α contain for the moment only the $\frac{1}{2}+$ and $\frac{5}{2}+$ states. The octet with spin $J^P = \frac{5}{2}+$ is formed by the above-mentioned N-resonance and particles with mass $m_\Lambda = 1815$ MeV, $m_\Sigma = 1910$ MeV, $m_\Xi = 2030$ MeV. Figure 23 shows one of the octet trajectories—the trajectory Λ_α.

Figure 23 also shows exchange-degenerate trajectories of those just mentioned. For example, the trajectory Λ_α passes through the resonances $\Lambda\,(J^P = \frac{3}{2}-, m = 1520$ MeV) and $\Lambda\,(J^P = \frac{7}{2}-, m = 2100$ MeV).

The Pomeranchuk trajectory occupies a special place among all Regge trajectories. This trajectory passes through an integer at 0 but has positive signature, so the point $\alpha_P(0) = 1$ cannot correspond to a particle. It seems, in fact, that it has no resonances whatever on it; at least, none has been observed. While nonvacuum trajectories describe the asymptotic behavior of two-particle charge exchange scattering (which vanishes as $s \to \infty$), the Pomeranchuk trajectory describes the asymptotic behavior of elastic scattering for small angles and (for the total cross-section) all inelastic processes. The source of this trajectory is not clear at present.

Figures 22 and 23 illustrate the following characteristic features of trajectories.

1. To a good approximation, trajectories are linear not only in a small region around $t \approx 0$, but for a finite interval of t as well:

$$\alpha(t) = \alpha(0) + \alpha'(0)t. \tag{69}$$

2. The slope $\alpha'(0)$ of all trajectories is approximately the same (with the exception of th Pomeranchuk trajectory):

$$\alpha'(0) \approx 0.9 \text{ GeV}^{-2}. \tag{7}$$

Reference 189 gives some typical intercepts and slopes for various Regge trajectories $\alpha(t)$ More recent reviews should be consulted for precise parameters.[170] The Pomeranchuk trajectory seems to have considerably smaller slope of about $\frac{1}{4}$ GeV^{-2}.[190]

As we saw in § 13.4 the values of $\alpha(t)$ at $t = 0$ directly determine the asymptotic behavior of the forward scattering amplitude as $s \to \infty$: $F(s, 0) \approx s^{\alpha(0)}$. The intercept $\alpha(0)$ characterizes the role of the trajectory in the asymptotic behavior of the amplitude. A comparison

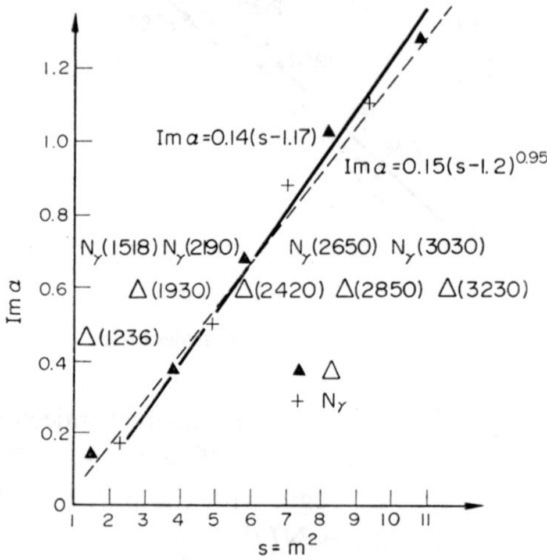

FIG. 24. Imaginary part of Regge trajectory function, Im $\alpha(s)$, versus $s = m^2$, for Δ_δ and N_γ trajectories.

of the values of $\alpha(0)$ shows that, after the Pomeranchuk trajectory, the greatest effect on the asymptotic behavior arises from the mesonic trajectories ϱ, ω, A_2, and P' and the baryon trajectories Δ_δ, Σ_δ, and N_α.

Knowing the trajectory slope $\alpha'(0)$ and the experimental total resonance widths Γ, one may calculate Im α at the resonance points via the formula Im $\alpha = \alpha'(0)\Gamma$.† The graph of Im α as a function of $s = m^2$ is approximately linear (Fig. 24) above the elastic threshold; in Fig. 24 the solid line is the straight line Im $\alpha = 0.14\,(s-1.17)$, and the dotted line is the curve Im $\alpha = 0.15\,(s-1.2)^{0.95}$.

Exchange degeneracy

By exchange degeneracy we mean the overlapping of trajectories with opposite signature.[191] When exchange degeneracy is present, resonances with the same (or even different) internal quantum numbers lie on trajectories spaced by one unit of spin, not by two as in the general theory (§ 13.3). For example, the trajectories ϱ, A_2, ω, and P' = f all nearly coincide with one another.

The occurrence of signature in characterizing trajectories is related to the fact that the analytic continuation in complex l is possible only for the separate Gribov–Froissart

† See eqn. (44) for the definition of Γ in this context.

amplitudes $a^+(l, s)$ and $a^-(l, s)$ (see § 12.4). We thus return to the definition (12.46) of these amplitudes. In the t-channel

$$\left.\begin{aligned} a^{\pm}(l, s) &= \frac{1}{\pi B}\int_{z_0}^{\infty}[A_s(s(t), z)\pm A_u(u(t), z)]\, Q_l(z)\, dz, \\ z &= 1 + \frac{2s}{t-4m^2}\,. \end{aligned}\right\} \tag{71}$$

From the definition of $a^{\pm}$ it follows that the sign $\pm$ has no meaning for the separate poles α^+ or α^- if either A_s or A_u does not contribute to the Regge asymptotic behavior $s^{\alpha^{\pm}(t)}$. We shall assume that (with the possible exclusion of the Pomeranchuk trajectory α_P) trajectories are always associated with resonances. If, for example, there are no resonances of the type r in the t-channel, then the trajectory $\alpha_r(t)$ does not exist.

The absorptive parts A_s and A_u in the physical regions of the s- and u-channels may be calculated using the unitarity condition (see § 12.3). Since we shall relate the existence of the trajectories $\alpha_n(s)$ and $\alpha_{n'}(u)$ to the existence of resonances in these channels, it is sufficient to use the resonance approximation in determining the contributions of A_s and A_u to the Regge asymptotic behavior. Let us replace the multi-particle states $|n\rangle$ in eqns (12.13) and (12.15) by resonances. Then it is clear from expression (71) for $a^{\pm}$ that if the s-channel contains no resonances, the contribution of A_s to the Regge term vanishes and the trajectories $\alpha^{\pm}(t)$ will be degenerate: $\alpha^{\pm}(t) = \alpha^-(t)$. Both trajectories $\alpha^+(t)$ and $\alpha^-(t)$ will be due only to resonances in the u-channel.

Analogously, the trajectories $\alpha^+(t)$ and $\alpha^-(t)$ will be degenerate if there are no resonances in the u-channel and, consequently, A_u does not contribute to the Regge asymptotic behavior. These cases ($A_s = 0$ or $A_u = 0$) differ with respect to the signs of the residues at the Regge poles.

Thus degeneracy of trajectories with respect to signature, or exchange degeneracy, signifies the absence of resonances in one of the crossed channels.

The experimentally observed trajectory degeneracies (see Fig. 22 and 23) may be explained by the fact that no exotic resonances exist (see § 10.3). Let us assume that only those resonances are observed whose quantum numbers are equal to the quark–antiquark system $q\bar{q}$ (mesons) or the three-quark system qqq (baryons). The allowed SU_3 multiplets will then be singlets and octets (for mesons) and singlets, octets, and decimets (for baryons).

Let us consider the scattering process $\pi^+ + \pi^- \to \pi^+ + \pi^-$. In this case the s- and t-channels are identical. They may contain resonances beginning with ϱ ($I = 1$, odd spin) and f ($I = 0$, even spin), lying on the trajectories ϱ, $\sigma(\varrho) = -1$, and f, $\sigma(\mathrm{f}) = +1$. The u-channel is the reaction $\pi^+ + \pi^+ \to \pi^+ + \pi^+$, so that the resonances in the u-channel must be exotic (isospin $I = 2$). The absence of exotic resonances means that the trajectories ϱ and f must coincide:

$$\varrho = \mathrm{f}. \tag{72}$$

One may also arrive at this conclusion by considering the reaction $\pi^- + K^+ \to \pi^- + K^+$. In this case, (72) follows from the fact that the resonances in the u-channel would have the quantum numbers of the $\pi^- K^-$ system and would be exotic.

In the process $K^+ + p \to K^+ + p$ the quantum numbers of the K^+p system (isospin $I = 1$, hypercharge $Y = 2$) do not belong to the "allowed" baryon SU_3 multiplets, so that the s-channel can contain only exotic resonances. Consequently, the trajectories in the t- and u-channels must be degenerate. In the t-channel, $p+\bar{p} \to K^+ + K^-$, several trajectories $\alpha^{\pm}(t)$ are possible: two trajectories of positive signature (the isotriplet A_2 and the isosinglet f) and two trajectories of negative signature (the isotriplet ϱ and the isosinglet ω), as well as the isosinglet trajectories φ and f′. However the last two, φ and f′ are not important since they couple weakly the to nucleon–antinucleon channel. The degeneracy due to $A_s \approx 0$ means that the $p+\bar{p}$ combinations of the trajectories $(\varrho+\omega)$ and (A_2+f) coincide.

A similar analysis may be performed for the t-channel $n+\bar{n} \to K^+ + K^-$ of the reaction $K^+ + n \to K^+ + n$, since the $K^+ + n$-system also has exotic quantum numbers. The exchange degeneracy in this case entails the degeneracy of the $(n+\bar{n})$ combinations of the trajectories $(\varrho+\omega)$ and (A_2+f).

Thus the absence of exotic resonances in the s-channels of both reactions $K^+ + p$ and $K^+ + n$ leads to the separate equality of the trajectories with isospin 0 and 1: $\varrho = A_2$ and $\omega = f$. But, from the $\pi^+\pi^-$ reaction, it was found above that $\varrho = f$. Consequently, the absence of exotic resonances explains the degeneracy of the meson trajectories:

$$\varrho = A_2 = f = \omega.$$

While exotic resonances may indeed exist, their contribution to absorptive part of amplitudes for reactions with the usual particles is small in comparison with the allowed resonances. We shall return to the question of exotic resonances in § 14.4.

Regge trajectories and particle classification

In contrast to the classification of particles with respect to internal symmetry groups like SU_3, the classification of particles in Regge families is primarily an approach to the explanation of the particle spectrum. In SU_3 classification the first step consists of the assignment of quantum numbers and the mere explanation of the level splitting.

Another distinctive feature of the classification of particles into Regge families is the treatment of particles on trajectories as composite objects. For isospin or SU_3 classification of particles the dynamical complexity or elementarity of particles is not discussed; the internal symmetry groups (like the Poincaré group) lead to a kinematic classification of particles whose elementarity is identified with the irreducibility of the representation. In the classification of particles into Regge families one may distinguish two types of particles: bound states and resonances. Particles of the first type lie (in mass) below the two-particle threshold, while particles of the second type lie above this threshold. Hence bound-state particles are stable while resonant particles are unstable. Particles of both types are composite objects. Thus, when classifying particles in Regge families we encounter for the first time the question of whether there exist truly elementary particles (which do not lie on Regge trajectories) or whether all particles have the same nature and are composite. Elementary particles may be associated with "elementary" poles of the Gribov–Froissart amplitude with $l \leqslant N_{\min} \approx 1$ [see (32) and the discussion of the third term in (32)].

Comparison with experiment shows that a treatment of all strongly interacting particles as composite is fully self-consistent. Such particles, like the nucleon, pion, and kaon, all lie on trajectories. In particular, the nucleon lies on a trajectory with resonances in the πN system of spins $\frac{5}{2}^+$, $\frac{9}{2}^+$, ..., and is the first particle $\frac{1}{2}^+$ of this Regge family.

A remarkable regularity of trajectories is their universal linearity and almost universal slope (with the exception of the Pomeranchuk trajectory). The linearity in the mass–spin relation $m^2(J)$, and the parallel nature of trajectories for all particles and resonances, may have a deeper basis (for example, see Chapter 14). Assuming that the linearity of trajectories continues to hold in the experimentally observable mass range, we may predict the existence of new resonances. The slow growth in mass of the resonance widths Γ (or the imaginary part of the trajectories), allows one to use the concept of a resonance at high masses (see Fig. 24).

If one recalls the isospin symmetry of the strong interactions, one sees that trajectories and Regge families must exist as isomultiplets, and, in the limit of SU_3 symmetry, in the form of SU_3 multiplets.

CHAPTER 14

DUALITY AND THE VENEZIANO MODEL

REGGE behavior is a characteristic feature of amplitudes at high energies; resonances govern the properties of amplitudes at low energies. Now, Regge trajectories, which describe the asymptotic behavior of the amplitude in the s-channel, may themselves (by analytic continuation into the crossed-channel region) pass through resonances in crossed channels. (The sole exception seems to be the Pomeranchuk trajectory.) Thus (if we do not consider diffraction scattering and the Pomeranchuk trajectory), the physics of elementary particles is to an important degree determined by resonances. The behavior of the amplitude at high energies, however, cannot be thought of separately from its behavior at low energies. The values of the amplitude in these regions are connected with one another by finite energy sum rules (FESR). Consequently, one may describe amplitudes approximately either using Regge poles or using resonances at low energies. Such a "dual" approach is described in the present chapter.

As an example of a dual amplitude we shall consider the Veneziano model, many of whose consequences are in remarkable accord with experiment. We shall perform several illustrative calculations using the Veneziano model. We shall not discuss multi-particle Veneziano amplitudes, detailed problems of unitarization, and other problems of the theory (e.g., its generalization to the case of fermions).

§ 14.1. Finite energy sum rules

Knowledge of the asymptotic behavior of the amplitude allows us to estimate dispersion integrals and to obtain sum rules connecting the low-energy region of the amplitude with its asymptotic behavior.[192–195] The derivation of sum rules is based on the Cauchy theorem according to which the integral over a closed contour C of a function $\varepsilon(\nu)$ regular inside the contour is equal to zero:

$$\int_C \varepsilon(\nu)\, d\nu = 0. \tag{1}$$

Let us consider the scattering amplitude $F(\nu, t)$ for spinless equal-mass particles as a function of $\nu \equiv (s-u)/2m$ and t. In the complex ν-plane, the function $F(\nu, t)$ has singularities along the real ν-axis for $\nu > 0$ associated with the s-channel, and singularities for $\nu < 0$ associated with the u-channel. The physical region of the s- and u-channels corresponds to $F(\nu+i\varepsilon, t)$ and $F(-\nu-i\varepsilon, t)$. The discontinuities of the function $F(\nu, t)$ across the cuts $\nu > 0$, $t < 0$, are equal to

$$\left.\begin{aligned} F(\nu+i0, t)-F(\nu-i0, t) &= 2i \operatorname{Im} F(\nu, t), \qquad \nu > 0; \\ F(-\nu-i0, t)-F(-\nu+i0, t) &= 2i \operatorname{Im} F(-\nu, t), \qquad \nu > 0. \end{aligned}\right\} \tag{2}$$

Here Im $F(\nu, t)$ and Im $F(-\nu, t)$ are the values of the imaginary part of the amplitude in the physical regions of the s- and u-channels, i.e. Im $F(\nu, t)$ is chosen along the right-hand cut, while Im $F(-\nu, t)$ is chosen along the left-hand cut in the ν-plane.

Let us first introduce a sum rule for the amplitude $F^-(\nu, t)$ antisymmetric under $\nu \to -\nu$ (and having negative signature). Let the asymptotic form $F^-_{as}(\nu, t)$ for $\nu \to \pm\infty$ be determined by a finite number of Regge poles $\alpha_i(t)$ (i = 1, 2, ..., n). According to (13.38), we may write F^-_{as} in the form

$$F^-_{as}(\nu, t) = \sum_{i=1}^{n} \frac{b_i^-(t)\,\eta^-(\alpha(t))}{\Gamma(\alpha_i(t)+1)}\,\nu^{\alpha_i(t)} = \sum_{i=1}^{n} \frac{b_i^-(t)}{-\Gamma(\alpha_i+1)\sin\pi\alpha_i}\,[(-\nu)^{\alpha_i}-\nu^{\alpha_i}], \tag{3}$$

where b is expressed in terms of the function $\tilde{\beta}$ introduced earlier [see (13.37)] as follows:

$$b_i^-(t) = \pi B(2\alpha_i+1)\,\Gamma(\alpha_i+1)\,\tilde{\beta}_i^-(t). \tag{4}$$

All $b_i^-(t)$ are real for $t < 0$. From (3) it is easily seen that the analytic properties of $F^-_{as}(\nu, t)$ in the ν-plane are the same as for the function $F^-(\nu, t)$. Both have right- and left-hand cuts in ν since the function ν^α has a cut for $\nu < 0$, while the function $(-\nu)^\alpha$ has a cut for $\nu > 0$.

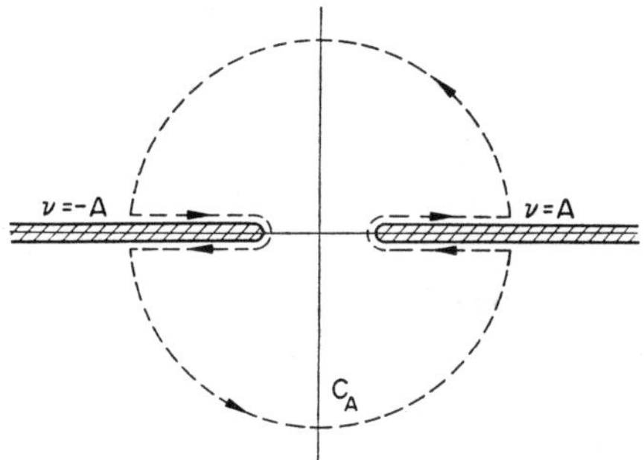

FIG. 25. Contour of integration in complex ν-plane used in writing finite-energy sum rules.

Let us chose the contour C in the integral (1) to be the contour C_A shown in Fig. 25, which has already been used in deriving dispersion relations (§ 12.3), but now let the radius A remain finite. The integral over the circle C_A of $F^-(\nu, t)$ may be rather large, since asymptotically $F^-_{as}(\nu, t) \sim \nu^{\alpha(t)} f(t)$, $\alpha \leqslant 1$. Let us thus apply the Cauchy theorem (1) not to $F(\nu, t)$ but to the function

$$\varepsilon(\nu, t) = F^-(\nu, t) - F^-_{as}(\nu, t) = -\varepsilon(-\nu, t),$$

where $F^-_{as}(\nu, t)$ is given by formulae (3). The function $F^-_{as}(\nu, t)$ has the same crossing-symmetry properties (under $\nu \to -\nu$) as $F^-(\nu, t)$, and also satisfies the relations (2). According to (3), for $\nu > 0$,

$$\operatorname{Im} F^-_{as}(\nu, t) = \sum_i \frac{b_i^-(t)}{\Gamma(\alpha_i+1)}\,\nu^{\alpha_i(t)}. \tag{5}$$

Let us now choose the boundary value $\nu = A$ so that (a) for $\nu \geqslant A$ one may use $F^{-}_{as}(\nu, t)$ instead of $F_{as}(\nu, t)$, and (b) the integral over the circle C_A

$$\int_{C_A} \varepsilon(\nu, t)\, d\nu \approx 0$$

may be neglected in comparison with $\int_0^A \operatorname{Im} F(\nu, t)\, d\nu$. In other words, we shall approximate the asymptotic behavior of the function $F^{-}(\nu, t)$ by a sum over the Regge poles $F^{-}_{as}(\nu, t)$ up to terms of order $\nu^{-1-\varepsilon}$, $\varepsilon > 0$. This, in turn, assumes that one may extend the Regge hypothesis in the l-plane to the region Re $l < -\frac{1}{2}$. In the case of potential scattering, the integration contour in the l-plane [see (13.27)] indeed may be moved to the left of the line Re $l = \frac{1}{2}$ (see ref. 159).

Under these assumption there remain in (1) only the integrals over both sides of the real axis, and, as a result of the properties of $\varepsilon(\nu, t)$,

$$\int_0^A \operatorname{Im} \varepsilon(\nu, t)\, d\nu = 0.$$

Consequently, for the amplitude $F^{-}(\nu, t)$, one has the sum rule

$$\frac{1}{A}\int_0^A \operatorname{Im} F^{-}(\nu, t)\, d\nu = \sum_i \frac{b_i^{-}(t) A^{\alpha_i}}{(\alpha_i+1)\Gamma(\alpha_i+1)}. \tag{6}$$

In the case of the amplitude $F^{+}(\nu, t)$ symmetric under the crossing-interchange $s \leftrightarrow u$, or $\leftrightarrow -\nu$, an analogous sum rule may be written for the antisymmetric combination $F^{+}(\nu, t)/\nu$:

$$\int_0^A \frac{d\nu}{\nu} \operatorname{Im} F^{+}(\nu, t)\, d\nu = \sum_i \frac{b_i^{+}(t)}{\Gamma(\alpha_i+1)} \frac{A^{\alpha_i}}{\alpha_i} + \pi F^{+}(0, t). \tag{7}$$

In (7) the signature of the trajectories is positive.

Since the energy interval $0 < \nu < A$ is finite, the sum rules also may be written for even "moments" of the antisymmetric amplitude $F^{-}(\nu, t)$:

$$S_n = \frac{1}{A^{n+1}} \int_0^{A_n} \nu^n \operatorname{Im} F^{-}(\nu, t) = \sum_i \frac{b_i^{-}(t) A_n^{\alpha_i}}{(\alpha_i+n+1)\Gamma(\alpha_i+1)}. \tag{8}$$

Here the number of Regge poles which must be included in $F^{-}_{as}(\nu, t)$, and the radius of the large radius of the large circle A_n, will depend on the (even) power n. The function $F^{-}_{as}(\nu, t)$ approximates $F(\nu, t)$ to order $\nu^{-n-1-\varepsilon}$, since the condition

$$\int_{C_{A_n}} \nu^n \{F(\nu, t) - F_{as}(\nu, t)\}\, d\nu \sim 0 \tag{9}$$

must hold. The number of Regge poles in (3) will grow as the power n in (8) and (9) grows.

For the symmetric amplitude $F^+(\nu, t)$, eqn. (8) is replaced by

$$\int_0^A \frac{\operatorname{Im} F^+(\nu, t)}{\nu^{n+1}} d\nu = \sum_i \frac{b_i^+(t)}{\Gamma(\alpha_i+1)} \frac{A^{\alpha_i-n}}{(\alpha_i-n)} + \begin{cases} \pi F^{+(n)}(0, t), & n \geqslant 0, \\ 0, & n < 0, \end{cases} \tag{10}$$

where n is an even integer, and $F^{(n)}$ is the nth derivative of F with respect to ν.

In the FESR (8) and (10), the integral on the left-hand side is over the low-energy region in the s-channel while the right-hand side contains the Regge trajectories $\alpha(t)$ in the t-channel. In these formulae $t \leqslant 0$. The sum rules (8) are consistency conditions imposed by the analyticity of the amplitude and by the asymptotic Regge form assumed in (9). The choice of the cut-off value $\nu_{max} = A$ in (6) and (10) is determined by the onset of the asymptotic régime, which, however, still has not reliably been established. The choice $A \approx 2$ GeV yields good results in a number of cases.

The direct use of (8) and (9) involves the correlation of experimental data with one another—the scattering phase shifts and the asymptotic parameters of the scattering amplitude. Using the phase shifts, one may find b_i and α_i, and, inversely, given b_i and α_i one may refine the low-energy scattering parameters.[195]

However, the significance of FESR does not involve merely the correlation of scattering data with one another. The pictures of scattering at low and high energies are significantly different. At low energies the total cross-sections have resonance-like behavior, so that the characteristic features of the imaginary part of the amplitude at low energies are determined by resonances. At high energies the basic features of scattering are described in the Regge-pole language. If one replaces Im F in the integrand by the contribution of resonances, FESR will be an approximate relation between the integrated contribution of resonances (in the s- and u-channels) in the region from 0 to $\nu = A$, and the contribution to Im F from the Regge poles $\alpha(t)$ at $\nu = A$. Consequently, FESR unify two different ways of describing scattering approximately—the resonance approximation (at low energies) and the Regge-pole model (at high energies). By virtue of analyticity one may continue the functions $\alpha^{\pm}(t)$ and $b^{\pm}(t)$ from the region $t \leqslant 0$ into the region $t > 4m^2$, where t has the meaning of an energy variable in the t-channel. For $t > 4m^2$ the trajectories $\alpha(t)$ (on the right-hand side of the FESR) pass through resonances in the t-channel, and, consequently, the low-energy resonances in the s-channel in Im F are connected via FESR with the low-energy resonances in the t-channel in $\alpha(t)$; the direct-channel resonances (the left-hand side of FESR) thus "build" the Regge trajectory in the crossed channel (the right-hand side of FESR). In turn, the resonances in the t-channel (in FESR for the t-channel) give rise to the trajectory $\alpha(s)$ with resonances in the s-channel (for $s < 0$) on the right-hand side. Of course, Im F in the low-energy region is not determined solely by resonances; there exists "background" scattering. However, the connection between resonances in both channels will be preserved if, for example, one of the trajectories α does not contain resonances but corresponds to background scattering and describes the phenomenological effect of inelastic processes.

The analysis of experimental data via FESR indicates[196, 197] that the Pomeranchuk trajectory is "built" (in the sense of FESR) by the nonresonant background of the s-channel amplitude, while the remaining trajectories $\alpha(t)$ are associated with resonances in the s-channel. In other words, if one represents the amplitude $F(\nu, t)$ as the sum of "background" (the diffractive part $F_P(\nu, t)$) and "resonances" (the nondiffractive part $F_{res}(\nu, t)$):

$$F = F_P + F_{res},$$

then, in the sum rule, (n is odd and positive)

$$\int_0^A \nu^n \operatorname{Im} F_P^+(\nu, t)\, d\nu \sim b_P(t) \frac{A^{\alpha_P+1+n}}{(\alpha_P+1+n)\Gamma(\alpha_P+1)} \tag{11}$$

the right-hand side will contain only α_P. For the resonance part the sum rule will have the form

$$\left.\begin{aligned} &\int_0^A \nu^n \operatorname{Im}(F_{res} - F_{as})\, d\nu \approx 0, \\ &\int_0^A \nu^n \operatorname{Im} F_{res}^{\pm}(\nu, t)\, d\nu \approx \sum_{i \neq P} b_i(t) \frac{A^{\alpha_i^{\pm}+1+n}}{(\alpha_i^{\pm}(t)+1+n)\,\Gamma(\alpha_i^{\pm}+1)}, \end{aligned}\right\} \tag{12}$$

where n is positive and odd for F^+, and n is nonnegative and even for F^-. The right-hand side of the second equation depends only on the trajectories $\alpha_i(t)$, on which resonances in the t-channel are situated. Equation (11) also signifies that the Pomeranchuk trajectory $\alpha_P(t)$ determines the asymptotic behavior of elastic scattering in the s-channel independently of whether there are resonances in this channel.

To illustrate FESR, let us consider the difference of total cross-sections

$$\sigma'_{tot} = \tfrac{1}{2}[\sigma_{tot}(\pi^- p) - \sigma_{tot}(\pi^+ p)].$$

The asymptotic behavior of σ'_{tot} as $\nu \to \infty$ is well described by a single trajectory $\alpha_\varrho(t)$ which also determines the asymptotic behavior of the charge-exchange scattering $\pi^- p \to \pi^0 n$ at $t = \text{const} < 0$. By virtue of the optical theorem, the value of σ'_{tot} is related to the imaginary part of the difference of the forward scattering amplitudes $\operatorname{Im} F'(\nu, 0)$, where $F' = \frac{1}{2}(F(\pi^- p) - F(\pi^+ p))$; in our terminology $F'(s, t)$ is the resonance amplitude: $F' = F_{res}^-$ (there is no Pomeranchuk pole in the asymptotic behavior). The asymptotic amplitude corresponding to F'_{res} will obviously contain only the trajectory α_ϱ: $F_{as}^- = F_\varrho$.

We will now write the FESR (12) with $n = 0$ for the case of forward scattering:

$$\int_0^A \operatorname{Im}(F_{res}^-(\nu, 0) - F_\varrho(\nu, 0))\, d\nu \approx 0,$$

and insert in it the function $\operatorname{Im} F^-(\nu, 0)$, found by measuring $\sigma_{tot}(\pi^- p)$ and $\sigma_{tot}(\pi^+ p)$, and the extrapolation to small ν of the function $\operatorname{Im} F_\varrho(\nu, 0)$. Graphs of these functions are shown in Fig. 26. Not only is the average of the Regge term $\operatorname{Im} F_\varrho$ equal approximately to the average of $\operatorname{Im} F^-$, but, as a consequence of its monotonicity, the Regge term $\operatorname{Im} F_\varrho$ itself at low energies is an average of $\operatorname{Im} F^-$, which has resonant behavior.

FESR also allow one to view the problem of exchange degeneracy from another point of view. We found earlier (see § 13.5) that in the resonance approximation the trajectories

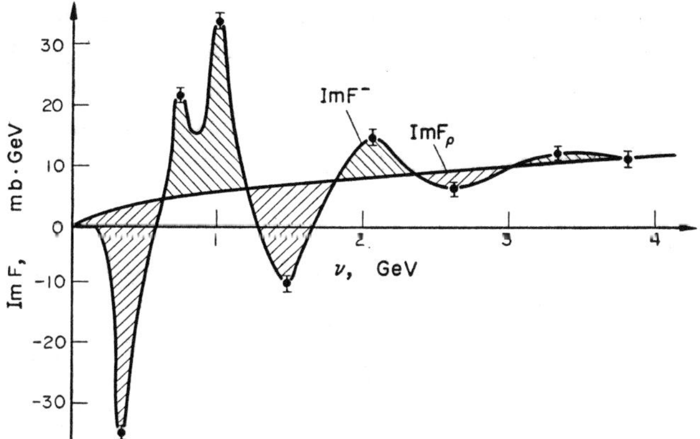

FIG. 26. Imaginary part of F^- in πN scattering from $\sigma_{tot}(\pi^- p) - \sigma_{tot}(\pi^+ p)$ (wavy line) and corresponding ϱ exchange term Im F_ϱ (smooth line).

of the t-channel would be degenerate if there were no resonances in the s- or u-channel and, consequently, Im $F = 0$ in these channels. Here the left-hand side of the FESR (12) depends on the Regge trajectories and residues; it vanishes if the contributions of the individual terms cancel one another. For this to occur, not only the trajectories but also the residues (up to a sign) must be identical. This is consistent with the general formula (13.71) for the amplitude $a^{\pm}(l, t)$ according to which, in the resonance approximation, $a^+ = a^-$ in the absence of resonances in the u-channel, and $a^+ = -a^-$ if there are no resonances in the s-channel. Thus FESR lead to the equality (up to a sign) of the residues $b^+(t) = \pm b^-(t)$ for exchange-degenerate trajectories $\alpha^+(t) = \alpha^-(t)$.

§ 14.2. Duality. Duality diagrams

The FESR (12) for the "resonance" amplitude F_{res}, and the example we have just considered (Fig. 26), suggest a new approximate approach to the construction of amplitudes based on the existence of resonances and the concept of Regge trajectories. This "dual" approach consists in passing from integral equations of the type (12) (for $n = 0$) to local ones.

Let us assume that the integral over the low-energy part of the spectrum coming from Im F_{res} in (12) may be replaced by an approximate sum of contributions of resonances (in the s-channel). This is the well-known resonant approximation in the dispersion approach (see § 12.3). From dispersion relations, it is clear that the amplitude $F_{res}(s, t, u)$ may then be written as a sum of pole terms in the s-channel up to the asymptotic region in s. If there are no resonances in the u-channel, then

$$F_{res}(s, t) \approx \sum_n \frac{c_n(t)}{s - s_n}, \qquad s_n > 0.$$

In this approximation we neglect the nonresonant values of absorptive parts of the amplitude. But dispersion relations describe the amplitude fully in principle (if the absorptive parts are known).

On the other hand, according to (13.32), this same "resonant" amplitude F_{res} may be written in terms of Regge-pole terms (without the Pomeranchuk pole) and an integral along the line Re $l = -\frac{1}{2}$ in the complex l-plane (and also terms arising from the cuts). Let us neglect this integral (and the contribution from the cuts) and assume that the amplitude F_{res} may be written approximately in the form of a sum of contributions from a finite number of Regge-pole terms in the t-channel:

$$F_{\text{res}}(s, t) \approx F_{\text{res}}^{\text{R}}(s, t) = \sum_n \beta_n(t)\, s^{\alpha_n(t)}\, \eta(\alpha_n(t)).$$

This form of the amplitude is used to describe the asymptotic behavior of the resonant amplitude $F_{\text{res}}^{\text{as}}$ in (12).

In a dual approach it is assumed that not only the integral formula (12), but also $F_{\text{res}}(\nu, t) \approx F_{\text{res}}^{\text{R}}(\nu, t)$, holds for each value of ν. In this approximate treatment, the amplitude is described using either resonances in one channel or Regge poles in the crossed channel. These descriptions supplement and complement one another; they are called dual.[198–200] The duality of the descriptions of amplitudes using Regge trajectories in the t-channel and resonances in the s-channel is shown symbolically in Fig. 27.

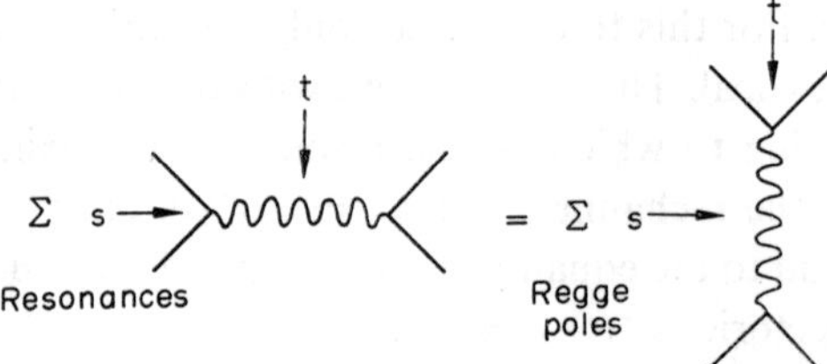

FIG. 27. Symbolic expression of duality. The relation is understood to hold only for imaginary parts, and when suitably averaged over s.

Let us enumerate the assumptions underlying duality (if there are no resonances in the u-channel). The dual description refers to the resonant amplitude $F_{\text{res}} = F - F_{\text{P}}$ which remains after separating out the diffractive part associated with the Pomeranchuk pole from the full amplitude F [see (11)].

1. In the whole physical region of the s-channel, the resonant amplitude is

$$F_{\text{res}}(s, t) \approx \sum_n \frac{c_n(t)}{s - s_n} \approx \sum_i \beta_i(t) s^{\alpha_i(t)}. \tag{13}$$

The number of resonances s_n must be infinite since in this case the sum of pole terms must give Regge asymptotic behavior for large s. By virtue of crossing symmetry, this approximate equality holds for resonances in the t-channel and Regge poles in the s-channel.

2. The sum of resonance poles of the amplitude in the t-channel is an analytic function which is approximately equal for $t < 0$ to a sum over Regge poles in the t-channel:

$$\sum_k \frac{c_k'(s)}{t - t_k} \approx \sum_i \beta_i(t) s^{\alpha_i(t)}. \tag{14}$$

The resonances and Regge trajectories appear asymmetrically in formulae (13) and (14) in the sense that the number of trajectories taken into account must be finite while the number of resonances is infinite. The finiteness of the number of Regge poles in (13) and (14) is related to the fact that the sum over all Regge poles diverges.[(201)] The total number of Regge trajectories, of course, will be infinite, since in each partial wave $a_l(s)$ there must exist an infinite number of resonances.

3. In addition to these two assumptions we shall also use the experimental linearity of trajectories and the positivity of their slope in the $(\mathrm{Re}\,\alpha(t), t)$-plane:

$$\mathrm{Re}\,\alpha = \mathrm{Re}\,\alpha(0)+\alpha'(0)t,$$

where the slope $\alpha'(0) > 0$ is the same for all trajectories (§ 13.5). In order that the trajectory may contain an infinite number of resonances, it must grow indefinitely.

The assumption of infinitely rising straight-line trajectories means that

$$\sum_k \frac{c_k'(s)}{t-t_k} = \sum_i \sum_l \frac{f_i(l, s)}{l-\alpha_i(t)}, \tag{15}$$

where the right-hand side contains a sum over the Regge trajectories in (13) and (14) and over the infinite number of resonances on each of these trajectories. By virtue of crossing symmetry, an analogous equation will hold, obtained from (15) by the substitution $t \leftrightarrow s$.

Up to now we have assumed that there are no resonances in the u-channel. To take these resonances into account in (15)–(17) presents no problem. Since the absorptive part in the u-channel A_u will be nonzero in this case, the dispersion relation in s requires the s-channel amplitude F_{res} to contain additional terms $c_n'/(u-u_n)$ (in the resonance approximation). The representation of the amplitude $F_{\mathrm{res}}^{\mathrm{R}}$ in terms of t-channel Regge trajectories maintains its form.

The case in which one of the channels is lacking resonances is important for dual theories since the absence of resonances entails the degeneracy of trajectories in crossed channels, and the number of parameters is reduced significantly. There thus arises the possibility of the self-consistent calculation of resonance parameters lying on trajectories, on the one hand, and encountered in the direct channel on the other.

Duality diagrams

Missing or exotic resonances (i.e. resonances whose contribution to FESR is greatly suppressed) have specific quantum numbers (see § 10.2). Exotic resonances do not exist in the simplest quark model, where the quantum numbers of particles are the same as in the nonrelativistic quark combinations $q\bar{q}$ (mesons) and qqq (baryons). Thus in the quark model one may introduce duality diagrams[(202)], which illustrate the restrictions arising from the dual relation between channels for meson–meson and meson–baryon scattering.

In a duality diagram each baryon is depicted by three quark lines going in the same direction, while each meson is represented by one quark and one antiquark line, i.e. by two lines in opposite directions (Fig. 28). The channels connected by duality to one another correspond to diagrams of the type shown in Fig. 28a without intersecting lines. The allowed resonances may be found in the intermediate quark states, i.e. by cutting the diagram in half in the direction of the channel of interest. The imaginary part of the amplitude is equal to the sum of the resonant contributions in the given channel.

The diagram contains intersecting quark lines (Fig. 28b) if only exotic resonances can occur in either the s- or t-channel. Passing to the meson–meson duality diagram for the u- and t-channels with no intersections, we see that resonances must be lacking in the s-channel. In this case the imaginary part of the amplitude F_{res} in the s-channel is equal to zero for large s and only the u- and t-channels are "dually" related to one another.

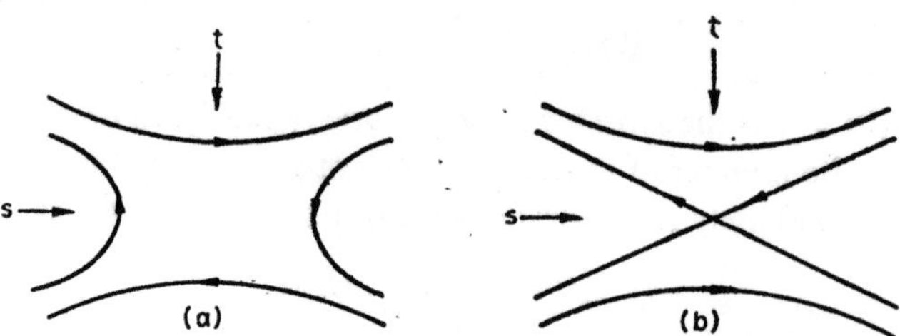

FIG. 28. Duality diagrams for meson–meson scattering. (a) Planar diagram, giving rise to an amplitude with nonzero imaginary part. (b) Nonplanar diagram, giving rise to a real amplitude.

Duality diagrams contain information in addition to that related to asymptotic SU_3 and the absence of exotic states. By assuming in the quark model that only the combinations $q\bar{q}$ and qqq are possible, we thereby reject more complicated combinations of the type $qq\bar{q}\bar{q}$ and $qqq(\bar{q}q)$, even if they give singlet and octet states.

To clarify the assumptions at the root of duality diagrams, let us consider the meson–meson scattering process $a+b \to \bar{c}+\bar{d}$, where the particles a, b, c, and d belong to the pseudo-scalar octet P (see § 10.1). Let us write the Regge amplitude in the s-channel (in the variable $\nu = (s-u)/2m$), analytically continued to small s, taking into account the factorizability of residues (13.52) and excluding the Pomeranchuk pole:[196, 202]

$$F_{res}(a+b \to \bar{c}+\bar{d}) \to \sum_E g_{aEc}(t)\, g_{b\tilde{E}d}(t) \frac{-\sigma_E - \exp[-i\pi\alpha_E(t)]}{\sin \pi\alpha_E(t)} \left(\frac{\nu}{\nu_0}\right)^{\alpha_E(t)}. \tag{16}$$

In an SU_3 invariant theory, the Reggeon E is an SU_3 multiplet. The interactions g_{aEc} and $g_{b\tilde{E}d}$ of the trajectories E (or the Reggeons) with the particles must be SU_3-invariant; they must be constructed by the same rules § 10.3 applying to the interaction Lagrangians.

Let E stand for the vector $V(J^P = 1^-)$ and the tensor $T(J^P = 2^+)$ nonets of trajectories with signature $\sigma_V = -1$, $\sigma_T = +1$. We shall choose nonets and not octets in view of the mixing of the neutral term of the octet with an SU_3 singlet (see § 10.2). The interactions (Lagrangians) of the Reggeons $E = V, T$ with the meson octet P have the form

$$g_{P_1TP_2} = \sqrt{2}\gamma_{PT} \mathrm{Tr}\,(P_1\{T, P_2\}) \tag{17}$$

$$g_{P_1VP_2} = i\sqrt{2}\gamma_{PV}\, \mathrm{Tr}\,(P_1[V, P_2]), \tag{18}$$

where γ_{PT}, γ_{VT} are the coupling constants of the pseudo-scalar octet P with the trajectories of the T- and V-type, and P, T, and V are 3 by 3 matrices of the multiplets (see § 10.1). The interaction g_{PTP} is described by pure D-type coupling, while g_{PVP} is described by pure

F-type coupling (see § 10.3), since F-coupling in (17) and D-coupling in (18) would violate charge-conjugation invariance. For the same reason, in (17), one cannot exclude the interaction of the P-mesons with a singlet component of the tensor nonet: $\mathrm{Tr}(PP)\,\mathrm{Tr}\,T$. The absence of the decay $f' \to \pi\pi$ indicates that this term is suppressed, and it is omitted here.

Let the s-channel be exotic, i.e. let it have the quantum numbers of exotic resonances, and, consequently,

$$\mathrm{Im}\, F_{\mathrm{res}}(P_a+P_b \to P_{\bar{c}}+P_{\bar{d}}) = 0, \quad s > 4m^2, \quad t \approx 0. \tag{19}$$

Here $P_a \ldots P_{\bar{d}}$ are those members of the pseudo-scalar octet P for which the systems (P_a+P_b) and $(P_{\bar{c}}+P_{\bar{d}})$ are exotic. e.g. $\pi^+\pi^+$. Equation (19) is the starting point in proving the dual nature of duality diagrams. We insert the Regge amplitude (16) as F_{res} into (19) and then find the conditions imposed by the vanishing of its imaginary part. We also use the experimental data (see § 13.5) according to which parts of the V- and T-multiplets are degenerate:

$$\left.\begin{aligned} &\alpha_{P'} = \alpha_{A_2} = \alpha_\omega = \alpha_\varrho \equiv \alpha_0, \\ &\alpha_{K^*} = \alpha_{K^{**}} \equiv \alpha_1, \quad \alpha_{f'} = \alpha_\varphi \equiv \alpha_2. \end{aligned}\right\} \tag{20}$$

Moreover, we take account of the experimental fact that $\gamma_{PT} = \gamma_{PV} \equiv \gamma_P$.

After inserting these experimental relations into (16), the Regge nondiffractive part of the amplitude F_{res} may be written in the form

$$F_{\mathrm{res}}(a+b \to \bar{c}+\bar{d}) \sim 4\gamma_P^2\{[(P_{c,i}{}^j P_{a,j}{}^k P_{b,k}{}^t P_{d,t}{}^i)+(P_{a,i}{}^j P_{c,j}{}^k P_{d,k}{}^t P_{b,t}{}^i)]$$
$$\times[i-\cot \pi\alpha_i^k(t)]+[(P_{c,i}{}^j P_{a,j}{}^k P_{d,k}{}^t P_{b,t}{}^i)+(P_{a,i}{}^j P_{c,j}{}^k P_{b,k}{}^t P_{d,t}{}^i)]\,[-\csc \pi\alpha_i^k(t)]\}\left(\frac{s}{s_0}\right)^{\alpha_i^k(t)}, \tag{21}$$

where $P_i{}^j$ and α_i^j are the elements of the matrices P and α, and, according to (20), $a_\beta^{\beta'} = \alpha_0$ for $\beta, \beta' = 1, 2$; $\alpha_\varrho^3 = \alpha_3^\varrho = \alpha_1$ for $\varrho = 1, 2$; and $\alpha_3^3 = \alpha_2$.

Expression (21) is the basis of the construction of duality diagrams (Fig. 28) for mesons. The indices a of the particles P_a correspond to the initial and final states of the diagram, while the quark lines correspond to summation over quark indices in (21) (an upper index refers to an antiquark and a lower index to a quark). The "allowed" diagram (Fig. 28a) is just a symbolic representation of the coefficient of $(-\cot \pi\alpha+i)$ in (21). In fact, if one transforms this coefficient into the form

$$(P_d P_c)_i{}^j\,(P_a P_b)_j{}^i+(P_c P_d)_i{}^j\,(P_b P_a)_j{}^i, \tag{22}$$

then it becomes obvious that it does not contain intermediate states in the multiplets $\underline{10}$, $\underline{10}^*$, and $\underline{27}$. The "antidual" diagram with intersecting lines (Fig. 28b) is the coefficient of $-\csc \pi\alpha$. The first term in (21) is nearly purely imaginary if one sets $\alpha \approx \frac{1}{2}$, while the second term is real. The vanishing of the coefficient (22) of $(-\cot \pi\alpha+i)$ in (21) when the diagrams of Fig. 28a cannot be drawn means, according to (19), the absence or cancellation of resonances in the s-channel. Of course, for exotic intermediate states in the s-channel, the coefficient (22) vanishes automatically.

A similar analysis applies to meson–baryon scattering. In the case of baryon–baryon and baryon–antibaryon scattering, the construction of duality diagrams encounters difficulties unless exotic states are permitted in the baryon–antibaryon system.[203]

§ 14.3. The Veneziano model

Veneziano[204] proposed a model dual amplitude based on straight-line trajectories and narrow resonances. We shall thus assume $\alpha(s) = \alpha(0)+\alpha'(0)\, s$, $\text{Im}\, \alpha(s) = 0$ for the trajectories $\alpha(s)$. Analogous expressions are assumed for the trajectories $\alpha(t)$ and $\alpha(u)$.

Let us consider for simplicity the scattering process $\pi^+ + \pi^- \to \pi^+ + \pi^-$, where there are only two dual channels s and t. In the u-channel the process is $\pi^+ + \pi^+ \to \pi^+ + \pi^+$, so that the resonances in the u-channel can exist only in exotic states ($I = 2$), which we neglect. The processes are the same in the s- and t-channels, and the amplitude $F_{\text{res}}(\pi^+ + \pi^- \to \pi^+ + \pi^-)$ must be symmetric in s and t. The poles in both channels include resonances on the ϱ- and f-trajectories with $I^G = 0^+$ and 1^+. The absence of resonances with $I = 2$ entails the exchange degeneracy of the ϱ- and f-trajectories, $\alpha_\varrho = \alpha_f$. Thus the resonances on the combined $\alpha_{\varrho f}$ trajectory must be separated by one unit of spin.

The Veneziano amplitude $V(s, t)$ is a function of the trajectories, and for the process $\pi^+ + \pi^- \to \pi^+ + \pi^-$ has the form

$$V(s, t) = -\lambda \frac{\Gamma(1-\alpha(s))\, \Gamma(1-\alpha(t))}{\Gamma(1-\alpha(s)-\alpha(t))}, \quad (\lambda > 0), \tag{23}$$

where the trajectory $\alpha = \alpha_{\varrho f}$ with $\alpha(0) \approx \frac{1}{2}$ is taken as linear, λ is a constant, and $\Gamma(z)$ is the γ-function

$$\Gamma(z) = \int_0^\infty e^{-t} t^{z-1}\, dt, \quad \text{Re}\, z > 0,$$

$$\Gamma(z+1) = z\Gamma(z).$$

If one introduces the Euler β-function,[205]

$$\text{B}(x, y) = \int_0^1 u^{x-1}(1-u)^{y-1}\, du = \frac{\Gamma(x)\, \Gamma(y)}{\Gamma(x+y)}, \tag{24}$$

the amplitude (23) will be equal to

$$V(s, t) = -\lambda(1-\alpha(s)-\alpha(t))\, \text{B}(1-\alpha(s), 1-\alpha(t)). \tag{25}$$

The function $\Gamma(z)$ has poles for integral nonpositive $z = 0, -1, -2. \ldots$ The residue of $\Gamma(z)$ at the pole $z = -n$ $(n \geqslant 0)$ is equal to

$$\lim_{z \to -n} \Gamma(z)\, (z+n) = (-1)^n \frac{1}{n!}. \tag{26}$$

Thus when $t \leqslant 0$, $s > 0$, the amplitude (23) describes resonances in the s-channel with mass $s = m_J^2$ for $\alpha(m_J^2) = J = 1, 2, \ldots$, while for $s \leqslant 0$, $t > 0$, it describes these same resonances in the t-channel. In the region $s > 0$, $t > 0$, the residues at the poles in s, considered as functions of t, do not have poles in t, since at this point the denominator also has a pole.

For $|z| \to \infty$ $(|\arg z| < \pi - \varepsilon, \ \varepsilon > 0)$

$$\frac{\Gamma(z+a)}{\Gamma(x+b)} = z^{a-b}\left(1+\frac{1}{z}(a-b)(a+b-1)\right). \tag{27}$$

If one sets $z = -\alpha(s) = -\alpha(0) - \alpha' s$, then (27) will hold for

$$|s| \to \infty, \quad |\arg s| < \pi - \varepsilon,$$

i.e. in the whole complex s-plane, except along the positive real axis, where there are poles (if $\operatorname{Im} \alpha' = 0$). Using this fact, we introduce an imaginary part into α', i.e. we set

$$\alpha(s) = \alpha(0) + \alpha'(s), \quad \operatorname{Im} \alpha' = \text{const} = \varepsilon > 0.$$

Then we obtain from (27) the Regge asymptotic behavior

$$V(s, t) \sim \lambda \Gamma(1-\alpha(t))\,(-\alpha' s)^{\alpha(t)}, \quad (s \to \infty). \tag{28}$$

By introducing an imaginary part of α' in deriving (28), we have imitated the "smearing out" of resonances in order to compare $V(s, t)$ with a real amplitude. Analogously,

$$V(s, t) \sim \lambda \Gamma(1-\alpha(s))\,(-\alpha' t)^{\alpha(s)}, \quad (t \to \infty). \tag{29}$$

From these expressions, it follows that the trajectory slope α' characterizes the energy scale in such models.

The crossing symmetry of the Veneziano amplitude is clear from its form (23). One may also show[204] that the Veneziano amplitude satisfies the sum rules of § 14.1, so that the integral over the moments of $V(s, t)$ in $v = s - u$, on the one hand, is equal to the "weighted" sum of the Regge terms, and, on the other hand, this sum comes entirely from the resonance contributions of $V(s, t)$.

Thus the Veneziano amplitude satisfies all conditions required of a dual amplitude. Resonances and Regge behavior are both determined in $V(s, t)$ by the trajectories and by the form of the amplitude. The concept of a trajectory α is basic to the dual Veneziano amplitude. Let us consider the behavior of the amplitude $V(s, t)$ near a pole of the s-channel $\alpha(m_J^2) = J = 1, 2. \ldots$ According to (26),

$$\Gamma(1-\alpha(s)) \sim \frac{(-1)^J}{(J-1)!\,\alpha'(s-m_J^2)}, \tag{30}$$

so that, near a pole, $V(s, t) \approx c_J/(s - m_J^2)$, where

$$c_J = \lambda \frac{1}{\alpha'}(1-\alpha(t)-J)\frac{\alpha(t)\,(\alpha(t)+1) \ldots [\alpha(t)+(J-2)]}{(J-1)!}$$

$$\equiv \lambda \frac{1}{\alpha'}(1-J-\alpha(t))\binom{\alpha(t)+(J-2)}{(J-1)}. \tag{31}$$

By virtue of the linearity of the trajectories, c_J is a polynomial of degree J in t. In the physical region of the s-channel, t contains $\cos\theta$ linearly, so that in (31) one encounters all Legendre polynomials up to $P_J(\cos\theta)$. Consequently, a pole in the s-channel at $\alpha(m_J^2) = J$

refers not only to a particle with spin J but also to particles with $J-1, J-2, \ldots, 0$ ("daughters"). Thus each "parent" trajectory with $\alpha(m_J^2) = J$ will be accompanied by a series of daughter trajectories $\tilde{\alpha}^{(n)}$ with $\tilde{\alpha}^{(n)}(m_J^2) = J-n$. In Fig. 29 the daughter trajectories are parallel to the "parent" trajectory with spacing $\Delta J = 1$. The number of daughter trajectories $\tilde{\alpha}^{(n)}$ is infinite, since there are an infinite number of resonances on the parent trajec-

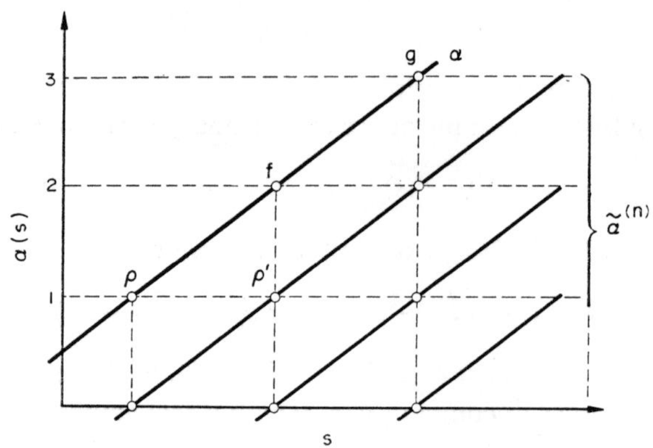

FIG. 29. Daughter trajectories in the Veneziano model, shown as lines parallel to the leading (degenerate) ϱf trajectories.

tory. (In factorizable multi-particle dual amplitudes, one also finds a high degree of degeneracy among daughters.) The Veneziano amplitude (23) may be represented in the form of a sum

$$V(s, t) = -\lambda \sum_{J=1}^{\infty} \binom{\alpha(t)+J}{J-1} \frac{1-\alpha(s)-\alpha(t)}{J-\alpha(s)}, \tag{32}$$

which, by virtue of $\alpha(s) = \alpha(0)+\alpha' s$ and $\alpha(m_J^2) = J$, is an expansion in terms of poles in the s-channel. As a consequence of the (s, t) symmetry of $V(s, t)$ a similar sum may be written in the t-channel. Thus an infinite set of poles in the s-channel generates a set of poles in the t-channel and vice versa, so that the descriptions of the amplitudes from the point of view of poles (resonances) and trajectories are equivalent. However, eqn. (23) for the amplitude is not unique, since one may add terms of the type

$$V'(s, t) = -\lambda' \frac{\Gamma(m-\alpha(s))\,\Gamma(n-\alpha(t))}{\Gamma(m+n+p-\alpha(s)-\alpha(t))}, \tag{33}$$

$$(p \leq 0, \quad -p \leq m, \quad -p \leq n, \quad m, n \geq 1),$$

without changing the basic properties of the amplitude. In contrast to (23), the first pole of (33) associated with the trajectory $\alpha(s)$ now appears at $\alpha(s) = m$, while the first pole of (33) in $\alpha(t)$ lies at $\alpha(t) = n$. The asymptotic behavior of (33), instead of (28) and (29), has the form

$$\left.\begin{aligned} V'(s, t) &\sim s^{\alpha(t)-n-p}, \quad s \to \infty, \\ V'(s, t) &\sim t^{\alpha(t)-m-p}, \quad t \to \infty. \end{aligned}\right\} \tag{34}$$

Terms of the type (33) are called satellite terms. In defining the Veneziano amplitude via the β-function (25), the indeterminancy with respect to terms of the type (33) corresponds to the fact that if one modifies the integrand of the β-function in (25) by a function $f(x)$ regular in the segment $0 \leqslant x \leqslant 1$, so that

$$B \to B' = \int_0^1 f(u)\, u^{-\alpha(s)-1}(1-u)^{-\alpha(t)-1}\, du, \tag{35}$$

the dual properties of the amplitude (25) do not change. An expansion of the function f in a series leads to a sum of satellite terms:

$$B' = f(0)B + \sum_{m,\,n} \sum_{p=0} c_{n,\,m,\,p} \frac{\Gamma(n-\alpha(s))\,\Gamma(m-\alpha(t))}{\Gamma(n+m+p-\alpha(s)-\alpha(t))}, \tag{36}$$

where $n > 0$, $m > 0$, $p_0 = -\min\,(n, m)$, and $f_0 = -\lambda$. The term with $m = 1$, $n = 1$, $p = -1$ is excluded from the sum.

Using several satellite terms one may eliminate unwanted daughter trajectories.[206] The full Veneziano amplitude in all cases, however, is determined only up to satellite terms.

Another important shortcoming of the Veneziano model is its inconsistency with unitarity. Since the trajectories α are real, the poles lie at real s and t. This contradicts unitarity, according to which resonance poles must lie on the second sheet. When the imaginary parts of the poles vanish (i.e. when the total width of the resonances vanishes) the residue at the pole (i.e. the partial width) must also vanish.

The Veneziano amplitude (23) is written for spinless particles. Its generalization to the case of fermions encounters difficulties.

§ 14.4. Some applications of the Veneziano model

Resonance widths

The Veneziano model contains only a few parameters, so that its predictions may be checked easily. In the case of $\pi^+\pi^- \to \pi^+\pi^-$ scattering, the residues (31) of the amplitude (23) depend on a single constant λ. Now the residue at a pole corresponding to a resonance in the s- or t-channel is characterized by a constant for the interaction of the resonance with the $\pi^+\pi^-$ system for zero momentum transfer, i.e. by a decay width of the resonance into $\pi^+\pi^-$. In the Veneziano amplitude (23), the constant λ is the same for all particles in the trajectory α, and thus determines (together with α') the relative magnitude of the partial widths $\Gamma_{\pi\pi}$ of all resonances lying on α.

A phenomenological interaction of the $\pi\pi$ system with the ϱ-meson may be written in the form

$$G_{\varrho\pi\pi}\varepsilon^{\mu}(\mathbf{q}_1+\mathbf{q}_2, \sigma)\,(q_1-q_2)_{\mu},$$

where ε^{μ} is the polarization vector of the ϱ-meson. The residue at the ϱ-meson pole of the amplitude is then equal to

$$-G^2_{\varrho\pi\pi}4\mathbf{q}\cdot\mathbf{q}', \quad \mathbf{q}_1 = -\mathbf{q}_2 = \mathbf{q},$$

where $\mathbf{q}$ and $\mathbf{q}'$ are the initial and final momenta of the π^+ particles in the center of mass system. On the other hand, the contribution to (23) from the term with $J = 1$ is equal to

$-\lambda\alpha' t$, which also contains a daughter trajectory contribution passing through the point $J = 0$ at $t = m^2_{J=1}$. The contribution of the ϱ-meson itself is equal to $-2\lambda \boldsymbol{q}\cdot\boldsymbol{q}'$, from which we may find λ:

$$\lambda = 2G^2_{\varrho\pi\pi}. \tag{37}$$

$G_{\varrho\pi\pi}$ is known from the width of the ϱ-meson: $\Gamma_\varrho \approx 100$ MeV. The width Γ_f of the next resonance may be determined by (31) in terms of Γ_ϱ and the trajectory slope $\alpha' \approx 0.9\ \text{GeV}^{-2}$, which may be found independently in terms of the resonances (see § 13.5). Equation (23) cannot be strictly accurate at m_f, however, since it predicts an unobserved $J^P = 1^-$ resonance (ϱ') at this mass with an appreciable $\pi\pi$ coupling.

ππ-scattering[207]

Let us construct the isospin amplitudes $T^{(I)}$ for $\pi\pi$ scattering from the amplitude (23). Since pions obey symmetric statistics, the amplitudes $T^{(I)}$ ($I = 0, 2$) in the s-channel are symmetric under $t \leftrightarrow u$, while the amplitude $T^{(1)}$ is antisymmetric. Consequently, in the s-channel,

$$\left.\begin{aligned} T_s^{(0)} &= \beta(V_{\varrho\varrho}(s,t)+V_{\varrho\varrho}(s,u))+\gamma V_{\varrho\varrho}(t,u),\\ T_s^{(1)} &= a(V_{\varrho\varrho}(s,t)-V_{\varrho\varrho}(s,u)),\\ T_s^{(2)} &= V_{\varrho\varrho}(t,u). \end{aligned}\right\} \tag{38}$$

The relation between the coefficients a, β, γ may be obtained from the condition that exotic resonances be absent. For the amplitude (38), this means that in the t-channel $\pi^+ + \pi^+ \to \pi^+ + \pi^+$ the amplitude with $I_t = 2$ is equal to zero (I_t is the isospin in the t-channel). Expressing the amplitude T_t in the t-channel in terms of the amplitudes T_s in the s-channel (38) and the crossing matrix X_{ts}

$$T_t^{(2)} = \sum_I (X_{ts})^{2,I}\, T_s^{(I)} = \gamma V_{\varrho\varrho}(s,u), \tag{39}$$

one may easily find $\gamma = -\frac{1}{2}a$, $\beta = +\frac{3}{2}a$. Moreover, the amplitude $T_t^{(0)}$ cannot have a pole at $\alpha(t) = 1$, since the trajectory $\alpha_\varrho(t)$ is isovector, while the resonances on the α_f trajectory begin with $J = 2$:

$$[\text{residue of } T_t^{(0)}] = 0 \quad \text{at} \quad \alpha(t) = 1, \tag{40}$$

yielding $a = 1$. Thus all three $\pi\pi$ amplitudes (38) depend on one constant λ.

Let one of the pions have zero mass and its momentum also approach zero, while the remaining pions stay on the mass shell: $p_i^2 = m_\pi^2$ ($i = 2, 3, 4$). Then the invariant variables are equal to $s = t = u = m_\pi^2$. One may attempt to apply the Veneziano formula (23) for the $\pi\pi$ scattering amplitude off the mass shell without any modifications. It turns out that one then obtains results coinciding with those of current algebra, which are examined in the following chapter.

As will be shown in Chapter 15, when $p_{1\mu} \to 0$, one has

$$F(m_\pi^2, m_\pi^2, m_\pi^2) = 0, \quad p_{1\mu}(\pi) = 0 \tag{41}$$

[see the Adler consistency condition (15.74)]. In order that the amplitude $W_{\varrho\varrho}(s, t, u)$

satisfy the condition (41), the relation

$$\alpha_\varrho(m_\pi^2) = \tfrac{1}{2}. \tag{42}$$

must hold. Describing the ϱ-trajectory in the form $\alpha_\varrho(s) = \frac{1}{2}+\alpha'(s-m_\pi^2)$,

$$\alpha' = \frac{1}{2(m_\varrho^2-m_\pi^2)} = 0.88\ \text{GeV}^{-2}. \tag{43}$$

Since $m_\pi^2 \ll m_\varrho^2$,

$$\alpha' \approx \frac{1}{2m_\varrho^2}. \tag{44}$$

This formula agrees well with experiment.

Quantization of trajectories[208]

The Adler consistency condition (41) refers to all processes with one soft pion:

$$\pi + A \to B + c.$$

The amplitude for such a process also must vanish when the pion momentum approaches zero: $p_\mu(\pi) \to 0$. In the Veneziano model this condition leads to mass "quantization."

We shall examine below the application of the Veneziano model to the scattering of particles with spin. The scattering amplitudes may be first expanded in the usual manner (see §§ 7.5 and 11.3) in terms of invariant amplitudes, and these invariant amplitudes may be described in the Veneziano form. As was pointed out earlier, the application of the Veneziano model to fermions encounters difficulties (the appearance of unphysical states). Nonetheless we shall formally apply the Veneziano model to the scattering of fermions as well. We thus arrive at interesting results agreeing with experiment. This may indicate the unimportance of the difficulties just mentioned with respect to the quantization of trajectories. Thus let us consider a typical term in the Veneziano-type amplitude for the process $\pi + A \to B + c$. It has the form

$$W_{XY}(s, t) \sim \frac{\Gamma(k+J_A-\alpha_X(s))\,\Gamma(l+J_B-\alpha_Y(t))}{\Gamma(n+J_A+J_B-\alpha_X(s)-\alpha_Y(t))}, \tag{45}$$

where J_A and J_B are the spins of the external particles A and B; the integers l, k, and n ($n \leqslant l+k$) characterize the resonances on the trajectories $\alpha_X(s)$ and $\alpha_Y(t)$. Trajectory X has internal quantum numbers and parity of the $(\pi+A)$ system, while trajectory Y corresponds to the $(\pi+B)$ system. The normality of the trajectory X must coincide with that of the system (A+pion with zero orbital momentum), and consequently, must be opposite to the normality of particle A. Analogously, the normality of the Y-trajectory must be opposite to the normality of B [see (13.68)].

Let us assume that the vanishing of the amplitude $F(\pi+A \to B+c)$ when $p_\mu(\pi) = 0$ (or $s = m_A^2$, $t = m_B^2$, $u = m_c^2$) does not arise from the mutual cancellation of terms of the type (45) for any A, B, and c. Then each term (45) must be separately equal to zero, which is possible if the argument N_{AB} of the γ-function in the denominator is integral and $N_{AB} \leqslant 0$, i.e.

$$\alpha_X(m_A^2)+\alpha_Y(m_B^2) = J_A+J_B+N_{AB}, \tag{46}$$

where the number N_{AB} may be considered as characteristic of the particles A and B. In the case of the scattering process $\pi+A \to A+c'$, one must have

$$\alpha_X(m_A^2) = J_A+\frac{N_{AA}}{2}, \tag{47}$$

while in the case of $\pi+B \to B+c''$ the condition takes the form

$$\alpha_Y(m_B^2) = J_B+\frac{N_{BB}}{2}.$$

Consequently, the number N_{AB} obeys the rule

$$2N_{AB} = N_{AA}+N_{BB}; \tag{48}$$

all numbers of the type N_{AB}, N_{BB} must be either odd or even. From $\pi\pi$ scattering it is known that for $A = \pi$ and $X = \varrho$ the quantity (47) is equal to $\alpha_\varrho(m_\pi^2) = \frac{1}{2}$ [formula (42)], i.e. $N_{\pi\pi}$ is odd. Consequently, for all particles A the number N_{AA} must be odd. The mass of particle A in the condition (47) may be expressed in terms of the parameters of its trajectory α_A and its spin J_A:

$$m_A^2 = \frac{J_A-\alpha_A(0)}{\alpha_A'}. \tag{49}$$

Then (47) becomes

$$a_X(0)+\frac{\alpha_X'}{\alpha_A'}(J_A-\alpha_A(0)) = J_A+\frac{N_{AA}}{2}. \tag{50}$$

Now this relation must hold also for a particle a of spin J_A on a parallel daughter trajectory α_a which lies one unit below α_A. Consequently, $\alpha_X'/\alpha_A' = (\frac{1}{2})(N_{AA}-N_{aa})$, which is an integer. But if we consider a different scattering process in which the scattered ("external") particles lie on the parent or daughter trajectories X, while the particle A is one of the resonances in the s-channel, then in (50) one should also interchange $X \leftrightarrow A$, and the ratio α_A'/α_X' would also be an integer. Consequently

$$\alpha_A' = \alpha_X', \tag{51}$$

so that the trajectories α_X and α_A are parallel. The condition (50) now connects the intercepts of the trajectories:

$$\alpha_X(0)-\alpha_A(0) = \tfrac{1}{2}N_{AA} \qquad (N_{AA} \text{ an odd integer}). \tag{52}$$

Formulae (51) and (52) express the "quantization" of trajectories in the Veneziano model; if a particle on the trajectory α_X may decay into a pion and a particle of opposite normality lying on the trajectory α_A, then the trajectories α_A and α_X are parallel, while the difference of their intercepts (52) must be half an odd integer.

Using the approximate formula $\alpha' \approx \frac{1}{2}m_\varrho^2$ or $m_\pi^2 \approx 0$, we may write (49) and (52) in the form of a relation between the masses of the particles X and A:

$$m_X^2 = m_A^2+2m_\varrho^2\left(J_X-J_A-\frac{N}{2}\right) \tag{53}$$

with a mass "quantum" m_ϱ^2. Choosing as pairs (X, A) the particles (A_1, ϱ), (Δ, N), $(\mathrm{K}^*, \mathrm{K})$, (A_2, η), we obtain the mass formulae

$$\left.\begin{aligned} m_{\mathrm{A}_1}^2 &= 2m_\varrho^2, & m_\Delta^2 &= m_{\mathrm{N}}^2 + m_\varrho^2, \\ m_{\mathrm{K}^*}^2 &= m_{\mathrm{K}}^2 + m_\varrho^2, & m_{\mathrm{A}_2}^2 &= 3m_\varrho^2 + m_\eta^2. \end{aligned}\right\} \tag{54}$$

The first of formulae (54) was obtained historically from the spectral sum rules of Weinberg (see § 15.5). In the case of the baryon $\mathrm{Y}_1^* = \Sigma$ (1385) in the decimet, two types of pairs (X, A), are possible, specifically $(\mathrm{Y}_1^*, \Lambda)$ and (Y_1^*, Σ). Here (53) gives

$$m_{\mathrm{Y}_1^*}^2(1385) = \begin{pmatrix} m_\Sigma^2 \\ m_\Lambda^2 \end{pmatrix} + m_\varrho^2, \tag{55}$$

which, in fact holds for the average $\frac{1}{2}(m_\Sigma^2 + m_\Lambda^2)$. One must, however, bear in mind that the mass formulae for the octet and for the decimet (see § 10.1) are different, which perhaps demands the introduction of satellite terms. Moreover, the Veneziano model has so far not been satisfactorily generalized to the case of fermions.

Thus despite its oversimplifications (narrow resonances in all channels) and difficulties (satellite terms and unitarity), the dual Veneziano model evidently captures many features of hadron dynamics in a remarkable way. One might imagine that the Veneziano amplitude is some sort of Born approximation to the real amplitude. Much work has been devoted to the development of the Veneziano model and its generalizations to the case of multi-particle amplitudes (see the reviews of refs. 209–211).

CHAPTER 15

ELECTROMAGNETIC AND WEAK CURRENTS. CURRENT ALGEBRA

As in the case of interactions, currents may be subdivided into strong, electromagnetic, and weak. Strong currents were introduced earlier in § 11.1 as sources of hadronic fields. The possibility of using the reduction formula to introduce weak currents is doubtful since, on the one hand, the presence of zero-mass particles disturbs a series of proofs in axiomatic theory, and, on the other hand, we do not at present know of fields which could be sources of weak currents. We shall consider the weak currents $j_\mu(x)$ as local vector or axial vector operators serving as a primary basis for internal symmetry groups (see Part III).

The current algebra of Gell-Mann starts with the assumption that the spatial integrals of the densities of weak currents

$$\int d^3x j^0(x^0, \boldsymbol{x})$$

are equal to the generators of some internal symmetry group (exact or approximate). To some extent, current algebra uses both the group-theoretic and dispersion approaches described in earlier chapters. The relations between group generators are replaced by local relations between current densities. The matrix elements of currents may be expressed in terms of form factors whose behavior is analyzed using ideas of the dispersion approach.

In contrast to strong currents, weak and electromagnetic currents are observable (in principle). This property is related to the smallness of their coupling constant with electromagnetic and lepton fields, allowing one to apply perturbation theory. In other words, the weak and electromagnetic lepton currents can play the role of "test charges" in studying the structure of the strong interactions without disturbing them.

This chapter also contains a section on the violation of CP invariance, since this question is related to problems of the weak interactions which we discuss in connection with current algebra.

§ 15.1. Electromagnetic and weak currents

The interaction of systems with the electromagnetic field $A^\mu(x)$ is described by the Lagrangian

$$\mathcal{L}(x) = -ej_\mu(x)\, A^\mu(x). \tag{1}$$

The electromagnetic current $j_\mu = j_\mu^{\mathrm{h}} + j_\mu^{\mathrm{l}}$ consists of hadronic j_μ^{h} and leptonic j_μ^{l} parts. The electric charge

$$Q = e\int j_0(x)\,d^3x \tag{2}$$

is conserved: $[Q, P_\mu] = 0$, $[Q, S] = 0$, which entails the current conservation equation

$$\partial^\mu j_\mu(x) = 0. \tag{3}$$

The electromagnetic current is neutral:

$$[B, j_\mu] = [L, j_\mu] = [Y, j_\mu] = [I_3, j_\mu] = 0.$$

Under SU_3 rotations, the hadronic current $j_\mu^{\rm h}$ transforms like the hadronic electric charge, i.e. as a sum of $v_{\mu 8}$ and $v_{\mu 3}$:

$$j_\mu^{\rm h}(x) = \left(v_{\mu 3}(x) + \frac{1}{\sqrt{3}}\, v_{\mu 8}(x)\right), \tag{4}$$

where $v_{\mu a}(x)$ is the octet of vector currents ($a = 1 \ldots 8$). In terms of U-spin multiplets, the electromagnetic hadron current (4) is a U-spin singlet.

Formula (4) for the hadronic current means that the spatial integrals of the densities of the v_{03} and v_{08} define the generators of the group SU_3—the isospin component I_3 and the hypercharge Y:

$$\int v_{03}(x)\,d^3x = I_3, \qquad \frac{2}{\sqrt{3}}\int v_{08}(x)\,d^3x = Y. \tag{5}$$

The quantities I_3 and Y are conserved separately, if one neglects the weak interactions. In this case, the currents $v_{\mu 3}$ and $v_{\mu 8}$ are also conserved in the sense of (3):

$$\partial^\mu v_{\mu 3}(x) = 0, \qquad \partial^\mu v_{\mu 8}(x) = 0. \tag{6}$$

The electromagnetic interaction is invariant separately with respect to the operations C, P, and T. From (1) and from the transformation properties of A_μ (see Chapter 6) it then follows that the electromagnetic current and the separate currents $v_{\mu 3}$ and $v_{\mu 8}$ are first-class currents (see § 11.4):

$$\mathrm{C} j_\mu \mathrm{C}^{-1} = -j_\mu, \qquad \mathrm{C} v_{\mu 3}\mathrm{C}^{-1} = -v_{\mu 3}, \qquad \mathrm{C} v_{\mu 8}\mathrm{C}^{-1} = -v_{\mu 8}. \tag{7}$$

The matrix elements of the currents then may be found from lepton–hadron scattering experiments and electro- and photoproduction of hadrons. The matrix elements of the current between single-particle states (vertex parts, see § 11.4)

$$\Gamma_\mu(p', \lambda'; p, \lambda) = \langle \boldsymbol{p}', \lambda' \,|\, j_\mu(0) \,|\, \boldsymbol{p}, \lambda\rangle \tag{8}$$

appear, for example, in the amplitude $T(\mathrm{e^-h \to e^-h})$ for the scattering of an electron on a hadron:

$$T(\mathrm{e^-h \to e^-h}) = \frac{2\alpha}{(2\pi)^2}\,\bar{u}(k_1)\,\gamma_\mu u(k_2)\,\frac{1}{k^2 - i\varepsilon}\,\Gamma^\mu(p', \lambda'; p, \lambda). \tag{9}$$

The matrix elements of the current between one- and two-particle states appear in the amplitudes for electro- and photoproduction, also calculated to first (electromagnetic) order:

$$\gamma + a \to b + c, \quad \mathrm{e} + a \to \mathrm{e} + c + d.$$

In the pole approximation, when the main contribution to these processes comes from a resonance with the quantum numbers of the $(b+c)$ or $(c+d)$ system, these matrix elements may be expressed in terms of one-particle matrix elements of the type $\Gamma_\mu(p', \lambda'; p, \lambda)$.

The vertex parts Γ_μ contain effects of the strong interactions. By virtue of current conservation, eqn. (3),

$$(p'^\mu - p^\mu)\Gamma_\mu(p', \lambda'; p, \lambda) = 0. \tag{10}$$

The electric charge (2) of a hadron h may be expressed in terms of the value of Γ_0 at zero momentum transfer:

$$\langle \boldsymbol{p}', \lambda'; \mathrm{h} | Q | \boldsymbol{p}, \lambda; \mathrm{h} \rangle = 2p_0 \delta(\boldsymbol{p} - \boldsymbol{p}')\, \delta_{\lambda\lambda'} Q_\mathrm{h} = (2\pi)^3\, \delta(\boldsymbol{p} - \boldsymbol{p}')\, \Gamma_0(p', \lambda'; p, \lambda). \tag{11}$$

The vertex functions Γ_μ may be expanded in terms of invariant-form factors (see § 11.4), whose dependence on invariant momentum variables is determined only by the dynamics of the process.

The matrix elements of the current and the form factors are usually considered to lowest electromagnetic order, so that one need not worry about renormalization of the charge Q by electromagnetic effects. Let us see whether the charge Q_h of hadrons is influenced by strong interactions. If, for simplicity, we pass to normalized proton states $|\overline{p, \lambda}\rangle$, with $\langle \overline{p, \lambda} | \overline{p, \lambda} \rangle = \delta_{\lambda\lambda'}$, then instead of (11) we may write the electric charge of the proton in the form

$$Q_\mathrm{p} = \langle \overline{\boldsymbol{p}, \lambda} | Q | \overline{\boldsymbol{p}, \lambda} \rangle.$$

The unrenormalized ("bare") charge Q_p^0 appears in the phase transformation of the interpolating proton field Ψ, i.e. in the commutator

$$[Q, \Psi] = -Q_\mathrm{p}^0 \Psi. \tag{12}$$

But the matrix element of the left-hand side of this relation between the vacuum and a one-proton state

$$\langle 0 | [Q, \Psi] | \overline{\boldsymbol{p}, \lambda} \rangle = -\sum_n \langle 0 | \Psi | \bar{n} \rangle \langle \bar{n} | Q | \overline{\boldsymbol{p}, \lambda} \rangle = -\langle 0 | \Psi | \overline{\boldsymbol{p}, \lambda} \rangle\, Q_\mathrm{p}$$

may contain only the proton as an intermediate state $|\bar{n}\rangle$, since, being a conserved operator, Q must be diagonal in all states. Consequently, $Q_\mathrm{p} = Q_\mathrm{p}^0$, and the strong interactions do not renormalize the electric charge. This result is in fact contained in (11) as well, since it is directly connected with the normalization condition for the electric form factor $F_1(0) = 1$ (see § 11.4).

While the properties of the electromagnetic interactions are well-established and quantum electrodynamics is a self-consistent theory, the weak interactions are treated at present on a purely phenomenological level. The Hamiltonian density for the weak interactions may be written in the form of the product of a weak current by a weak current:

$$H_w(x) = \frac{G}{\sqrt{2}} \left(J_\mu^{w+}(x)\, J^{\mu w}(x) + J^{\mu w}(x)\, J_\mu^{w+}(x) \right), \tag{13}$$

and calculations performed using perturbation theory. Sensible results are thus obtained only to first order. In other words, for the weak interactions, only the form of the amplitude for the (weak) transition

$$\langle a|T|b\rangle \approx \langle a|H_w(0)|b\rangle, \tag{14}$$

is established, but the way to calculate higher-order processes in G, as well as to take account of unitarity, is not indicated.

From experiment it is clear that there do not exist excited lepton states, at least up to a couple of GeV. Consequently, in this region of energy, one should not expect resonant phenomena, and the weak interaction constant G is the only parameter characterizing the order of magnitude of the transition amplitude.

In view of the smallness of the constant $G \approx 10^{-5}\, m_{\text{p}}^2$, the restriction to first order in G looks quite well-founded from a practical point of view. But, of course, the absence of a self-consistent theory should not in any way be considered an advantage of the present treatment of weak interactions.

The total weak current J_μ^{w} may be written as a sum of a leptonic current l_μ and a hadronic weak current J_μ:

$$J_\mu^w = l_\mu + J_\mu. \tag{15}$$

The effective Hamiltonian (13), with the current (15), describes a set of different processes which include purely leptonic processes (the term $l_\mu^+ l^\mu$), leptonic decays of hadrons (the term $l^\mu J_\mu^+ + \text{H.c.}$) and nonleptonic decays of hadrons (the term $J_\mu^+ J^\mu$).

The form (13), with an overall constant G for weak processes of all types, presupposes the universality of G.

We shall assume in this chapter (through § 15.5) that H_{w} conserves CP; we thereby exclude the weak processes which violate CP invariance (the decays of K_L^0 mesons; see § 15.6).

The effective Lagrangian for the decay of the muon and the β-decay of the neutron

$$\mu^- \to \text{e}^- + \nu_\mu + \bar{\nu}_\text{e}, \quad \text{n} \to \text{p} + \text{e}^- + \bar{\nu}_\text{e}$$

was proposed by Sudarshan and Marshak[(212)] and Feynman and Gell-Mann.[(213)] In their CP-invariant Lagrangian, the weak currents have the form

$$l_\lambda = \bar{\nu}_\mu \gamma_\lambda (1+\gamma_5)\mu + \bar{\nu}_\text{e} \gamma_\lambda (1+\gamma_5)\text{e}, \tag{16}$$

$$J_\lambda = \bar{\text{p}} \gamma_\lambda (1-\gamma_5)\text{n}, \tag{17}$$

where the field of each (free) particle is denoted by its symbol. The difference in sign in front of γ_5 in (16) and (17) is related to the different choice of particles and antiparticles in the two cases. In (16) the particles are taken as the negatively charged e^- and μ^-, while in (17) they are the neutron n and the positively charged proton p.

The expression (16) for the lepton current may be checked by comparing the calculated muon decay rate

$$\Gamma(\mu^- \to \text{e}^- \nu_\mu \bar{\nu}_\text{e}) \sim |\langle \text{e}^- \nu_\mu \bar{\nu}_\text{e} |\, l_\lambda^+ l^\lambda \,| \mu^- \rangle|^2 \tag{18}$$

with experiment. Both the angular distribution and the polarization of the electrons are very closely described using the current (16), so that the form (16) for the lepton current may be considered well established at low energies.

The lepton current (16) has the following characteristic features: (a) it transforms as a vector under transformations of the homogeneous Lorentz group (and does not contain tensor or scalar parts); (b) the current l_μ consists only of parts carrying electric charge $Q = +1$, while all parts of the current l_μ^+ carry the charge $Q = -1$; (c) the vector current $\bar{\nu}_\mu \gamma_\lambda \mu + \ldots$ and the axial vector current $-\bar{\nu}_\mu \gamma_\lambda \gamma_5 \mu + \ldots$ appear in the total current (16) with equal weights. For this reason, the theory of the weak interactions with currents (16) and (17) is conventionally called the $(V-A)$ theory. It is obvious that a different choice of particles would change the $(V-A)$ theory into a $(V+A)$ theory. The $(V-A)$ structure of the current (16) corresponds to a maximal violation of P-invariance for the weak interactions of leptons.†

The weak hadronic current (17) describes only the β-decay of the neutron. Many other weak decays are observed experimentally, (e.g., the decay of the pion $\pi^- \to \mu^- + \bar{\nu}_\mu$, the decays $K^- \to \mu^- + \bar{\nu}_\mu$, $\Sigma^+ \to \Lambda + e^+ + \nu_e$, etc., see Table A.1, p. 359), so that (17) is only one part of the total hadronic weak current $J_\lambda(x)$. Moreover, (17) parametrizes the matrix element of the weak current $\langle p | J_\lambda(x) | n \rangle$ only for vanishing momentum transfer $p_\lambda(n) - p_\lambda(p) \approx 0$. In the general case, the weak hadronic current $J_\lambda(x)$ must be considered simply as a local quantity characterized by selection rules.

The matrix elements of the current J_λ may be found from leptonic decays of hadrons whose amplitudes, according to (14), contain J_λ linearly, since the leptonic and hadronic parts of the amplitude factorize. Thus the study of the amplitude for β-decay of the neutron allows one to obtain information about the matrix element $\langle p | J_\lambda(x) | n \rangle$ and to check (17) for the weak current. The amplitude for the decay $\pi^+ \to \mu^+ + \nu_\mu$

$$T(\pi^+ \to \mu^+ \nu_\mu) \sim \frac{G}{\sqrt{2}} \langle \mu^+ \nu_\mu | l_\lambda(0) | 0 \rangle \langle 0 | J^{\lambda +}(0) | \pi^+ \rangle$$

gives information about one of the simplest matrix elements of the weak hadronic current—the matrix element of J_λ between the vacuum and the single-pion state. The decay $\pi^+ \to \pi^0 + e^+ + \nu_e$ is associated with the matrix elements of the current between single-pion states:

$$\langle \pi^0 | J_\lambda^+(0) | \pi^+ \rangle$$

Formula (17) for the β-decay weak current contains valuable information about the transformation properties of the current: like the lepton current, the current $\bar{p}\gamma_\lambda(1-\gamma_5)n$ consists of vector and axial vector parts.

Under isospin rotations the vector part of the current (17) behaves as the $(1+i2)$ component of an isovector with $I_3 = 1$:

$$\bar{p}\gamma_\lambda n = \tfrac{1}{2}\bar{N}\gamma_\lambda(\tau_1 + i\tau_2)N, \tag{19}$$

where N denotes the nucleon field. In the simple form (17) or (19) for the β-decay current, the current (19) is assumed to be a component of the same isovector $\frac{1}{2}\bar{N}\gamma_\lambda\tau_k N$ as the isovector part of the electromagnetic current $\frac{1}{2}\bar{N}\gamma_\lambda\tau_3 N$ (the hypothesis of the isovector nature of the weak vector current[213, 214].)

To go beyond the specific process of neutron β-decay and the assumption of zero-momentum transfer, one must consider the current $J_\mu(x)$ as a local quantity characterized

† See translator's note, p. 358.

by its transformation properties instead of using the explicit expression (19) for the hadron current in terms of the fields N(x) and $\overline{\text{N}}$(x). For transitions that do not change strangeness or hypercharge, a natural generalization of (19) is the expression

$$J_\lambda(\Delta Y = 0) = [v_{\lambda 1} - a_{\lambda 1} + i(v_{\lambda 2} - a_{\lambda 2})], \tag{20}$$

where $v_{\lambda k}$ and $a_{\lambda k}$ ($k = 1, 2, 3$) are the vector and axial vector isotriplets of hadronic currents. The expression (20) is well supported experimentally.

Formula (20) assumes that the $(V-A)$ structure of the weak currents holds both for baryonic and for mesonic decays. In fact, for $\pi^+ \to \mu^+ + \nu_\mu$, the decay amplitude is determined by the pseudo-vector part of the current a_λ^+ in $\langle 0 | J_\lambda^+ | \pi^+ \rangle$, while for $\pi^+ \to \pi^0 + e^+ + \nu_e$ the probability is determined by the contribution of the vector part v_λ in $\langle \pi^0 | J_\lambda^+ | \pi^+ \rangle$.

The isovector current $v_{\lambda k}$ is a quantity whose two charged components

$$\left.\begin{aligned} v_{\lambda +} &= (v_{\lambda 1} + i v_{\lambda 2}), & I_3 &= 1, \\ v_{\lambda -} &= (v_{\lambda 1} - i v_{\lambda 2}), & I_3 &= -1, \end{aligned}\right\} \tag{21}$$

occur in the weak currents J_λ and J_λ^+, while the neutral component $v_{\lambda 3}$ appears in the electromagnetic hadron current (4). This means that the invariant isovector form factor must be the same for the electromagnetic and weak interactions.

In particular, for the nucleon, the form factors F_1^V and F_2^V [eqn. (12.73)] determine the matrix elements of the currents both for the electromagnetic and for the weak interactions. For the current $v_{\lambda +}$, eqn. (12.73) and the hypothesis of the isovector nature of the current with components $v_{\lambda \pm}$ and $v_{\lambda 3}$ yield:

$$\langle \boldsymbol{p}_2, \sigma_2; \text{N} | v_{\lambda +}(0) | \boldsymbol{p}_1, \sigma_1; \text{N} \rangle$$
$$= \frac{1}{(2\pi)^3} \bar{u}(\boldsymbol{p}_2, \sigma_2)(\tau_1 + i\tau_2) \tfrac{1}{2} [\gamma_\mu F_1^V - i\sigma_{\mu\nu} q^\nu F_2^V] u(\boldsymbol{p}_1, \sigma_1), \quad (q = p_1 - p_2). \tag{22}$$

The form factors F_1^V and F_2^V may be observed in weak decays. The term with F_2^V in (22) is sometimes called "weak magnetism."[215]

The isovector nature of the current $v_{\lambda k}$ also assumes, by isospin symmetry, that if one of the components of $v_{\lambda k}$ is conserved all other components must be conserved. The conservation of the third component: $\partial^\lambda v_{\lambda 3} = 0$, following from the properties of the electromagnetic current, hence entails[214]

$$\partial^\lambda v_{\lambda k}(x) = 0 \qquad (k = 1, 2, 3). \tag{23}$$

By virtue of (23) the hypothesis of the isovector nature of the current $v_{\lambda k}$ is sometimes called the hypothesis of conserved vector current (CVC).

The electromagnetic interaction violates conservation of vector current, i.e., it violates (23) for $k = 1, 2$. However, the electromagnetic interaction usually is not taken into account when studying the properties of the weak currents J_λ in the Hamiltonian (14). Thus the weak hadronic currents in (14) may be treated in the limit of isospin symmetry, i.e. for conserved vector currents.

In analogy with the electromagnetic current, conservation of vector current entails the absence of renormalization of the "vector" coupling constant in the hadron–lepton Hamiltonian. This means that if we write this Hamiltonian, using (20), in the form

$$\frac{G}{\sqrt{2}}[(v_{\lambda+}-a_{\lambda+})l^{\lambda}+(v_{\lambda-}-a_{\lambda-})l^{\lambda+}], \tag{24}$$

the one-particle matrix elements of the vector part of (24) in the limit of zero-momentum transfer become the matrix elements of free fields (contained in v_λ) with the same coupling constant G. In other words, in (22) one will have $F_1^V(0) = 1$, just as in the case of $v_{\lambda 3}$ (the electromagnetic field), so that the strong interactions do not change the isospin of the states considered. We may write (24) in the form

$$\frac{1}{\sqrt{2}}(G_V^0 v_{\lambda+}l^{\lambda}+G_A^0 a_{\lambda+}l^{\lambda}+\text{H.c.}), \tag{25}$$

introducing explicitly the unrenormalized vector G_V^0 and axial G_A^0 coupling constants. In view of the conservation of vector current, the first term in (25) for β-decay (with zero-momentum transfer) is equivalent to

$$\frac{1}{\sqrt{2}}\left(G_V\bar{\text{N}}\gamma_\lambda \tfrac{1}{2}(\tau_1+i\tau_2)\text{N}l^{\lambda}\right),$$

containing the free field $\text{N}(x)$ and $G_V = G_V^0 F_1^V(0) = G_V^0$. As we shall see below in § 15.3, the axial vector coupling constant G_A^0 undergoes renormalization, so that $G_A \neq G_A^0 = G_V$.

The isovector nature of the current (23) entails a selection rule with respect to isospin for leptonic decays:

$$|\Delta \boldsymbol{I}| = 1. \tag{26}$$

This rule, however, is hard to check since, aside from β-decay, it gives clear-cut predictions only for neutrino reactions. In the case of nuclear β-decay, one must also take into account the violation of isospin invariance by Coulomb fields.

For isospin multiplets, the G-parity operator is more convenient than charge conjugation. Hence the classification of isotriplets of currents into first- and second-class currents is performed using the operator GP. For Hermitian first-class currents,

$$\text{GP}(v_{0k}-a_{0k})(\text{GP})^{-1} = -v_{0k}-a_{0k} \qquad (k = 1, 2, 3), \tag{27}$$

while second-class currents have opposite GP parity. The current (20) is first-class. With an accuracy of $\approx 5\%$, experiment indicates the absence of second-class currents.

There are both strangeness-conserving and strangeness-violating leptonic and nonleptonic decays of hadrons. Hence the total weak hadron current is equal to the sum of (20) and the current $J_\lambda(\Delta Y \neq 0)$, which changes hypercharge:

$$J_\lambda = J_\lambda(\Delta Y=0)+J_\lambda(\Delta Y \neq 0).$$

The current $J_\lambda(\Delta Y \neq 0)$ may be reconstructed on the basis of selection rules under the assumption that all weak processes (both with and without leptons) are generated by the Lagrangian

$$\frac{G}{\sqrt{2}}(J_\mu^+ + l_\mu^+)(J^\mu + l^\mu).$$

The leptonic decays of hadrons (whose amplitudes contain the current $J_\lambda(\Delta Y \neq 0)$ linearly) satisfy the following empirical selection rules:

$$\Delta Y = \Delta Q, \quad \Delta Y = \pm 1, \tag{28}$$

$$|\Delta \boldsymbol{I}| = \tfrac{1}{2}, \tag{29}$$

where ΔQ is the change in the electric charge of the hadrons. Rule (29) holds as well for nonleptonic decays with $\Delta Y \neq 0$. Isospin invariance is violated by electromagnetic interactions, so that rule (29) can hold only up to electromagnetic corrections.

Violations of the rules (28) and (29) are related to one another. From the Gell-Mann–Nishijima formula it follows that

$$\Delta I_3 = \Delta(Q-Y)+\tfrac{1}{2}\Delta Y.$$

If the rule $\Delta(Q-Y) = 0$ is violated for leptonic decays with $\Delta Y = \pm 1$, then the rule $|\Delta \boldsymbol{I}| = \frac{1}{2}$ is also violated. In the case of nonleptonic decays, where $\Delta Q = 0$, the rule $|\Delta \boldsymbol{I}| = \frac{1}{2}$ is violated if $|\Delta Y| > 1$.

The rule $|\Delta Y| = 1$ rests on the absence of other decays for all available cases, such as

$$\Xi^- \to \mathrm{n}+\pi, \quad \Xi^- \to \mathrm{n}+\mathrm{e}^-+\bar{\nu}_\mathrm{e}, \quad \Omega^- \to \Sigma^-+\pi^0.$$

The rule $\Delta Y = \Delta Q$ is borne out by the absence of the decays $\Sigma^+ \to \mathrm{n}+\mathrm{e}^++\nu_\mathrm{e}$, $\mathrm{K}^+ \to \pi^++\pi^++\mathrm{e}^-+\bar{\nu}_\mathrm{e}$, etc.

The rule $|\Delta \boldsymbol{I}| = \frac{1}{2}$ is also deduced from the analysis of multi-particle decays. The degree of its violation may be characterized by the ratio

$$\xi = \frac{\text{amplitude with } |\Delta \boldsymbol{I}| = \frac{3}{2}}{\text{amplitude with } |\Delta \boldsymbol{I}| = \frac{1}{2}}$$

(under the assumption that the amplitude with $|\Delta \boldsymbol{I}| > \frac{3}{2}$ is equal to zero). In the case of $\mathrm{K}^+ \to \pi^++\pi^0$ and $\mathrm{K}^0 \to \pi+\pi$, this ratio is equal to $\xi \simeq 4\times 10^{-2}$. The selection rules (28) and (29) show that the current $J_\lambda(\Delta Y \neq 0)$ is a charged isospinor, so that for $\Delta Q = +1$ its properties are the same as for the $(4+i5)$ component of the octet.

From the analysis of decays, it also follows that the current $J_\lambda(\Delta Y \neq 0)$ contains both a vector and an axial vector part. In particular, two decays are known: $\mathrm{K}^\pm \to \mu^\pm + \begin{Bmatrix}\nu_\mu \\ \bar{\nu}_\mu\end{Bmatrix}$ and $\mathrm{K}^\pm \to \pi^0+\mu^\pm+\begin{Bmatrix}\nu_\mu \\ \bar{\nu}_\mu\end{Bmatrix}$, the first of which is connected with the axial vector part of J_λ and the second with the vector part. Hence the current $J_\lambda(\Delta Y \neq 0)$ may be written naturally in the form

$$J_\lambda(\Delta Y \neq 0) \sim (v_{\lambda 4}-a_{\lambda 4})+i(v_{\lambda 5}-a_{\lambda 5}), \tag{30}$$

introducing the octet of vector $v_{\lambda a}$ and axial vector $a_{\lambda a}$ currents, $a = 1 \ldots 8$.

In (30) the "strength" of the current, or the overall factor, remains undetermined. Experiments show that the universality of G does not hold for $J_\lambda(\Delta Y \neq 0)$ (the sign $\sim$ in (30) cannot be replaced by $=$). The comparison of the processes $\mathrm{K}^\pm \to \pi^0+\mathrm{e}^\pm+\begin{Bmatrix}\nu_\mathrm{e} \\ \bar{\nu}_\mathrm{e}\end{Bmatrix}$ and

$\pi^{\pm} \to \pi^0 + e^{\pm} + \begin{Bmatrix} \nu_e \\ \bar{\nu}_e \end{Bmatrix}$ shows that processes with $\Delta Y \neq 0$ are suppressed approximately 20 times in comparison with processes that do not change hypercharge.

The universality of the constant G is restored using the hypothesis of Cabibbo,[216] according to which all three types of weak interactions (ll^+, lJ^+, and JJ^+) have the same coupling constant, but the weak hadronic current is modified by the introduction of the Cabibbo angle θ:

$$J_\lambda = [(v_{\lambda 1} - a_{\lambda 1}) + i(v_{\lambda 2} - a_{\lambda 2})] \cos\theta + [(v_{\lambda 4} - a_{\lambda 4}) + i(v_{\lambda 5} - a_{\lambda 5})] \sin\theta. \tag{31}$$

The hypothesis of Cabibbo is confirmed by experiment for a value $\theta = 15°$.

Thus the hadronic current (31), together with the leptonic current (16), is the total weak current, which appears in the effective Hamiltonian

$$\frac{G}{\sqrt{2}} (J^\lambda + l^\lambda)(J_\lambda^+ + l_\lambda^+) \tag{32}$$

with a universal constant G; here the unrenormalized vector and axial vector constants are now equal to $G_V^0 = G\cos\theta$ and $G_A^0 = G\sin\theta$. The Hamiltonian (32) explains all selection rules except for the empirical rule $|\Delta \boldsymbol{I}| = \frac{1}{2}$ in strangeness-changing nonleptonic decays. According to (32), nonleptonic decays must also contain transitions with $|\Delta \boldsymbol{I}| = \frac{3}{2}$ which, however, are suppressed by factors as yet not understood. From the point of view of the group SU_3, the terms in (32) with $|\Delta \boldsymbol{I}| = \frac{3}{2}$ are contained in a $\underline{27}$-plet, while the remaining terms in (32) belong to an octet. For this reason the hypothesis of the suppression of $\underline{27}$-plet terms [within the framework of the theory with the Hamiltonian (32)] is usually called the "hypothesis of octet dominance."

§ 15.2. The Gell-Mann algebra of densities and charges. The groups $SU_2 \times SU_2$ and $SU_3 \times SU_3$

Let us form integral quantities—the charges $V_k(x_0)$—from the densities of isovector currents $v_{0k}(x)$:

$$V_k(x_0) = \int v_{0k}(\boldsymbol{x}, x_0)\, d^3x \qquad (k = 1, 2, 3). \tag{33}$$

Since the isovector currents are conserved, $\partial^\mu v_\mu = 0$, the charges (33) do not depend on time:

$$i[P_0, V_k] = \int \partial_0 V_k(x_0) = (\partial_0 v_{0k} + \operatorname{div} \boldsymbol{v}_k)\, d^3x = 0. \tag{34}$$

Here we have added the vector current $\boldsymbol{v}_k$ across a surface at infinity, which is equal to 0. But, according to (4) and (5), the component V_3 appears in the Gell-Mann–Nishijima formula and must be identified with I_3. Hence the isovector V_k may coincide with isospin

$$V_k = I_k, \tag{35}$$

and, consequently, the components (35) may be taken to satisfy the commutation relations:

$$\left.\begin{aligned} [I_k, I_j] &= i\varepsilon_{kjl} I_l, \\ \left[\int v_{0k}(\boldsymbol{x}, x_0)\, d^3x, \int v_{0j}(\boldsymbol{x}', x_0)\, d^3x'\right] &= i\varepsilon_{kji} \int v_{0i}(\boldsymbol{x}, x_0)\, d^3x. \end{aligned}\right\} \tag{36}$$

The conditions (35) and (36) define the choice of normalization of the currents $v_{\mu k}$.

The axial charges $A_k(x_0)$ are determined in terms of the integral over the axial densities

$$A_k(x_0) \int a_{0k}(\boldsymbol{x}, x_0)\, d^3x \qquad (k = 1, 2, 3). \tag{37}$$

Since $a_{0k}(x)$ is an isovector, the commutator of the charges (37) with the generators of the isospin group I_k must be equal to

$$[A_k(x_0), I_j] = i\varepsilon_{kjl}A_l(x_0). \tag{38}$$

The axial current, in general, is not conserved: $\partial^\mu a_\mu \neq 0$, and the axial charges $A_k(x_0)$ cannot be constants of the motion:

$$[P_0, A_k(x_0)] = 0.$$

Let us try to close the algebra of the charges I_k and $A_l(x_0)$ at equal times. For this, we must postulate a commutation relation between the components $A_k(x_0)$ and $A_l(x_0)$ that does not introduce new quantities. As an additional condition, we shall demand that the charges I_k and A_l may be interpreted as generators of an extended compact Lie group. This condition leads to a unique solution:

$$[A_l(x_0), A_k(x_0)] = i\varepsilon_{lkj}I_j. \tag{39}$$

Since A_l is a pseudo-scalar, the right-hand side of (39) must be a scalar and is either expressible in terms of I_i or equal to zero. Zero on the right-hand side of (39) leads to the noncompact group $SU_2 \times T_3$ (isospin rotations with translations), where $A_l(x_0)$ plays the role of three-dimensional momentum. The choice of the opposite sign on the right-hand side of (39) would lead to the noncompact group $SL(2, c)$, where A_k would be similar to the generators of Lorentz transformations N_k. The relation (39) uniquely normalizes the axial charges A_k.

Formulae (36), (38), and (39) are the commutation relations for the generators of the group $SU_2 \times SU_2$. If one introduces the operators

$$V_k^{\pm} = \tfrac{1}{2}(I_k \pm A_k) \qquad (k = 1, 2, 3), \tag{40}$$

the sets V_k^+ and V_k^- will commute with one another:

$$[V_k^+, V_j^-] = 0, \tag{41}$$

and separately form SU_2 algebrae:

$$[V_k^{\pm}, V_j^{\pm}] = i\varepsilon_{kjl}V_l^{\pm}. \tag{42}$$

The relations (36) may be obtained in nearly any quantum-field theory model. Equation (38) follows from the assumption of the isovector nature of the axial-vector currents. Relation (39) was postulated by Gell-Mann;[(217)] it can be obtained in several models.

Let us introduce formulae of the type (36), (38), and (39) for the group $SU_3 \times SU_3$ in the quark model with currents

$$v_{\mu a}(\boldsymbol{x}, x_0) = \tfrac{1}{2}\bar{q}\gamma_\mu\lambda_a q, \tag{43}$$

$$a_{\mu a}(\boldsymbol{x}, x_0) = \tfrac{1}{2}\bar{q}\gamma_\mu\gamma_5 q, \tag{44}$$

where the operators $q(\boldsymbol{x}, x_0)$ and $q^+(\boldsymbol{x}, x_0)$ obey the usual equal-time commutation relations

$$\{q_{\alpha k}(\boldsymbol{x}, x_0), q^+_{\beta i}(\boldsymbol{x}', x_0)\} = \delta_{\alpha\beta}\delta_{ik}\delta(\boldsymbol{x}-\boldsymbol{x}')$$
$$(\alpha, \beta = 1, 2, 3, 4;\ i, k = 1, 2, 3). \tag{45}$$

If O_1 and O_2 are spin-SU_3 matrices of the type $\frac{1}{2}\gamma_5\lambda_a$, $\frac{1}{2}\lambda_a$, then, according to (45),

$$[q^+(\boldsymbol{x}, x_0)\, O_1 q(\boldsymbol{x}, x_0),\ q^+(\boldsymbol{x}', x_0)\, O_2 q(\boldsymbol{x}', x_0)] = q^+(\boldsymbol{x}, x_0)\, [O_1, O_2]\, q(\boldsymbol{x}, x_0)\, \delta(\boldsymbol{x}-\boldsymbol{x}'). \tag{46}$$

It then follows that, in the quark model, the current densities (43) and (44) satisfy the commutation relations

$$[v_{0a}(\boldsymbol{x}, x_0), v_{0b}(\boldsymbol{x}', x_0)] = if_{abc}v_{0c}(\boldsymbol{x}, x_0)\, \delta(\boldsymbol{x}-\boldsymbol{x}'), \tag{47}$$
$$[a_{0a}(\boldsymbol{x}, x_0), v_{0b}(\boldsymbol{x}', x_0)] = if_{abc}a_{0c}(\boldsymbol{x}, x_0)\, \delta(\boldsymbol{x}-\boldsymbol{x}'), \tag{48}$$
$$[a_{0a}(\boldsymbol{x}, x_0), a_{0b}(\boldsymbol{x}', x_0)] = if_{abc}v_{0c}(\boldsymbol{x}, x_0)\, \delta(\boldsymbol{x}-\boldsymbol{x}'). \tag{49}$$

For the octets of the vector and axial vector charges

$$F_a(x_0) = \int_{x_0} d^3x v_0(x_0, \boldsymbol{x}), \qquad A_a(x_0) = \int_{x_0} d^3x a_0(x_0, \boldsymbol{x}) \tag{50}$$

relations (47)–(49) lead to the algebra of $SU_3 \times SU_3$:

$$[F_a(x_0), F_b(x_0)] = if_{abc}F_c(x_0), \tag{51}$$
$$[F_a(x_0), A_b(x_0)] = if_{abc}A_c(x_0), \tag{52}$$
$$[A_a(x_0), A_b(x_0)] = if_{abc}F_c(x_0), \tag{53}$$

where we have retained the argument x_0 in conserved charges for uniformity. The charges F_a are generators of the usual group SU_3. If we introduce the quantities

$$F^\pm_a = \tfrac{1}{2}(F_a \pm A_a), \tag{54}$$

they will satisfy the commutation relations for the generators of two independent SU_3 groups:

$$[F^+_a, F^-_b] = 0, \tag{55}$$
$$[F^\pm_a, F^\pm_b] = if_{abc}F^\pm_c. \tag{56}$$

The representations of the group $SU_3 \times SU_3$ may be designated by $(\underline{n}^+, \underline{n}^-)$, where $n^\pm$ is the dimension of the representation of SU_3 of the subgroup with generators $\frac{1}{2}(F_a \pm A_a)$. The charges associated with the weak and electromagnetic hadronic currents belong to the reducible representation $(\underline{8}, \underline{1})+(\underline{1}, \underline{8})$.

The Gell-Mann commutation relations of currents and charges are nonlinear; they thereby allow one to establish the scale of the charges. The vector interaction constant G_V is determined by the leptonic decay $a \to b+1$, where 1 is a lepton. The strong interactions affect the constant G_V, inasmuch as they determine the vector form-factor $\langle \boldsymbol{k}_2, b | v_{\lambda, 1+i2} | \boldsymbol{k}_1, a\rangle$. The renormalization of the constant G_V is determined by the value of this form-factor at $(k_2 - k_1) = 0$ so that, for example, for the β-decay of the neutron,

$$G \cos\theta \langle \boldsymbol{k}_2, \mathrm{p} | v_{\lambda, 1+i2}(0) | \boldsymbol{k}_1, \mathrm{n}\rangle = G_V \langle \boldsymbol{k}_2, \mathrm{p} | \bar{\mathrm{p}}(0)\, \gamma_\lambda \tau_{1+i2} \mathrm{n}(0) | \boldsymbol{k}_1, \mathrm{n}\rangle, \tag{57}$$

where p(x) and n(x) are the free proton and neutron fields. For $\lambda = 0$, the left-hand side of (57) is proportional to the expectation value of the charge I_{1+i2}:

$$\langle \boldsymbol{k}_2, \mathrm{p} | I_{1+i2} | \boldsymbol{k}_1, \mathrm{n} \rangle = (2\pi)^3\, \delta(\boldsymbol{k}_1 - \boldsymbol{k}_2) \langle \boldsymbol{k}_1, \mathrm{p} | v_{0,\,1+i2}(0) | \boldsymbol{k}_1, \mathrm{n} \rangle = \delta(\boldsymbol{k}_1 - \boldsymbol{k}_2) 2k_0 . \quad (58)$$

The last relation in (58) is a consequence of the fact that I_k is a conserved generator of the isospin group [satisfying the commutation relations (36)], and thus the operators I_k act within the (p, n) multiplet, while the values of the matrix elements $I_k^{\pm}$ follow from (36). From (57) and (58) one sees that

$$G_V = G \cos \theta \quad (59)$$

—the vector constant is not renormalized by the strong interactions nor are the charges I_k.

In the limit of exact SU_3 symmetry, the current $v_{\mu a}$ ($a = 4, 5$) are also conserved, while the values of F_a remain constants of the motion. From the commutation relations (51) between these charges and $F_k = I_k$ ($k = 1, 2, 3$), it is clear that in the limit of SU_3 symmetry the charges F_4, F_5 are also not renormalized. If one allows for violation of SU_3 symmetry, the charges F_4 and F_5 are renormalized only in second order in the SU_3-violating interaction (the Ademollo–Gatto theorem[(218)]).

The scale of the axial vector charges A_a is fixed by the commutation relations (52) and (53) and is connected with the vector charges. Since the axial currents are not conserved, one should expect important effects of renormalization of the weak axial charges. The value of the renormalized axial isovector charge is calculated in § 15.4.

§ 15.3. Partial conservation of axial current

While the vector isotriplet current is conserved, the axial current cannot be conserved. To see this, consider the decay

$$\pi^{\pm} \to \mu^{\pm} + \nu_\mu(\bar{\nu}_\mu).$$

The probability amplitude for this decay

$$\frac{G}{\sqrt{2}} \langle \mu\nu | J_\lambda^+ l^\lambda | \pi \rangle$$

contains the simplest matrix element of the weak hadron current (p_λ is the pion momentum)

$$\langle 0 | J_\lambda^+(0) | \boldsymbol{p}, \pi^+ \rangle = \langle 0 | a_{\lambda -}(0) | \boldsymbol{p}, \pi^+ \rangle,$$
$$a_{\lambda -} = a_{\lambda 1} - i a_{\lambda 2}, \quad (60)$$

in which only the axial part contributes since the pion is a pseudo-scalar particle. From considerations of covariance, the matrix element (60) is determined by one constant f_π:

$$\langle 0 | a_{\lambda -}(0) | \boldsymbol{p}; \pi^+ \rangle = \frac{i f_\pi}{(2\pi)^{3/2}} p_\lambda . \quad (61)$$

The constant f_π may be found from the decay of the pion: $f_\pi \approx 135$ MeV (if G is found from other experiments). Calculating the divergence of the current $a_{\lambda -}$,

$$\langle 0 | \partial^\lambda a_{\lambda -}(0) | \boldsymbol{p}, \pi^+ \rangle = i \langle 0 | [p_\lambda, a_{\lambda -}] | \boldsymbol{p}, \pi^+ \rangle = \frac{f_\pi m_\pi^2}{(2\pi)^{3/2}} . \quad (62)$$

If one were to set $\partial^\lambda a_\lambda = 0$, then, according to (62), one would have to simultaneously consider a pion to be a massless particle, $m_\pi = 0$, or to take $f_\pi = 0$ and forbid the decay $\pi \to \mu\nu$.

The isovector field $\partial_\lambda a^\lambda(x)$ has the same quantum numbers as the pion fields $\pi_{\text{in}}(x)$ and $\pi_{\text{out}}(x)$, and the matrix element of this field between the vacuum and a one-pion state is non-zero. If the field $\partial_\lambda a^\lambda(x)$ is also irreducible, it satisfies the condition of the theorem of Haag *et al.*,[219–221] according to which such a field $\partial_\lambda a^\lambda(x)$ may be used as an interpolating pion field $c\pi(x)$, and

$$\partial_\lambda a^\lambda(x) \to c_\pi \pi_{\text{in}}(x), \quad x_0 \to \mp\infty,$$

where the limit is understood in the sense of weak convergence (see § 11.1).

The hypothesis of partial conservation of axial current (PCAC), relates the divergence of the axial current to the interpolating pion field:[222, 223]

$$\partial_\mu a^\mu(x) = c_\pi \pi(x). \tag{63}$$

Equation (63) defines the continuation of the pion field off the mass shell $p^2 = m_\pi^2$.

A different continuation would be given, for example, by the field

$$\pi'(x) = \pi(x) + \lambda(\partial_\mu \partial^\mu - m_\pi^2)\, \pi(x), \tag{64}$$

coinciding with $\pi(x)$ asymptotically: $\pi'_{\substack{\text{in}\\\text{out}}} = \pi_{\substack{\text{in}\\\text{out}}}$. One of the possible fields (64) is the canonical field $\pi(x)$, satisfying the usual equal-time commutation relations. For the field (63), the canonical commutation relations hold only in model theories (the σ-model[224]). Since the matrix elements of the current a_μ are observable in principle, the PCAC hypothesis (63) implies that the corresponding matrix elements of the pion field off the mass shell are also observable.

Comparing (62) and (63), we find

$$c_\pi = m_\pi^2 f_\pi \approx m_\pi^2 \times 135 \text{ MeV}. \tag{65}$$

As an example of the application of the PCAC hypothesis, let us consider the matrix element of the axial current between proton and neutron states contained in the amplitude for the β-decay of the neutron. From invariance considerations, it follows that here there are three independent form factors, G_1, G_2, G_3 (see § 11.4), which we shall associate with the vectors $\gamma_5\gamma_\lambda$, $q_\lambda\gamma_5 = (\text{p}_\text{n} - \text{p}_\text{p})_\lambda\gamma_5$ and $\sigma_{\lambda\nu}q^\nu\gamma_5$:

$$G \cos\theta \langle \boldsymbol{p}_2, \sigma_2; \text{p} \,|\, a_{\lambda 1} + i a_{\lambda 2} \,|\, \boldsymbol{p}_1, \sigma_1; \text{n} \rangle = \frac{1}{(2\pi)^3} \bar{u}_\text{N}(\boldsymbol{p}_2, \sigma_2)\, \tau_+ \{\gamma_5\gamma_\lambda G_1(q^2) - q_\lambda\gamma_5 G_2(q^2) + i\sigma_{\lambda\varrho}q^\varrho\gamma_5 G_3(q^2)\}\, u_\text{N}(\boldsymbol{p}_1, \sigma_1), \tag{66}$$

where we have used the isovector nature of the current a_λ and have introduced on the right-hand side the components of the isovector $\bar{u}_\text{N}\frac{1}{2}\tau_k u_\text{N}$ (u_N is an eight-component nucleon Dirac spinor). The hermiticity of the operator $a_{\lambda k}$ demands that G_1 and G_2 be real and G_3 be purely imaginary. The current $i\bar{u}\tau_k\sigma_{\lambda\varrho}q^\varrho\gamma_5 u/2$ is a second-class current with respect to the GP transformation (31). The presence of a current with an imaginary form factor G_3 violates invariance of the theory with respect to time-reversal T (see § 5.4). Hence one must set

$$G_3 = 0. \tag{67}$$

For $q^2 = 0$, the right-hand side of (66) becomes

$$\frac{G_1(0)}{(2\pi)^3}\,\bar{u}_N(p_2, \sigma_2)\,\tau_+\gamma_\lambda\gamma_5 u_N(p_1, \sigma_1), \tag{68}$$

since the form factors cannot have poles at $q^2 = 0$: such poles would correspond to charged particles of zero mass.

Expression (68) differs from the formula for the matrix element of the axial current

$$\frac{G\cos\theta}{(2\pi)^3}\,\bar{u}_N(p_2, \sigma_2)\,\tau_+\gamma_\lambda\gamma_5 u_N(p_1, \sigma_1)$$

(applicable when one neglects the effects of strong interactions) by a factor $G_1(0)/G\cos\theta$. In other words, the effective (renormalized) axial weak coupling constant is equal to

$$G_A \equiv G_1(0) \approx 1.25G\cos\theta, \tag{69}$$

with the number 1.25 (more precisely, 1.250 ± 0.009) obtained from experiment.[8] Instead of (17), we should now write the β-decay current in the form

$$J_\lambda = \bar{\text{p}}\gamma_\lambda(1 - 1.25\gamma_5)\text{n}\cos\theta. \tag{70}$$

The Goldberger–Treiman relation

Let us find a relation between G_A and the pion parameters m_π and f_π. The matrix element of the divergence of the axial current is equal to

$$\cos\theta G\langle p_2;\,\text{p}\,|\,\partial^\lambda(a_{\lambda 1} + ia_{\lambda 2})\,|\,p_1;\,\text{n}\rangle = \frac{i}{(2\pi)^3}\,\bar{u}_\text{p}(p_2)\,\{(m_\text{p} + m_\text{n})\,G_1(q^2) + q^2 G_2(q^2)\}\,\gamma_5 u_\text{n}(p_1). \tag{71}$$

Now, according to (63), the left-hand side of (71) is equal to

$$G\cos\theta\langle p_2;\,\text{p}\,|\,\pi(0)\,|\,p_1;\,\text{n}\rangle m_\pi^2 f_\pi.$$

This quantity may be related to the pion form-factor of the nucleon $K_{\text{NN}\pi}(q^2)$, which is defined by

$$\frac{1}{(2\pi)^3}\,ig_{\text{NN}\pi}\bar{u}(p_2)\,\gamma_5\tau_k u(p_1)\,K_{\text{NN}\pi}(q^2) = \langle p_2,\,\text{N}\,|\,J_{5k}\,|\,p_1;\,\text{N}\rangle, \tag{72}$$

where $q_{\text{NN}\pi}$ is the pion–nucleon coupling constant, $g^2_{\text{NN}\pi}/4\pi \approx 14.6$, while $J_{5k}(x) = (-\Box + m_\pi^2)\,\pi_k(x)$ is the pion current.

The form factor $K_{\text{NN}\pi}(q^2)$ is normalized by the condition $K_{\text{NN}\pi}(m_\pi^2) = 1$, where the point $q^2 = m_\pi^2$ corresponds to the pion pole in the annihilation channel.

Thus a simple combination of axial form factors may be expressed in terms of the pion form factor of the nucleon and the constant $g_{NN\pi}$:

$$\frac{2m_N G_1(q^2)+q^2G_2(q^2)}{2G\cos\theta} = f_\pi m_\pi^2 \frac{g_{NN\pi}}{\sqrt{2}} \frac{K_{NN\pi}(q^2)}{m_\pi^2-q^2}.$$

Setting $q^2 = 0$, we obtain the Goldberger–Treiman relation[225]

$$\frac{G_1(0)}{G\cos\theta} \equiv \frac{G_A}{G\cos\theta} = \frac{f_\pi}{\sqrt{2}m_N} g_{NN\pi}K_{NN\pi}(0). \tag{73}$$

Equation (73) allows one to calculate $K_{NN\pi}(0)$ in terms of the known experimental values for $g_A(0) \approx 1.25$ and $f_\pi \approx 135$ MeV, giving $K_{NN\pi}(0) \cong 0.91$. This value differs very little from $K_{NN\pi}(m_\pi^2) = 1$, and the function $K_{NN\pi}(q^2)$ probably varies monotonically in the interval $0 \leqslant q^2 \leqslant m_\pi^2$. For this reason, one usually sets $K_{NN\pi} \approx 1$, writing formula (73) in the form ($\cos\theta \approx 1$)

$$\frac{G_A}{G} = \frac{f_\pi}{\sqrt{2}m_N} g_{NN\pi}.$$

The accuracy of this relation clearly cannot exceed 10 per cent.

If one passes to exact $SU_2\times SU_2$ symmetry, where the axial currents are conserved:

$$\partial^\lambda a_\lambda = 0,$$

relation (71) implies:

$$2m_N G_1(q^2) = -q^2 G_2(q^2).$$

To determine the axial constant for the case of $SU_2\times SU_2$ symmetry, one must study the limit of the right-hand side as $q^2 \to 0$. According to (62), conservation of axial current suggests two possibilities for the treatment of the pion: (1) $m_\pi = 0, f_\pi \neq 0$, (2) $f_\pi = 0, m_\pi \neq 0$. If $m_\pi = 0$, the form factor $G_2(q^2)$ acquires a pion pole at $q^2 = 0$, so that

$$G_1(0) = -\frac{1}{2m_N} \lim_{q^2\to 0} q^2 G_2(q^2) \neq 0.$$

If $m_\pi \neq 0$, then $G_2(q^2)$ does not have a pole corresponding to charged particles with $m^2 = q^2 = 0$, and $q^2G_2(q^2) \to 0$ for $q^2 \to 0$. In this case, the axial constant is nonzero only if the mass of the nucleon vanishes: $m_N = 0$. The nucleon states $|\boldsymbol{p}, \lambda; N\rangle$ of definite helicity $\pm\lambda$ thus form a basis for the two simplest representations $V^+ = \frac{1}{2}$, $V^- = 0$, and $V^+ = 0$, $V^- = \frac{1}{2}$ of the group $SU_2^{(+)}\times SU_2^{(-)}$.

The Adler consistency condition[226]

Another important consequence of the PCAC hypothesis is the vanishing (under certain conditions) of the hadron scattering amplitude $b \to c+\pi$ if the momentum q_μ of the pion approaches zero: $q_\mu \to 0$; here b and c stand for arbitrary hadron states.

Let us consider the amplitude for this process and use the reduction formula (11.14):

$$\langle \pi c|S-1|b\rangle = \delta^4(p_c-p_b+q)\langle \pi c|T|b\rangle = -\frac{1}{(2\pi)^{3/2}}\int d^4x e^{iqx}K(x)\langle c|\pi(x)|b\rangle,$$

where $\pi(x)$ is the interpolating pion field and q_μ is the momentum of the pion in the final state. The field $\pi(x)$ may be expressed in terms of $\partial_\mu a^\mu$ using the PCAC hypothesis (63):

$$\langle \pi c \,|\, S-1 \,|\, b\rangle = -\frac{1}{c_\pi(2\pi)^{3/2}}\int d^4x e^{iqx}K(x)\langle c\,|\,\partial_\mu a^\mu(x)\,|\,b\rangle$$
$$= \frac{iq_\mu}{c_\pi(2\pi)^{3/2}}\int d^4x e^{iqx}K(x)\langle c\,|\,a^\mu(x)\,|\,b\rangle,$$

since $p_b = p_c + q$. Consequently, the amplitude may be nonzero at $q_\mu = 0$ only when the "scattering" amplitude of the axial current $b \to c+a$ possesses a pole at this point. When such poles are absent, we obtain the Adler consistency condition:

$$\lim_{q_\mu \to 0} \langle \pi c \,|\, T \,|\, b\rangle = 0. \tag{74}$$

This condition has already been used in § 14.4.

§ 15.4. Renormalization of the axial vector coupling constant

The axial vector weak coupling constant G_A may be defined by eqns. (66), (68) and (69), or

$$G_V\langle \boldsymbol{k}, \sigma_1;\, \mathrm{p}\,|\, a_{01}+ia_{02}\,|\,\boldsymbol{k}, \sigma_2;\, \mathrm{n}\rangle = G_A\langle \boldsymbol{k}, \sigma_1;\, \mathrm{p}\,|\,\bar{p}(x)\,\gamma_4\gamma_5 n(x)\,|\,\boldsymbol{k}, \sigma_2;\, \mathrm{n}\rangle$$
$$= \frac{G_A}{(2\pi)^3}\,\bar{u}(\boldsymbol{k}', \sigma_1)\,\gamma_4\gamma_5 u(\boldsymbol{k}, \sigma_2), \tag{75}$$

where G_V is the vector interaction constant, and p(x) and n(x) are the free proton and neutron fields. Our problem is to calculate G_A/G_V, i.e. to derive an expression which, evaluated using independent experimental data, can give the coefficient 1.25 in (69) and (70).

To calculate G_A, we use the equal-time commutation relations (39) for axial charges

$$[A_+(x_0),\, A_-(x_0)] = 2I_3 \qquad (A_\pm = A_1 \pm iA_2). \tag{76}$$

We set $x_0 = 0$ and write $A(x_0) = A$. Let us find the matrix element of both sides of (76) between proton states $|\,\boldsymbol{k}, \sigma;\, \mathrm{p}\rangle$ and $|\,\boldsymbol{k}', \sigma';\, \mathrm{p}\rangle$. Averaging over the proton spin,

$$\tfrac{1}{2}\sum_{\sigma,\sigma'}\langle \boldsymbol{k}, \sigma;\, \mathrm{p}\,|\,2I_3\,|\,\boldsymbol{k}', \sigma';\, \mathrm{p}\rangle = 2k_0\delta(\boldsymbol{k}-\boldsymbol{k}'). \tag{77}$$

for the right-hand side of (76). On the left-hand side of (76) the matrix element of the product of axial charges A_+A_- may be expanded in a complete set of intermediate states $|\,\alpha\rangle$:

$$\tfrac{1}{2}\sum_{\sigma,\sigma'}\langle \boldsymbol{k}, \sigma;\, \mathrm{p}\,|\,A_+A'_-\,|\,\boldsymbol{k}', \sigma';\, \mathrm{p}\rangle = \tfrac{1}{2}\sum_{(\mathrm{n})}\sum_{\sigma,\sigma'}\langle \boldsymbol{k}, \sigma;\, \mathrm{p}\,|\,A_+\,|\,(\mathrm{n})\rangle$$
$$\times\langle(\mathrm{n})\,|\,A_-\,|\,\boldsymbol{k}', \sigma';\, \mathrm{p}\rangle + \tfrac{1}{2}\sum_{\alpha\neq(\mathrm{n});\,\sigma,\sigma'}\langle \boldsymbol{k}, \sigma;\, \mathrm{p}\,|\,A_+\,|\,\alpha\rangle\langle\alpha\,|\,A_-\,|\,\boldsymbol{k}', \sigma', \mathrm{p}\rangle \equiv C_1 + C_2(+\,-), \tag{78}$$

where we have separated out the sum C_1 over single-particle (in the present case, neutron) states n. The second sum $C_2(+\,-)$ contains the multi-particle states $|\,\alpha\rangle$. The term A_-A_+ in (76) does not yield a one-particle contribution of the type C_1 (there are no doubly charged particles decaying only as a consequence of the weak interaction). The relation obtained from (76) thus has the form

$$2k_0\delta(\boldsymbol{k}-\boldsymbol{k}') = C_1+C_2(+\,-)-C_2(-\,+). \tag{79}$$

The contribution of the neutron state C_1 may be expressed in terms of the axial vector coupling constant G_A. According to (37), the operator A_k does not change the momentum, and each of the matrix elements in (78) contains a δ-function of the momentum of the states:

$$\langle a|A_k|b\rangle = (2\pi)^3\,\delta(\boldsymbol{p}_a-\boldsymbol{p}_b)\,\langle a|a_{0k}|b\rangle. \tag{80}$$

Now, by virtue of (75) for one-particle states, the matrix element of the current a_{0k} (for equal momenta) is proportional to G_A. Inserting (75) and (80) into (78), we obtain for C_1 the expression

$$C_1 = 2\left(\frac{G_A}{G_V}\right)^2\frac{\boldsymbol{k}^2}{k_0}\,\delta(\boldsymbol{k}-\boldsymbol{k}'). \tag{81}$$

The relation (79) may now be written symbolically in the form

$$1-\left(\frac{G_A}{G_V}\right)^2\frac{\boldsymbol{k}^2}{k_0^2} = \frac{1}{2k_0}\left(C_2'(+\,-)-C_2'(-\,+)\right), \tag{79'}$$

where C_2' designates the contribution of the multi-particle states in (78) without the momentum δ-function.

From (79) it is clear that the renormalization of G_A is significantly related to multi-particle states. If the symmetry $SU_2\times SU_2$ were exact and the axial charges A_k were conserved, while the nucleon belonged to an irreducible representation of the group $SU_2\times SU_2$, the charges A_k, like the generators of this group, could not connect the nucleon with multi-particle states. Consequently, in the case of $SU_2\times SU_2$ symmetry, the right-hand side of (79′) would vanish. Assuming that the neutron and proton are the only single-particle states, we thereby exclude parity-doubling and thus must set $m_N = 0$ (see § 15.3). The relation (79′) would then lead to the absence of renormalization: $G_A = G_V$.

If the mass of the nucleon is nonzero and the symmetry $SU_2\times SU_2$ is broken, then (79′) allows one in principle to find the ratio G_A/G_V as a measure of the breaking of the symmetry. The relation (79′) depends on the frame of reference and is most easily visualized in the infinite momentum frame of the nucleon, where $\boldsymbol{k}^2/k_0^2 \approx 1$. Then the left-hand side contains only $1-G_A^2/G_V^2$ and its deviation from unity is wholly determined by the multi-particle contributions C_2'.

The contribution of multi-particle states may be found using the PCAC hypothesis (63). We first transform the matrix element of the axial charge A_k in such a way as to replace A_k by the divergence $\partial_\mu a_k^\mu$:

$$\langle\alpha|A|\boldsymbol{k},\sigma;\mathrm{p}\rangle = \frac{\langle\alpha|\partial_0 A|\boldsymbol{k},\sigma;\mathrm{p}\rangle}{i(p_{\alpha 0}-k_0)} = \frac{\langle\alpha|\int d^3x\,\partial^\mu a_\mu|\boldsymbol{k},\sigma;\mathrm{p}\rangle}{i\omega_\alpha}$$

$$= \frac{C_\pi}{\sqrt{2}}\,\frac{1}{i\omega_\alpha}\langle\alpha|\int d^3x\pi(x)|\boldsymbol{k},\sigma;\mathrm{p}\rangle, \quad (\omega_\alpha = p_{\alpha 0}-k_0). \tag{82}$$

To express C_2 in terms of observable quantities, let us introduce the pion current related to the strong interactions in place of the field $\pi(x)$:

$$\langle\alpha|\int d^3x\pi(x)|\boldsymbol{k},\sigma;\mathrm{p}\rangle = (2\pi)^3\,\delta^3(\boldsymbol{p}_\alpha-\boldsymbol{k})\,\langle\alpha|\pi(0)|\boldsymbol{k},\sigma;\mathrm{p}\rangle$$
$$= (2\pi)^3\,\delta(\boldsymbol{p}_\alpha-\boldsymbol{k})\,\frac{\langle\alpha|J_\pi(0)|\boldsymbol{k},\mathrm{p}\rangle}{(p_\alpha-k)^2-m_\pi^2}, \tag{83}$$

$$(\Box-m_\pi^2)\pi(x) = J_\pi(x).$$

On the energy-momentum surface $p_\alpha = k+p_\pi$, the matrix element of the pion current $\langle\alpha|J_\pi(0)|\boldsymbol{k},\sigma;\mathrm{p}\rangle$ determines the amplitude for the process $\pi+\mathrm{p}\to\alpha$, if $|\alpha\rangle\equiv|\alpha,\text{out}\rangle$.

Inserting (82) and (83) into $C_2(+-)$,

$$C_2(+-) = \delta(\boldsymbol{k}-\boldsymbol{k}')\,\frac{c_\pi^2(2\pi)^6}{4}\sum_{n_\alpha,\,\sigma_\alpha}\int\delta(\boldsymbol{p}_\alpha-\boldsymbol{k})\,\delta(p_1^2-m_1^2)$$
$$\ldots\,\delta(p_{n_\alpha}^2-m_{n_\alpha}^2)\,\frac{|\langle\boldsymbol{p}_\alpha,\sigma_\alpha;\alpha|J_\pi(0)|\boldsymbol{k},\sigma;\mathrm{p}\rangle|^2}{\omega_\alpha^2[(p_\alpha-k)^2-m_\pi^2]^2}$$
$$= \delta(\boldsymbol{k}-\boldsymbol{k}')\,\frac{c_\pi^2(2\pi)^6}{4}\sum_{n_\alpha,\,\sigma_\alpha}\int dR_{n_\alpha}(k+q)\,\frac{|\langle\boldsymbol{p}_\alpha,\sigma_\alpha;\alpha|J_\pi(0)|\boldsymbol{k},\sigma;\mathrm{p}\rangle|^2}{\omega_\alpha^2[(p_\alpha-k)^2-m_\pi^2]^2}, \tag{84}$$

where n_α is the number of particles in the state α, while σ_α is their polarization state. The invariant phase volume dR_n is determined by (7.32), where we have introduced the vector $q^\mu = (\omega_\alpha, 0, 0, 0)$. The matrix element of the current in C_2 can be related to the total cross-section for the scattering of soft pions ($m_\pi^2 = 0$) on protons. For the scattering

$$\pi+\mathrm{p}\to\alpha$$

from the conservation of energy-momentum $q+k = p_\alpha$ (q is the pion momentum) when $q^2 = 0$, we obtain in the centre of mass system

$$M = \sqrt{s} = q_0+k_0, \quad q_0 = \frac{M^2-m_\mathrm{N}^2}{2M}.$$

The scattering cross-section $d\sigma(\pi+\mathrm{p}\to\alpha)$ is equal (see § 7.3) to

$$d\sigma = \frac{2(2\pi)^{10}}{M^2-m_\mathrm{N}^2}\,|\langle\alpha|T|\pi\mathrm{p}\rangle|^2\,dR_\alpha(q+k), \tag{85}$$

where, according to (11.14),

$$\langle\alpha|T|\pi\mathrm{p}\rangle = -(2\pi)^{-3/2}\,\langle\alpha|J_\pi(0)|\boldsymbol{k},\sigma;\mathrm{p}\rangle. \tag{86}$$

The total cross-section σ_{tot} may be found from (85) by summing Σ_α over all possible states α with given energy $\sqrt{s}$. If one introduces the notation $\sigma_{\text{tot}}(\sqrt{s}, m_\pi^2)$, stressing the dependence of σ_{tot} on the pion mass, (85) yields $\sigma_{\text{tot}}(\sqrt{s}, 0)$.

Let us pass in (84) to the infinite-momentum frame, where $k_0\to\infty$, and insert expressions (85) and (86), obtaining

$$C_2(+-) = \delta(\boldsymbol{k}-\boldsymbol{k}')\,\frac{c_\pi^2}{m_\pi^4}\int_{m_\mathrm{N}+m_\pi}^{\infty} dW\,\frac{W}{W^2-m_\mathrm{N}^2}\,\sigma_{\text{tot}}^{-}(W, 0), \tag{87}$$

where $\sigma_{tot}^{-}(W, 0)$ is the total-cross section for scattering of soft negative pions on a proton. The term $C_2(-\ +)$ is calculated analogously, and contains the total cross-section $\sigma_{tot}^{+}(W, 0)$ for the scattering of soft positive pions on a proton:

$$C_2(-+) = \delta(\boldsymbol{k}-\boldsymbol{k}')\frac{c_\pi^2}{m_\pi^4}\int_{m_N+m_\pi}^{\infty} dW \frac{W}{W^2-m_N^2}\,\sigma_{tot}^{+}(W, 0). \tag{88}$$

The pion constant c_π contains G_A/G_V. Using the Goldberger–Treiman formula (73), we may, with the help of (81)–(88), transform (79) into the form[227, 228]

$$1-\frac{G_V^2}{G_A^2} = \frac{4m_N^2}{(K_{NN\pi}(0)g_{NN\pi})^2}\,\frac{1}{\pi}\int_{m_N+m_\pi}^{\infty} dW \frac{W}{W^2-m_N^2}\,(\sigma_{tot}^{+}(W, 0)-\sigma_{tot}^{-}(W, 0)). \tag{89}$$

The Adler–Weisberger relation (89) allows one to calculate the renormalization of the weak axial vector coupling constant G_A in terms of the scattering cross-sections and the constants $K_{NN\pi}$, $g_{NN\pi}$, associated with the strong interactions. It is assumed, of course, that the experimental cross-sections $\sigma_{tot}^{\pm}(W, m_\pi^2)$ may be extrapolated to $m_\pi^2 = 0$. In the low-energy region, σ_{tot}^{+} dominates over σ_{tot}^{-} as a consequence of the Δ^{++} resonance. As the energy increases, the difference $\sigma_{tot}^{+}-\sigma_{tot}^{-}$ decreases, and eventually becomes negative. Calculations give $|G_A/G_V| \approx 1.24$[227] and $|G_A/G_V| \approx 1.16$,[228] values very close to the experimental number 1.25 if one bears in mind the approximate nature ($\sim 10\%$) of the Goldberger–Treiman formula.

§ 15.5. Asymptotic chiral symmetry and spectral sum rules

If chiral symmetry $SU_2 \times SU_2$ were exact, then, as we saw in § 15.3, the pion would have to be massless. Hence one might expect chiral symmetry to become more and more exact with increasing energy, while for low energies its violations would be very significant. The latter is certainly the case since, for example, the baryon octet B cannot be associated with another octet B' of opposite parity. The demand of asymptotic symmetry, when combined with an assumption about the asymptotic behavior of vacuum expectation values of products of currents, allows one to obtain strict limitations on spectral functions or on pole approximations to the masses of particles. These restrictions are known as the Weinberg spectral sum rules.[229]

Let us consider the vacuum expectation value of the product of two currents $j_{\mu k}$, $j_{\nu j}$, where k, j refer to isospin:

$$\Delta_{\mu\nu}(x) = \langle 0|j_\mu(x)j_\nu(0)|0\rangle. \tag{90}$$

The Fourier transform of (90) is

$$\Delta_{\mu\nu}(p) = \int e^{ipx}\,\Delta_{\mu\nu}(x)\,d^4x = (2\pi)^4 \sum_n \int dR_n(p)\,\langle 0|j_\mu|n\rangle\langle n|j_\nu|0\rangle, \tag{91}$$

where we have introduced a complete set of intermediate states, the second integral in (91) is over n-particle invariant phase space, and $j_\mu \equiv j_\mu(0)$.

The currents j_μ may, in general, not be conserved: $\partial^\mu j_\mu \neq 0$. Hence the matrix elements of the current in (91) may include states $|n\rangle$ both with spin 1 and with spin zero. Specifically, for the matrix element of a current between the vacuum and a state with spin J and mass $m_J \neq 0$,

$$-m_J^2 J(J+1)\langle 0|j_\mu|m_J, J\rangle = \langle 0|[w^2, j_\mu]|m_J, J\rangle = 2\langle 0|g_{\mu\nu}\Box j_\nu - \partial_\mu\partial_\nu j^\nu|m_J, J\rangle. \quad (92)$$

For the "transverse" current j_μ^t with $\partial^\mu j_\mu^t = 0$, the matrix element in (92) is nonzero only when $J = 1$. If $[w^2, j_\mu] = 0$, i.e. for the "longitudinal" current j_ν^l, satisfying the condition $\Box j_\mu^l - \partial_\mu \partial^\nu j_\nu^l = 0$, one must have $J = 0$ in the matrix element of the current in (92).

Consequently, for a nonconserved current, the sum in (91) decomposes into two parts, in which the states $|n\rangle$ have spin 1 and spin 0:

$$\Delta_{\mu\nu}(p) = (2\pi)^4 \Big\{ \sum_n \delta^4(p-p_n)\langle 0|j_\mu^t|n, J=1\rangle\langle n, J=1|j_\nu^t|0\rangle + \sum_n \delta^4(p-p_n)\langle 0|j_\mu^l|n, J=0\rangle\langle n, J=0|j_\nu^l|0\rangle \Big\}. \quad (93)$$

Each of these parts may be expressed in terms of an invariant spectral function $\varrho^{(J)}(p^2)$, so that

$$\Delta_{\mu\nu}(p) = 2\pi\left(g_{\mu\nu} - \frac{1}{p^2}p_\mu p_\nu\right)\varrho^{(1)}(p^2)\,\theta(p_0) + \frac{2\pi}{p^2}p_\mu p_\nu \varrho^{(0)}(p^2)\,\theta(p_0). \quad (94)$$

From (93) and (94) it follows that the spectral functions $\varrho^{(1)}$ and $\varrho^{(0)}$ are nonnegative:

$$\varrho(p^2) \geqslant 0, \quad p^2 > 0, \quad p_0 > 0, \quad (95)$$

and the threshold for $\varrho(p^2)$ is the lowest mass $m_n^2 = p^2$ of the state $|n\rangle$. Returning to coordinate space, we obtain a spectral representation of the Källen–Lehmann type:

$$\Delta_{\mu\nu}(x) = g_{\mu\nu}\int dm^2 \varrho^{(1)}(m^2)\,\Delta^+(x; m^2) - \partial_\mu\partial_\nu \int \frac{dm^2}{m^2}(\varrho^{(1)}(m^2) + \varrho^{(0)}(m^2))\,\Delta^+(x, m^2), \quad (96)$$

where $\Delta^+(x, m^2)$ is the vacuum expectation value of the product of two free fields with spin 0 and mass m^2:

$$\Delta^+(x, m^2) = \frac{1}{(2\pi)^3}\int d^4p e^{-ipx}\theta(p_0)\,\delta(p^2 - m^2). \quad (97)$$

Equation (96) implies a spectral representation for the vacuum expectation value of the value of the time-ordered product of two currents:

$$\Delta_{F\mu\nu}(x) = \langle 0|T(j_\mu(x)j_\nu(0))|0\rangle = \langle 0|\theta(x_0)[j_\mu(x), j_\nu(0)]|0\rangle + \langle 0|j_\nu(0)\,j_\mu(x)|0\rangle. \quad (98$$

The result is

$$\Delta_{F\mu\nu}(x) = g_{\mu\nu}\int dm^2 \varrho^{(1)}(m^2)\,\Delta_F(x, m^2) + \int \frac{dm^2}{m^2}(\varrho^{(1)} + \varrho^{(0)})\,\partial_\mu\partial_\nu \Delta_F(x, m^2) + g_{\mu 0}g_{\nu 0}\delta^4(x)\int \frac{dm^2}{m^2}(\varrho^{(1)} + \varrho^{(0)}), \quad (99)$$

where

$$\Delta_F(x, m^2) = i\langle 0|T(\varphi(x)\varphi(0))|0\rangle$$

is the Feynman propagator for a free field with spin 0, and the last term (the "Schwinger" term) appears upon differentiating $\theta(x_0)$ and noting that

$$\partial_0(\Delta^+(x)-\Delta^+(-x))\Big|_{x_0=0} = -i\delta^3(x). \tag{100}$$

We note that only the difference between $\Delta_{F\mu\nu}$ and the Schwinger term transforms as a tensor.

The assumption of asymptotic symmetry of vacuum expectation values (99) for axial and vector currents may be formulated in the following manner.[230] Let

$$\Delta^V_{F\mu\nu}(q) = \int d^4x e^{iqx}\langle 0|T(v_\mu(x)\,v_\nu(0))|0\rangle, \tag{101}$$

$$\Delta^A_{F\mu\nu}(q) = \int d^4x e^{iqx}\langle 0|T(a_\mu(x)\,a_\nu(0))|0\rangle \tag{102}$$

be the Fourier components of the vacuum expectation values of T-products of currents, In the case of exact $SU_2\times SU_2$ symmetry, one must have $\Delta^V_{F\mu\nu}(q) = \Delta^A_{F\mu\nu}(q)$ for all values of q. Asymptotic chiral symmetry means that

$$\lim_{q^2\to\infty} g^{\mu\nu}(\tilde{\Delta}^V_{F\mu\nu}(q)-\tilde{\Delta}^A_{F\mu\nu}(q)) = 0, \tag{103}$$

where $\tilde{\Delta}_{F\mu\nu}(q)$ is the covariant part of the propagator, not containing the Schwinger term. In the spectral representation (99) for the axial vector current, it is convenient to separate out the pion contribution from $\varrho_A^{(0)}$. Bearing in mind formula (61) for $\langle 0|a_\mu|\pi\rangle$, we find that this contribution to $\Delta^A_{F\mu\nu}(q)$ is equal to

$$\frac{f_\pi^2 q_\mu q_\nu}{q^2-m_\pi^2+i\varepsilon}. \tag{104}$$

Then the difference between the vector and axial vector propagators (103) is

$$\left\{\int \frac{dm^2(\varrho_V^{(1)}-\varrho_A^{(1)})}{q^2-m^2-i\varepsilon} - q^2\int \frac{\varrho_V^{(1)}-\varrho_A^{(1)}-\varrho_A^{(0)}}{m^2(q^2-m^2+i\varepsilon)} - \frac{f_\pi^2}{q^2-m_\pi^2+i\varepsilon}\right\}, \tag{105}$$

where ϱ_V and ϱ_A are the vector and axial vector spectral densities defined in (93) and (94). In view of the conservation of the vector current and the absence of particles with zero mass, $\varrho_V^{(0)} = 0$, and $\varrho_V \equiv \varrho_V^{(1)}$ contains only contributions from particles with spin 1.

As $q^2 \to \infty$, the first term in (105) vanishes, if we consider the expression

$$C = \int dm^2(\varrho_V - \varrho_A^{(1)}).$$

to be finite.

We note that the change in the order of integration used above is possible if the Schwinger term integral is finite, or

$$\int_0^\infty dm^2 \frac{\varrho(m^2)}{m^2} < \infty.$$

The limit of the second term in (105) is easily calculated if one can interchange the order of integration and the passage to the limit. Under this assumption we obtain from (103) and (104) the first Weinberg sum rule:[(229)]

$$\int_0^\infty \frac{\varrho_V(m^2)-\varrho_A(m^2)}{m^2}\, dm^2 = f_\pi^2. \tag{106}$$

If one makes the further assumption that $C = 0$, i.e. that asymptotically not only (103) but also

$$\lim_{q^2\to\infty} q^2 \int \frac{dm^2(\varrho_V-\varrho_A^{(1)})}{q^2-m^2-i\varepsilon} = 0, \tag{107}$$

holds, the spectral densities must satisfy the second sum rule:[(229)]

$$\int_0^\infty (\varrho_V(m^2)-\varrho_A^{(1)}(m^2))\, dm^2 = 0. \tag{108}$$

Formulae (106) and (108) are restrictions imposed by asymptotic chiral $SU_2\times SU_2$ symmetry and by the assumption (107) of the rapid vanishing of the density $\varrho_V-\varrho_A^{(1)}$.

The application of the sum rules (106) and (108) is based on the assumption that only sharp maxima in the region of resonances are important in the spectral densities ϱ_A and ϱ_V. Taking into account only the ϱ-meson contribution to ϱ_V and the A_1-meson contribution to $\varrho_A^{(1)}$, and neglecting infinitesimally narrow resonances, one may set

$$\varrho_V(m^2) = G_\varrho^2\delta(m^2-m_\varrho^2), \tag{109}$$

$$\varrho_A^{(1)}(m^2) = G_{A_1}^2\delta(m^2-m_{A_1}^2), \tag{110}$$

where the constants G_ϱ and G_{A_1} may be found from the leptonic decays of ϱ and A_1 e.g. $\varrho\to e^+e^-,\ \mu^+\mu^-$:

$$\left.\begin{aligned} \langle 0|v_\mu(0)|\varrho\rangle &= \frac{\varepsilon_\mu(k_\varrho,\lambda)}{(2\pi)^{3/2}}G_\varrho, \\ \langle 0|a_\mu(0)|\mathrm{A}_1\rangle &= \frac{\varepsilon_\mu(k_{A_1},\lambda)}{(2\pi)^{3/2}}G_{A_1}. \end{aligned}\right\} \tag{111}$$

The sum rules then lead to $G_{\mathrm{A}_1}^2 = G_\varrho^2$ and to

$$G_\varrho^2\left(\frac{1}{m_\varrho^2}-\frac{1}{m_{A_1}^2}\right) = f_\pi^2, \tag{112}$$

relating the masses of the ϱ and the A_1 to the decay constants G_ϱ and f_π.

From the decay $\varrho\to 2\pi$, it is clear that G_ϱ and f_π are proportional to one another:[(231–233)]

$$G_\varrho^2 \approx 2m_\varrho^2 f_\pi^2. \tag{113}$$

Introducing this empirical value of G_ϱ^2 into (112), we obtain a formula for the ratio of mass of the ϱ and the A_1:

$$m_{\mathrm{A}_1} = \sqrt{2}m_\varrho, \tag{114}$$

which agrees well with experiment ($m_{A_1} \approx 1100$ MeV, $m_\varrho \approx 770$ MeV). In the case of exact $SU_2 \times SU_2$ symmetry, the masses m_ϱ and m_{A_1} would have to be equal.

The idea of asymptotic symmetry is easily extended to the chiral group $SU_3 \times SU_3$.

The vector strange currents are conserved only in the limit of SU_3. Hence if one allows for the effects of SU_3 violation (which are comparable with the effects of chiral $SU_2 \times SU_2$ violation; see § 13.3), one should consider the currents $v_{\mu a}$ $(a = 4, 5, 6, 7)$ to be nonconserved. The scalar part of the spectral density $\varrho_V^{(0)}$ will then be nonzero. If one assumes that the contribution to $\varrho_V^{(0)}$ is connected with scalar K-mesons ($\varkappa$-mesons), $J^P = 0^+$, then, for example,

$$\langle 0 | v_{\mu 4} + i v_{\mu 5} | q; \varkappa^+ \rangle = \frac{F_\varkappa q_\mu}{(2\pi)^{3/2}}. \tag{115}$$

The nonconservation of the axial currents $a_{\mu a}$ $(a = 4, 5, 6, 7)$ is related (via PCAC) to the existence of K-mesons:

$$\langle 0 | a_{\mu 4} + i a_{\mu 5} | q; \mathrm{K}^+ \rangle = \frac{F_{\mathrm{K}} q_\mu}{(2\pi)^{3/2}}. \tag{116}$$

Thus in the case of asymptotic $SU_3 \times SU_3$, in addition to the difference (103) one should also consider the difference

$$\Delta^{\mathrm{K}^*}_{F\mu\nu}(q) - \Delta^{\mathrm{K}_A}_{F\mu\nu}(q) \equiv F(q^2) g_{\mu\nu} + G(q^2) q_\mu q_\nu, \tag{117}$$

$$F(q^2) = \int dm^2 \frac{\varrho_{\mathrm{K}^*}(m^2) - \varrho_{\mathrm{K}_A}(m^2)}{q^2 - m^2 - i\varepsilon}, \tag{118}$$

$$G(q^2) = \int dm^2 \frac{\varrho_{\mathrm{K}^*}(m^2) - \varrho_{\mathrm{K}_A}(m^2)}{m^2(q^2 - m^2 - i\varepsilon)} + \frac{F_\varkappa^2}{q^2 - m_\varkappa^2 - i\varepsilon} - \frac{F_{\mathrm{K}}^2}{q^2 - m_{\mathrm{K}}^2 - i\varepsilon}, \tag{119}$$

where ϱ_{K^*} is the vector density, ϱ_{K_A} is the axial vector density, and the contribution of the scalar densities is limited to the $\varkappa$ and K terms in (119).

The assumption that the difference (117) vanishes asymptotically as $q^2 \to \infty$ gives the first Weinberg sum rule:

$$\int dm^2 \frac{\varrho_{\mathrm{K}^*}(m^2) - \varrho_{\mathrm{K}_A}(m^2)}{m^2} = F_{\mathrm{K}}^2 - F_\varkappa^2. \tag{120}$$

The second sum rule, which also does not agree badly with experiment, may be derived[234] under the additional assumption that $q^2 F(q^2)$ becomes a constant in the limit $q^2 \to \infty$ while $q^2(g^{\mu\nu} q_\mu q_\nu G(q^2))$ vanishes asymptotically. Then, from (119) and the first sum rule (120), the second sum rule follows:

$$\int_0^\infty dm^2 \left(\varrho_{\mathrm{K}^*}(m^2) - \varrho_{\mathrm{K}_A}(m^2)\right) = m_{\mathrm{K}}^2 F_{\mathrm{K}}^2 - m_\varkappa^2 F_\varkappa^2. \tag{121}$$

The first resonances contributing to the spectral densities ϱ_{K^*} and ϱ_{K_A} are $\mathrm{K}^*(J^P = 1^-$, $m^{\mathrm{K}^*} \approx 892$ MeV) and $\mathrm{K}_A(J^P = 1^+$, $m_{\mathrm{K}_A} \approx 1300$ MeV). The scalar particle $\varkappa$ has a mass $m_\varkappa \approx 1100$–1400 MeV and a width of several hundred MeV, but has not been identified conclusively.[8, 235]

If chiral $SU_3 \times SU_3$ becomes exact at large energies, SU_3 also becomes exact. We shall thus assume that as $q^2 \to \infty$ the difference between the following propagators vanishes:[(230)]

$$\lim_{q^2 \to \infty} [\tilde{\Delta}^V_{F\mu\nu}(q) - \tilde{\Delta}^{K^*}_{F\mu\nu}(q)] = 0, \tag{122}$$

where the intermediate states in Δ^V and Δ^{K^*} are the particles ϱ and K^* of the same octet V, for example

$$\Delta^{K^*}_{F\mu\nu}(q) = \int d^4x e^{-iqx} \langle 0 | T(v_{\mu 4-i5}(x)\, v_{\nu 4+i5}(0)) | 0 \rangle. \tag{123}$$

The condition (122) is equivalent to the following relation for spectral densities:

$$\int_0^\infty \frac{\varrho^V(m^2) - \varrho^{K^*}(m^2)}{m^2} dm^2 = 0. \tag{124}$$

If only the first resonances ϱ and K^* contribute to the functions ϱ^V and ϱ^{K^*}:

$$\varrho^V(m^2) = G_\varrho^2 \delta(m^2 - m_\varrho^2), \quad \varrho^{K^*}(m^2) = G_{K^*}^2 \delta(m^2 - m_{K^*}^2),$$

we find a relation between the decay constants G_ϱ and G_{K^*} and masses:

$$G_\varrho^2 m_{K^*}^2 = m_\varrho^2 G_{K^*}^2. \tag{125}$$

Relations of the type (114) and (125), arising in the one-resonance approximation to spectral densities, are, of course, very rough in nature. The fact that they hold fairly well indicates the dominant role of the low-energy region (and the resonance peaks) in integrals over masses. Other applications of current algebra may be found in various reviews.[(16, 236–239)]

§ 15.6. Violation of CP invariance

The neutral kaons K^0 and $\bar{K}^0$ decay in the form of two particles—the "long-lived" kaon K_L with lifetime $\tau_L \cong 5 \times 10^{-8}$ sec and the "short-lived" kaon K_S with lifetime $\tau_S \cong 0.9 \times 10^{-10}$ sec (see Table A.1, p. 359). In a CP-invariant theory the particles K_S and K_L have definite CP parity and are described by the respective combinations K_1^0 and K_2^0 (see § 6.3):

$$\left. \begin{aligned} | K_1^0 \rangle &= -\frac{i}{\sqrt{2}} \{ | K^0 \rangle - | \bar{K}^0 \rangle \}, \\ | K_2^0 \rangle &= \frac{1}{\sqrt{2}} \{ | K^0 \rangle + | \bar{K}^0 \rangle \}, \\ CP | K^0_{\substack{1\\2}} \rangle &= \pm | K^0_{\substack{1\\2}} \rangle, \end{aligned} \right\} \tag{126}$$

if the phase is chosen so that $CP | K^0 \rangle = - | \bar{K}^0 \rangle$. All kaon states in (126) and in the rest of this section will be considered in their rest frame.

CP symmetry of a theory is equivalent to the presence of a selection rule with respect to CP parity. In particular, the particle K_2^0 (identified with K_L) cannot decay into two pions, since the two-pion state (in the c.m.s.) has positive CP parity: $CP | \pi^+ \pi^- \rangle = + | \pi^+ \pi^- \rangle$.

In 1964 Christenson, *et al.* (see ref. 82) discovered the decay $K_L \to 2\pi$ and thus showed that CP invariance was only an approximate property of the weak interactions. To avoid misunderstandings we shall call any interaction which violates CP symmetry superweak (not associating this name, however, with any particular model of CP violation). Further experiments (see refs. 240, 244) confirmed the existence of CP-noninvariant $K_L \to 2\pi$ decays. CP-noninvariant lepton decays $K_L \to \pi^{\pm} l^{\mp} \nu$ have also been observed.[241, 242, 244]

Although many hypotheses were advanced to explain CP noninvariance, a unique theory of the superweak interaction still has not been constructed. This is partly due to the fact that the known experimental data still do not allow one to determine all the necessary parameters of CP violation. To present clearly the problem any future theory must face, we shall consider in this section a phenomenological analysis of CP-noninvariant decays.[15, 243]

Parameters

In experiments on two-pion K-decays the following quantities are measured:

$$\eta_{+-} \equiv |\eta_{+-}| e^{i\varphi_{+-}} = \frac{\langle \pi^+\pi^- | T | K_L \rangle}{\langle \pi^+\pi^- | T | K_S \rangle}, \tag{127}$$

$$\eta_{00} \equiv |\eta_{00}| e^{i\varphi_{00}} = \frac{\langle \pi^0\pi^0 | T | K_L \rangle}{\langle \pi^0\pi^0 | T | K_S \rangle}. \tag{128}$$

Here $\langle \pi\pi | T | K \rangle$ is the amplitude for the decay of a kaon at rest into two pions. The quantities η_{+-} and η_{00} are parameters of CP violation; under CP symmetry $\eta_{+-} = \eta_{00} = 0$.

Instead of two-pion states with a given type of particles, it is more convenient to use states with definite isospin. Two pions in a state with total spin equal to zero can only be in the (symmetric) isospin states with $I = 0$ and $I = 2$. Thus (see § 8.2)

$$\left.\begin{aligned} |\pi^+\pi^-\rangle &= \sqrt{\tfrac{2}{3}}\,|0\rangle + \sqrt{\tfrac{1}{3}}\,|2\rangle, \\ |\pi^0\pi^0\rangle &= \sqrt{\tfrac{1}{3}}\,|0\rangle - \sqrt{\tfrac{2}{3}}\,|2\rangle, \end{aligned}\right\} \tag{129}$$

where $|I\rangle$ denotes the two-pion state (in the c.m.s.) with isospin $I = 0, 2$ and $I_3 = 0$.

After inserting (129) into (127) and (128) it is best to pass to other (theoretical) parameters of CP violation ε_0 and ε_2, which appear in the theoretical analysis of K_L decays:

$$\varepsilon_0 = \frac{\langle 0 | T | K_L \rangle}{\langle 0 | T | K_S \rangle}, \quad \varepsilon_2 = \frac{1}{2} \frac{\langle 2 | T | K_L \rangle}{\langle 0 | T | K_S \rangle}. \tag{130}$$

The connection between the parameters η_{+-}, η_{00} and ε_0, ε_2 is given by

$$\left(1 + \frac{1}{\sqrt{2}}\omega\right)\eta_{+-} = \varepsilon_0 + \varepsilon_2, \quad (1 - \sqrt{2}\omega)\eta_{00} = \varepsilon_0 - 2\varepsilon_2, \tag{131}$$

where ω characterizes the isospin content of the two-pion channel for the decay of K_S:

$$\omega = \frac{\langle 2 | T | K_S \rangle}{\langle 0 | T | K_S \rangle}. \tag{132}$$

The numerator in (132) refers to the decay of K_S with $\Delta I = \frac{3}{2}$, while the denominator refers to the decay of K_S with $\Delta I = \frac{1}{2}$. The parameter ω thus describes the violation of the $\Delta I = \frac{1}{2}$ rule.

Information on ω may be obtained from experiments on K_S decays. Experimentally one may define the quantity

$$R = \frac{\Gamma(K_S \to \pi^0\pi^0)}{\Gamma(K_S \to \pi^0\pi^0)+\Gamma(K_S \to \pi^+\pi^-)} = \frac{1}{3}\,\frac{|1-\sqrt{2}\omega|^2}{1+|\omega|^2} \tag{133}$$

($\Gamma(a \to b)$ is the partial width for the decay of particle into the state b). If the rule $\Delta I = \frac{1}{2}$ were to hold ($\omega = 0$), then one would have $R = \frac{1}{3}$. Experimentally, $R = 0.312 \pm 0.003$; this allows one to assume that $|\omega^2| \ll 1$, and we find $\text{Re}\,\omega \cong 2\times 10^{-2}$.

Consequently,

$$\eta_{+-} \cong \varepsilon_0+\varepsilon_2, \quad \eta_{00} \cong \varepsilon_0 - 2\varepsilon_2. \tag{134}$$

The latest experimental data are:[244]

$$\begin{aligned} |\eta_{+-}| &= (2.279\pm 0.025)\times 10^{-3}, \\ \varphi_{+-} &= (45.0\pm 1.3)^\circ, \\ |\eta_{00}| &= (2.30\pm 0.10)\times 10^{-3}, \\ \varphi_{00} &= (48.3\pm 13.0)^\circ. \end{aligned}$$

States of K_L *and* K_S *mesons*

Let us turn to the theoretical description of K_L decays. The neutral kaons K^0 and $\overline{K}^0$ are produced in the strong interactions, so that the states $|K^0\rangle$ and $|\overline{K}^0\rangle$ are eigenstates of the strong interaction hamiltonian H_0+H_S. The particles K^0 and $\overline{K}^0$ possess only hypercharge ($Y = +1$ and $Y = -1$); the remaining charges of these particles are equal to zero. With the "inclusion" of the weak H_w and "superweak" H_{sw} interactions, the particles K^0 and $\overline{K}^0$ become unstable and decay. The interaction $H_w + H_{sw}$ does not conserve hypercharge, and, by connecting K^0 and $\overline{K}^0$ with a continuous spectrum of decay states, leads to the transitions $K^0 \leftrightarrow \overline{K}^0$. The existence of such transitions means that it is not the K^0 and $\overline{K}^0$ particles but their superpositions $|K_L\rangle$ and $|K_S\rangle$ which decay:

$$|K_L\rangle = p|K^0\rangle - q|\overline{K}^0\rangle, \quad |K_S\rangle = r|K^0\rangle + s|\overline{K}^0\rangle. \tag{135}$$

In (135), all states are relative to the c.m.s. The complex coefficients p, q, r, s satisfy the normalization conditions

$$|p^2|+|q^2| = |r|^2+|s|^2 = 1.$$

To see which superpositions (135) describe the K_L and K_S particles, we consider the effective transition matrix $\hat{T}$ between the states K^0 and $\overline{K}^0$. (The usual Hamiltonian H_w has no matrix elements between K^0 and $\overline{K}^0$ states: $\langle K^0|H_w|K^0\rangle = \langle \overline{K}^0|H_w|\overline{K}^0\rangle = \langle K^0|H_w|\overline{K}^0\rangle = 0$.) The matrix $\hat{T}$ acts in the space of (K^0, $\overline{K}^0$)-states (in the c.m.s.):

$$\hat{T} = \begin{pmatrix} \langle K^0|T|K^0\rangle & \langle K^0|T|\overline{K}^0\rangle \\ \langle \overline{K}^0|T|K^0\rangle & \langle \overline{K}^0|T|\overline{K}^0\rangle \end{pmatrix} \equiv \begin{pmatrix} M & A \\ B & \overline{M} \end{pmatrix}; \tag{136}$$

the explicit form of the transition amplitude $\langle \ldots | T | \ldots \rangle$ is unimportant for the moment. We only take account of the fact that the theory must be invariant with respect to the total reflection θ = CPT. Then (see § 6.1) the diagonal elements in (136) must be equal:

$$\langle K^0 | T | K^0 \rangle = \langle \overline{K}^0 | T | \overline{K}^0 \rangle,$$

or $M = \overline{M}$.

The physical states $|K_L\rangle$ and $|K_S\rangle$, which decay as a result of the interaction $H_w + H_{sw} = H'$, are distinguished from other possible superpositions of $|K^0\rangle$ and $|\overline{K}^0\rangle$ states by the condition that the K_L and K_S particles have definite mass and lifetime. In other words, the states (135) must diagonalize the matrix (136).

The diagonalization of the matrix (136) is elementary. The complex eigenvalues of the matrix (136) are equal to

$$\lambda = M \pm \sqrt{AB},$$

and the coefficients in (135) must satisfy the relations

$$|s|^2 = |q|^2 = \frac{|B|}{|A|+|B|}, \quad |p|^2 = |r|^2 = \frac{|A|}{|B|+|A|},$$

$$r = \sqrt{\frac{A}{B}}\, s, \quad p = \sqrt{\frac{A}{B}}\, q.$$

Choosing the relative phase factors s and q so that $s = q$,

$$|K_L\rangle = p|K^0\rangle - q|\overline{K}^0\rangle, \quad |K_S\rangle = p|K^0\rangle + q|\overline{K}^0\rangle. \tag{137}$$

Since the states $|K_L\rangle$ and $|K_S\rangle$ are normalized, they depend on only one complex parameter p/q. Instead of p/q one often uses the parameter

$$\varepsilon = \frac{p-q}{p+q}, \tag{138}$$

whose deviation from the value $\varepsilon = 0$ is related to CP asymmetry. Indeed, if the interaction $H' = H_w + H_{sw}$ were CP symmetric (i.e. if $A = B$), then one would have $|p|^2 = |q|^2 = \frac{1}{2}$, and in this case the states $|K_S\rangle = |K_1^0\rangle$ and $|K_L\rangle = |K_2^0\rangle$ would be orthogonal (see § 6.3). But the scalar product

$$\langle K_L | K_S \rangle = |p|^2 - |q|^2 = \frac{2\,\mathrm{Re}\,\varepsilon}{1+|\varepsilon|^2} \tag{139}$$

is determined by the parameter ε, which thus characterizes the degree of violation of CP symmetry. We note that, according to (139), the quantity $\langle K_L | K_S \rangle$ is real (in a CPT-invariant theory).

The unitarity condition

Since the particles K_L and K_S have definite masses and lifetimes, they behave under a time translation as

$$U(\tau, 0)|K_S\rangle = e^{-iM_S\tau}|K_S\rangle, \quad U(\tau, 0)|K_L\rangle = e^{-iM_L\tau}|K_L\rangle. \tag{140}$$

The complex quantities M_S and M_L are the well-known combinations of mass m and decay width Γ:

$$M_{S,L} = m_{S,L} - \frac{i}{2}\Gamma_{S,L}. \tag{141}$$

The formulae (140) assume that the exponential decay law holds.

Let $|\Psi(\tau)\rangle$ be a nonstationary state which can connect with the state $|f\rangle$. According to the unitarity condition, the rate of decrease of the norm of the state $|\Psi\rangle$ is equal to the probability for the transition from $|\Psi\rangle$ to any possible state $|f\rangle$:

$$-\left.\frac{d\langle\Psi|\Psi\rangle}{d\tau}\right|_{\tau=0} = \sum_f |\langle f|T|\Psi\rangle|^2. \tag{142}$$

Let us choose the state $|\Psi(\tau)\rangle$ to be an arbitrary superposition of $|K_S(\tau)\rangle$ and $|K_L(\tau)\rangle$:

$$|\Psi(\tau)\rangle = ae^{-iM_S\tau}|K_S\rangle + be^{-iM_L\tau}|K_L\rangle.$$

Inserting this expression into (142), we find, comparing coefficients of the independent combinations $a\,b^*$, a^*b, etc.,

$$\Gamma_S = \sum_f |\langle f|T|K_S\rangle|^2, \tag{143}$$

$$\Gamma_L = \sum_f |\langle f|T|K_L\rangle|^2, \tag{144}$$

$$i(M_L^* - M_S)\langle K_L|K_S\rangle = \sum_f \langle f|T|K_L\rangle^* \langle f|T|K_S\rangle. \tag{145}$$

Equation (145) allows one to obtain an upper bound on the value of $\langle K_L|K_S\rangle$. Using the triangle inequality, we find[245]

$$|M_L^* - M_S|\langle K_L|K_S\rangle \leqslant (\Gamma_L\Gamma_S)^{1/2}. \tag{146}$$

Now the mass difference $m_L - m_S$ and the widths Γ_L, Γ_S are known from experiment:

$$m_L - m_S = (0.479 \pm 0.002)\Gamma_S, \quad \Gamma_L = 1.7\times 10^{-3}\Gamma_S. \tag{147}$$

The bound (146) then gives

$$\langle K_L|K_S\rangle \leqslant 0.06,$$

showing that the violation of CP symmetry must be relatively small. Experimentally, the charge asymmetry in the decays $K_L \to \pi^{\pm} l^{\mp}\nu$ (l is a lepton e or μ) shows that $\langle L|S\rangle \approx 10^{-3}$. Consequently, formula (139) may be rewritten in the approximate form:

$$\langle K_L|K_S\rangle = \begin{cases} 2\,\mathrm{Re}\,\varepsilon, \\ 3.3\times 10^{-3}, \end{cases} \tag{148}$$

neglecting quantities of order ε^2.

Let us turn to the study of the unitarity relations (145). If one separates out the most important terms in the sum over the states $|f\rangle$, one obtains an approximate unitarity relation of practical significance. For an energy equal to the mass of the K-meson, such states will be the states $|\pi\pi\rangle$, $|\pi l\nu\rangle$, $|\pi\pi\pi\rangle$ (here $l = e, \mu$). The states with photons, of the type

$|\pi\pi\gamma\rangle$, $|\pi\pi\gamma\gamma\rangle$, etc., may be discarded, since their total contribution will be roughly $\alpha = \frac{1}{137}$ times smaller than that of the two-pion contribution.

The two-pion part of the sum over $|f\rangle$ in (145) is easily expressed in terms of the parameters η_{+-}, η_{00}, and R [eqns. (127), (128), (133)] and the probability Γ_S for the two-pion decay of K_S [formula 143)]:

$$\sum_{(2\pi)} \langle 2\pi | T | K_S\rangle^* \langle 2\pi | T | K_L\rangle = [(1-R)\eta_{+-} + R\eta_{00}]\Gamma_S = 10^{-3}\Gamma_S.$$

The numerical estimate of this expression was found by substituting the experimental data for R, η_{+-}, and η_{00}.

An upper bound on the contribution of the three-pion states $|f\rangle = |3\pi\rangle$ in (145) is easily established if one uses the triangle inequality and the experimental values for the three-pion widths:

$$\Gamma(K_L \to 3\pi) = 6\times 10^{-4}\Gamma_S, \quad \Gamma(K_S \to 3\pi) < 10^{-4}\Gamma_S.$$

Then

$$\sum_{(3\pi)} \langle 3\pi | T | K_S\rangle^* \langle 2\pi | T | K_L\rangle \leq (\Gamma(K_S \to 3\pi)\Gamma(K_L \to 3\pi))^{1/2} \approx 2.5\times 10^{-4}\Gamma_S,$$

and, consequently, the three-pion states do not contribute significantly to (145).

To evaluate the contribution of the lepton states $|\pi l \nu\rangle$, let us assume that the $\Delta Q = \Delta Y$ rule holds in decays of K^0 and $\bar{K}^0$ mesons. Then only the (hypothetical) decays $K^0 \to \pi^- l^+ \nu_l$, and $\bar{K}^0 \to \pi^+ l^- \bar{\nu}_l$ are possible, so that

$$\sum_{(l\pi\nu)} \langle l\pi\nu | T | K^0\rangle^* \langle l\pi\nu | T | \bar{K}^0\rangle = 0.$$

Here one sums over the leptons $l = e^{\pm}$, $\mu^{\pm}$ and all possible $(l\pi\nu)$ -states. On the other hand, the sum

$$\sum_{(l\pi\nu)} |\langle l\pi\nu | T | K^0\rangle|^2 = \Gamma(K^0 \to l\pi\nu)$$

is the probability of the leptonic decay K^0. By virtue of the CPT invariance of the theory, the probability $\Gamma(K^0 \to l\pi\nu)$ is equal to the probability for the decay $\Gamma(\bar{K}^0 \to \bar{l}\bar{\pi}\bar{\nu})$ of an antiparticle $\bar{K}^0$ into the antiparticles $\bar{l}\bar{\pi}\bar{\nu}$.

Let us now insert the expressions (135), (139) for $|K_L\rangle$ and $|K_S\rangle$ into the lepton terms of the unitarity relation (145). We write

$$\Gamma(K_{S,L} \to (l)) = \sum_{l=e,\,\mu} [\Gamma(K_{S,L} \to \pi^- l^+ \nu) + \Gamma(K_{S,L} \to \pi^+ l^- \bar{\nu})].$$

It is easy to check using (135) and (139) that

$$\Gamma(K_S \to (l)) = \Gamma(K_L \to (l)) = \Gamma(K^0 \to l^+\pi^-\nu),$$

$$\sum_{(l\pi\nu)} \langle l\pi\nu | T | K_S\rangle^* \langle l\pi\nu | T | K_L\rangle = \langle K_L | K_S\rangle \Gamma(K^0 \to l^+\pi^-\nu) \approx 10^{-6}\Gamma_S.$$

For a numerical estimate we have used the previous model (148) for $\langle K_L | K_S\rangle$ and the experimental value $\Gamma(K_L \to (l)) \approx 10^{-3}\Gamma_S$. Hence the lepton terms of the sum in (145) are negligibly small in comparison with the main (two-pion) terms.

Thus on the right-hand side of the unitarity relation (145) one may omit all terms in the sum over $|f\rangle$ except the two-pion ones. On the left-hand side of (145) one may neglect Γ_L in comparison with Γ_S. Moreover, we shall set $R = \frac{1}{3}$ instead of the experimental value $R \approx 0.31$ (i.e. we shall neglect transitions with $\Delta I = \frac{3}{2}$). The approximate unitarity relation may then be written in the form

$$i\left(m_S - m_L - \frac{i}{2}\Gamma_S\right)\langle \mathrm{K}_S | \mathrm{K}_L\rangle = \frac{1}{3}\Gamma_S(2\eta_{+-}+\eta_{00}) = \varepsilon_0 \Gamma_S. \tag{149}$$

But $\langle \mathrm{K}_S | \mathrm{K}_L\rangle$ is real, allowing one to find the phase of the parameter ε_0:

$$\frac{\mathrm{Im}\ \varepsilon_0}{\mathrm{Re}\ \varepsilon_0} = \frac{2(m_S - m_L)}{\Gamma_S} \approx 0.96, \tag{150}$$

since the right-hand side contains quantities directly observable in experiment.

Relations between CP-*violating parameters*

Now that one knows the states $|\mathrm{K}_S\rangle$ and $|\mathrm{K}_L\rangle$, one can find the relations between experimental and theoretical parameters of CP violation.

Inserting eqns. (135) into the definition of the parameters (130) and (132) introduces the transition amplitudes $\langle I|T|\mathrm{K}^0\rangle$ and $\langle I|T|\overline{\mathrm{K}}^0\rangle$, labeled by the total isospin I of the two-pion system. By virtue of the invariance of the theory with respect to the reflection $\theta = $ CPT, these amplitudes are related to one another.

Let us find this relation.

To first order in perturbation theory (with respect to $H'(x^0) = H_{\mathrm{w}}(x^0)+H_{\mathrm{sw}}(x^0)$), the transition amplitude is equal to the corresponding matrix element of $H'(0)$. As a consequence of θ-invariance, $\theta H'(0)\theta^{-1} = H'(0)$. We then have (see §§ 6.1 and 6.4)

$$\begin{aligned}\langle I|T|\mathrm{K}^0\rangle &= \langle I(\pi\pi), \text{out}\,|H'(0)|\mathrm{K}^0\rangle = \langle \theta I(\pi,\pi), \text{out}\,|\theta H'(0)|\mathrm{K}^0\rangle^* \\ &= \langle I(\pi\pi), \text{in}\,|H'(0)|\overline{\mathrm{K}}^0\rangle^* = \langle \overline{\mathrm{K}}^0|H'(0)|I, \text{in}\rangle.\end{aligned}$$

The matrix element on the right-hand side may be expanded in terms of a complete set of states $|n, \text{out}\rangle$. For an energy equal to the energy of the kaon at rest, only the two-pion state can contribute appreciably, since it is the only possible state allowed under the strong interactions alone. The three-pion states $|\pi\pi\pi, \text{out}\rangle$ do not contribute if G-parity is conserved. Electromagnetic transitions to the states $|\pi\pi\gamma, \text{out}\rangle$ may be neglected in view of the smallness of their contribution. Then one may set

$$\langle I(\pi\pi), \text{out}\,|I(\pi\pi), \text{in}\rangle = e^{2i\delta_I},$$

where δ_I is the phase of elastic pion–pion scattering for an energy $E = m_{\mathrm{K}}$ (in the c.m.s.) in a state with isospin I. Consequently,

$$\langle I|T|\mathrm{K}^0\rangle = \langle I|T|\overline{\mathrm{K}}^0\rangle^* e^{2i\delta_I}.$$

This formula allows one to separate out the effects of the strong interactions (the phase δ_I)

and to introduce the decay amplitudes A_0, A_2 characterizing the weak interactions:

$$\langle 0|T|\mathrm{K}^0\rangle = ie^{i\delta_0}A_0\,, \quad \langle 0|T|\overline{\mathrm{K}}{}^0\rangle = -ie^{i\delta_0}A_0^*\,, \tag{151}$$

$$\langle 2|T|\mathrm{K}^0\rangle = ie^{i\delta_2}A_2, \quad \langle 2|T|\overline{\mathrm{K}}{}^0\rangle = -ie^{i\delta_2}A_2^*\,. \tag{152}$$

The strongly interacting particle states $|\mathrm{K}^0\rangle$ and $|\overline{\mathrm{K}}{}^0\rangle$ are defined up to a phase $e^{i\gamma_\alpha}$. Let us redefine these states so that the amplitude A_0 is real (the Wu-Yang[246] convention):

$$\mathrm{Im}\, A_0 = 0.$$

(We thus no longer take $\mathrm{CP}|\mathrm{K}^0\rangle = -|\overline{\mathrm{K}}{}^0\rangle$.)

For this set of phases, the parameter ε_0 coincides with the parameter ε characterizing the nonorthogonality of the states $|\mathrm{K}_L\rangle$ and $|\mathrm{K}_S\rangle$:

$$\varepsilon_0 = \frac{\langle 0|T|\mathrm{K}_L\rangle}{\langle 0|T|\mathrm{K}_S\rangle} = \frac{pA_0-qA_0^*}{pA_0+qA_0^*} = \frac{p-q}{p+q} = \varepsilon.$$

The expressions for the parameters ε_2 and ω take the form

$$\varepsilon_2 = \frac{1}{\sqrt{2}}\frac{1}{A_0}(\varepsilon\,\mathrm{Re}\,A_2+i\,\mathrm{Im}\,A_2)e^{i(\delta_2-\delta_0)} \approx \frac{i}{\sqrt{2}}\frac{\mathrm{Im}\,A_2}{A_0}e^{i(\delta_2-\delta_0)}, \tag{153}$$

$$\omega = \frac{1}{A_0}(\mathrm{Re}\,A_2+i\varepsilon\,\mathrm{Im}\,A_2)e^{i(\delta_2-\delta_0)} \approx \frac{\mathrm{Re}\,A_2}{A_0}e^{i(\delta_2-\delta_0)}. \tag{154}$$

In passing to the approximate formulae for ε_2 and ω, we used the smallness of the quantity $|\varepsilon| \approx 10^{-3}$ [see (148)]. The parameter ε depends only on the states $|\mathrm{K}_L\rangle$ and $|\mathrm{K}_S\rangle$, but not on the decay amplitudes A_0 and A_2. The real part of ε was determined from experiments on the charge asymmetry of the leptonic decays of K_L, while the imaginary part of ε may now be estimated from the unitarity condition (150) and the fact that $\varepsilon = \varepsilon_0$.

The parameter ε_2 describes CP violation in the decay amplitudes, while the parameter ω describes the breaking of the $\Delta I = \frac{1}{2}$ rule.

As is clear from (153), if the rule $\Delta I = \frac{1}{2}$ holds, so that $A_2 = 0$, CP violation will be related only to properties of the states K_L and K_S (the parameter ε). In this case the parameters η_{+-} and η_{00} are equal: $\eta_{+-} = \eta_{00}$. Present experiments are consistent with this situation.

Translator's note to Chapter 15

The recent observation of weak neutral currents in neutrino reactions[247] indicates that the V—A theory mentioned in this chapter is most likely only a low-energy limit of a more general theory which has the potential of unifying the weak and electromagnetic interactions. The neutral weak currents do not appear to be of the V—A form. These new and exciting developments do not obviate the considerations of the present chapter.

APPENDIX

TABLE A.1[8]

Particle	$I^G(J^P)C$	Mass (MeV)	Mean life (sec)	Major decay channels, (%)		
γ	0, $1(1^-)^-$	$0(<2)10^{-21}$	Stable			
ν_e	$J=\frac{1}{2}$	0(< 60 eV)	Stable			
ν_μ		0(< 1.2)				
e	$J=\frac{1}{2}$	0.5110034 ±0.0000014	Stable (> 2×10^{21} *y*)			
μ	$J=\frac{1}{2}$	105.65948 ±0.00035	2.1994×10^{-6} ±0.0006	$e\nu\bar{\nu}$	100	
$\pi^\pm$	$1^-(0^-)$	139.5688 ±0.0064	2.6030×10^{-8} ±0.0023	$\mu\nu$	~ 100	
π^0	$1^-(0^-)^+$	134.9645 ±0.0074	0.84×10^{-16} ±0.10	$\gamma\gamma$	98.83	±0.05
				γe^+e^-	1.17	±0.05
$K^\pm$	$\frac{1}{2}(0^-)$	493.707 ±0.037	1.2371×10^{-8} ±0.0026	$\mu\nu$	63.54	±0.19
				$\pi\pi^0$	21.12	±0.17
				$\pi\pi^-\pi^+$	5.59	±0.03
				$\pi\pi^0\pi^0$	1.73	±0.05
				$\mu\pi^0\nu$	3.20	±0.09
				$e\pi^0\nu$	4.82	±0.05
K^0	$\frac{1}{2}(0^-)$	497.70 ±0.13	50% K_S, 50% K_L			
K_S^0	$\frac{1}{2}(0^-)$		0.886×10^{-10} ±0.008	$\pi^+\pi^-$	68.77	±0.26
				$\pi^0\pi^0$	31.19	
K_L^0	$\frac{1}{2}(0^-)$		5.181×10^{-8} ±0.041	$\pi^0\pi^0\pi^0$	21.3	±0.6
				$\pi^+\pi^-\pi^0$	11.9	±0.4
				$\pi\mu\nu$	27.5	±0.5
				$\pi e\nu$	39.0	±0.6
				$\pi e\nu\gamma$	1.3	±0.8
				$\pi^+\pi^-$	0.177	±0.018
				$\pi^0\pi^0$	0.093	±0.019
η	$0^+(0^-)^+$	548.8 ±0.6	$\Gamma=(2.63\pm0.58)$ keV	$\gamma\gamma$	38.0	±1.0
			Neutral decays 71.1%	$\pi^0\gamma\gamma$	3.1	±1.1
				$3\pi^0$	30.0	±1.1
			Charged decays 28.9%	$\pi^+\pi^-\pi^0$	23.9	±0.6
				$\pi^+\pi^-\gamma$	5.0	±0.1
p	$\frac{1}{2}(\frac{1}{2}^+)$	938.2796 ±0.0027	Stable (> 2×10^{28} *y*)			

Table A.1 (continued)

Particle	$I^G(J^P)C$	Mass (MeV)	Mean life (sec)	Major decay channels, (%)	
n	$\frac{1}{2}(\frac{1}{2}^+)$	939.5731 $\pm$0.0027	918$\pm$14	$pe^-\nu$	100
Λ	$0(\frac{1}{2}^+)$	1115.60$\pm$0.05	2.578×10^{-10} $\pm$0.021	$n\pi^0$	$\begin{matrix}(64.2 \\ (35.8\end{matrix}\pm0.5$
Σ^+	$1(\frac{1}{2}^+)$	1189.37$\pm$0.06	0.800×10^{-10} $\pm$0.006	$p\pi^0$ $n\pi^+$	51.6$\pm$0.7 48.4$\pm$0.7
Σ^0	$1(\frac{1}{2}^+)$	1192.48$\pm$0.10	$< 1.0\times10^{-14}$	$\Lambda\gamma$	$\approx$ 100
Σ^-	$1(\frac{1}{2}^+)$	1197.35$\pm$0.06	1.482×10^{-10} $\pm$0.017	$n\pi^-$	$\approx$ 100
Ξ^0	$\frac{1}{2}(\frac{1}{2}^+)$	1314.9 $\pm$0.6	2.96×10^{-10} $\pm$0.12	$\Lambda\pi^0$	100
Ξ^-	$\frac{1}{2}(\frac{1}{2}^+)$	1321.29$\pm$0.14	1.652×10^{-10} $\pm$0.023	$\Lambda\pi^-$	$\approx$ 100
Ω^-	$0(\frac{3}{2}^+)$	1672.2 $\pm$0.4	$1.3^{+0.3}_{-0.2}\times10^{-10}$	$\Xi^0\pi^-$ $\Xi^-\pi^0$ ΛK^-	

TABLE A.2

Particle	$I^G(J^P)C$	Mass M (MeV)	Full width Γ (MeV)	Major decay channels (%)	
$\pi^\pm$ (140) π^0 (135)	$1^-(0^-)+$	139.57 134.96	0.0 7.8 eV $\pm$0.9 eV	See Table A. 1	
η (549)	$0^-(0^-)+$	548.8 $\pm$0.6	2.63 KeV $\pm$0.58 KeV	All neutral $\pi^+\pi^-\pi^0+\pi^+\pi^-\gamma$ See Table A.1	71 29
ε	$0^+(0^+)+$	$\lesssim$ 700	$\gtrsim$ 600	$\pi\pi$	
ϱ (770)	$1^+(1^-)-$	770 $\pm$10	150 $\pm$10	$\pi\pi$	$\approx$ 100
ω (783)	$0^-(1^-)-$	782.7 $\pm$0.6	10 $\pm$0.4	$\pi^+\pi^-\pi^0$ $\pi^+\pi^-$ $\pi^0\gamma$	90.0$\pm$0.6 1.3$\pm$0.3 8.7$\pm$0.5
η' or χ^0 (958)	$0^+(0^-)+$	957.6 $\pm$0.3	$<$ 1	$\eta\pi\pi$ $\pi^+\pi^-\gamma$ (mainly $\varrho^0\gamma$) $\gamma\gamma$	70.6$\pm$2.5 27.4$\pm$2.2 1.9$\pm$0.3
δ (970)	$1^-(0^+)+$	~ 976 $\pm$10	50 $\pm$20	$\eta\pi$	
S* (993)	$0^+(0^+)+$	~ 993 $\pm$5	40$\pm$8	$\pi\pi$ $K\bar{K}$	 dominant
Φ (1019)	$0^-(1^-)-$	1019.7 $\pm$0.3	4.2 $\pm$0.2	K^+K^- K_LK_S $\pi^+\pi^-\pi^0$ (incl. $\varrho\pi$) $\eta\gamma$	46.6$\pm$2.5 34.6$\pm$2.2 15.8$\pm$1.5 3.0$\pm$1.1

Table A.2 (continued)

Particle	$I^G(J^P)C$	Mass M (MeV)	Full width Γ (MeV)	Major decay channels (%)	
A_1 (1100)	$1^-(1^+)+$	~ 1100	~ 300	$\varrho\pi$	~ 100
Broad enhancement in the $J^P = 1^+$ $\varrho\pi$ partial wave; not a Breit–Wigner resonance.					
B (1235)	$1^+(1^+)-$	1237 ±10	120 ±20	$\omega\pi$	only mode seen
f (1270)	$0^+(2^+)+$	1270 ±10	170 ±20	$\pi\pi$ $2\pi^+2\pi^-$ $K\bar{K}$	83±5 4±1 4±3
D (1285)	$0^+(A)+$	1286 ±10	30 ±20	$K\bar{K}\pi$ $\eta\pi\pi$ $2\pi^+2\pi^-$ (prob. $\varrho^0\pi^+\pi^-$)	
A_2 (1310)	$1^-(2^+)+$	1310 ±10	100 ±10	$\varrho\pi$ $\eta\pi$ $\omega\pi\pi$ $K\bar{K}$ η' (958) π	71.5±1.8 15.2±1.2 8.6±1.8 4.7±0.6 < 1
E (1420)	$0^+(A)+$	1416 ±10	60 ±20	$K\bar{K}\pi$ $\eta\pi\pi$	~ 40 ~ 60
f′ (1514)	$0^+(2^+)+$	1516 ±3	40 ±10	$K\bar{K}$	only mode seen
F_1 (1540)	1(A)	1540 ±5	40 ±15	$K^*\bar{K}+\bar{K}^*K$	only mode seen
ϱ' (1600)	$1^+(1^-)-$	~ 1600	~ 400	4π [$\varrho\pi\pi$ $\pi\pi$	dominant seen] possibly seen
Resonance interpretation uncertain.					
A_3 (1640)	$1^-(2^-)+$	~1640	~ 300	$f\pi$	~ 100
Broad enhancement in the $J^P = 2^-$ $f\pi$ partial wave; not a Breit–Wigner resonance.					
ω (1675)	$0^-(N)-$	1666 ±10	142 ±20	$\varrho\pi$ 3π 5π	seen possibly seen possibly seen
g (1680)	$1^+(3^-)-$	1686 ±20	180 ±30	2π 4π $K\bar{K}$ $K\bar{K}\pi$ (incl. K^*K)	~ 26±5 ~ 70 ~ 2 ~ 3
$K^\pm$ (494) K^0 (498)	$\frac{1}{2}(0^-)$	493.71 497.70		See Table A. 1	
K^* (892)$^\pm$ ($m^0-m^\pm$ = 6.1 ± 1.5 MeV)	$\frac{1}{2}(1^-)$	892.2 ±0.5	49.8 ±1.1	$K\pi$ $K\pi\pi$	≈ 100 < 0.2
K_N	$\frac{1}{2}(0^+)$	δ is near 90°, with slow variation, in mass region 1200–1400 MeV.			
Q { K_A (1240) or C	$\frac{1}{2}(1^+)$	1242 ±10 seen in $\bar{p}p$ at rest	127 ±25	$K\pi\pi$ [$K^*\pi$	only mode seen large]
Q { K_A (1280–1400)	$\frac{1}{2}(1^+)$	1280–1400		[$K\varrho$ [$K(\pi\pi)_{l=0}$	seen] possibly seen]

(*Table A. 2, continued*)

Particle	$I^G(J^P)C$	Mass M (MeV)	Full width Γ (MeV)	(Major decay channels %)	
K_N (1420)	$\frac{1}{2}(2^+)$	1421 ± 5	100 ± 10	$K\pi$ $K^*\pi$ $K\varrho$ $K\omega$ $K\eta$	55.0 ± 2.7 29.5 ± 2.5 9.2 ± 2.4 4.4 ± 1.7 2.0 ± 2.0
L (1770)	$\frac{1}{2}$(A)	1765 ± 10	140 ± 50	$K\pi\pi$ $K\pi\pi\pi$	dominant seen

Other states which require further confirmation are listed in ref. 8.

TABLE A.3

Particle	$I(J^P)$	Mass (MeV)	Full width Γ (MeV)	Major decay channels (%)	
p n	$\frac{1}{2}(\frac{1}{2}^+)$	938.3 939.6		See Table A. 1	
N (1470)	$\frac{1}{2}(\frac{1}{2}^+)\ P'_{11}$	~ 1470	165–300	$N\pi$ $N\pi\pi$	60 35
N (1520)	$\frac{1}{2}(\frac{3}{2}^-)\ D'_{13}$	1510–1540	105–150	$N\pi$ $N\pi\pi$	55 ~ 45
N (1535)	$\frac{1}{2}(\frac{1}{2}^-)\ S'_{11}$	1500–1600	50–160	$N\pi$ $N\eta$ $N\pi\pi$	35 55 ~ 10
N (1670)	$\frac{1}{2}(\frac{5}{2}^-)\ D'_{15}$	1670–1685	115–175	$N\pi$ $N\pi\pi$	40 60
N (1688)	$\frac{1}{2}(\frac{5}{2}^+)\ F'_{15}$	1680–1690	105–180	$N\pi$ $N\pi\pi$	60 40
N (1700)	$\frac{1}{2}(\frac{1}{2}^-)\ S''_{11}$	1665–1765	100–300	$N\pi$ $N\pi\pi$	55 25
N (1780)	$\frac{1}{2}(\frac{1}{2}^+)\ P''_{11}$	1650–1860	50–350	$N\pi$ $N\pi\pi$	~ 20 > 40
N (1860)	$\frac{1}{2}(\frac{3}{2}^+)\ P_{13}$	1770–1860	180–330	$N\pi$ $N\pi\pi$	25 > 50
N (2190)	$\frac{3}{2}(\frac{7}{2}^-)\ G_{17}$	2000–2260	150–325	$N\pi$ $N\pi\pi$	25
N (2220)	$\frac{1}{2}(\frac{9}{2}^+)\ H_{19}$	2200–2245	260–330	$N\pi$	15
N (2650)	$\frac{1}{2}(?^-)$	~ 2650	~ 360	$N\pi$	
N (3030)	$\frac{1}{2}(?)$	~ 3030	~ 400	$N\pi$	
Δ (1236)	$\frac{3}{2}(\frac{3}{2}^+)\ P'_{33}$	1230–1236	110–122	$N\pi$ $N\gamma$	99.4 0.72–0.74
Δ (1650)	$\frac{3}{2}(\frac{1}{2}^-)\ S_{31}$	1615–1695	140–200	$N\pi$ $N\pi\pi$	~ 30 ~ 70

Table A. 3 (continued)

Particle	$I(J^P)$	Mass (MeV)	Full width Γ MeV	Major decay channels (%)
Δ (1670)	$\frac{3}{2}(\frac{3}{2}^-)$ D_{33}	1650–1720	190–270	Nπ ~15 Nππ > 60
Δ (1890)	$\frac{3}{2}(\frac{5}{2}^+)$ F_{35}	1840–1920	140–350	Nπ ~17 Nππ > 50
Δ (1910)	$\frac{3}{2}(\frac{1}{2}^+)$ P_{31}	1780–1935	200–340	Nπ ~25 Nππ ?
Δ (1950)	$\frac{3}{2}(\frac{7}{2}^+)$ F_{37}	1930–1980	170–270	Nπ 40 Nππ > 25
Δ (2420)	$\frac{3}{2}(\frac{11}{2}^+)$	2320–2450	250–350	Nπ 11 Nππ > 20
Δ (2850)	$\frac{3}{2}(?^+)$	~ 2850	~ 400	Nπ
Δ (3230)	$\frac{3}{2}(?)$	~ 3230	~ 440	Nπ
Λ	$0(\frac{1}{2}^+)$	1115.6		See Table A.1
Λ (1405)	$0(\frac{1}{2}^-)$ S'_{01}	1405 ±5	40 ±10	Σπ 100
Λ (1520)	$0(\frac{3}{2}^-)$ D'_{03}	1518 ±2	16 ±2	N$\bar{K}$ 45 ±1 Σπ 41 ±1 Λππ 10 ±0.5 Σππ 0.8±0.1
Λ (1670)	$0(\frac{1}{2}^-)$ S''_{01}	1660–1680	23–40	N$\bar{K}$ 15–35 Λη 15–25 Σπ 30–50
Λ (1690)	$0(\frac{3}{2}^-)$ D''_{03}	1690 ±10	30–70	N$\bar{K}$ 20–30 Σπ 30–50 Λππ < 25 Σππ < 25
Λ (1815)	$0(\frac{5}{2}^+)$ F'_{05}	1820 ±5	70–100	N$\bar{K}$ ~61 Σπ ~11 Σ (1385)π 15–20
Λ (1830)	$0(\frac{5}{2}^-)$ D'_{05}	1810–1840	70–120	N$\bar{K}$ ~ 10 Σπ 20–60 Λη ~2
Λ (2100)	$0(\frac{7}{2}^-)$ G_{07}	2090–2120	60–140	N$\bar{K}$ ~30 Σπ ~ 5 Λη < 2 Ξ < 3 Λω < 3
Λ (2350)	0(?)	~ 2350	140–320	N$\bar{K}$
Λ (2585)	0(?)	~ 2585	~ 300	N$\bar{K}$
Σ	$1(\frac{1}{2}^+)$	(+)1189.4 (0)1192.5 (−)1197.4		See Table A.1

Table A.3 (continued)

Particle	$I(J^P)$	Mass (MeV)	Full width Γ MeV	Major decay channels (%)	
Σ (1385)	$1(\frac{3}{2}^+)$ P'_{13}	(+)1383±1 (−)1387±1	(+)34±2 (−)42±5	$\Lambda\pi$ $\Sigma\pi$	88±2 12±2
Σ (1670)	$1(\frac{3}{2}^-)D'_{13}$	1670 ±10	35–60	$N\bar{K}$ $\Sigma\pi$ $\Lambda\pi$ $\Sigma\pi\pi$ $\Lambda\pi\pi$	~ 8 30−60 ~ 12 seen
Σ (1750)	$1(\frac{1}{2}^-)$ S'_{11}	1700–1790	50–100	$N\bar{K}$ $\Lambda\pi$ $\Sigma\eta$	12−45
Σ (1765)	$1(\frac{5}{2}^-)$ D_{15}	1765 ±5	~ 120	$N\bar{K}$ $\Lambda\pi$ $\Lambda(1520)\,\pi$ $\Sigma(1385)\,\pi$ $\Sigma\pi$	~ 41 ~ 13 ~ 15 ~ 10 ~ 1
Σ (1915)	$1(\frac{5}{2}^+)$ F'_{15}	1900–1930	50–120	$N\bar{K}$ $\Lambda\pi$ $\Sigma\pi$	~ 14 ~ 6 ~ 6
Σ (1940)	$1(\frac{3}{2}^-)D''_{13}$	1865–1950	120–280	$N\bar{K}$ $\Lambda\pi$ $\Sigma\pi$	~ 21 ~ 4 < 7
Σ (2030)	$1(\frac{7}{2}^+)$ F_{17}	~ 2030	120–170	$N\bar{K}$ $\Lambda\pi$ $\Sigma\pi$ ΞK	~ 20 ~ 20 ~ 4 < 2
Σ (2250)	1(?)	~ 2250	100–230	$N\bar{K}$ $\Sigma\pi$ $\Lambda\pi$	
Σ (2455)	(1?)	~ 2455	~ 120	$N\bar{K}$	
Σ (2620)	(1?)	~ 2620	~ 175	$N\bar{K}$	
Ξ	$\frac{1}{2}(\frac{1}{2}^+)$	(0)1314.9 (−)1321.3		See Table A. 1	
Ξ (1530)	$\frac{1}{2}(\frac{3}{2}^+)$ P_{13}	(0)1531.6±0.4 (0) (−)1535.0±0.6 (−)	9.1±0.5 10.6±2.6	$\Xi\pi$	100
Ξ (1820)	$\frac{1}{2}$(?)	1795–1870	12–100	$\Lambda\bar{K}$ $\Xi\pi$ $\Xi(1530)\pi$ $\Sigma\bar{K}$	
Ξ (1940)	$\frac{1}{2}$(?)	1894–1961	40–140	$\Xi\pi$ $\Xi(1530)\pi$	
Ω^-	$0(\frac{3}{2}^+)$	1672.5	See Table A.1		

TABLE A.5 ISOSCALAR FACTORS FOR THE GROUP SU_3 [117]

According to eqn. (9.73) the Clebsch–Gordan coefficients of SU_3 are the product of SU_2 Clebsch–Gordan coefficients and isoscalar factors. These isoscalar factors

$$\begin{pmatrix} n_1 & n_2 & \Big| & n \\ I_1Y_1 & I_2Y_2 & \Big| & IY \end{pmatrix}$$

also depend on the isospins I_1, I_2 and hypercharges Y_1, Y_2 of the particles in the SU_3 multiplets $\underline{n}_1$ and $\underline{n}_2$ whose product $\underline{n}_1 \times \underline{n}_2$ we are studying. Two products are considered below: $\underline{8} \times \underline{8}$ and $\underline{10} \times \underline{8}$, both of which arise in meson–baryon scattering. In the product $\underline{n}_1 \times \underline{n}_2$ the first representation $\underline{n}_1$ refers to the baryon.

States which occur in the product $\underline{8} \times \underline{8}$ are shown in Fig. 30, while those in the product $\underline{10} \times \underline{8}$ are shown in Fig. 31. Fig. 30 is the weight diagram of the $\underline{27}$-plet, whose single states are denoted by dots, doubled ones by

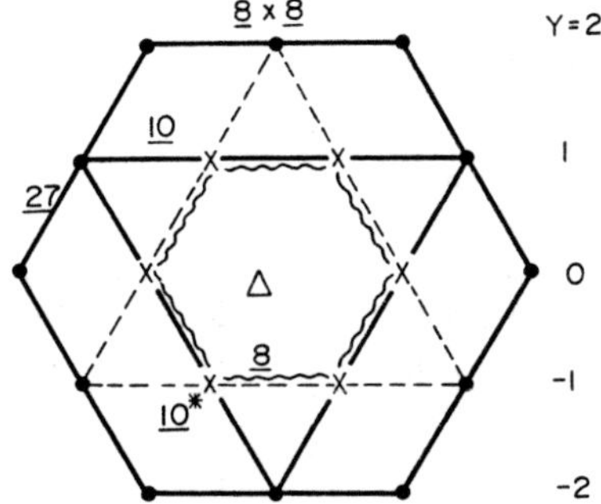

FIG. 30. States arising from the product $\underline{8} \times \underline{8}$. (See text.)

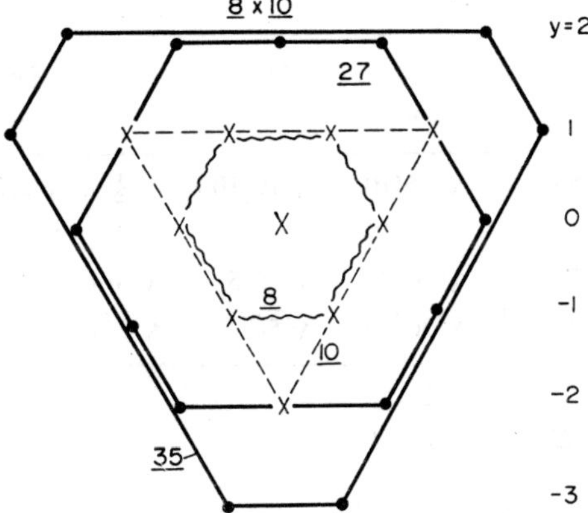

FIG. 31. States arising from the product $\underline{10} \times \underline{8}$. (See text.)

crosses, and tripled one by a triangle. The weight diagrams of the $\underline{10}$, $\underline{10}^*$, and the octet are also shown on this figure, by solid, dashed, and wavy lines respectively. Figure 31 is the weight diagram of the $\underline{35}$-plet whose doubled states are denoted by crosses. The weight diagrams of the $\underline{27}$, $\underline{10}$, and octet are also shown.

Each table refers to a given combination of I and Y, written over the table. The left-hand column contains combinations of states of the initial multiplets leading to given I and Y. Each successive column gives the soscalar factors for a given resulting representation $\underline{n}$. The sign $\pm$ under the value of $\underline{n}$ gives the factor $\xi_1(n, in_1, n_2) = \pm 1$.

Table A.5 (continued)

$$\underline{8}\times\underline{8} = \underline{27}+\underline{10}+\underline{10}^*+\underline{8}_S+\underline{8}_A+\underline{1}$$

N: $Y = 1$, $I = \frac{1}{2}$

	$\underline{27}$ +	$\underline{8}_D$ +	$\underline{8}_F$ −	$\underline{10}^*$ −
Nπ	$\sqrt{5}/10$	$3\sqrt{5}/10$	$\frac{1}{2}$	$-\frac{1}{2}$
ΣK	$-\sqrt{5}/10$	$-3\sqrt{5}/10$	$\frac{1}{2}$	$-\frac{1}{2}$
Nη	$3\sqrt{5}/10$	$-\sqrt{5}/10$	$\frac{1}{2}$	$\frac{1}{2}$
ΛK	$3\sqrt{5}/10$	$-\sqrt{5}/10$	$-\frac{1}{2}$	$-\frac{1}{2}$

Δ: $Y = 1$, $I = \frac{3}{2}$

	$\underline{27}$ +	$\underline{10}$ −
Nπ	$\sqrt{2}/2$	$-\sqrt{2}/2$
ΣK	$\sqrt{2}/2$	$\sqrt{2}/2$

Ξ: $Y = -1$, $I = \frac{1}{2}$

	$\underline{27}$ +	$\underline{8}_D$ +	$\underline{8}_F$ +	$\underline{10}$ −
$\Xi\pi$	$-\sqrt{5}/10$	$-3\sqrt{5}/10$	$\frac{1}{2}$	$\frac{1}{2}$
$\Sigma\overline{\text{K}}$	$\sqrt{5}/10$	$3\sqrt{5}/10$	$\frac{1}{2}$	$\frac{1}{2}$
$\Xi\eta$	$3\sqrt{5}/10$	$-\sqrt{5}/10$	$-\frac{1}{2}$	$\frac{1}{2}$
$\Lambda\overline{\text{K}}$	$3\sqrt{5}/10$	$-\sqrt{5}/10$	$\frac{1}{2}$	$-\frac{1}{2}$

Ξ: $Y = -1$, $I = \frac{3}{2}$

	$\underline{27}$ +	$\underline{10}^*$ −
$\Xi\pi$	$\sqrt{2}/2$	$-\sqrt{2}/2$
$\Sigma\overline{\text{K}}$	$\sqrt{2}/2$	$\sqrt{2}/2$

Λ : $Y = 0$, $I = 0$

	$\underline{27}$ +	$\underline{8}_D$ +	$\underline{1}$ +	$\underline{8}_F$ −
N$\overline{\text{K}}$	$\sqrt{15}/10$	$\sqrt{10}/10$	$\frac{1}{2}$	$\sqrt{2}/2$
ΞK	$-\sqrt{15}/10$	$-\sqrt{10}/10$	$-\frac{1}{2}$	$\sqrt{2}/2$
$\Sigma\pi$	$-\sqrt{10}/20$	$-\sqrt{15}/5$	$\sqrt{6}/4$	0
$\Lambda\eta$	$3\sqrt{30}/20$	$-\sqrt{5}/5$	$-\sqrt{2}/4$	0

Σ: $Y = 0$, $I = 1$

	$\underline{27}$ +	$\underline{8}_D$ +	$\underline{8}_F$ −	$\underline{10}$ −	$\underline{10}^*$ −
N$\overline{\text{K}}$	$\sqrt{5}/5$	$-\sqrt{30}/10$	$\sqrt{6}/6$	$-\sqrt{6}/6$	$\sqrt{6}/6$
ΞK	$\sqrt{5}/5$	$-\sqrt{30}/10$	$-\sqrt{6}/6$	$\sqrt{6}/6$	$-\sqrt{6}/6$
$\Sigma\pi$	0	0	$\sqrt{6}/3$	$\sqrt{6}/6$	$-\sqrt{6}/6$
$\Sigma\eta$	$\sqrt{30}/10$	$\sqrt{5}/5$	0	$\frac{1}{2}$	$\frac{1}{2}$
$\Lambda\pi$	$\sqrt{30}/10$	$\sqrt{5}/5$	0	$-\frac{1}{2}$	$-\frac{1}{2}$

Table A.5 (continued)

$\underline{10} \times \underline{8} = \underline{35} + \underline{27} + \underline{10} + \underline{8}$

Σ: $Y = 0$, $I = 1$

	$\underline{35}$ +	$\underline{27}$ −	$\underline{10}$ −	$\underline{8}$ +
$\Sigma\pi$	$\sqrt{3}/6$	$-3\sqrt{5}/10$	$\sqrt{3}/3$	$-\sqrt{30}/15$
$\Sigma\eta$	$\sqrt{2}/2$	$\sqrt{30}/10$	0	$-\sqrt{5}/5$
ΞK	$\sqrt{3}/3$	$-\sqrt{5}/5$	$-\sqrt{3}/3$	$\sqrt{30}/50$
$\Delta\bar{K}$	$\sqrt{3}/6$	$\sqrt{5}/10$	$\sqrt{3}/3$	$2\sqrt{30}/15$

$Y = 0$, $I = 2$

	$\underline{35}$ +	$\underline{27}$ −
$\Sigma\pi$	$\sqrt{3}/2$	$-\frac{1}{2}$
$\Delta\bar{K}$	$\frac{1}{2}$	$\sqrt{3}/2$

Ξ: $Y = -1$, $I = \frac{1}{2}$

	$\underline{35}$ +	$\underline{27}$ −	$\underline{10}$ −	$\underline{8}$ +
$\Xi\pi$	$\frac{1}{4}$	$-7\sqrt{5}/20$	$\sqrt{2}/4$	$-\sqrt{5}/5$
$\Xi\eta$	$\frac{3}{4}$	$3\sqrt{5}/20$	$-\sqrt{2}/4$	$-\sqrt{5}/5$
ΩK	$\sqrt{2}/4$	$-3\sqrt{10}/20$	$-\frac{1}{2}$	$\sqrt{10}/5$
$\Sigma\bar{K}$	$\frac{1}{2}$	$\sqrt{5}/10$	$\sqrt{2}/2$	$\sqrt{5}/5$

Ξ: $Y = -1$, $I = \frac{3}{2}$

	$\underline{35}$ +	$\underline{27}$ −
$\Xi\pi$	$\sqrt{2}/2$	$-\sqrt{2}/2$
$\Sigma\bar{K}$	$\sqrt{2}/2$	$\sqrt{2}/2$

N: $Y = 1$, $I = \frac{1}{2}$

	$\underline{27}$ −	$\underline{8}$ +
$\Delta\pi$	$-\sqrt{5}/5$	$-2\sqrt{5}/5$
ΣK	$-2\sqrt{5}/5$	$\sqrt{5}/5$

Δ: $Y = 1$, $I = \frac{3}{2}$

	$\underline{35}$ +	$\underline{27}$ −	$\underline{10}$ −
$\Delta\pi$	$\frac{1}{4}$	$-\sqrt{5}/4$	$\sqrt{10}/4$
$\Delta\eta$	$\sqrt{5}/4$	$\frac{3}{4}$	$\sqrt{2}/4$
ΣK	$\sqrt{10}/4$	$-\sqrt{2}/4$	$-\frac{1}{2}$

Ω: $Y = -2$, $I = 0$

	$\underline{35}$ +	$\underline{10}$ −
$\Omega\eta$	$\sqrt{2}/2$	$-\sqrt{2}/2$
$\Xi\bar{K}$	$\sqrt{2}/2$	$\sqrt{2}/2$

Ω: $Y = -2$, $I = 1$

	$\underline{35}$ +	$\underline{27}$ −
$\Omega\pi$	$\frac{1}{2}$	$-\sqrt{3}/2$
$\Xi\bar{K}$	$\sqrt{3}/2$	$\frac{1}{2}$

Λ: $Y = 0$, $I = 0$

	$\underline{27}$ −	$\underline{8}$ +
$\Sigma\pi$	$-\sqrt{10}/5$	$-\sqrt{15}/5$
ΞK	$-\sqrt{15}/5$	$\sqrt{10}/5$

TABLE A.6 CROSSING MATRICES FOR THE ISOSPIN GROUP $SU_2^{(142)}$

Each reaction is characterized by the isospins of the particles involved. The s-channel reaction is indicated above each table. The crossing matrix X_{st} has the matrix elements $(X_{st})_{I_s I_t}$, where I_s and I_t are the values of the isospins in the s- and t-channels. Amplitudes with different isospins belonging to the same channel form a column vector.

Notation:

		$I_{t1}\ I_{t2}\ \ldots$	$I_{u1}\ I_{u2}\ \ldots$	
X_{st}	I_{s1}, I_{s2}, $\vdots$	$(X_{st})_{I_s I_t}$	$(X_{su})_{I_s I_u}$	X_{su}
X_{ut}	I_{u1}, I_{u2}, $\vdots$	$(X_{ut})_{I_u I_t}$		

$\frac{1}{2}+\frac{1}{2}' \to \frac{1}{2}''+\frac{1}{2}'''$

		0	1	0	1	
X_{st}	0	$-\frac{1}{2}$	$-\frac{3}{2}$	$\frac{1}{2}$	$\frac{3}{2}$	X_{su}
	1	$-\frac{1}{2}$	$\frac{1}{2}$	$-\frac{1}{2}$	$\frac{1}{2}$	
X_{ut}	0	$\frac{1}{2}$	$-\frac{3}{2}$			
	1	$-\frac{1}{2}$	$-\frac{1}{2}$			

Table A.6 (continued)

$\frac{1}{2}+\frac{1}{2}' \to 1+1'$

		$\frac{1}{2}$	$\frac{3}{2}$	$\frac{1}{2}$	$\frac{3}{2}$	
X_{st}	0	$-\sqrt{\frac{2}{3}}$	$-2\sqrt{\frac{2}{3}}$	$-\sqrt{\frac{2}{3}}$	$-2\sqrt{\frac{2}{3}}$	X_{su}
	1	$-\frac{2}{3}$	$\frac{2}{3}$	$\frac{2}{3}$	$-\frac{2}{3}$	
X_{ut}	$\frac{1}{2}$	$-\frac{1}{3}$	$\frac{4}{3}$			
	$\frac{3}{2}$	$\frac{2}{3}$	$\frac{1}{3}$			

$\frac{1}{2}+\frac{1}{2}' \to \frac{3}{2}+\frac{3}{2}'$

		1	2	1	2	
X_{st}	0	$-3\sqrt{2}/4$	$-5\sqrt{2}/4$	$3\sqrt{2}/4$	$5\sqrt{2}/4$	X_{su}
	1	$-\sqrt{10}/4$	$\sqrt{10}/4$	$-\sqrt{10}/4$	$\sqrt{10}/4$	
X_{ut}	1	$\frac{1}{4}$	$-\frac{5}{4}$			
	2	$-\frac{3}{4}$	$-\frac{1}{4}$			

$1+1' \to 1''+1'''$

		0	1	2	0	1	2	
X_{st}	0	$\frac{1}{3}$	1	$\frac{5}{3}$	$\frac{1}{3}$	1	$\frac{5}{3}$	X_{su}
	1	$\frac{1}{3}$	$\frac{1}{2}$	$-\frac{5}{6}$	$-\frac{1}{3}$	$-\frac{1}{2}$	$\frac{5}{6}$	
	2	$\frac{1}{3}$	$-\frac{1}{2}$	$\frac{1}{6}$	$\frac{1}{3}$	$-\frac{1}{2}$	$\frac{1}{6}$	
X_{ut}	0	$\frac{1}{3}$	-1	$\frac{5}{3}$				
	1	$-\frac{1}{3}$	$\frac{1}{2}$	$\frac{5}{6}$				
	2	$\frac{1}{3}$	$\frac{1}{2}$	$\frac{1}{6}$				

Table A.6 (continued)

$1+1' \rightarrow \frac{3}{2}+\frac{3}{2}'$

		$\frac{1}{2}$	$\frac{3}{2}$	$\frac{5}{2}$	$\frac{1}{2}$	$\frac{3}{2}$	$\frac{5}{2}$	
X_{st}	0	$1/\sqrt{3}$	$2/\sqrt{3}$	$\sqrt{3}$	$1/\sqrt{3}$	$2/\sqrt{3}$	$\sqrt{3}$	X_{su}
	1	$\sqrt{10}/6$	$2\sqrt{10}/15$	$-3/\sqrt{10}$	$-\sqrt{10}/6$	$-2\sqrt{10}/15$	$3/\sqrt{10}$	
	2	$1/\sqrt{6}$	$-4\sqrt{6}/15$	$\sqrt{6}/10$	$1/\sqrt{6}$	$-4\sqrt{6}/15$	$\sqrt{6}/10$	
X_{ut}	$\frac{1}{2}$	$\frac{1}{6}$	$-\frac{2}{3}$	$\frac{3}{2}$				
	$\frac{3}{2}$	$-\frac{1}{3}$	$\frac{11}{15}$	$\frac{3}{5}$				
	$\frac{5}{2}$	$\frac{1}{2}$	$\frac{2}{5}$	$\frac{1}{10}$				

$\frac{1}{2}+\frac{3}{2} \rightarrow \frac{3}{2}'+\frac{3}{2}''$

		1	2	1	2	
X_{st}	1	$\frac{1}{2}$	$\sqrt{5}/2$	$-\frac{1}{2}$	$\sqrt{5}/2$	X_{su}
	2	$3\sqrt{5}/10$	$\frac{1}{2}$	$-3\sqrt{5}/10$	$\frac{1}{2}$	
X_{ut}	1	$-\frac{1}{2}$	$\sqrt{5}/2$			
	2	$3\sqrt{5}/10$	$\frac{1}{2}$			

$\frac{3}{2}+\frac{3}{2}' \rightarrow \frac{3}{2}''+\frac{3}{2}'''$

		0	1	2	3	0	1	2	3	
X_{st}	0	$-\frac{1}{4}$	$-\frac{3}{4}$	$-\frac{5}{4}$	$-\frac{7}{4}$	$\frac{1}{4}$	$\frac{3}{4}$	$\frac{5}{4}$	$\frac{7}{4}$	X_{su}
	1	$-\frac{1}{4}$	$-\frac{11}{20}$	$-\frac{1}{4}$	$\frac{21}{20}$	$-\frac{1}{4}$	$-\frac{11}{20}$	$-\frac{1}{4}$	$\frac{21}{20}$	
	2	$-\frac{1}{4}$	$-\frac{3}{20}$	$\frac{3}{4}$	$-\frac{7}{20}$	$\frac{1}{4}$	$\frac{3}{20}$	$-\frac{3}{4}$	$\frac{7}{20}$	
	3	$-\frac{1}{4}$	$\frac{9}{20}$	$-\frac{1}{4}$	$\frac{1}{0}$	$-\frac{1}{4}$	$\frac{9}{20}$	$-\frac{1}{4}$	$\frac{1}{20}$	
X_{ut}	0	$\frac{1}{4}$	$-\frac{3}{4}$	$\frac{5}{4}$	$-\frac{7}{4}$					
	1	$-\frac{1}{4}$	$\frac{11}{20}$	$-\frac{1}{4}$	$-\frac{21}{20}$					
	2	$\frac{1}{4}$	$-\frac{3}{20}$	$-\frac{3}{4}$	$-\frac{7}{20}$					
	3	$-\frac{1}{4}$	$-\frac{9}{20}$	$-\frac{1}{4}$	$-\frac{1}{20}$					

TABLE A.7. CROSSING MATRICES FOR THE GROUP SU_3 (142)

The s-channel reaction $\underline{n}_1+\underline{n}_2 \rightarrow \underline{n}_3+\underline{n}_4$ is indicated above each table, where $\underline{n}_i$ is an SU_3 multiplet. If $\underline{n}_s$ and $\underline{n}_t$ are the allowed multiplets in the s- and t-channels, the crossing matrix has the matrix elements $(X_{st})_{\underline{n}_s\underline{n}_t}$. Amplitudes with different $\underline{n}$ belonging to the same channel form a column vector. The notation $\underline{8}_{SS}$, $\underline{8}_{SA}$, $\underline{8}_{AA}$, $\underline{8}_{AS}$ is explained in § 10.3.

Notation:

		$\underline{n}_{t1}\underline{n}_{t2}$ ...
X_{st}	$\underline{n}_{s1}$ $\underline{n}_{s2}$ . . .	$(X_{st})_{\underline{n}_s\underline{n}_t}$

$\underline{8}+\underline{8}' \rightarrow \underline{8}''+\underline{8}'''$

		$\underline{1}$	$\underline{8}_{AS}$	$\underline{8}_{SA}$	$\underline{8}_{SS}$	$\underline{8}_{AA}$	$\underline{10}$	$\underline{10}^*$	$\underline{27}$
$X_{st} = X_{ts}$	$\underline{1}$	$\frac{1}{8}$	0	0	1	1	$\frac{5}{4}$	$\frac{5}{4}$	$\frac{27}{8}$
	$\underline{8}_{AS}$	0	$-\frac{1}{2}$	$-\frac{1}{2}$	0	0	$\sqrt{5}/4$	$-\sqrt{5}/4$	0
	$\underline{8}_{SA}$	0	$-\frac{1}{2}$	$-\frac{1}{2}$	0	0	$-\sqrt{5}/4$	$\sqrt{5}/4$	0
	$\underline{8}_{SS}$	$\frac{1}{8}$	0	0	$-\frac{3}{10}$	$\frac{1}{2}$	$-\frac{1}{2}$	$-\frac{1}{2}$	$\frac{27}{40}$
	$\underline{8}_{AA}$	$\frac{1}{8}$	0	0	$\frac{1}{2}$	$\frac{1}{2}$	0	0	$-\frac{9}{8}$
	$\underline{10}$	$\frac{1}{8}$	$1/\sqrt{5}$	$-1/\sqrt{5}$	$-\frac{2}{5}$	0	$\frac{1}{4}$	$\frac{1}{4}$	$-\frac{9}{40}$
	10^*	$\frac{1}{8}$	$-1/\sqrt{5}$	$1/\sqrt{5}$	$-\frac{2}{5}$	0	$\frac{1}{4}$	$\frac{1}{4}$	$-\frac{9}{40}$
	$\underline{27}$	$\frac{1}{8}$	0	0	$\frac{1}{5}$	$-\frac{1}{3}$	$-\frac{1}{12}$	$-\frac{1}{12}$	$\frac{7}{40}$
X_{su}	$\underline{1}$	$\frac{1}{8}$	0	0	1	1	$\frac{5}{4}$	$\frac{5}{4}$	$\frac{27}{8}$
	$\underline{8}_{AS}$	0	$\frac{1}{2}$	$\frac{1}{2}$	0	0	$-\sqrt{5}/4$	$\sqrt{5}/4$	0
	$\underline{8}_{SA}$	0	$-\frac{1}{2}$	$-\frac{1}{2}$	0	0	$-\sqrt{5}/4$	$\sqrt{5}/4$	0
	$\underline{8}_{SS}$	$\frac{1}{8}$	0	0	$-\frac{3}{10}$	$\frac{1}{2}$	$-\frac{1}{2}$	$-\frac{1}{2}$	$\frac{27}{40}$
	$\underline{8}_{AA}$	$-\frac{1}{8}$	0	0	$-\frac{1}{2}$	$-\frac{1}{2}$	0	0	$\frac{9}{8}$
	$\underline{10}$	$-\frac{1}{8}$	$-1/\sqrt{5}$	$1/\sqrt{5}$	$\frac{2}{5}$	0	$-\frac{1}{4}$	$-\frac{1}{4}$	$\frac{9}{40}$
	$\underline{10}^*$	$-\frac{1}{8}$	$1/\sqrt{5}$	$-1/\sqrt{5}$	$\frac{2}{5}$	0	$-\frac{1}{4}$	$-\frac{1}{4}$	$\frac{9}{40}$
	$\underline{27}$	$\frac{1}{8}$	0	0	$\frac{1}{5}$	$-\frac{1}{3}$	$-\frac{1}{12}$	$-\frac{1}{12}$	$\frac{7}{40}$

Table A.7 (continued)

		$\underline{1}$	$\underline{8}_{AS}$	$\underline{8}_{SA}$	$\underline{8}_{SS}$	$\underline{8}_{AA}$	$\underline{10}$	$\underline{10}^*$	$\underline{27}$
$X_{tu} = X_{ut}$	$\underline{1}$	$\frac{1}{8}$	0	0	1	-1	$-\frac{5}{4}$	$-\frac{5}{4}$	$\frac{27}{8}$
	$\underline{8}_{AS}$	0	$-\frac{1}{2}$	$\frac{1}{2}$	0	0	$\sqrt{5}/4$	$-\sqrt{5}/4$	0
	$\underline{8}_{SA}$	0	$\frac{1}{2}$	$-\frac{1}{2}$	0	0	$\sqrt{5}/4$	$-\sqrt{5}/4$	0
	$\underline{8}_{SS}$	$\frac{1}{8}$	0	0	$-\frac{3}{10}$	$-\frac{1}{2}$	$\frac{1}{2}$	$\frac{1}{2}$	$\frac{27}{40}$
	$\underline{8}_{AA}$	$-\frac{1}{8}$	0	0	$-\frac{1}{2}$	$\frac{1}{2}$	0	0	$\frac{9}{8}$
	$\underline{10}$	$-\frac{1}{8}$	$1/\sqrt{5}$	$1/\sqrt{5}$	$\frac{2}{5}$	0	$\frac{1}{4}$	$\frac{1}{4}$	$\frac{9}{40}$
	$\underline{10}^*$	$-\frac{1}{8}$	$-1/\sqrt{5}$	$-1/\sqrt{5}$	$\frac{2}{5}$	0	$\frac{1}{4}$	$\frac{1}{4}$	$\frac{9}{40}$
	$\underline{27}$	$\frac{1}{8}$	0	0	$\frac{1}{5}$	$\frac{1}{3}$	$\frac{1}{12}$	$\frac{1}{12}$	$\frac{7}{40}$
X_{us}	$\underline{1}$	$\frac{1}{8}$	0	0	1	-1	$-\frac{5}{4}$	$-\frac{5}{4}$	$\frac{27}{8}$
	$\underline{8}_{AS}$	0	$\frac{1}{2}$	$-\frac{1}{2}$	0	0	$-\sqrt{5}/4$	$\sqrt{5}/4$	0
	$\underline{8}_{SA}$	0	$\frac{1}{2}$	$-\frac{1}{2}$	0	0	$\sqrt{5}/4$	$-\sqrt{5}/4$	0
	$\underline{8}_{SS}$	$\frac{1}{8}$	0	0	$-\frac{3}{10}$	$-\frac{1}{2}$	$\frac{1}{2}$	$\frac{1}{2}$	$\frac{27}{40}$
	$\underline{8}_{AA}$	$\frac{1}{8}$	0	0	$\frac{1}{2}$	$-\frac{1}{2}$	0	0	$-\frac{9}{8}$
	$\underline{10}$	$\frac{1}{8}$	$-1/\sqrt{5}$	$-1/\sqrt{5}$	$-\frac{2}{5}$	0	$-\frac{1}{4}$	$-\frac{1}{4}$	$-\frac{9}{40}$
	$\underline{10}^*$	$\frac{1}{8}$	$1/\sqrt{5}$	$1/\sqrt{5}$	$-\frac{2}{5}$	0	$-\frac{1}{4}$	$-\frac{1}{4}$	$-\frac{9}{40}$
	$\underline{27}$	$\frac{1}{8}$	0	0	$\frac{1}{5}$	$\frac{1}{3}$	$\frac{1}{12}$	$\frac{1}{12}$	$\frac{7}{40}$

$\underline{8}+\underline{8}' \to \underline{8}''+\underline{10}$

		$\underline{8}_S$	$\underline{8}_A$	$\underline{10}^*$	$\underline{27}$
$X_{st} = X_{ts}$	$\underline{8}_S$	$\frac{2}{5}$	$1/\sqrt{5}$	$\frac{1}{4}\sqrt{2}$	$\frac{27}{20}$
	$\underline{8}_A$	$1/\sqrt{5}$	0	$\sqrt{\frac{5}{2}}/2$	$-9\sqrt{5}/20$
	$\underline{10}$	$\sqrt{2}/5$	$\sqrt{\frac{2}{5}}$	$-\frac{1}{2}$	$-\frac{9}{20}\sqrt{2}$
	$\underline{27}$	$\frac{2}{5}$	$-\frac{2}{15}\sqrt{5}$	$-\frac{1}{6}\sqrt{2}$	$\frac{1}{10}$

		$\underline{8}_S$	$\underline{8}_A$	$\underline{10}$	$\underline{27}$
X_{su}	$\underline{8}_S$	$-\frac{2}{5}$	$1/\sqrt{5}$	$-\frac{1}{4}\sqrt{2}$	$\frac{27}{20}$
	$\underline{8}_A$	$1/\sqrt{5}$	0	$\sqrt{\frac{5}{2}}/2$	$9\sqrt{5}/20$
	$\underline{10}$	$\sqrt{2}/5$	$-\sqrt{\frac{2}{5}}$	$-\frac{1}{2}$	$\frac{9}{20}\sqrt{2}$
	$\underline{27}$	$-\frac{2}{5}$	$-\frac{2}{15}\sqrt{5}$	$\frac{1}{6}\sqrt{2}$	$\frac{1}{10}$

Table A.7 (continued)

		$\underline{8}_S$	$\underline{8}_A$	$\underline{10}^*$	$\underline{27}$
X_{tu}	$\underline{8}_S$	$-\frac{2}{5}$	$-\frac{1}{5}$	$\frac{1}{4}\sqrt{2}$	$\frac{27}{20}$
	$\underline{8}_A$	$1/\sqrt{5}$	0	$-\sqrt{\frac{5}{2}}/2$	$9\sqrt{5}/20$
	$\underline{10}$	$\sqrt{2}/5$	$\sqrt{\frac{2}{5}}$	$\frac{1}{2}$	$\frac{9}{20}\sqrt{2}$
	$\underline{27}$	$-\frac{2}{5}$	$\frac{2}{15}\sqrt{5}$	$-\frac{1}{6}\sqrt{2}$	$\frac{1}{10}$

$\underline{8}+\underline{8}' \rightarrow \underline{10}+\underline{10}'$

		$\underline{8}$	$\underline{27}$	$\underline{8}$	$\underline{27}$	
X_{st}	$\underline{10}^*$	$2\sqrt{2}/5$	$9\sqrt{2}/10$	$-2\sqrt{2}/5$	$-9\sqrt{2}/10$	X_{su}
	$\underline{27}$	$2\sqrt{\frac{2}{5}}/3$	$-\sqrt{\frac{2}{5}}$	$2\sqrt{\frac{2}{5}}/3$	$-\sqrt{\frac{2}{5}}$	
		$\underline{8}$	$\underline{27}$	$\underline{10}^*$	$\underline{27}$	
X_{tu}	$\underline{8}$	$\frac{1}{5}$	$-\frac{9}{5}$	$1/\sqrt{2}$	$9\sqrt{10}/20$	X_{ts}
$= X_{ut}$	$\underline{27}$	$-\frac{8}{15}$	$-\frac{1}{5}$	$\sqrt{2}/3$	$-\sqrt{10}/5$	

$\underline{8}+\underline{10} \rightarrow \underline{8}'+\underline{10}'$

		$\underline{1}$	$\underline{8}_S$	$\underline{8}_A$	$\underline{27}$
X_{st}	$\underline{8}$	$\frac{1}{20}\sqrt{5}$	$\sqrt{2}/5$	$\sqrt{\frac{2}{5}}$	$9\sqrt{7}/20$
	$\underline{10}$	$\frac{1}{20}\sqrt{5}$	$\frac{3}{10}\sqrt{2}$	$1/\sqrt{10}$	$-9\sqrt{7}/20$
	$\underline{27}$	$\frac{1}{20}\sqrt{5}$	$-\frac{3}{10}\sqrt{2}$	$\frac{1}{3}\sqrt{10}$	$-\sqrt{7}/20$
	$\underline{35}$	$\frac{1}{20}\sqrt{5}$	$\frac{3}{10}\sqrt{2}$	$-1/\sqrt{10}$	$\frac{9}{20}\sqrt{7}$
		$\underline{8}$	$\underline{10}^*$	$\underline{27}$	$\underline{35}$
X_{su}	$\underline{8}$	$\frac{1}{5}$	$\frac{1}{2}$	$\frac{9}{20}$	$\frac{7}{4}$
	$\underline{10}$	$-\frac{2}{5}$	$-\frac{3}{4}$	$\frac{9}{40}$	$\frac{7}{8}$
	$\underline{27}$	$-\frac{2}{15}$	$\frac{1}{12}$	$-\frac{37}{40}$	$\frac{7}{24}$
	$\underline{35}$	$\frac{2}{5}$	$-\frac{1}{4}$	$-\frac{9}{40}$	$\frac{1}{8}$

REFERENCES

1. S. Gasiorowicz, *Elementary Particle Physics*, Wiley, New York, 1966.
2. G. Källen, *Elementary Particle Physics*, Addison-Wesley, Reading, Mass., 1964.
3. R. E. Marshak and E. C. G. Sudarshan, *Introduction to Elementary Particle Physics*, Interscience, New York, 1961.
4. K. Nishijima, *Fundamental Particles*, W. A. Benjamin, New York, 1963.
5. N. N. Bogolyubov and D. V. Shirkov, *Vvedenie v teoriyu kvantovannykh poleĭ*, Gostekhizdat, 1957. (*Introduction to the Theory of Quantized Fields*, translated by G. M. Volkoff, Interscience, New York, 1959).
6. V. B. Berestetskiĭ, E. M. Lifshitz, and L. P. Pitaevskiĭ, *Relyativistskaya kvantovaya teoriya*, part I, "Nauka", 1968. (*Relativistic Quantum Theory*, translated by J. B. Sykes and J. S. Bell, Pergamon Press, Oxford and New York, 1971.)
7. S. S. Schweber, *An Introduction to Relativistic Quantum Field Theory*, Row, Peterson, Evanston, Ill., 1961.
8. V. Chaloupka *et al.* (Particle Data Group), *Phys. Lett.* **50B**, S1 No.1 (1974).
9. *Lectures in Theoretical Physics*, vol. VIIA, *Lorentz Group* (ed. W. E. Brittin and A. O. Barut), University of Colorado Press, Boulder, 1965.
10. *Group Theoretical Concepts and Methods in Elementary Particle Physics* (Istanbul Summer School of Theoretical Physics, 1962) (ed. by F. Gürsey), Gordon & Breach, New York, 1964.
11. A. O. Barut, *The Theory of the Scattering Matrix for the Interactions of Fundamental Particles*, Macmillan, New York, 1967.
12. A. M. Baldin, V. I. Gol'danskiĭ, and I. L. Rozental', *Kinematika yadernykh reaktsiĭ*, Fizmatgiz, 1959. (*Kinematics of Nuclear Reactions*, translated by Ronald F. Peierls, Oxford University Press, London, 1961.)
13. J. Werle, *Relativistic Theory of Reactions*, North-Holland, Amsterdam; Wiley, Interscience, New York, 1966.
14. L. B. Okun', *Slaboe vzaimodeĭstvie elementarnykh chastits*, Fizmatgiz, 1963. (*Weak Interactions of Elementary Particles*, translated by Z. Lerman, translation ed. I. Meroz, Jerusalem, Israel Program for Scientific Translations, 1965.)
15. T. D. Lee and C. S. Wu, *Ann. Rev. Nucl. Sci.* **15,** 381 (1965); **16,** 471 (1966).
16. J. Bernstein, *Elementary Particles and their Currents*, W. H. Freeman, San Francisco, 1968.
17. N. F. Nelipa, *Vvedenie v teoriyu sil'no vzaimodeistvuyushchykh chastits*, Atomizdat, 1970. (*Introduction to the Theory of Strongly Interacting Particles.*)
18. E. Wigner, *Group Theory and its Application to the Quantum Mechanics of Atomic Spectra* (translated from German by J. J. Griffin), Academic Press, New York, 1959.
19. M. Hamermesh, *Group Theory and its Application to Physical Problems*, Addison-Wesley, Reading, Mass., 1964.
20. D. P. Zhelobenko, *Kompaktnye gruppy Lie i ikh predstavleniya*, "Nauka", 1970. (*Compact Lie Groups and their Representations.*)
21. G. C. Wick, E. P. Wigner, and A. S. Wightman, *Phys. Rev.* **88,** 101 (1952); **D1,** 3267 (1970).
22. G. C. Hegerfeldt, K. Kraus, and E. P. Wigner, *J. Math. Phys.* **9,** 2029 (1968).
23. E. P. Wigner, *Ann. Math.* **40,** 149 (1939).
24. V. Bargmann, *Ann. Math.* **59,** 1 (1954).
25. V. A. Fock, *Z. Phys.* **75,** 622 (1932).
26. R. M. F. Houtappel, H. van Dam, and E. P. Wigner, *Rev. Mod. Phys.* **37,** 595 (1965).
27. W. Brenig and R. Haag, *Fortschr. Phys.* **7,** 183 (1959).

28. M. FIERZ, *Helv. phys. acta* **12,** 3 (1939).
29. W. PAULI, in *Niels Bohr and the Development of Physics* (ed. W. PAULI, L. ROSENFELD, and V. WEISSKOPF, McGraw-Hill, New York; Pergamon Press, London, 1955, p. 30.
30. G. LÜDERS and B. ZUMINO, *Phys. Rev.* **110,** 1450 (1958).
31. N. BURGOYNE, *Nuovo Cim.* **8,** 607 (1958).
32. G. F. DELL'ANTONIO, *Ann. Phys. NY* **16,** 153 (1961).
33. M. FROISSART and J. R. TAYLOR, *Phys. Rev.* **153,** 1636 (1967).
34. T. G. TRIPPE *et al., Phys. Lett.* **28B,** 203 (1968).
35. A. BARBARO-GALTIERI *et al.*, in *Proceedings of the International Conference on $\pi\pi$ Scattering and Associated Topics, Florida State University, March 28–30, 1973*, American Institute of Physics, New York 1973.
36. N. S. KRYLOV and V. A. FOCK, *Zh. éksp. teor. Fiz.* **17,** 93 (1947).
37. A. BARBARO-GALTIERI, in *Advances in Particle Physics* (ed. R. L. COOL and R. E. MARSHAK), Interscience, New York, 1968, vol. 2, p. 175.
38. E. J. CARTAN, *Leçons sur la Theorie des Spineurs*, Hermann, Paris 1938. (*The Theory of Spinors*, foreword by Raymond Straeter, MIT Press, Cambridge, Mass., 1966.)
39. YU. B. RUMER, *Spinornyĭ analiz*, ONTI, 1936. (*Spinor Analysis.*)
40. M. A. NAĬMARK, *Lineĭnye predstavleniya gruppy Lorentza*, Fizmatgiz, 1958. (*Linear Representations of the Lorentz Group*, translated by ANN SWINFEN and O. J. MARSTRAND, translation ed. H. K. FARAHAT, Pergamon Press, Oxford and New York, 1964.
41. W. L. BADE and H. JEHLE, *Rev. Mod. Phys.* **25,** 714 (1953).
42. O. N. WILLIAMS, Lawrence Radiation Laboratory preprint UCRL-11113, 1963 (unpublished).
43. G. W. MACKEY, *Ann. Math.* **55,** 101 (1952).
44. YU. M. SHIROKOV, *Zh. éksp. teor. Fiz.* **21,** 748 (1951); *Dokl. Akad. Nauk* **94,** 857 (1954), Lektsii po osnovaniyam relyativistskoĭ kvantovoĭ teorii, Novosibirsk, 1964, (*Lectures on the Foundations of Relativistic Quantum Theory.*)
45. C. FRONSDAL, *Phys. Rev.* **113,** 1367 (1959).
46. L. L. FOLDY, *Phys. Rev.* **102,** 568 (1956).
47. J. S. LOMONT and H. E. MOSES, *J. Math. Phys.* **3,** 405 (1962).
48. A. CHAKRABARTI, *J. Math. Phys.* **5,** 922 (1964).
49. I. S. SHAPIRO, *Dokl. Akad. Nauk.* **106,** 647 (1956) (*Sov. Phys., Dokl.* **1,** 91 (1956)).
50. CHOU KUANG-CHAO and L. G. ZASTAVENKO, *Zh. éksp. teor. Fiz.* **35,** 1417 (1958) (*Sov. Phys., JETP* **8,** 990 (1959)).
51. YU. V. NOVOZHILOV and E. V. PROKHVATILOV, *Teor. mat. Fiz.* **1,** 101 (1969) (*Theor. Math. Phys.* **1,** 78 1969)).
52. H. JOOS, *Fortschr. Phys.* **10,** 65 (1962).
53. H. P. STAPP, *Phys. Rev.* **125,** 2139 (1962).
54. S. WEINBERG, *Phys. Rev.* **133,** B1318 (1964).
55. D. L. WEAVER, C. L. HAMMER, and R. H. GOOD, Jr., *Phys. Rev.* **135,** B241 (1964).
56. S. WEINBERG, *Phys. Rev.* **134,** B882 (1964).
57. D. ZWANZIGER, *Phys. Rev.* **133,** B1036 (1964).
58. M. JACOB and G. C. WICK, *Ann Phys. NY* **7,** 404 (1959).
59. V. I. RITUS, *Zh. éksp. teor. Fiz.* **40,** 352 (1961) (*Sov. Phys., JETP* **13,** 240 (1961)).
60. M. I. SHIROKOV, *Zh. éksp. teor. Fiz.* **40,** 1387 (1961) (*Sov. Phys., JETP* **13,** 975 (1961)).
61. A. J. MACFARLANE, *Rev. Mod. Phys.* **34,** 41 (1962).
62. G. C. WICK, *Ann. Phys. NY* **18,** 65 (1962).
63. P. A. M. DIRAC, *Proc. R. Soc. Lond.* A **155,** 447 (1936).
64. M. FIERZ, *Helv. phys. acta* **12,** 3 (1939).
65. M. FIERZ and W. PAULI, *Proc. R. Soc. Lond.* A **173,** 211 (1939).
66. R. J. DUFFIN, *Phys. Rev.* **54,** 1114 (1938).
67. N. KEMMER, *Proc. R. Soc. Lond.* A **173,** 91 (1939).
68. W. RARITA and J. SCHWINGER, *Phys. Rev.* **60,** 61 (1941).
69. V. BARGMANN and E. P. WIGNER, *Proc. Natn. Acad. Sci. USA* **34,** 211 (1948).
70. H. A. KRAMERS, F. J. BELINFANTE, and J. K. LUBAŃSKI, *Physica* **8,** 597 (1941).
71. H. J. BHABHA, *Rev. Mod Phys.* **17,** 200 (1945); **21,** 451 (1949).
72. HARISH – CHANDRA, *Phys. Rev.* **71,** 793 (1947); *Proc. R. Soc. Lond.* A, **192,** 195 (1948).
73. I. E. TAMM and V. L. GINZBURG, *Zh. éksp. teor. Fiz.* **17,** 227 (1947).
74. I. M. GEL'FAND and A. M. YAGLOM, *Zh. éksp. teor. Fiz.* **18,** 703 (1947).

75. D. L. PURSEY, *Ann. Phys. NY* **32,** 157 (1965).
76. W. K. TUNG, *Phys. Rev.* **156,** 1385 (1967).
77. S. J. CHANG, *Phys. Rev.* **148,** 1259 (1966).
78. L. FONDA and G. C. GHIRARDI, *Phys. Rev.* **175,** 2082 (1968).
79. S. TANI, *Phys. Rev.* **D2,** 980 (1970).
80. C. S. WU *et al.*, *Phys. Rev.* **105,** 1413 (1957).
81. T. D. LEE and C. N. YANG, *Phys. Rev.* **104,** 254 (1956).
82. J. H. CHRISTENSON *et al.*, *Phys. Rev. Lett.* **13,** 138 (1964).
83. J. SCHWINGER, *Phys. Rev.* **91,** 713 (1953); **94,** 1362 (1953).
84. G. LÜDERS, *K. danske Vidensk. Selsk. Mat. -Fys. Medd.* **28,** no. 5 (1954).
85. R. JOST, *Helv. phys. acta* **30,** 409 (1957).
86. R. F. STREATER and A. S. WIGHTMAN, *PCT, Spin and Statistics, and All That*, W. A. BENJAMIN, New York, 1964.
87. H. A. KRAMERS, *K. Akad. Amsterdam, Proc.* **33,** 953 (1930).
88. E. P. WIGNER, *Göttinger Nachr. Math. -Phys.* **5,** 546 (1932).
89. E. P. WIGNER, in the collection of ref. 10, p. 37.
90. T. D. LEE and G. C. WICK, *Phys. Rev.* **148,** 1385 (1966).
91. B. ZUMINO and D. ZWANZIGER, *Phys. Rev.* **164,** 1959 (1967).
92. L. D. LANDAU, *Nucl. Phys.* **3,** 127 (1957).
93. A. SALAM, *Nuovo Cim.* **5,** 299 (1957).
94. T. D. LEE, R. OEHME, and C. N. YANG, *Phys. Rev.* **106,** 340 (1957).
95. G. LÜDERS and B. ZUMINO, *Phys. Rev.* **106,** 385 (1957).
96. J. G. LAYTER *et al.*, *Phys. Rev. Lett.* **29,** 316 (1972).
97. J. K. BAIRD *et al.*, *Phys. Rev.* **179,** 1285 (1969).
98. W. B. DRESS *et al.*, *Phys. Rev.* **D7,** 3147 (1973).
99. M. GELL-MANN and A. PAIS, *Phys. Rev.* **97,** 1387 (1955).
100. A. PAIS and O. PICCIONI, *Phys. Rev.* **100,** 1487 (1955).
101. T. W. B. KIBBLE, *Phys. Rev.* **117,** 1159 (1960).
102. S. MANDELSTAM, *Phys. Rev.* **112,** 1344 (1958).
103. K. HEPP, *Helv. phys. acta* **36,** 355 (1963); **37,** 55 (1964).
104. A. O. BARUT, I. MUZINICH, and D. WILLIAMS, *Phys. Rev.* **130,** 442 (1963).
105. K. HARDENBERG, K.-H. MÜTTER, and W. R. THEIS, *Nuovo Cim.* **62A,** 385 (1969).
106. H. KLEINERT and K.-H. MÜTTER, *Nuovo Cim.* **63A,** 657 (1969).
107. M. D. SCADRON and H. F. JONES, *Phys. Rev.* **173,** 1734 (1968).
108. B. H. KELLETT, *Nuovo Cim.* **56A,** 1003 (1968).
109. C. de CALAN and R. STORA, CERN preprint TH. 1004, 1969 (unpublished).
110. P. CARRUTHERS, *Phys. Rev. Lett.* **18,** 353 (1967).
111. O. STEINMANN, *Phys. Lett.* **25B,** 234 (1967).
112. L. MICHEL, *Nuovo Cim.* **10,** 319 (1953).
113. M. GELL-MANN, *Phys. Rev.* **125,** 1067 (1962); California Institute of Technology report CTSL-20, 1961, reprinted in *The Eightfold Way* by M. GELL-MANN and Y. NE'EMAN, W. A. BENJAMIN, New York, 1964.
114. Y. NE'EMAN, *Nucl. Phys.* **26,** 222 (1961).
115. R. E. BEHRENDS *et al.*, *Rev. Mod. Phys.* **34,** 1 (1962).
116. M. GELL-MANN and Y. NE'EMAN, *The Eightfold Way*, W. A. BENJAMIN, New York, 1964.
117. J. J. DE SWART, *Rev. Mod. Phys.* **35,** 916 (1963).
118. S. SAKATA, *Progr. Theor. Phys.* **16,** 686 (1956).
118a. M. BOTT-BODENHAUSEN *et al.*, *Phys. Lett.* **40B,** 693 (1972).
119. S. OKUBO, *Progr. Theor. Phys.* **27,** 949 (1962).
120. J. J. SAKURAI, *Phys. Rev. Lett.* **9,** 472 (1962).
121. S. MESHKOV, C. A. LEVINSON, and H. J. LIPKIN, *Phys. Rev. Lett.* **10,** 361 (1963).
122. YU. B. RUMER and A. I. FET, *Teoriya unitarnoĭ simmetrii*, "Nauka", 1970. (*Theory of Unitary Symmetry.*)
123. NGUEN VAN HIEU, *Lektsii po teorii unitarnoĭ simmetrii elementarnykh chastits*, Atomizdat, 1967 (*Lectures on the Theory of Unitary Symmetry of Elementary Particles*); the collection *Fizika vysokikh energiĭ i teoriya elementarnykh chastits*, "Naukova dumka", Kiev, 1967. (*High Energy Physics and Elementary Particle Theory.*)
124. A. BARBARO-GALTIERI, in *Proceedings of the XVI International Conference on High Energy Physics*,

Batavia, Illinois, September 6–13, 1972 (*ed.* J. D. Jackson, A. Roberts, and R. Donaldson), National Accelerator Laboratory, Batavia, Ill., 1972, Vol. 1, p. 159; N. P. SAMIOS, M. GOLDBERG, and B. T. MEADOWS, Hadrons and SU(3): a critical review, Brookhaven National Laboratory report BNL–17851, May 1, 1973, *Rev. Mod. Phys.* **46**, 49 (1974).

125. M. GELL-MANN, *Phys. Lett.* **8,** 214 (1964); G. ZWEIG, CERN reports TH.401 and TH.412, 1964 (unpublished).
126. O. W. GREENBERG, *Phys. Rev. Lett.* **13,** 598 (1964); M. Y. HAN and Y. NAMBU, *Phys. Rev.* **139,** B1006 (1965); M. GELL-MANN, in *Acta phys. austriaca*, Suppl. *IX*, 733 (1972); H. J. LIPKIN, *Phys. Lett.* **45B,** 267 (1973).
127. L. O'RAIFEARTAIGH, *Phys. Rev. Lett.* **14,** 332 (1965); *Phys. Rev.* **139B,** 1052 (1965); R. JOST, *Helv. phys. acta* **39,** 369 (1966).
128. F. GÜRSEY and L. A. RADICATI, *Phys. Rev. Lett.* **13,** 173 (1964).
129. B. SAKITA, *Phys. Rev.* **136,** B1756 (1964).
130. E. P. WIGNER, *Phys. Rev.* **51,** 106 (1937).
131. R. F. DASHEN and M. GELL-MANN, *Phys. Rev. Lett.* **17,** 340 (1966); M. GELL-MANN, *Acta phys. austriaca*, Suppl. IX, 733 (1972).
132. H. J. MELOSH, thesis, California Institute of Technology, 1973 (unpublished); *Phys. Rev.* **D9,** 1095 (1974); F. GILMAN, M. KUGLER, and S. MESHKOV, *Phys. Rev.* **D9,** 715 (1974); A. HEY and T. WEYERS, *Phys. Lett.* **48B,** 69 (1974).
133. M. A. B. BÉG, B. W. LEE, and A. PAIS, *Phys. Rev. Lett.* **13,** 514 (1964); B. SAKITA, *Ibid.* 643 (1964).
134. H. LEHMANN, K. SYMANZIK, and W. ZIMMERMANN, *Nuovo Cim.* **1,** 205 (1955); **6,** 319 (1957).
135. H. BROS, H. EPSTEIN, and V. GLASER, *Commun. Math. Phys.* **1,** 240 (1965).
136. Y. HARA, *Phys. Rev.* **136,** B507 (1964).
137. L.-L. C. WANG, *Phys. Rev.* **142,** 1187 (1966).
138. T. L. TRUEMAN and G. C. WICK, *Ann. Phys. NY* **26,** 322 (1964).
139. E. LEADER, *Phys. Rev.* **166,** 1599 (1968).
140. H. F. JONES, *Nuovo Cim.* **50A,** 814 (1967).
141. G. COHEN-TANNOUDJI, A. MOREL, and H. NAVELET, *Ann. Phys. NY* **46,** 239 (1968).
142. C. REBBI and R. SLANSKY, *Rev. Mod. Phys.* **42,** 68 (1970).
143. J. J. DE SWART, *Rev. Mod. Phys.* **35,** 916 (1963).
144. L. DURAND, III, P. DE CELLES, and R. MARR, *Phys. Rev.* **126,** 1882 (1962).
145. J. D. BJORKEN and J. D. WALECKA, *Ann. Phys. NY* **38,** 35 (1966).
146. W. R. THEIS and P. H. HERTEL, *Nuovo Cim.* **66A,** 152 (1970).
147. M. D. SCADRON, *Phys. Rev.* **165,** 1640 (1968).
148. E. V. PROKHVATILOV, *Teor. mat. Fiz.* **9,** 65 (1971) (*Theor. Math. Phys.* **9,** 983 (1971)).
149. N. N. BOGOLIUBOV, B. V. MEDVEDEV, and M. K. POLIVANOV, *Voprosy teorii dispersionnykh soothosheniĭ*, Fizmatgiz, 1958. (*Questions in the Theory of Dispersion Relations*, translated by S. G. Brush, Lawrence Radiation Laboratory translation UCRL — 499 (L), unpublished.) J. HILGEVOORD, *Dispersion Relations and Causal Description: An Introduction to Dispersion Relations in Field Theory*, North-Holland, Amsterdam, 1960.
150. D. OLIVE, *Phys. Rev.* **135,** B745 (1964).
151. R. J. EDEN *et al.*, *The Analytic S-Matrix*, Cambridge University Press, Cambridge, 1966.
152. G. F. CHEW, *The Analytic W-Matrix: A Basis for Nuclear Democracy*, W. A. BENJAMIN, New York, 1966.
153. R. J. EDEN, *High Energy Collisions of Elementary Particles*, Cambridge University Press, Cambridge, 1967.
154. L. D. LANDAU, *Nucl. Phys.* **13,** 181 (1959).
155. M. GELL-MANN, M. L. GOLDBERGER, and W. E. THIRRING, *Phys. Rev.* **95,** 1612 (1954); M. L. GOLDBERGER, *Phys. Rev.* **99,** 979 (1955).
156. G. HÖHLER, G. EBEL, and J. GIESECKE, Z. *Phys.* **180,** 430 (1964). More recently: G. HÖHLER and H. P. JAKOB, Tables of pion–nucleon forward amplitudes, Karlsruhe University preprint, 1972.
157. V. N. GRIBOV, *Zh. éksp. teor. Fiz.* **41,** 667, 1962 (1961) (*Sov. Phys., – JETP* **14,** 478, 1395 (1962)).
158. M. FROISSART, in Proceedings of the La Jolla Conference on Weak and Strong Interactions, 1961 (unpublished).
159. E. J. SQUIRES, *Complex Angular Momenta and Particle Physics*, W. A. BENJAMIN, New York, 1963.
160. M. L. GOLDBERGER, *Dispersion Theory and Elementary Particles*, Wiley, New York, 1961.
161. M. L. GOLDBERGER and K. M. WATSON, *Collision Theory*, Wiley, New York, 1964.
162. G. BARTON, *Introduction to Dispersion Techniques in Field Theory*, W. A. BENJAMIN, New York, 1965.

163. V. B. Berestetskiĭ, *Usp. fiz. Nauk.* **76,** 25 (1962) (*Sov. Phys., Usp.* **5,** 7 (1962)).
164. R. Omnès, *Nuovo Cim.* **8,** 316 (1958).
165. G. Bellettini *et al., Phys. Lett.* **14,** 164 (1965).
166. W. Galbraith *et al., Phys. Rev.* **138,** B913 (1965); K. J. Foley et al., *Phys. Rev. Lett.* **19,** 330, 857 (1967).
167. J. V. Allaby *et al., Phys. Lett.* **30B,** 500 (1969); G. G. Beznogikh *et al., Ibid.* 270 (1969).
168. J. W. Chapman *et al., Phys. Rev. Lett.* **29,** 1686 (1972); G. Charlton *et al., Ibid.* 515 (1972); F. T. Dao *et al., Ibid.* 1627 (1972; R. H. Gustafson *et al., Ibid.* **32,** 441 (1974); W. F. Baker *et al.,* in *Proceedings of the XVII International Conference on High Energy Physics, London, 1–10, July 1974,* edited by J. R. Smith, Rutherford High Energy Laboratory, Chilton, England, 1974.
169. U. Amaldi *et al., Phys. Lett.* **44B,** 112 (1973); S. R. Amendolia *et al., Ibid.* 119 (1973).
170. V. Barger, in *Proceedings of the XVII International Conference on High Energy Physics, op. cit.*
171. O. W. Greenberg and F. E. Low, *Phys. Rev.* **124,** 2047 (1961).
172. M. Froissart, *Phys. Rev.* **123,** 1053 (1961).
173. A. Martin, *Nuovo Cim.* **42A,** 930 (1966); **44A,** 1219 (1966).
174. T. Kinoshita, J. J. Loeffel, and A. Martin, *Phys. Rev. Lett.* **10,** 460 (1963); *Phys. Rev.* **135,** **B**1464 (1964); A. A. Logunov, Nguyen van Hieu, and I. T. Todorov, *Usp. fiz. Nauk* **88,** 51 (1966) (*Sov. Phys., Usp.* **9,** 31 (1966)).
175. L. Łukaszuk and A. Martin, *Nuovo Cim.* **52A,** 122 (1967). For a review, see A. Martin, in *Elementary Processes at High Energy* (1970 International School of Subnuclear Physics, Erice, Italy) (ed. A. Zichichi, Academic Press, New York and London, 1971, p. 22.
176. A. K. Common and F. J. Yndurain, *Nucl. Phys.* **B26,** 167 (1971).
177. F. Cerulus and A. Martin, *Phys. Lett.* **8,** 80 (1964); J. D. Bessis, *Nuovo Cim.* **45,** 974 (1966).
178. A. Martin, in *Proceedings of the Fifth International Conference on High Energy Collisions, Stony Brook, New York, August 23–24, 1973* (ed. C. Quigg), American Institute of Physics, New York, 1973.
179. T. Regge, *Nuovo Cim.* **14,** 951 (1959); **18,** 947 (1960).
180. G. Chew and S. C. Frautschi, *Phys. Rev. Lett.* **5,** 580 (1960).
181. V. N. Gribov and I. Ya. Pomeranchuk, *Zh. éksp. teor Fiz.* **42,** 1141 (1962) (*Phys. Rev. Lett.* **8,** 343 (1962)); V. N. Gribov, ref. 157.
182. A. O. Barut and D. E. Zwanziger, *Phys. Rev.* **127,** 974 (1962).
183. A. O. Barut, *Phys. Rev.* **128,** 1959 (1962).
184. V. N. Gribov and I. Ya. Pomeranchuk, *Zh. eksp. teor, Fiz.* **43,** 308 (1962) (*Sov. Phys., – JETP* **16,** 220 (1963)).
185. S. Mandelstam, *Nuovo Cim.* **30,** 1113, 1148 (1963).
186. J. C. Polkinghorne, *J. Math. Phys.* **4,** 1396 (1963).
187. V. N. Gribov and I. Ya. Pomeranchuk, ref. 181; *Phys. Rev. Lett.* **8,** 412 (1962); M. Gell–Mann, *Ibid.* 263 (1962).
188. V. N. Gribov, *Zh. éksp. teor. Fiz.* **43,** 1529 (1962) (*Sov. Phys., JETP* **16,** 1080 (1963)); V. Singh, *Phys. Rev.* **129,** 1889 (1963).
188a. S. W. MacDowell, *Phys. Rev.* **116,** 774 (1959).
189. D. V. Shirkov, *Usp. fiz. Nauk* **102,** 87 (1970) (*Sov. Phys., Usp.* **13,** 599 (1971)).
190. V. Bartenev *et al., Phys. Rev. Lett.* **31,** 1088 (1973).
191. R. C. Arnold, *Phys. Rev. Lett.* **14,** 657 (1965).
192. K. Igi, *Phys. Rev. Lett.* **9,** 76 (1962).
193. A. Logunov, L. D. Soloviev, and A. N. Tavkhelidze, *Phys. Lett.* **24B,** 181 (1967).
194. K. Igi and S. Matsuda, *Phys. Rev. Lett.* **18,** 625 (1967).
195. R. Dolen, D. Horn, and C. Schmid, *Phys. Rev.* **166,** 1768 (1968).
196. P. G. O. Freund, *Phys. Rev. Lett.* **20,** 235 (1968); H. Harari, *Ibid.* 1395 (1968).
197. F. Gilman, H. Harari, and Y. Zarmi, *Phys. Rev. Lett.* **21,** 323 (1968); H. Harari and Y. Zarmi, *Phys. Rev.* **187,** 2230 (1969); Y. Zarmi, *Ibid.* **D4,** 3455 (1971).
198. C. Schmid, *Phys. Rev. Lett.* **20,** 628, 689, (1968).
199. G. F. Chew and A. Pignotti, *Phys. Rev. Lett.* **20,** 1078 (1968).
200. P. D. B. Collins, R. C. Johnson, and E. J. Squires, *Phys. Lett.* **27B,** 23 (1968).
201. M. Jacob, *Acta phys. austriaca,* suppl. VI, 215 (1969); M. Kugler, *Ibid.,* suppl. VII, 443 (1970).
202. H. Harari, *Phys. Rev. Lett.* **22,** 562 (1969); J. Rosner, *Ibid.* 689 (1969).
203. J. L. Rosner, *Phys. Rev. Lett.* **21,** 950, 1422 (E) (1968); H. J. Lipkin, *Nucl. Phys.* **B9,** 349 (1969).
204. G. Veneziano, *Nuovo Cim.* **57A,** 190 (1968).
205. N. N. Lebedev, *Spetsialnye funktsii i ikh prilozheniya,* Gostekhizdat, 1953. (*Special Functions and their*

Applications, translated by Richard A. SILVERMAN, Prentice-Hall, Englewood Cliffs, NJ, 1965.)
206. S. MANDELSTAM, *Phys. Rev. Lett.* **21,** 1724 (1968).
207. C. LOVELACE, *Phys. Lett.* **28B**, 264 (1968) T. SHAPIRO, *Phys Rev.* **179**, 1345 (1969).
208. M. ADEMOLLO, G. VENEZIANO, and S. WEINBERG, *Phys. Rev. Lett.* **22,** 83 (1969).
209. M. JACOB, in *Proceedings of the Lund International Conference on Elementary Particles, June 25 – July 1, 1969* (ed. G. VON DARDEL), Berlingska Boktryckeriet, Lund, Sweden, 1969, p. 125.
210. L. L. ENKOVSKIĬ, V. V. KUKHTIN, and V. P. SHELEST, Dual'nye resonansnye modeli, Inst. of Theor. Physics of the USSR Academy of Sciences preprint ITF-70-43, 1970 (unpublished).
211. V. ALESSANDRINI *et al.*, *Phys Reports* **1C**, 269 (1971); D. OLIVE, in *Proceedings of the XVII International Conference on High Energy Physis, op. cit.*
212. E. C. G. SUDARSHAN and R. E. MARSHAK, *Phys. Rev.* **109,** 1860 (1958).
213. R. P. FEYNMAN and M. GELL-MANN, *Phys. Rev.* **109,** 193 (1958).
214. S. S. GERSHTEĬN and YA. B. ZEL'DOVICH, *Zh. éksp. teor. Fiz.* **29,** 698 (1955) (*Sov. Phys., JETP* **2,** 576 (1956)).
215. M. GELL-MANN, *Phys. Rev.* **111,** 362 (1958).
216. N. CABIBBO, *Phys. Rev. Lett.* **10,** 531 (1963).
217. M. GELL-MANN, *Phys. Rev.* **125,** 1067 (1962); *Physics* **1,** 63 (1964).
218. M. ADEMOLLO and R. GATTO, *Phys. Rev. Lett.* **13,** 264 (1964).
219. R. HAAG, *Phys. Rev.* **112,** 669 (1958).
220. K. NISHIJIMA, *Phys. Rev.* **111,** 995 (1958).
221. W. ZIMMERMANN, *Nuovo Cim.* **10,** 597 (1958).
222. Y. NAMBU, *Phys. Rev. Lett.* **4,** 380 (1960).
223. CHOU KUANG-CHAU, *Zh. éksp. teor. Fiz.* **39,** 703 (1960) (*Sov. Phys., JETP* **12,** 492 (1961)).
224. M. GELL-MANN and M. LÉVY, *Nuovo Cim.* **16,** 705 (1960).
225. M. L. GOLDBERGER and S. B. TREIMAN, *Phys. Rev.* **110,** 1178 (1958).
226. S. L. ADLER, *Phys. Rev.* **137,** B1022 (1965).
227. S. L. ADLER, *Phys. Rev. Lett.* **14,** 1051 (1965); *Phys. Rev.* **140,** B736 (1965).
228. W. I. WEISBERGER, *Phys. Rev. Lett.* **14,** 1047 (1965); *Phys. Rev.* **143,** 1302 (1966).
229. S. WEINBERG, *Phys. Rev. Lett.* **18,** 507 (1967).
230. T. DAS, V. S. MATHUR, and S. OKUBO, *Phys. Rev. Lett.* **18,** 761 (1967).
231. K. KAWARABAYASHI and M. SUZUKI, *Phys. Rev. Lett.* **16,** 255 (1966); RIAZUDDIN and FAYYAZUDDIN, *Phys. Rev.* **147,** 1071 (1966).
232. C. H. ALBRIGHT, P. R. AUVIL, and N. G. DESHPANDE, *Nuovo Cim.* **52A**, 301 (1967).
233. D. A. GEFFEN, *Phys. Rev. Lett.* **19,** 770 (1967); S. A. BROWN and G. B. WEST, *Ibid.* 812 (1967). A modified relation follows from the more recent approach of ref. 132. See F. GILMAN and M. KUGLER, *Ibid.* **30,** 518 (1973).
234. C. L. COOK *et al.*, *Phys. Rev. Lett.* **20,** 295 (1968).
235. See refs. 34, 35.
236. S. L. ADLER and R. F. DASHEN, *Current Algebras and Applications to Particle Physics*, W. A. BENJAMIN, New York, 1968.
237. B. RENNER, *Current Algebras and their Applications*, Pergamon Press, Oxford and New York, 1968.
238. S. GASIOROWICZ and D. A. GEFFEN, *Rev. Mod. Phys.* **41,** 531 (1969).
239. A. I. VAĬNSHTEĬN and V. I. ZAKHAROV, *Usp. fiz. Nauk* **100,** 225 (1970) (*Sov. Phys., Usp.* **13,** 73 (1970)).
240. V. M. LOBASHEV, in *Proceedings of the XV International Conference on High Energy Physics, Kiev, USSR, August. 26–September 5, 1970* (ed. V. P. SHELEST, *et al.*), "Naukova Dumka", Kiev, 1972, p. 262.
241. D. DORFAN *et al.*, *Phys. Rev. Lett.* **19,** 987 (1967); S. BENNETT *et al.*, *Ibid.* 993 (1967).
242. J. MARX *et al.*, *Phys. Lett.* **32B,** 219 (1970); R. PICCIONI *et al.*, *Phys. Rev. Lett.* **29,** 1412 (1972); R. C. WEBB, thesis, Princeton University, 1972 (unpublished).
243. S. M. BILEN'KIĬ, in *Problemy fiziki elementarnykh chastits i atomnogo yadra*, Atomizdat, 1970, vol. 1, no. 1, p. 227 (*Sov. J. Particles Nuclei* **1,** 146 (1972)).
244. K. KLEINKNECHT in *Proceedings of the XVII International Conference in High Energy Physics, op cit.*
245. J. S. BELL. and J. STEINBERGER, in *Proceedings of the Oxford International Conference on Elementary Particles, September 19–25, 1965* (ed. by R. G. MOORHOUSE, A. E. TAYLOR, and T. R. WALSH), Rutherford High Energy Laboratory, Chilton, England, 1966, p. 195.
246. T. T. WU and C. N. YANG, *Phys. Rev. Lett.* **13,** 380 (1964).
247. D. CUNDY, in *Proceedings of the XVII International Conference on High Energy Physics, op. cit.*

INDEX

OTHER TITLES IN THE SERIES IN NATURAL PHILOSOPHY

Vol. 1. DAVYDOV—Quantum Mechanics
Vol. 2. FOKKER—Time and Space, Weight and Inertia
Vol. 3. KAPLAN—Interstellar Gas Dynamics
Vol. 4. ABRIKOSOV, GOR'KOV and DZYALOSHINSKII—Quantum Field Theoretical Methods in Statistical Physics
Vol. 5. OKUN'—Weak Interaction of Elementary Particles
Vol. 6. SHKLOVSKII—Physics of the Solar Corona
Vol. 7. AKHIEZER *et al.*—Collective Oscillations in a Plasma
Vol. 8. KIRZHNITS—Field Theoretical Methods in Many-body Systems
Vol. 9. KLMONTOVICH—The Statistical Theory of Non-equilibrium Processes in a Plasma
Vol. 10. KUIRTH—Introduction to Stellar Statistics
Vol. 11. CHALMERS—Atmospheric Electricity (2nd Edition)
Vol. 12. RENNER—Current Algebras and their Applications
Vol. 13. FAIN and KHANIN—Quantum Electronics, Volume 1—Basic Theory
Vol. 14. FAIN and KHANIN—Quantum Electronics, Volume 2—Maser Amplifiers and Oscillators
Vol. 15. MARCH—Liquid Metals
Vol. 16. HORI—Spectral Properties of Disordered Chains and Lattices
Vol. 17. SAINT JAMES, THOMAS and SARMA—Type II Superconductivity
Vol. 18. MARGENAU and KESTNER—Theory of Intermolecular Forces (2nd Edition)
Vol. 19. JANCEL—Foundations of Classical and Quantum Statistical Mechanics
Vol. 20. TAKAHASHI—An Introduction to Field Quantization
Vol. 21. YVON—Correlations and Entropy in Classical Statistical Mechanics
Vol. 22. PENROSE—Foundations of Statistical Mechanics
Vol. 23. VISCONTI—Quantum Field Theory, Volume 1
Vol. 24. FURTH—Fundamental Principles of Modern Theoretical Physics
Vol. 25. ZHELEZNYAKOV—Radioemission of the Sun and Planets
Vol. 26. GRINDLAY—An Introduction to the Phenomenological Theory of Ferroelectricity
Vol. 27. UNGER—Introduction to Quantum Electronics
Vol. 28. KOGA—Introduction to Kinetic Theory: Stochastic Processes in Gaseous Systems
Vol. 29. GALASIEWICZ—Superconductivity and Quantum Fluids
Vol. 30. CONSTANTINESCU and MAGYARI—Problems in Quantum Mechanics
Vol. 31. KOTKIN and SERBO—Collection of Problems in Classical Mechanics
Vol. 32. PANCHEV—Random Functions and Turbulence
Vol. 33. TALPE—Theory of Experiments in Paramagnetic Resonance
Vol. 34. TER HAAR—Elements of Hamiltonian Mechanics (2nd Edition)
Vol. 35. CLARKE and GRAINGER—Polarized Light and Optical Measurement
Vol. 36. HAUG—Theoretical Solid State Physics, Volume 1
Vol. 37. JORDAN and BEER—The Expanding Earth
Vol. 38. TODOROV—Analytical Properties of Feynman Diagrams in Quantum Field Theory
Vol. 39. SITENKO—Lectures in Scattering Theory
Vol. 40. SOBEL'MAN—Introduction to the Theory of Atomic Spectra
Vol. 41. ARMSTRONG and NICHOLLS—Emission, Absorption and Transfer of Radiation in Heated Atmospheres
Vol. 42. BRUSH—Kinetic Theory, Volume 3
Vol. 43. BOGOLYUBOV—A Method for Studying Model Hamiltonians

Vol. 44. Tsytovich—An Introduction to the Theory of Plasma Turbulence
Vol. 45. Pathria—Statistical Mechanics
Vol. 46. Haug—Theoretical Solid State Physics, Volume 2
Vol. 47. Nieto—The Titius–Bode Law of Planetary Distances: Its History and Theory
Vol. 48. Wagner—Introduction to the Theory of Magnetism
Vol. 49. Irvine—Nuclear Structure Theory
Vol. 50. Strohmeier—Variable Stars
Vol. 51. Batten—Binary and Multiple Systems of Stars
Vol. 52. Rousseau and Mathieu—Problems in Optics
Vol. 53. Bowler—Nuclear Physics
Vol. 54. Pomraning— The Equations of Radiation Hydrodynamics
Vol. 55. Belinfante—A Survey of Hidden-Variables Theories
Vol. 56. Scheibe—The Logical Analysis of Quantum Mechanics
Vol. 57. Robinson—Macroscopic Electromagnetism
Vol. 58. Gombás and Kisdi—Wave Mechanics and its Applications
Vol. 59. Kaplan and Tsytovich—Plasma Astrophysics
Vol. 60. Kovács and Zsoldos—Dislocations and Plastic Deformation
Vol. 61. Auvray and Fourrier—Problems in Electronics
Vol. 62. Mathieu—Optics
Vol. 63. Atwater—Introduction to General Relativity
Vol. 64. Müller—Quantum Mechanics: A Physical World Picture
Vol. 65. Bilenky—Introduction to Feynman Diagrams
Vol. 66. Vodar and Romand—Some Aspects of Vacuum Ultraviolet Radiation Physics
Vol. 67. Willett—Gas Lasers: Population Inversion Mechanisms
Vol. 68. Akhiezer—Plazma Electrodynamics, Volume 1—Linear Theory
Vol. 69. Glasby—The Nebular Variables
Vol. 70. Białynicki-Birula—Quantum Electrodynamics
Vol. 71. Karpman—Non-linear Waves in Dispersive Media
Vol. 72. Cracknell—Magnetism in Crystalline Materials
Vol. 73. Pathria—The Theory of Relativity
Vol. 74. Sitenko and Tartakovskiĭ—Lectures on the Theory of the Nucleus
Vol. 75. Belinfante—Measurement and Time Reversal in Objective Quantum
Vol. 76. Sobolev—Light Scattering in Planetary Atmospheres
Vol. 77. Novaković—The Pseudo-spin Method in Magnetism and Ferroelectricity